STUDENT'S SOLUTIONS MANUAL
MULTIVARIABLE

KELLY BARBER ELKA M. BLOCK FRANK PURCELL

UNIVERSITY CALCULUS
EARLY TRANSCENDENTALS
THIRD EDITION

Joel Hass
University of California, Davis

Maurice D. Weir
Naval Postgraduate School

George B. Thomas, Jr
Massachusetts Institute of Technology

with the assistance of

Christopher Heil
Georgia Institute of Technology

PEARSON

Boston Columbus Hoboken Indianapolis New York San Francisco
Amsterdam Cape Town Dubai London Madrid Milan Munich Paris Montreal Toronto
Delhi Mexico City São Paulo Sydney Hong Kong Seoul Singapore Taipei Tokyo

The author and publisher of this book have used their best efforts in preparing this book. These efforts include the development, research, and testing of the theories and programs to determine their effectiveness. The author and publisher make no warranty of any kind, expressed or implied, with regard to these programs or the documentation contained in this book. The author and publisher shall not be liable in any event for incidental or consequential damages in connection with, or arising out of, the furnishing, performance, or use of these programs.

Reproduced by Pearson from electronic files supplied by the author.

ISBN-13: 978-0-321-99985-6
ISBN-10: 0-321-99985-1

www.pearsonhighered.com

PEARSON

TABLE OF CONTENTS

CHAPTER 9   INFINITE SEQUENCES AND SERIES

9.1   SEQUENCES

1. $a_1 = \frac{1-1}{1^2} = 0, \ a_2 = \frac{1-2}{2^2} = -\frac{1}{4}, \ a_3 = \frac{1-3}{3^2} = -\frac{2}{9}, \ a_4 = \frac{1-4}{4^2} = -\frac{3}{16}$

3. $a_1 = \frac{(-1)^2}{2-1} = 1, \ a_2 = \frac{(-1)^3}{4-1} = -\frac{1}{3}, \ a_3 = \frac{(-1)^4}{6-1} = \frac{1}{5}, \ a_4 = \frac{(-1)^5}{8-1} = -\frac{1}{7}$

5. $a_1 = \frac{2}{2^2} = \frac{1}{2}, \ a_2 = \frac{2^2}{2^3} = \frac{1}{2}, \ a_3 = \frac{2^3}{2^4} = \frac{1}{2}, \ a_4 = \frac{2^4}{2^5} = \frac{1}{2}$

7. $a_1 = 1, \ a_2 = 1 + \frac{1}{2} = \frac{3}{2}, \ a_3 = \frac{3}{2} + \frac{1}{2^2} = \frac{7}{4}, \ a_4 = \frac{7}{4} + \frac{1}{2^3} = \frac{15}{8}, \ a_5 = \frac{15}{8} + \frac{1}{2^4} = \frac{31}{16}, \ a_6 = \frac{63}{32}, \ a_7 = \frac{127}{64}, \ a_8 = \frac{255}{128},$
    $a_9 = \frac{511}{256}, \ a_{10} = \frac{1023}{512}$

9. $a_1 = 2, \ a_2 = \frac{(-1)^2(2)}{2} = 1, \ a_3 = \frac{(-1)^3(1)}{2} = -\frac{1}{2}, \ a_4 = \frac{(-1)^4\left(-\frac{1}{2}\right)}{2} = -\frac{1}{4}, \ a_5 = \frac{(-1)^5\left(-\frac{1}{4}\right)}{2} = \frac{1}{8}, \ a_6 = \frac{1}{16}, \ a_7 = -\frac{1}{32},$
    $a_8 = -\frac{1}{64}, \ a_9 = \frac{1}{128}, \ a_{10} = \frac{1}{256}$

11. $a_1 = 1, \ a_2 = 1, \ a_3 = 1 + 1 = 2, \ a_4 = 2 + 1 = 3, \ a_5 = 3 + 2 = 5, \ a_6 = 8, \ a_7 = 13, \ a_8 = 21, \ a_9 = 34, \ a_{10} = 55$

13. $a_n = (-1)^{n+1}, \ n = 1, 2, \ldots$ 15. $a_n = (-1)^{n+1} n^2, \ n = 1, 2, \ldots$

17. $a_n = \frac{2^{n-1}}{3(n+2)}, \ n = 1, 2, \ldots$ 19. $a_n = n^2 - 1, \ n = 1, 2, \ldots$

21. $a_n = 4n - 3, \ n = 1, 2, \ldots$ 23. $a_n = \frac{3n+2}{n!}, \ n = 1, 2, \ldots$

25. $a_n = \frac{1+(-1)^{n+1}}{2}, \ n = 1, 2, \ldots$

27. $\lim\limits_{n \to \infty} \left(2 + (0.1)^n\right) = 2 \Rightarrow$ converges   (Theorem 5, #4)

29. $\lim\limits_{n \to \infty} \frac{1-2n}{1+2n} = \lim\limits_{n \to \infty} \frac{\left(\frac{1}{n}\right)-2}{\left(\frac{1}{n}\right)+2} = \lim\limits_{n \to \infty} \frac{-2}{2} = -1 \Rightarrow$ converges

31. $\lim\limits_{n \to \infty} \frac{1-5n^4}{n^4+8n^3} = \lim\limits_{n \to \infty} \frac{\left(\frac{1}{n^4}\right)-5}{1+\left(\frac{8}{n}\right)} = -5 \Rightarrow$ converges

33. $\lim\limits_{n \to \infty} \frac{n^2-2n+1}{n-1} = \lim\limits_{n \to \infty} \frac{(n-1)(n-1)}{n-1} = \lim\limits_{n \to \infty} (n-1) = \infty \Rightarrow$ diverges

35. $\lim\limits_{n \to \infty} \left(1 + (-1)^n\right)$ does not exist $\Rightarrow$ diverges

37. $\lim\limits_{n\to\infty} \left(\frac{n+1}{2n}\right)\left(1-\frac{1}{n}\right) = \lim\limits_{n\to\infty} \left(\frac{1}{2}+\frac{1}{2n}\right)\left(1-\frac{1}{n}\right) = \frac{1}{2} \Rightarrow$ converges

39. $\lim\limits_{n\to\infty} \frac{(-1)^{n+1}}{2n-1} = 0 \Rightarrow$ converges

41. $\lim\limits_{n\to\infty} \sqrt{\frac{2n}{n+1}} = \sqrt{\lim\limits_{n\to\infty} \frac{2n}{n+1}} = \sqrt{\lim\limits_{n\to\infty}\left(\frac{2}{1+\frac{1}{n}}\right)} = \sqrt{2} \Rightarrow$ converges

43. $\lim\limits_{n\to\infty} \sin\left(\frac{\pi}{2}+\frac{1}{n}\right) = \sin\left(\lim\limits_{n\to\infty}\left(\frac{\pi}{2}+\frac{1}{n}\right)\right) = \sin\frac{\pi}{2} = 1 \Rightarrow$ converges

45. $\lim\limits_{n\to\infty} \frac{\sin n}{n} = 0$ because $-\frac{1}{n} \leq \frac{\sin n}{n} \leq \frac{1}{n} \Rightarrow$ converges by the Sandwich Theorem for sequences

47. $\lim\limits_{n\to\infty} \frac{n}{2^n} = \lim\limits_{n\to\infty} \frac{1}{2^n \ln 2} = 0 \Rightarrow$ converges (using l'Hôpital's rule)

49. $\lim\limits_{n\to\infty} \frac{\ln(n+1)}{\sqrt{n}} = \lim\limits_{n\to\infty} \frac{\left(\frac{1}{n+1}\right)}{\left(\frac{1}{2\sqrt{n}}\right)} = \lim\limits_{n\to\infty} \frac{2\sqrt{n}}{n+1} = \lim\limits_{n\to\infty} \frac{\left(\frac{2}{\sqrt{n}}\right)}{1+\left(\frac{1}{n}\right)} = 0 \Rightarrow$ converges

51. $\lim\limits_{n\to\infty} 8^{1/n} = 1 \Rightarrow$ converges   (Theorem 5, #3)

53. $\lim\limits_{n\to\infty} \left(1+\frac{7}{n}\right)^n = e^7 \Rightarrow$ converges   (Theorem 5, #5)

55. $\lim\limits_{n\to\infty} \sqrt[n]{10n} = \lim\limits_{n\to\infty} 10^{1/n}\cdot n^{1/n} = 1\cdot 1 = 1 \Rightarrow$ converges   (Theorem 5, #3 and #2)

57. $\lim\limits_{n\to\infty} \left(\frac{3}{n}\right)^{1/n} = \frac{\lim\limits_{n\to\infty} 3^{1/n}}{\lim\limits_{n\to\infty} n^{1/n}} = \frac{1}{1} = 1 \Rightarrow$ converges   (Theorem 5, #3 and #2)

59. $\lim\limits_{n\to\infty} \frac{\ln n}{n^{1/n}} = \frac{\lim\limits_{n\to\infty} \ln n}{\lim\limits_{n\to\infty} n^{1/n}} = \frac{\infty}{1} = \infty \Rightarrow$ diverges   (Theorem 5, #2)

61. $\lim\limits_{n\to\infty} \sqrt[n]{4^n\, n} = \lim\limits_{n\to\infty} 4\sqrt[n]{n} = 4\cdot 1 = 4 \Rightarrow$ converges   (Theorem 5, #2)

63. $\lim\limits_{n\to\infty} \frac{n!}{n^n} = \lim\limits_{n\to\infty} \frac{1\cdot 2\cdot 3\cdots(n-1)(n)}{n\cdot n\cdot n\cdots n\cdot n} \leq \lim\limits_{n\to\infty}\left(\frac{1}{n}\right) = 0$ and $\frac{n!}{n^n} \geq 0 \Rightarrow \lim\limits_{n\to\infty}\frac{n!}{n^n} = 0 \Rightarrow$ converges

65. $\lim\limits_{n\to\infty} \frac{n!}{10^{6n}} = \lim\limits_{n\to\infty} \frac{1}{\left(\frac{(10^6)^n}{n!}\right)} = \infty \Rightarrow$ diverges   (Theorem 5, #6)

67. $\displaystyle\lim_{n\to\infty}\left(\tfrac{1}{n}\right)^{1/(\ln n)}=\lim_{n\to\infty}\exp\left(\tfrac{1}{\ln n}\ln\left(\tfrac{1}{n}\right)\right)=\lim_{n\to\infty}\exp\left(\tfrac{\ln 1-\ln n}{\ln n}\right)=e^{-1}\Rightarrow$ converges

69. $\displaystyle\lim_{n\to\infty}\left(\tfrac{3n+1}{3n-1}\right)^{n}=\lim_{n\to\infty}\exp\left(n\ln\left(\tfrac{3n+1}{3n-1}\right)\right)=\lim_{n\to\infty}\exp\left(\tfrac{\ln(3n+1)-\ln(3n-1)}{\frac{1}{n}}\right)=\lim_{n\to\infty}\exp\left(\tfrac{\frac{3}{3n+1}-\frac{3}{3n-1}}{\left(-\frac{1}{n^2}\right)}\right)$

$\displaystyle=\lim_{n\to\infty}\exp\left(\tfrac{6n^2}{(3n+1)(3n-1)}\right)=\exp\left(\tfrac{6}{9}\right)=e^{2/3}\Rightarrow$ converges

71. $\displaystyle\lim_{n\to\infty}\left(\tfrac{x^n}{2n+1}\right)^{1/n}=\lim_{n\to\infty}x\left(\tfrac{1}{2n+1}\right)^{1/n}=x\lim_{n\to\infty}\exp\left(\tfrac{1}{n}\ln\left(\tfrac{1}{2n+1}\right)\right)=x\lim_{n\to\infty}\exp\left(\tfrac{-\ln(2n+1)}{n}\right)=x\lim_{n\to\infty}\exp\left(\tfrac{-2}{2n+1}\right)$

$=xe^0=x,\ x>0\Rightarrow$ converges

73. $\displaystyle\lim_{n\to\infty}\tfrac{3^n\cdot 6^n}{2^{-n}\cdot n!}=\lim_{n\to\infty}\tfrac{36^n}{n!}=0\Rightarrow$ converges   (Theorem 5, #6)

75. $\displaystyle\lim_{n\to\infty}\tanh n=\lim_{n\to\infty}\tfrac{e^n-e^{-n}}{e^n+e^{-n}}=\lim_{n\to\infty}\tfrac{e^{2n}-1}{e^{2n}+1}=\lim_{n\to\infty}\tfrac{2e^{2n}}{2e^{2n}}=\lim_{n\to\infty}1=1\Rightarrow$ converges

77. $\displaystyle\lim_{n\to\infty}\tfrac{n^2\sin\left(\frac{1}{n}\right)}{2n-1}=\lim_{n\to\infty}\tfrac{\sin\left(\frac{1}{n}\right)}{\left(\frac{2}{n}-\frac{1}{n^2}\right)}=\lim_{n\to\infty}\tfrac{-\left(\cos\left(\frac{1}{n}\right)\right)\left(\frac{1}{n^2}\right)}{\left(-\frac{2}{n^2}+\frac{2}{n^3}\right)}=\lim_{n\to\infty}\tfrac{-\cos\left(\frac{1}{n}\right)}{-2+\left(\frac{2}{n}\right)}=\tfrac{1}{2}\Rightarrow$ converges

79. $\displaystyle\lim_{n\to\infty}\sqrt{n}\sin\left(\tfrac{1}{\sqrt{n}}\right)=\lim_{n\to\infty}\tfrac{\sin\left(\frac{1}{\sqrt{n}}\right)}{\frac{1}{\sqrt{n}}}=\lim_{n\to\infty}\tfrac{\cos\left(\frac{1}{\sqrt{n}}\right)\left(-\frac{1}{2n\sqrt[3]{2}}\right)}{-\frac{1}{2n\sqrt[3]{2}}}=\lim_{n\to\infty}\cos\left(\tfrac{1}{\sqrt{n}}\right)=\cos 0=1\Rightarrow$ converges

81. $\displaystyle\lim_{n\to\infty}\tan^{-1}n=\tfrac{\pi}{2}\Rightarrow$ converges

83. $\displaystyle\lim_{n\to\infty}\left(\left(\tfrac{1}{3}\right)^n+\tfrac{1}{\sqrt{2^n}}\right)=\lim_{n\to\infty}\left(\left(\tfrac{1}{3}\right)^n+\left(\tfrac{1}{\sqrt{2}}\right)^n\right)=0\Rightarrow$ converges   (Theorem 5, #4)

85. $\displaystyle\lim_{n\to\infty}\tfrac{(\ln n)^{200}}{n}=\lim_{n\to\infty}\tfrac{200(\ln n)^{199}}{n}=\lim_{n\to\infty}\tfrac{200\cdot199(\ln n)^{198}}{n}=\ldots=\lim_{n\to\infty}\tfrac{200!}{n}=0\Rightarrow$ converges

87. $\displaystyle\lim_{n\to\infty}\left(n-\sqrt{n^2-n}\right)=\lim_{n\to\infty}\left(n-\sqrt{n^2-n}\right)\left(\tfrac{n+\sqrt{n^2-n}}{n+\sqrt{n^2-n}}\right)=\lim_{n\to\infty}\tfrac{n}{n+\sqrt{n^2-n}}=\lim_{n\to\infty}\tfrac{1}{1+\sqrt{1-\frac{1}{n}}}=\tfrac{1}{2}\Rightarrow$ converges

89. $\displaystyle\lim_{n\to\infty}\tfrac{1}{n}\int_1^n\tfrac{1}{x}\,dx=\lim_{n\to\infty}\tfrac{\ln n}{n}=\lim_{n\to\infty}\tfrac{1}{n}=0\Rightarrow$ converges   (Theorem 5, #1)

91. Since a_n converges $\Rightarrow\lim_{n\to\infty}a_n=L\Rightarrow\lim_{n\to\infty}a_{n+1}=\lim_{n\to\infty}\tfrac{72}{1+a_n}\Rightarrow L=\tfrac{72}{1+L}\Rightarrow L(1+L)=72$

$\Rightarrow L^2+L-72=0\Rightarrow L=-9$ or $L=8$; since $a_n>0$ for $n\ge1\Rightarrow L=8$

93. Since a_n converges $\Rightarrow \lim_{n\to\infty} a_n = L \Rightarrow \lim_{n\to\infty} a_{n+1} = \lim_{n\to\infty} \sqrt{8+2a_n} \Rightarrow L = \sqrt{8+2L} \Rightarrow L^2 - 2L - 8 = 0$

$\Rightarrow L = -2$ or $L = 4$; since $a_n > 0$ for $n \geq 3 \Rightarrow L = 4$

95. Since a_n converges $\Rightarrow \lim_{n\to\infty} a_n = L \Rightarrow \lim_{n\to\infty} a_{n+1} = \lim_{n\to\infty} \sqrt{5a_n} \Rightarrow L = \sqrt{5L} \Rightarrow L^2 - 5L = 0 \Rightarrow L = 0$ or $L = 5$;

since $a_n > 0$ for $n \geq 1 \Rightarrow L = 5$

97. $a_{n+1} = 2 + \frac{1}{a_n}, \; n \geq 1, a_1 = 2.$ Since a_n converges $\Rightarrow \lim_{n\to\infty} a_n = L \Rightarrow \lim_{n\to\infty} a_{n+1} = \lim_{n\to\infty} \left(2 + \frac{1}{a_n}\right) \Rightarrow L = 2 + \frac{1}{L}$

$\Rightarrow L^2 - 2L - 1 = 0 \Rightarrow L = 1 \pm \sqrt{2}$; since $a_n > 0$ for $n \geq 1 \Rightarrow L = 1 + \sqrt{2}$

99. $1, 1, 2, 4, 8, 16, 32, \ldots = 1, 2^0, 2^1, 2^2, 2^3, 2^4, 2^5, \ldots \Rightarrow x_1 = 1$ and $x_n = 2^{n-2}$ for $n \geq 2$

101. (a) $f(x) = x^2 - 2$; the sequence converges to $1.414213562 \approx \sqrt{2}$

(b) $f(x) = \tan(x) - 1$; the sequence converges to $0.7853981635 \approx \frac{\pi}{4}$

(c) $f(x) = e^x$; the sequence $1, 0, -1, -2, -3, -4, -5, \ldots$ diverges

103. (a) If $a = 2n+1$, then $b = \left\lfloor \frac{a^2}{2} \right\rfloor = \left\lfloor \frac{4n^2+4n+1}{2} \right\rfloor = \left\lfloor 2n^2 + 2n + \frac{1}{2} \right\rfloor = 2n^2 + 2n, \; c = \left\lceil \frac{a^2}{2} \right\rceil = \left\lceil 2n^2 + 2n + \frac{1}{2} \right\rceil$

$= 2n^2 + 2n + 1$ and $a^2 + b^2 = (2n+1)^2 + \left(2n^2 + 2n\right)^2 = 4n^2 + 4n + 1 + 4n^4 + 8n^3 + 4n^2$

$= 4n^4 + 8n^3 + 8n^2 + 4n + 1 = \left(2n^2 + 2n + 1\right)^2 = c^2.$

(b) $\lim_{a\to\infty} \frac{\left\lfloor \frac{a^2}{2} \right\rfloor}{\left\lceil \frac{a^2}{2} \right\rceil} = \lim_{a\to\infty} \frac{2n^2+2n}{2n^2+2n+1} = 1$ or $\lim_{a\to\infty} \frac{\left\lfloor \frac{a^2}{2} \right\rfloor}{\left\lceil \frac{a^2}{2} \right\rceil} = \lim_{a\to\infty} \sin\theta = \lim_{\theta\to\frac{\pi}{2}} \sin\theta = 1$

105. (a) $\lim_{n\to\infty} \frac{\ln n}{n^c} = \lim_{n\to\infty} \frac{\left(\frac{1}{n}\right)}{cn^{c-1}} = \lim_{n\to\infty} \frac{1}{cn^c} = 0$

(b) For all $\epsilon > 0$, there exists an $N = e^{-(\ln \epsilon)/c}$ such that $n > e^{-(\ln \epsilon)/c} \Rightarrow \ln n > -\frac{\ln \epsilon}{c} \Rightarrow \ln n^c > \ln\left(\frac{1}{\epsilon}\right)$

$\Rightarrow n^c > \frac{1}{\epsilon} \Rightarrow \frac{1}{n^c} < \epsilon \Rightarrow \left|\frac{1}{n^c} - 0\right| < \epsilon \Rightarrow \lim_{n\to\infty} \frac{1}{n^c} = 0$

107. $\lim_{n\to\infty} n^{1/n} = \lim_{n\to\infty} \exp\left(\frac{1}{n} \ln n\right) = \lim_{n\to\infty} \exp\left(\frac{1}{n}\right) = e^0 = 1$

109. Assume the hypotheses of the theorem and let ϵ be a positive number. For all ϵ there exists an N_1 such that when $n > N_1$ then $|a_n - L| < \epsilon \Rightarrow -\epsilon < a_n - L < \epsilon \Rightarrow L - \epsilon < a_n$, and there exists an N_2 such that when $n > N_2$ then $|c_n - L| < \epsilon \Rightarrow -\epsilon < c_n - L < \epsilon \Rightarrow c_n < L + \epsilon.$ If $n > \max\{N_1, N_2\}$, then $L - \epsilon < a_n \leq b_n \leq c_n < L + \epsilon$ $\Rightarrow |b_n - L| < \epsilon \Rightarrow \lim_{n\to\infty} b_n = L.$

111. $a_{n+1} \geq a_n \Rightarrow \frac{3(n+1)+1}{(n+1)+1} > \frac{3n+1}{n+1} \Rightarrow \frac{3n+4}{n+2} > \frac{3n+1}{n+1} \Rightarrow 3n^2 + 3n + 4n + 4 > 3n^2 + 6n + n + 2 \Rightarrow 4 > 2;$

the steps are reversible so the sequence is nondecreasing; $\frac{3n+1}{n+1} < 3 \Rightarrow 3n+1 < 3n+3 \Rightarrow 1 < 3;$

the steps are reversible so the sequence is bounded above by 3

113. $a_{n+1} \leq a_n \Rightarrow \frac{2^{n+1}3^{n+1}}{(n+1)!} \leq \frac{2^n 3^n}{n!} \Rightarrow \frac{2^{n+1}3^{n+1}}{2^n 3^n} \leq \frac{(n+1)!}{n!} \Rightarrow 2 \cdot 3 \leq n+1$ which is true for $n \geq 5;$ the steps are reversible so

the sequence is decreasing after a_5, but it is not nondecreasing for all its terms; $a_1 = 6, \; a_2 = 18, \; a_3 = 36,$

$a_4 = 54, \; a_5 = \frac{324}{5} = 64.8 \Rightarrow$ the sequence is bounded from above by 64.8

115. $a_n = 1 - \frac{1}{n}$ converges because $\frac{1}{n} \to 0$ by Example 1; also it is a nondecreasing sequence bounded above by 1

117. $a_n = \frac{2^n - 1}{2^n} = 1 - \frac{1}{2^n}$ and $0 < \frac{1}{2^n} < \frac{1}{n};$ since $\frac{1}{n} \to 0$ (by Example 1) $\Rightarrow \frac{1}{2^n} \to 0,$ the sequence converges; also it is

a nondecreasing sequence bounded above by 1

119. $a_n = \left((-1)^n + 1 \right)\left(\frac{n+1}{n} \right)$ diverges because $a_n = 0$ for n odd, while for n even $a_n = 2\left(1 + \frac{1}{n} \right)$ converges to 2; it

diverges by definition of divergence

121. $a_n \geq a_{n+1} \Leftrightarrow \frac{1+\sqrt{2n}}{\sqrt{n}} \geq \frac{1+\sqrt{2(n+1)}}{\sqrt{n+1}} \Leftrightarrow \sqrt{n+1} + \sqrt{2n^2 + 2n} \geq \sqrt{n} + \sqrt{2n^2 + 2n} \Leftrightarrow \sqrt{n+1} \geq \sqrt{n}$ and $\frac{1+\sqrt{2n}}{\sqrt{n}} \geq \sqrt{2};$

thus the sequence is nonincreasing and bounded below by $\sqrt{2} \Rightarrow$ it converges

123. $\frac{4^{n+1} + 3^n}{4^n} = 4 + \left(\frac{3}{4} \right)^n$ so $a_n \geq a_{n+1} \Leftrightarrow 4 + \left(\frac{3}{4} \right)^n \geq 4 + \left(\frac{3}{4} \right)^{n+1} \Leftrightarrow \left(\frac{3}{4} \right)^n \geq \left(\frac{3}{4} \right)^{n+1} \Leftrightarrow 1 \geq \frac{3}{4}$ and $4 + \left(\frac{3}{4} \right)^n \geq 4;$ thus the

sequence is nonincreasing and bounded below by $4 \Rightarrow$ it converges

125. For a given ε, choose N to be any integer greater than $1/\varepsilon.$ Then for $n > N,$

$\left| \frac{\sin n}{n} - 0 \right| = \frac{|\sin n|}{n} \leq \frac{1}{n} < \frac{1}{N} < \varepsilon.$

127. Let $0 < M < 1$ and let N be an integer greater than $\frac{M}{1-M}.$ Then $n > N \Rightarrow n > \frac{M}{1-M} \Rightarrow n - nM > M$

$\Rightarrow n > M + nM \Rightarrow n > M(n+1) \Rightarrow \frac{n}{n+1} > M.$

129. The sequence $a_n = 1 + \frac{(-1)^n}{2}$ is the sequence $\frac{1}{2}, \frac{3}{2}, \frac{1}{2}, \frac{3}{2}, \ldots.$ This sequence is bounded above by $\frac{3}{2},$ but it

clearly does not converge, by definition of convergence.

131. Given an $\epsilon > 0,$ by definition of convergence there corresponds an N such that for all $n > N, \; |L_1 - a_n| < \epsilon$ and

$|L_2 - a_n| < \epsilon.$ Now $|L_2 - L_1| = |L_2 - a_n + a_n - L_1| \leq |L_2 - a_n| + |a_n - L_1| < \epsilon + \epsilon = 2\epsilon. \; |L_2 - L_1| < 2\epsilon$ says that the

difference between two fixed values is smaller than any positive number $2\epsilon.$ The only nonnegative number

smaller than every positive number is 0, so $|L_1 - L_2| = 0$ or $L_1 = L_2.$

133. $a_{2k} \to L \Leftrightarrow$ given an $\epsilon > 0$ there corresponds an N_1 such that $\left[2k > N_1 \Rightarrow |a_{2k} - L| < \epsilon\right]$. Similarly, $a_{2k+1} \to L \Leftrightarrow \left[2k+1 > N_2 \Rightarrow |a_{2k+1} - L| < \epsilon\right]$. Let $N = \max\{N_1, N_2\}$. Then $n > N \Rightarrow |a_n - L| < \epsilon$ whether n is even or odd, and hence $a_n \to L$.

135. (a) $f(x) = x^2 - a \Rightarrow f'(x) = 2x \Rightarrow x_{n+1} = x_n - \frac{x_n^2 - a}{2x_n} \Rightarrow x_{n+1} = \frac{2x_n^2 - (x_n^2 - a)}{2x_n} = \frac{x_n^2 + a}{2x_n} = \frac{\left(x_n + \frac{a}{x_n}\right)}{2}$

   (b) $x_1 = 2$, $x_2 = 1.75$, $x_3 = 1.732142857$, $x_4 = 1.73205081$, $x_5 = 1.732050808$; we are finding the positive number where $x^2 - 3 = 0$; that is, where $x^2 = 3, x > 0$, or where $x = \sqrt{3}$.

137-147. Example CAS Commands:

Mathematica: (sequence functions may vary):

    Clear[a, n]

    a[n_] := n^{1/n}

    first25= Table[N[a[n]],{n, 1, 25}]

    Limit[a[n], n → 8]

The last command (Limit) will not always work in Mathematica. You could also explore the limit by enlarging your table to more than the first 25 values.

If you know the limit (1 in the above example), to determine how far to go to have all further terms within 0.01 of the limit, do the following.

    Clear[minN, lim]

    lim= 1

    Do[{diff =Abs[a[n] – lim], If[diff <.01, {minN= n, Abort[]}]}, {n, 2, 1000}]

    minN

For sequences that are given recursively, the following code is suggested. The portion of the command a[n_]:=a[n] stores the elements of the sequence and helps to streamline computation.

    Clear[a, n]

    a[1]= 1;

    a[n_] := a[n]= a[n – 1] + (1/5)^{n-1}

    first25= Table[N[a[n]],{n, 1, 25}]

The limit command does not work in this case, but the limit can be observed as 1.25.

    Clear[minN, lim]

    lim= 1.25

    Do[{diff =Abs[a[n] – lim], If[diff <.01, {minN= n, Abort[]}]}, {n, 2, 1000}]

    minN

9.2   INFINITE SERIES

1. $s_n = \dfrac{a\left(1-r^n\right)}{(1-r)} = \dfrac{2\left(1-\left(\frac{1}{3}\right)^n\right)}{1-\left(\frac{1}{3}\right)} \Rightarrow \lim_{n\to\infty} s_n = \dfrac{2}{1-\left(\frac{1}{3}\right)} = 3$

3. $s_n = \dfrac{a\left(1-r^n\right)}{(1-r)} = \dfrac{1-\left(-\frac{1}{2}\right)^n}{1-\left(-\frac{1}{2}\right)} \Rightarrow \lim_{n\to\infty} s_n = \dfrac{1}{\left(\frac{3}{2}\right)} = \dfrac{2}{3}$

5. $\dfrac{1}{(n+1)(n+2)} = \dfrac{1}{n+1} - \dfrac{1}{n+2} \Rightarrow s_n = \left(\frac{1}{2}-\frac{1}{3}\right)+\left(\frac{1}{3}-\frac{1}{4}\right)+\dots+\left(\frac{1}{n+1}-\frac{1}{n+2}\right) = \frac{1}{2}-\frac{1}{n+2} \Rightarrow \lim_{n\to\infty} s_n = \frac{1}{2}$

7. $1-\frac{1}{4}+\frac{1}{16}-\frac{1}{64}+\dots$, the sum of this geometric series is $\dfrac{1}{1-\left(-\frac{1}{4}\right)} = \dfrac{1}{1+\left(\frac{1}{4}\right)} = \dfrac{4}{5}$

9. $\frac{7}{4}+\frac{7}{16}+\frac{7}{64}+\dots$, the sum of this geometric series is $\dfrac{\left(\frac{7}{4}\right)}{1-\left(\frac{1}{4}\right)} = \dfrac{7}{3}$

11. $(5+1)+\left(\frac{5}{2}+\frac{1}{3}\right)+\left(\frac{5}{4}+\frac{1}{9}\right)+\left(\frac{5}{8}+\frac{1}{27}\right)+\dots$, is the sum of two geometric series; the sum is

$\dfrac{5}{1-\left(\frac{1}{2}\right)} + \dfrac{1}{1-\left(\frac{1}{3}\right)} = 10+\frac{3}{2} = \frac{23}{2}$

13. $(1+1)+\left(\frac{1}{2}-\frac{1}{5}\right)+\left(\frac{1}{4}+\frac{1}{25}\right)+\left(\frac{1}{8}-\frac{1}{125}\right)+\dots$, is the sum of two geometric series; the sum is

$\dfrac{1}{1-\left(\frac{1}{2}\right)} + \dfrac{1}{1+\left(\frac{1}{5}\right)} = 2+\frac{5}{6} = \frac{17}{6}$

15. Series is geometric with $r=\frac{2}{5} \Rightarrow \left|\frac{2}{5}\right|<1 \Rightarrow$ Converges to $\dfrac{1}{1-\frac{2}{5}} = \dfrac{5}{3}$

17. Series is geometric with $r=\frac{1}{8} \Rightarrow \left|\frac{1}{8}\right|<1 \Rightarrow$ Converges to $\dfrac{\frac{1}{8}}{1-\frac{1}{8}} = \dfrac{1}{7}$

19. $0.\overline{23} = \displaystyle\sum_{n=0}^{\infty} \frac{23}{100}\left(\frac{1}{10^2}\right)^n = \dfrac{\left(\frac{23}{100}\right)}{1-\left(\frac{1}{100}\right)} = \dfrac{23}{99}$

21. $0.\overline{7} = \displaystyle\sum_{n=0}^{\infty} \frac{7}{10}\left(\frac{1}{10}\right)^n = \dfrac{\left(\frac{7}{10}\right)}{1-\left(\frac{1}{10}\right)} = \dfrac{7}{9}$

23. $0.0\overline{6} = \displaystyle\sum_{n=0}^{\infty} \left(\frac{1}{10}\right)\left(\frac{6}{10}\right)\left(\frac{1}{10}\right)^n = \dfrac{\left(\frac{6}{100}\right)}{1-\left(\frac{1}{10}\right)} = \dfrac{6}{90} = \dfrac{1}{15}$

25. $1.24\overline{123} = \frac{124}{100}+\displaystyle\sum_{n=0}^{\infty} \frac{123}{10^5}\left(\frac{1}{10^3}\right)^n = \frac{124}{100}+\dfrac{\left(\frac{123}{10^5}\right)}{1-\left(\frac{1}{10^3}\right)} = \frac{124}{100}+\dfrac{123}{10^5-10^2} = \frac{124}{100}+\frac{123}{99,900} = \frac{123,999}{99,900} = \frac{41,333}{33,300}$

27. $\displaystyle\lim_{n\to\infty} \frac{n}{n+10} = \lim_{n\to\infty} \frac{1}{1} = 1 \neq 0 \Rightarrow$ diverges

29. $\lim\limits_{n\to\infty} \frac{1}{n+4} = 0 \Rightarrow$ test inconclusive

31. $\lim\limits_{n\to\infty} \cos\frac{1}{n} = \cos 0 = 1 \neq 0 \Rightarrow$ diverges

33. $\lim\limits_{n\to\infty} \ln\frac{1}{n} = -\infty \neq 0 \Rightarrow$ diverges

35. $s_k = \left(1-\frac{1}{2}\right)+\left(\frac{1}{2}-\frac{1}{3}\right)+\left(\frac{1}{3}-\frac{1}{4}\right)+\ldots+\left(\frac{1}{k-1}-\frac{1}{k}\right)+\left(\frac{1}{k}-\frac{1}{k+1}\right)=1-\frac{1}{k+1} \Rightarrow \lim\limits_{k\to\infty} s_k = \lim\limits_{k\to\infty}\left(1-\frac{1}{k+1}\right)=1,$

 series converges to 1

37. $s_k = \left(\ln\sqrt{2}-\ln\sqrt{1}\right)+\left(\ln\sqrt{3}-\ln\sqrt{2}\right)+\left(\ln\sqrt{4}-\ln\sqrt{3}\right)+\ldots+\left(\ln\sqrt{k}-\ln\sqrt{k-1}\right)+\left(\ln\sqrt{k+1}-\ln\sqrt{k}\right)$

 $= \ln\sqrt{k+1}-\ln\sqrt{1} = \ln\sqrt{k+1} \Rightarrow \lim\limits_{k\to\infty} s_k = \lim\limits_{k\to\infty} \ln\sqrt{k+1} = \infty;$ series diverges

39. $s_k = \left(\cos^{-1}\left(\frac{1}{2}\right)-\cos^{-1}\left(\frac{1}{3}\right)\right)+\left(\cos^{-1}\left(\frac{1}{3}\right)-\cos^{-1}\left(\frac{1}{4}\right)\right)+\left(\cos^{-1}\left(\frac{1}{4}\right)-\cos^{-1}\left(\frac{1}{5}\right)\right)+\ldots$

 $+\left(\cos^{-1}\left(\frac{1}{k}\right)-\cos^{-1}\left(\frac{1}{k+1}\right)\right)+\left(\cos^{-1}\left(\frac{1}{k+1}\right)-\cos^{-1}\left(\frac{1}{k+2}\right)\right)=\frac{\pi}{3}-\cos^{-1}\left(\frac{1}{k+2}\right)$

 $\Rightarrow \lim\limits_{k\to\infty} s_k = \lim\limits_{k\to\infty}\left[\frac{\pi}{3}-\cos^{-1}\left(\frac{1}{k+2}\right)\right]=\frac{\pi}{3}-\frac{\pi}{2}=\frac{\pi}{6},$ series converges to $\frac{\pi}{6}$

41. $\frac{4}{(4n-3)(4n+1)} = \frac{1}{4n-3}-\frac{1}{4n+1} \Rightarrow s_k = \left(1-\frac{1}{5}\right)+\left(\frac{1}{5}-\frac{1}{9}\right)+\left(\frac{1}{9}-\frac{1}{13}\right)+\ldots+\left(\frac{1}{4k-7}-\frac{1}{4k-3}\right)+\left(\frac{1}{4k-3}-\frac{1}{4k+1}\right)=1-\frac{1}{4k+1}$

 $\Rightarrow \lim\limits_{k\to\infty} s_k = \lim\limits_{k\to\infty}\left(1-\frac{1}{4k+1}\right)=1$

43. $\frac{40n}{(2n-1)^2(2n+1)^2} = \frac{A}{(2n-1)}+\frac{B}{(2n-1)^2}+\frac{C}{(2n+1)}+\frac{D}{(2n+1)^2} = \frac{A(2n-1)(2n+1)^2+B(2n+1)^2+C(2n+1)(2n-1)^2+D(2n-1)^2}{(2n-1)^2(2n+1)^2}$

 $\Rightarrow A(2n-1)(2n+1)^2+B(2n+1)^2+C(2n+1)(2n-1)^2+D(2n-1)^2 = 40n$

 $\Rightarrow A(8n^3+4n^2-2n-1)+B(4n^2+4n+1)+C(8n^3-4n^2-2n+1)+D(4n^2-4n+1)=40n$

 $\Rightarrow (8A+8C)n^3+(4A+4B-4C+4D)n^2+(-2A+4B-2C-4D)n+(-A+B+C+D)=40n$

 $\Rightarrow \begin{cases} 8A+8C=0 \\ 4A+4B-4C+4D=0 \\ -2A+4B-2C-4D=40 \\ -A+B+C+D=0 \end{cases} \Rightarrow \begin{cases} A+C=0 \\ A+B-C+D=0 \\ -A+2B-C-2D=20 \\ -A+B+C+D=0 \end{cases} \Rightarrow \begin{cases} B+D=0 \\ 2B-2D=20 \end{cases} \Rightarrow 4B=20 \Rightarrow B=5$

 and $D=-5 \Rightarrow \begin{cases} A+C=0 \\ -A+5+C-5=0 \end{cases} \Rightarrow C=0$ and $A=0$. Hence,

 $\sum\limits_{n=1}^{k}\left[\frac{40n}{(2n-1)^2(2n+1)^2}\right] = 5\sum\limits_{n=1}^{k}\left[\frac{1}{(2n-1)^2}-\frac{1}{(2n+1)^2}\right] = 5\left(\frac{1}{1}-\frac{1}{9}+\frac{1}{9}-\frac{1}{25}+\frac{1}{25}-\ldots-\frac{1}{(2(k-1)+1)^2}+\frac{1}{(2k-1)^2}-\frac{1}{(2k+1)^2}\right)$

 $= 5\left(1-\frac{1}{(2k+1)^2}\right) \Rightarrow$ the sum is $\lim\limits_{n\to\infty} 5\left(1-\frac{1}{(2k+1)^2}\right)=5$

45. $s_k = \left(1 - \frac{1}{\sqrt{2}}\right) + \left(\frac{1}{\sqrt{2}} - \frac{1}{\sqrt{3}}\right) + \left(\frac{1}{\sqrt{3}} - \frac{1}{\sqrt{4}}\right) + \ldots + \left(\frac{1}{\sqrt{k-1}} + \frac{1}{\sqrt{k}}\right) + \left(\frac{1}{\sqrt{k}} - \frac{1}{\sqrt{k+1}}\right) = 1 - \frac{1}{\sqrt{k+1}}$

$\Rightarrow \lim_{k \to \infty} s_k = \lim_{k \to \infty} \left(1 - \frac{1}{\sqrt{k+1}}\right) = 1$

47. $s_k = \left(\frac{1}{\ln 3} - \frac{1}{\ln 2}\right) + \left(\frac{1}{\ln 4} - \frac{1}{\ln 3}\right) + \left(\frac{1}{\ln 5} - \frac{1}{\ln 4}\right) + \ldots + \left(\frac{1}{\ln (k+1)} - \frac{1}{\ln k}\right) + \left(\frac{1}{\ln (k+2)} - \frac{1}{\ln (k+1)}\right) = -\frac{1}{\ln 2} + \frac{1}{\ln (k+2)}$

$\Rightarrow \lim_{k \to \infty} s_k = -\frac{1}{\ln 2}$

49. convergent geometric series with sum $\dfrac{1}{1 - \left(\frac{1}{\sqrt{2}}\right)} = \dfrac{\sqrt{2}}{\sqrt{2} - 1} = 2 + \sqrt{2}$

51. convergent geometric series with sum $\dfrac{\left(\frac{3}{2}\right)}{1 - \left(-\frac{1}{2}\right)} = 1$

53. The sequence $a_n = \cos \dfrac{n\pi}{2}$ starting with $n = 0$ is $1, 0, -1, 0, 1, 0, -1, 0, \ldots$, so the sequence of partial sums for the given series is $1, 1, 0, 0, 1, 1, 0, 0, \ldots$ and thus the series diverges.

55. convergent geometric series with sum $\dfrac{1}{1 - \left(\frac{1}{e^2}\right)} = \dfrac{e^2}{e^2 - 1}$

57. convergent geometric series with sum $\dfrac{2}{1 - \left(\frac{1}{10}\right)} - 2 = \dfrac{20}{9} - \dfrac{18}{9} = \dfrac{2}{9}$

59. difference of two geometric series with sum $\dfrac{1}{1 - \left(\frac{2}{3}\right)} - \dfrac{1}{1 - \left(\frac{1}{3}\right)} = 3 - \dfrac{3}{2} = \dfrac{3}{2}$

61. $\lim_{n \to \infty} \dfrac{n!}{1000^n} = \infty \neq 0 \Rightarrow$ diverges

63. $\displaystyle\sum_{n=1}^{\infty} \dfrac{2^n + 3^n}{4^n} = \sum_{n=1}^{\infty} \dfrac{2^n}{4^n} + \sum_{n=1}^{\infty} \dfrac{3^n}{4^n} = \sum_{n=1}^{\infty} \left(\frac{1}{2}\right)^n + \sum_{n=1}^{\infty} \left(\frac{3}{4}\right)^n$; both $\displaystyle\sum_{n=1}^{\infty} \left(\frac{1}{2}\right)^n$ and $\displaystyle\sum_{n=1}^{\infty} \left(\frac{3}{4}\right)^n$ are geometric series, and both

converge since $r = \frac{1}{2} \Rightarrow \left|\frac{1}{2}\right| < 1$ and $r = \frac{3}{4} \Rightarrow \left|\frac{3}{4}\right| < 1$, respectively $\Rightarrow \displaystyle\sum_{n=1}^{\infty} \left(\frac{1}{2}\right)^n = \dfrac{\frac{1}{2}}{1 - \frac{1}{2}} = 1$ and $\displaystyle\sum_{n=1}^{\infty} \left(\frac{3}{4}\right)^n = \dfrac{\frac{3}{4}}{1 - \frac{3}{4}} = 3$

$\Rightarrow \displaystyle\sum_{n=1}^{\infty} \dfrac{2^n + 3^n}{4^n} = 1 + 3 = 4$ by Theorem 8, part (1)

65. $\displaystyle\sum_{n=1}^{\infty} \ln\left(\frac{n}{n+1}\right) = \sum_{n=1}^{\infty} [\ln(n) - \ln(n+1)]$

$\Rightarrow s_k = [\ln(1) - \ln(2)] + [\ln(2) - \ln(3)] + [\ln(3) - \ln(4)] + \ldots + [\ln(k-1) - \ln(k)] + [\ln(k) - \ln(k+1)] = -\ln(k+1)$

$\Rightarrow \lim_{k \to \infty} s_k = -\infty, \Rightarrow$ diverges

67. convergent geometric series with sum $\frac{1}{1-\left(\frac{e}{\pi}\right)} = \frac{\pi}{\pi-e}$

69. $\sum_{n=0}^{\infty} (-1)^n x^n = \sum_{n=0}^{\infty} (-x)^n$; $a = 1$, $r = -x$; converges to $\frac{1}{1-(-x)} = \frac{1}{1+x}$ for $|x| < 1$

71. $a = 3$, $r = \frac{x-1}{2}$; converges to $\frac{3}{1-\left(\frac{x-1}{2}\right)} = \frac{6}{3-x}$ for $-1 < \frac{x-1}{2} < 1$ or $-1 < x < 3$

73. $a = 1$, $r = 2x$; converges to $\frac{1}{1-2x}$ for $|2x| < 1$ or $|x| < \frac{1}{2}$

75. $a = 1$, $r = -(x+1)$; converges to $\frac{1}{1+(x+1)} = \frac{1}{2+x}$ for $|x+1| < 1$ or $-2 < x < 0$

77. $a = 1$, $r = \sin x$; converges to $\frac{1}{1-\sin x}$ for $x \neq (2k+1)\frac{\pi}{2}$, k an integer

79. (a) $\sum_{n=-2}^{\infty} \frac{1}{(n+4)(n+5)}$ (b) $\sum_{n=0}^{\infty} \frac{1}{(n+2)(n+3)}$ (c) $\sum_{n=5}^{\infty} \frac{1}{(n-3)(n-2)}$

81. (a) one example is $\frac{1}{2} + \frac{1}{4} + \frac{1}{8} + \frac{1}{16} + \ldots = \frac{\left(\frac{1}{2}\right)}{1-\left(\frac{1}{2}\right)} = 1$

    (b) one example is $-\frac{3}{2} - \frac{3}{4} - \frac{3}{8} - \frac{3}{16} - \ldots = \frac{\left(-\frac{3}{2}\right)}{1-\left(\frac{1}{2}\right)} = -3$

    (c) one example is $1 - \frac{1}{2} - \frac{1}{4} - \frac{1}{8} - \frac{1}{16} - \ldots = 1 - \frac{\left(\frac{1}{2}\right)}{1-\left(\frac{1}{2}\right)} = 0$

83. Let $a_n = b_n = \left(\frac{1}{2}\right)^n$. Then $\sum_{n=1}^{\infty} a_n = \sum_{n=1}^{\infty} b_n = \sum_{n=1}^{\infty} \left(\frac{1}{2}\right)^n = 1$, while $\sum_{n=1}^{\infty} \left(\frac{a_n}{b_n}\right) = \sum_{n=1}^{\infty} (1)$ diverges.

85. Let $a_n = \left(\frac{1}{4}\right)^n$ and $b_n = \left(\frac{1}{2}\right)^n$. Then $A = \sum_{n=1}^{\infty} a_n = \frac{1}{3}$, $B = \sum_{n=1}^{\infty} b_n = 1$ and $\sum_{n=1}^{\infty} \left(\frac{a_n}{b_n}\right) = \sum_{n=1}^{\infty} \left(\frac{1}{2}\right)^n = 1 \neq \frac{A}{B}$.

87. Since the sum of a finite number of terms is finite, adding or subtracting a finite number of terms from a series that diverges does not change the divergence of the series.

89. (a) $\frac{2}{1-r} = 5 \Rightarrow \frac{2}{5} = 1 - r \Rightarrow r = \frac{3}{5}$; $2 + 2\left(\frac{3}{5}\right) + 2\left(\frac{3}{5}\right)^2 + \ldots$

    (b) $\frac{\left(\frac{13}{2}\right)}{1-r} = 5 \Rightarrow \frac{13}{10} = 1 - r \Rightarrow r = -\frac{3}{10}$; $\frac{13}{2} - \frac{13}{2}\left(\frac{3}{10}\right) + \frac{13}{2}\left(\frac{3}{10}\right)^2 - \frac{13}{2}\left(\frac{3}{10}\right)^3 + \ldots$

91. $s_n = 1 + 2r + r^2 + 2r^3 + r^4 + 2r^5 + \ldots + r^{2n} + 2r^{2n+1}$, $n = 0, 1, \ldots$

    $\Rightarrow s_n = \left(1 + r^2 + r^4 + \ldots + r^{2n}\right) + \left(2r + 2r^3 + 2r^5 + \ldots + 2r^{2n+1}\right) \Rightarrow \lim_{n \to \infty} s_n = \frac{1}{1-r^2} + \frac{2r}{1-r^2} = \frac{1+2r}{1-r^2}$,

    if $|r^2| < 1$ or $|r| < 1$

93. (a) After 24 hours, before the second pill: $300e^{(-0.12)(24)} \approx 16.840$ mg; after 48 hours, the amount present after 24 hours continues to decay and the dose taken at 24 hours has 24 hours to decay, so the amount present is $300e^{(-0.12)(48)} + 300e^{(-0.12)(24)} \approx 0.945 + 16.840 = 17.785$ mg.

   (b) The long-run quantity of the drug is $300 \sum_{1}^{\infty} \left(e^{(-0.12)(24)} \right)^n = 300 \dfrac{e^{(-0.12)(24)}}{1 - e^{(-0.12)(24)}} \approx 17.84$ mg.

95. (a) The endpoint of any closed interval remaining at any stage of the construction will remain in the Cantor set, so some points in the set include $0, \frac{1}{27}, \frac{2}{27}, \frac{1}{9}, \frac{2}{9}, \frac{7}{27}, \frac{8}{27}, \frac{1}{3}, \frac{2}{3}, \frac{7}{9}, \frac{8}{9}, 1$.

   (b) The lengths of the intervals removed are:

   Stage 1: $\dfrac{1}{3}$

   Stage 2: $\dfrac{1}{3}\left(1 - \dfrac{1}{3}\right) = \dfrac{2}{9}$

   Stage 3: $\dfrac{1}{3}\left(1 - \dfrac{1}{3} - \dfrac{2}{9}\right) = \dfrac{4}{27}$ and so on.

   Thus the sum of the lengths of the intervals removed is $\sum_{n=1}^{\infty} \dfrac{1}{3}\left(\dfrac{2}{3}\right)^{n-1} = \dfrac{1}{3} \cdot \dfrac{1}{1-(2/3)} = 1$.

9.3   THE INTEGRAL TEST

1. $f(x) = \frac{1}{x^2}$ is positive, continuous, and decreasing for $x \ge 1$; $\int_1^\infty \frac{1}{x^2} dx = \lim_{b \to \infty} \int_1^b \frac{1}{x^2} dx = \lim_{b \to \infty} \left[-\frac{1}{x} \right]_1^b$

   $= \lim_{b \to \infty} \left(-\frac{1}{b} + 1 \right) = 1 \Rightarrow \int_1^\infty \frac{1}{x^2} dx$ converges $\Rightarrow \sum_{n=1}^{\infty} \frac{1}{n^2}$ converges

3. $f(x) = \frac{1}{x^2+4}$ is positive, continuous, and decreasing for $x \ge 1$; $\int_1^\infty \frac{1}{x^2+4} dx = \lim_{b \to \infty} \int_1^b \frac{1}{x^2+4} dx$

   $= \lim_{b \to \infty} \left[\frac{1}{2} \tan^{-1} \frac{x}{2} \right]_1^b = \lim_{b \to \infty} \left(\frac{1}{2} \tan^{-1} \frac{b}{2} - \frac{1}{2} \tan^{-1} \frac{1}{2} \right) = \frac{\pi}{4} - \frac{1}{2} \tan^{-1} \frac{1}{2} \Rightarrow \int_1^\infty \frac{1}{x^2+4} dx$ converges

   $\Rightarrow \sum_{n=1}^{\infty} \frac{1}{n^2+4}$ converges

5. $f(x) = e^{-2x}$ is positive, continuous, and decreasing for $x \ge 1$; $\int_1^\infty e^{-2x} dx = \lim_{b \to \infty} \int_1^b e^{-2x} dx$

   $= \lim_{b \to \infty} \left[-\frac{1}{2} e^{-2x} \right]_1^b = \lim_{b \to \infty} \left(-\frac{1}{2e^{2b}} + \frac{1}{2e^2} \right) = \frac{1}{2e^2} \Rightarrow \int_1^\infty e^{-2x} dx$ converges $\Rightarrow \sum_{n=1}^{\infty} e^{-2n}$ converges

7. $f(x) = \frac{x}{x^2+4}$ is positive and continuous for $x \ge 1$, $f'(x) = \frac{4-x^2}{\left(x^2+4\right)^2} < 0$ for $x > 2$, thus f is decreasing for

   $x \ge 3$; $\int_3^\infty \frac{x}{x^2+4} dx = \lim_{b \to \infty} \int_3^b \frac{x}{x^2+4} dx = \lim_{b \to \infty} \left[\frac{1}{2} \ln\left(x^2+4\right) \right]_3^b = \lim_{b \to \infty} \left(\frac{1}{2} \ln\left(b^2+4\right) - \frac{1}{2} \ln(13) \right) = \infty \Rightarrow \int_3^\infty \frac{x}{x^2+4} dx$

   diverges $\Rightarrow \sum_{n=3}^{\infty} \frac{n}{n^2+4}$ diverges $\Rightarrow \sum_{n=1}^{\infty} \frac{n}{n^2+4} = \frac{1}{5} + \frac{2}{8} + \sum_{n=3}^{\infty} \frac{n}{n^2+4}$ diverges

9. $f(x) = \frac{x^2}{e^{\sqrt[3]{x}}}$ is positive and continuous for $x \geq 1$, $f'(x) = \frac{-x(x-6)}{3e^{\sqrt[3]{x}}} < 0$ for $x > 6$, thus f is decreasing for $x \geq 7$;

$$\int_7^\infty \frac{x^2}{e^{\sqrt[3]{x}}} dx = \lim_{b \to \infty} \int_7^b \frac{x^2}{e^{\sqrt[3]{x}}} dx = \lim_{b \to \infty} \left[-\frac{3x^2}{e^{\sqrt[3]{x}}} - \frac{18x}{e^{\sqrt[3]{x}}} - \frac{54}{e^{\sqrt[3]{x}}} \right]_7^b = \lim_{b \to \infty} \left(\frac{-3b^2 - 18b - 54}{e^{b/3}} + \frac{327}{e^{7/3}} \right) = \lim_{b \to \infty} \left(\frac{3(-6b-18)}{e^{b/3}} \right) + \frac{327}{e^{7/3}}$$

$$= \lim_{b \to \infty} \left(\frac{-54}{e^{b/3}} \right) + \frac{327}{e^{7/3}} = \frac{327}{e^{7/3}} \Rightarrow \int_7^\infty \frac{x^2}{e^{\sqrt[3]{x}}} dx \text{ converges} \Rightarrow \sum_{n=7}^\infty \frac{n^2}{e^{n/3}} \text{ converges}$$

$$\Rightarrow \sum_{n=1}^\infty \frac{n^2}{e^{n/3}} = \frac{1}{e^{1/3}} + \frac{4}{e^{2/3}} + \frac{9}{e^1} + \frac{16}{e^{4/3}} + \frac{25}{e^{5/3}} + \frac{36}{e^2} + \sum_{n=7}^\infty \frac{n^2}{e^{n/3}} \text{ converges}$$

11. converges; a geometric series with $r = \frac{1}{10} < 1$

13. diverges; by the nth-Term Test for Divergence, $\lim_{n \to \infty} \frac{n}{n+1} = 1 \neq 0$

15. diverges; $\sum_{n=1}^\infty \frac{3}{\sqrt{n}} = 3 \sum_{n=1}^\infty \frac{1}{\sqrt{n}}$, which is a divergent p-series with $p = \frac{1}{2}$

17. converges; a geometric series with $r = \frac{1}{8} < 1$

19. diverges by the Integral Test: $\int_2^n \frac{\ln x}{x} dx = \frac{1}{2}\left(\ln^2 n - \ln 2 \right) \Rightarrow \int_2^\infty \frac{\ln x}{x} dx \to \infty$

21. converges; a geometric series with $r = \frac{2}{3} < 1$

23. diverges; $\sum_{n=0}^\infty \frac{-2}{n+1} = -2 \sum_{n=0}^\infty \frac{1}{n+1}$, which diverges by the Integral Test

25. diverges; $\lim_{n \to \infty} a_n = \lim_{n \to \infty} \frac{2^n}{n+1} = \lim_{n \to \infty} \frac{2^n \ln 2}{1} = \infty \neq 0$

27. diverges; $\lim_{n \to \infty} \frac{\sqrt{n}}{\ln n} = \lim_{n \to \infty} \frac{\left(\frac{1}{2\sqrt{n}} \right)}{\left(\frac{1}{n} \right)} = \lim_{n \to \infty} \frac{\sqrt{n}}{2} = \infty \neq 0$

29. diverges; a geometric series with $r = \frac{1}{\ln 2} \approx 1.44 > 1$

31. converges by the Integral Test: $\int_3^\infty \frac{\left(\frac{1}{x} \right)}{(\ln x)\sqrt{(\ln x)^2 - 1}} dx$; $\left[u = \ln x, \, du = \frac{1}{x} dx \right]$

$$\to \int_{\ln 3}^\infty \frac{1}{u\sqrt{u^2-1}} du = \lim_{b \to \infty} \left[\sec^{-1}|u| \right]_{\ln 3}^b = \lim_{b \to \infty} \left[\sec^{-1} b - \sec^{-1}(\ln 3) \right] = \lim_{b \to \infty} \left[\cos^{-1}\left(\frac{1}{b} \right) - \sec^{-1}(\ln 3) \right]$$

$$= \cos^{-1}(0) - \sec^{-1}(\ln 3) = \frac{\pi}{2} - \sec^{-1}(\ln 3) \approx 1.1439$$

33. diverges by the nth-Term Test for divergence; $\lim_{n \to \infty} n \sin\left(\frac{1}{n} \right) = \lim_{n \to \infty} \frac{\sin\left(\frac{1}{n} \right)}{\left(\frac{1}{n} \right)} = \lim_{x \to 0} \frac{\sin x}{x} = 1 \neq 0$

35. converges by the Integral Test: $\int_1^\infty \frac{e^x}{1+e^{2x}}\,dx;$ $\left[u = e^x,\ du = e^x dx \right]$

$\to \int_e^\infty \frac{1}{1+u^2}\,du = \lim_{n\to\infty} \left[\tan^{-1} u \right]_e^b = \lim_{b\to\infty} \left(\tan^{-1} b - \tan^{-1} e \right) = \frac{\pi}{2} - \tan^{-1} e \approx 0.35$

37. converges by the Integral Test: $\int_1^\infty \frac{8\tan^{-1} x}{1+x^2}\,dx;$ $\left[\begin{array}{l} u = \tan^{-1} x \\ du = \frac{dx}{1+x^2} \end{array} \right] \to \int_{\pi/4}^{\pi/2} 8u\,du = \left[4u^2 \right]_{\pi/4}^{\pi/2} = 4\left(\frac{\pi^2}{4} - \frac{\pi^2}{16} \right) = \frac{3\pi^2}{4}$

39. converges by the Integral Test: $\int_1^\infty \operatorname{sech} x\,dx = 2 \lim_{b\to\infty} \int_1^b \frac{e^x}{1+(e^x)^2}\,dx = 2 \lim_{b\to\infty} \left[\tan^{-1} e^x \right]_1^b$

$= 2 \lim_{b\to\infty} \left(\tan^{-1} e^b - \tan^{-1} e \right) = \pi - 2\tan^{-1} e \approx 0.71$

41. $\int_1^\infty \left(\frac{a}{x+2} - \frac{1}{x+4} \right) dx = \lim_{b\to\infty} \left[a \ln|x+2| - \ln|x+4| \right]_1^b = \lim_{b\to\infty} \left[\ln\frac{(b+2)^a}{b+4} - \ln\left(\frac{3^a}{5} \right) \right] = \lim_{b\to\infty} \ln\frac{(b+2)^a}{b+4} - \ln\left(\frac{3^a}{5} \right);$

$\lim_{b\to\infty} \frac{(b+2)^a}{b+4} = a \lim_{b\to\infty} (b+2)^{a-1} = \begin{cases} \infty, & a>1 \\ 1, & a=1 \end{cases} \Rightarrow$ the series converges to $\ln\left(\frac{5}{3} \right)$ if $a=1$ and diverges to ∞ if

$a>1$. If $a<1$, the terms of the series eventually become negative and the Integral Test does not apply. From that point on, however, the series behaves like a negative multiple of the harmonic series, and so it diverges.

43. (a)

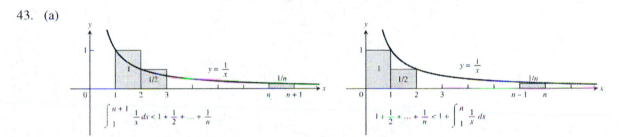

(b) There are $(13)(365)(24)(60)(60) \left(10^9 \right)$ seconds in 13 billion years; by part (a) $s_n \le 1 + \ln n$ where

$n = (13)(365)(24)(60)(60) \left(10^9 \right) \Rightarrow s_n \le 1 + \ln\left((13)(365)(24)(60)(60) \left(10^9 \right) \right)$

$= 1 + \ln(13) + \ln(365) + \ln(24) + 2\ln(60) + 9\ln(10) \approx 41.55$

45. Yes. If $\sum_{n=1}^\infty a_n$ is a divergent series of positive numbers, then $\left(\frac{1}{2} \right) \sum_{n=1}^\infty a_n = \sum_{n=1}^\infty \left(\frac{a_n}{2} \right)$ also diverges and $\frac{a_n}{2} < a_n$.

There is no "smallest" divergent series of positive number: for any divergent series $\sum_{n=1}^\infty a_n$ of positive

numbers $\sum_{n=1}^\infty \left(\frac{a_n}{2} \right)$ has smaller terms and still diverges.

47. (a) Both integrals can represent the area under the curve $f(x) = \frac{1}{\sqrt{x+1}}$, and the sum s_{50} can be considered an

approximation of either integral using rectangles with $\Delta x = 1$. The sum $s_{50} = \sum_{n=1}^{50} \frac{1}{\sqrt{n+1}}$ is an overestimate

of the integral $\int_1^{51} \frac{1}{\sqrt{x+1}}\,dx$. The sum s_{50} represents a left-hand sum (that is, the we are choosing the left-hand endpoint of each subinterval for c_i) and because f is a decreasing function, the value of f is a maximum at the left-hand endpoint of each subinterval. The area of each rectangle overestimates the true area, thus $\int_1^{51} \frac{1}{\sqrt{x+1}}\,dx < \sum_{n=1}^{50} \frac{1}{\sqrt{n+1}}$. In a similar manner, s_{50} underestimates the integral $\int_0^{50} \frac{1}{\sqrt{x+1}}\,dx$. In this case, the sum s_{50} represents a right-hand sum and because f is a decreasing function, the value of f is a minimum at the right-hand endpoint of each subinterval. The area of each rectangle underestimates the true area, thus $\sum_{n=1}^{50} \frac{1}{\sqrt{n+1}} < \int_0^{50} \frac{1}{\sqrt{x+1}}\,dx$. Evaluating the integrals we find $\int_1^{51} \frac{1}{\sqrt{x+1}}\,dx = \left[2\sqrt{x+1}\right]_1^{51}$

$$= 2\sqrt{52} - 2\sqrt{2} \approx 11.6 \text{ and } \int_0^{50} \frac{1}{\sqrt{x+1}}\,dx = \left[2\sqrt{x+1}\right]_0^{50} = 2\sqrt{51} - 2\sqrt{1} \approx 12.3. \text{ Thus, } 11.6 < \sum_{n=1}^{50} \frac{1}{\sqrt{n+1}} < 12.3.$$

(b) $s_n > 1000 \Rightarrow \int_1^{n+1} \frac{1}{\sqrt{x+1}}\,dx = \left[2\sqrt{x+1}\right]_1^{n+1} = 2\sqrt{n+1} - 2\sqrt{2} > 1000 \Rightarrow n > \left(500 + \sqrt{2}\right)^2 - 1 \approx 251414.2$

$\Rightarrow n \geq 251415.$

49. We want $S - s_n < 0.01 \Rightarrow \int_n^\infty \frac{1}{x^3}\,dx < 0.01 \Rightarrow \int_n^\infty \frac{1}{x^3}\,dx = \lim_{b\to\infty} \int_n^b \frac{1}{x^3}\,dx = \lim_{b\to\infty} \left[-\frac{1}{2x^2}\right]_n^b = \lim_{b\to\infty} \left(-\frac{1}{2b^2} + \frac{1}{2n^2}\right)$

$= \frac{1}{2n^2} < 0.01 \Rightarrow n > \sqrt{50} \approx 7.071 \Rightarrow n \geq 8 \Rightarrow S \approx s_8 = \sum_{n=1}^{8} \frac{1}{n^3} \approx 1.195$

51. $S - s_n < 0.00001 \Rightarrow \int_n^\infty \frac{1}{x^{1.1}}\,dx < 0.00001 \Rightarrow \int_n^\infty \frac{1}{x^{1.1}}\,dx = \lim_{b\to\infty} \int_n^b \frac{1}{x^{1.1}}\,dx = \lim_{b\to\infty} \left[-\frac{10}{x^{0.1}}\right]_n^b = \lim_{b\to\infty} \left(-\frac{10}{b^{0.1}} + \frac{10}{n^{0.1}}\right)$

$= \frac{10}{n^{0.1}} < 0.00001 \Rightarrow n > 1000000^{10} \Rightarrow n > 10^{60}$

53. Let $A_n = \sum_{k=1}^{n} a_k$ and $B_n = \sum_{k=1}^{n} 2^k a_{(2^k)}$, where $\{a_k\}$ is a nonincreasing sequence of positive terms converging to 0. Note that $\{A_n\}$ and $\{B_n\}$ are nondecreasing sequences of positive terms. Now,

$B_n = 2a_2 + 4a_4 + 8a_8 + \ldots + 2^n a_{(2^n)}$

$= 2a_2 + \left(2a_4 + 2a_4\right) + \left(2a_8 + 2a_8 + 2a_8 + 2a_8\right) + \ldots + \underbrace{\left(2a_{(2^n)} + 2a_{(2^n)} + \ldots + 2a_{(2^n)}\right)}_{2^{n-1}\,\text{terms}}$

$\leq 2a_1 + 2a_2 + \left(2a_3 + 2a_4\right) + \left(2a_5 + 2a_6 + 2a_7 + 2a_8\right) + \ldots + \left(2a_{(2^{n-1})} + 2a_{(2^{n-1}+1)} + \ldots + 2a_{(2^n)}\right) = 2A_{(2^n)}$

$\leq 2\sum_{k=1}^{\infty} a_k$. Therefore if $\sum a_k$ converges, then $\{B_n\}$ is bounded above $\Rightarrow \sum 2^k a_{(2^k)}$ converges. Conversely,

$A_n = a_1 + \left(a_2 + a_3\right) + \left(a_4 + a_5 + a_6 + a_7\right) + \ldots + a_n < a_1 + 2a_2 + 4a_4 + \ldots + 2^n a_{(2^n)} = a_1 + B_n < a_1 + \sum_{k=1}^{\infty} 2^k a_{(2^k)}.$

Therefore, if $\sum_{k=1}^{\infty} 2^k a_{(2^k)}$ converges, then $\{A_n\}$ is bounded above and hence converges.

55. (a) $\int_2^\infty \frac{dx}{x(\ln x)^p}$; $\left[u = \ln x, \, du = \frac{dx}{x}\right] \to \int_{\ln 2}^\infty u^{-p}\,du = \lim_{b\to\infty}\left[\frac{u^{-p+1}}{-p+1}\right]_{\ln 2}^b = \lim_{b\to\infty}\left(\frac{1}{1-p}\right)\left[b^{-p+1} - (\ln 2)^{-p+1}\right]$

$= \begin{cases} \frac{1}{p-1}(\ln 2)^{-p+1} & p > 1 \\ \infty, & p < 1 \end{cases} \Rightarrow$ the improper integral converges if $p > 1$ and diverges if $p < 1$. For $p = 1$:

$\int_2^\infty \frac{dx}{x \ln x} = \lim_{b\to\infty}\left[\ln(\ln x)\right]_2^b = \lim_{b\to\infty}\left[\ln(\ln b) - \ln(\ln 2)\right] = \infty$, so the improper integral diverges if $p = 1$.

(b) Since the series and the integral converge or diverge together, $\sum_{n=2}^\infty \frac{1}{n(\ln n)^p}$ converges if and only if $p > 1$.

57. (a) From Fig. 10.11 (a) in the text with $f(x) = \frac{1}{x}$ and $a_k = \frac{1}{k}$, we have $\int_1^{n+1} \frac{1}{x}\,dx \le 1 + \frac{1}{2} + \frac{1}{3} + \dots + \frac{1}{n}$

$\le 1 + \int_1^n f(x)\,dx \Rightarrow \ln(n+1) \le 1 + \frac{1}{2} + \frac{1}{3} + \dots + \frac{1}{n} \le 1 + \ln n \Rightarrow 0 \le \ln(n+1) - \ln n \le \left(1 + \frac{1}{2} + \frac{1}{3} + \dots + \frac{1}{n}\right) - \ln n \le 1$.

Therefore the sequence $\left\{\left(1 + \frac{1}{2} + \frac{1}{3} + \dots + \frac{1}{n}\right) - \ln n\right\}$ is bounded above by 1 and below by 0.

(b) From the graph in Fig. 10.11 (b) with $f(x) = \frac{1}{x}, \frac{1}{n+1} < \int_n^{n+1} \frac{1}{x}\,dx = \ln(n+1) - \ln n$

$\Rightarrow 0 > \frac{1}{n+1} - [\ln(n+1) - \ln n] = \left(1 + \frac{1}{2} + \frac{1}{3} + \dots + \frac{1}{n+1} - \ln(n+1)\right) - \left(1 + \frac{1}{2} + \frac{1}{3} + \dots + \frac{1}{n} - \ln n\right)$.

If we define $a_n = 1 + \frac{1}{2} = \frac{1}{3} + \frac{1}{n} - \ln n$, then $0 > a_{n+1} - a_n \Rightarrow a_{n+1} < a_n \Rightarrow \{a_n\}$ is a decreasing

sequence of nonnegative terms.

59. (a) $s_{10} = \sum_{n=1}^\infty \frac{1}{n^3} = 1.97531986$; $\int_{11}^\infty \frac{1}{x^3}\,dx = \lim_{b\to\infty}\int_{11}^b x^{-3}\,dx = \lim_{b\to\infty}\left[-\frac{x^{-2}}{2}\right]_{11}^b = \lim_{b\to\infty}\left(-\frac{1}{2b^2} + \frac{1}{242}\right) = \frac{1}{242}$

and $\int_{10}^\infty \frac{1}{x^3}\,dx = \lim_{b\to\infty}\int_{10}^b x^{-3}\,dx = \lim_{b\to\infty}\left[-\frac{x^{-2}}{2}\right]_{10}^b = \lim_{b\to\infty}\left(-\frac{1}{2b^2} + \frac{1}{200}\right) = \frac{1}{200}$

$\Rightarrow 1.97531986 + \frac{1}{242} < s < 1.97531986 + \frac{1}{200} \Rightarrow 1.20166 < s < 1.20253$

(b) $s = \sum_{n=1}^\infty \frac{1}{n^3} \approx \frac{1.20166 + 1.20253}{2} = 1.202095$; error $\le \frac{1.20253 - 1.20166}{2} = 0.000435$

61. The total area will be $\sum_{n=1}^\infty \frac{1}{n}\cdot\frac{1}{n} - \frac{1}{n+1} = \sum_{n=1}^\infty \frac{1}{n^2} - \frac{1}{n(n+1)}$. The p-series $\sum_{n=1}^\infty \frac{1}{n^2}$ converges to $\frac{\pi^2}{6}$ and

$\sum_{n=1}^\infty \frac{1}{n(n+1)}$ converges to 1 (see Example 5). Thus we can write the area as the difference of these two values,

or $\frac{\pi^2}{6} - 1 \approx 0.64493$.

9.4   COMPARISON TESTS

1. Compare with $\sum_{n=1}^\infty \frac{1}{n^2}$, which is a convergent p-series since $p = 2 > 1$. Both series have nonnegative terms for

$n \ge 1$. For $n \ge 1$, we have $n^2 \le n^2 + 30 \Rightarrow \frac{1}{n^2} \ge \frac{1}{n^2 + 30}$. Then by Comparison Test, $\sum_{n=1}^\infty \frac{1}{n^2 + 30}$ converges.

3. Compare with $\displaystyle\sum_{n=2}^{\infty}\frac{1}{\sqrt{n}}$, which is a divergent p-series since $p=\frac{1}{2}\leq 1$. Both series have nonnegative terms for

    $n\geq 2$. For $n\geq 2$, we have $\sqrt{n}-1\leq\sqrt{n}\Rightarrow\frac{1}{\sqrt{n-1}}\geq\frac{1}{\sqrt{n}}$. Then by Comparison Test, $\displaystyle\sum_{n=2}^{\infty}\frac{1}{\sqrt{n-1}}$ diverges.

5. Compare with $\displaystyle\sum_{n=1}^{\infty}\frac{1}{n^{3/2}}$, which is a convergent p-series since $p=\frac{3}{2}>1$. Both series have nonnegative terms for

    $n\geq 1$. For $n\geq 1$, we have $0\leq\cos^2 n\leq 1\Rightarrow\frac{\cos^2 n}{n^{3/2}}\leq\frac{1}{n^{3/2}}$. Then by Comparison Test, $\displaystyle\sum_{n=1}^{\infty}\frac{\cos^2 n}{n^{3/2}}$ converges.

7. Compare with $\displaystyle\sum_{n=1}^{\infty}\frac{\sqrt{5}}{n^{3/2}}$. The series $\displaystyle\sum_{n=1}^{\infty}\frac{1}{n^{3/2}}$ is a convergent p-series since $p=\frac{3}{2}>1$, and the series $\displaystyle\sum_{n=1}^{\infty}\frac{\sqrt{5}}{n^{3/2}}$

    $=\sqrt{5}\displaystyle\sum_{n=1}^{\infty}\frac{1}{n^{3/2}}$ converges by Theorem 8 part 3. Both series have nonnegative terms for $n\geq 1$. For $n\geq 1$,

    we have $n^3\leq n^4\Rightarrow 4n^3\leq 4n^4\Rightarrow n^4+4n^3\leq n^4+4n^4=5n^4\Rightarrow n^4+4n^3\leq 5n^4+20=5\left(n^4+4\right)$

    $\Rightarrow\frac{n^4+4n^3}{n^4+4}\leq 5\Rightarrow\frac{n^3(n+4)}{n^4+4}\leq 5\Rightarrow\frac{n+4}{n^4+4}\leq\frac{5}{n^3}\Rightarrow\sqrt{\frac{n+4}{n^4+4}}\leq\sqrt{\frac{5}{n^3}}=\frac{\sqrt{5}}{n^{3/2}}$. Then by Comparison Test, $\displaystyle\sum_{n=1}^{\infty}\sqrt{\frac{n+4}{n^4+4}}$
    converges.

9. Compare with $\displaystyle\sum_{n=1}^{\infty}\frac{1}{n^2}$, which is a convergent p-series since $p=2>1$. Both series have positive terms for

    $n\geq 1$. $\displaystyle\lim_{n\to\infty}\frac{a_n}{b_n}=\lim_{n\to\infty}\frac{\frac{n-2}{n^3-n^2+3}}{1/n^2}=\lim_{n\to\infty}\frac{n^3-2n^2}{n^3-n^2+3}=\lim_{n\to\infty}\frac{3n^2-4n}{3n^2-2n}=\lim_{n\to\infty}\frac{6n-4}{6n-2}=\lim_{n\to\infty}\frac{6}{6}=1>0$. Then by Limit

    Comparison Test, $\displaystyle\sum_{n=1}^{\infty}\frac{n-2}{n^3-n^2+3}$ converges.

11. Compare with $\displaystyle\sum_{n=2}^{\infty}\frac{1}{n}$, which is a divergent p-series since $p=1\leq 1$. Both series have positive terms for $n\geq 2$.

    $\displaystyle\lim_{n\to\infty}\frac{a_n}{b_n}=\lim_{n\to\infty}\frac{\frac{n(n+1)}{(n^2+1)(n-1)}}{1/n}=\lim_{n\to\infty}\frac{n^3+n^2}{n^3-n^2+n-1}=\lim_{n\to\infty}\frac{3n^2+2n}{3n^2-2n+1}=\lim_{n\to\infty}\frac{6n+2}{6n-2}=\lim_{n\to\infty}\frac{6}{6}=1>0$. Then by Limit

    Comparison Test, $\displaystyle\sum_{n=2}^{\infty}\frac{n(n+1)}{(n^2+1)(n-1)}$ diverges.

13. Compare with $\displaystyle\sum_{n=1}^{\infty}\frac{1}{\sqrt{n}}$, which is a divergent p-series since $p=\frac{1}{2}\leq 1$. Both series have positive terms for $n\geq 1$.

    $\displaystyle\lim_{n\to\infty}\frac{a_n}{b_n}=\lim_{n\to\infty}\frac{\frac{5^n}{\sqrt{n}\cdot 4^n}}{1/\sqrt{n}}=\lim_{n\to\infty}\frac{5^n}{4^n}=\lim_{n\to\infty}\left(\frac{5}{4}\right)^n=\infty$. Then by Limit Comparison Test, $\displaystyle\sum_{n=1}^{\infty}\frac{5^n}{\sqrt{n}\cdot 4^n}$ diverges.

15. Compare with $\displaystyle\sum_{n=2}^{\infty} \frac{1}{n}$, which is a divergent p-series, since $p = 1 \leq 1$. Both series have positive terms for $n \geq 2$.

$$\lim_{n\to\infty} \frac{a_n}{b_n} = \lim_{n\to\infty} \frac{\frac{1}{\ln n}}{1/n} = \lim_{n\to\infty} \frac{n}{\ln n} = \lim_{n\to\infty} \frac{1}{1/n} = \lim_{n\to\infty} n = \infty. \text{ Then by Limit Comparison Test, } \sum_{n=2}^{\infty} \frac{1}{\ln n} \text{ diverges.}$$

17. diverges by the Limit Comparison Test (part 1) when compared with $\displaystyle\sum_{n=1}^{\infty} \frac{1}{\sqrt{n}}$, a divergent p-series

$$\lim_{n\to\infty} \frac{\left(\frac{1}{2\sqrt{n}+\sqrt[3]{n}}\right)}{\left(\frac{1}{\sqrt{n}}\right)} = \lim_{n\to\infty} \frac{\sqrt{n}}{2\sqrt{n}+\sqrt[3]{n}} = \lim_{n\to\infty} \left(\frac{1}{2+n^{-1/6}}\right) = \frac{1}{2}$$

19. converges by the Direct Comparison Test; $\frac{\sin^2 n}{2^n} \leq \frac{1}{2^n}$, which is the nth term of a convergent geometric series

21. diverges since $\displaystyle\lim_{n\to\infty} \frac{2n}{3n-1} = \frac{2}{3} \neq 0$

23. converges by the Limit Comparison Test (part 1) with $\frac{1}{n^2}$, the nth term of a convergent p-series

$$\lim_{n\to\infty} \frac{\left(\frac{10n+1}{n(n+1)(n+2)}\right)}{\left(\frac{1}{n^2}\right)} = \lim_{n\to\infty} \frac{10n^2+n}{n^2+3n+2} = \lim_{n\to\infty} \frac{20n+1}{2n+3} = \lim_{n\to\infty} \frac{20}{2} = 10$$

25. converges by the Direct Comparison Test; $\left(\frac{n}{3n+1}\right)^n < \left(\frac{n}{3n}\right)^n = \left(\frac{1}{3}\right)^n$, the nth term of a convergent geometric series

27. diverges by the Direct Comparison Test; $n > \ln n \Rightarrow \ln n > \ln \ln n \Rightarrow \frac{1}{n} < \frac{1}{\ln n} < \frac{1}{\ln(\ln n)}$ and $\displaystyle\sum_{n=3}^{\infty} \frac{1}{n}$ diverges

29. diverges by the Limit Comparison Test (part 3) with $\frac{1}{n}$, the nth term of the divergent harmonic series:

$$\lim_{n\to\infty} \frac{\left[\frac{1}{\sqrt{n}\ln n}\right]}{\left(\frac{1}{n}\right)} = \lim_{n\to\infty} \frac{\sqrt{n}}{\ln n} = \lim_{n\to\infty} \frac{\left(\frac{1}{2\sqrt{n}}\right)}{\left(\frac{1}{n}\right)} = \lim_{n\to\infty} \frac{\sqrt{n}}{2} = \infty$$

31. diverges by the Limit Comparison Test (part 3) with $\frac{1}{n}$, the nth term of the divergent harmonic series:

$$\lim_{n\to\infty} \frac{\left(\frac{1}{1+\ln n}\right)}{\left(\frac{1}{n}\right)} = \lim_{n\to\infty} \frac{n}{1+\ln n} = \lim_{n\to\infty} \frac{1}{\left(\frac{1}{n}\right)} = \lim_{n\to\infty} n = \infty$$

33. converges by the Direct Comparison Test with $\frac{1}{n^{3/2}}$, the nth term of a convergent p-series $n^2 - 1 > n$ for

$$n \geq 2 \Rightarrow n^2(n^2-1) > n^3 \Rightarrow n\sqrt{n^2-1} > n^{3/2} \Rightarrow \frac{1}{n^{3/2}} > \frac{1}{n\sqrt{n^2-1}} \text{ or use Limit Comparison Test with } \frac{1}{n^2}.$$

35. converges because $\sum_{n=1}^{\infty}\frac{1-n}{n2^n} = \sum_{n=1}^{\infty}\frac{1}{n2^n} + \sum_{n=1}^{\infty}\frac{-1}{2^n}$ which is the sum of two convergent series: $\sum_{n=1}^{\infty}\frac{1}{n2^n}$ converges by

   the Direct Comparison Test since $\frac{1}{n2^n} < \frac{1}{2^n}$, and $\sum_{n=1}^{\infty}\frac{-1}{2^n}$ is a convergent geometric series

37. converges by the Direct Comparison Test: $\frac{1}{3^{n-1}+1} < \frac{1}{3^{n-1}}$, which is the nth term of a convergent geometric series

39. converges by Limit Comparison Test: compare with $\sum_{n=1}^{\infty}\left(\frac{1}{5}\right)^n$, which is a convergent geometric series with

   $|r| = \frac{1}{5} < 1$, $\lim_{n\to\infty}\frac{\left(\frac{n+1}{n^2+3n}\cdot\frac{1}{5^n}\right)}{(1/5)^n} = \lim_{n\to\infty}\frac{n+1}{n^2+3n} = \lim_{n\to\infty}\frac{1}{2n+3} = 0.$

41. diverges by Limit Comparison Test: compare with $\sum_{n=1}^{\infty}\frac{1}{n}$, which is a divergent p-series

   $\lim_{n\to\infty}\frac{\left(\frac{2^n-n}{n\cdot2^n}\right)}{1/n} = \lim_{n\to\infty}\frac{2^n-n}{2^n} = \lim_{n\to\infty}\frac{2^n\ln2-1}{2^n\ln2} = \lim_{n\to\infty}\frac{2^n(\ln2)^2}{2^n(\ln2)^2} = 1 > 0.$

43. converges by Comparison Test with $\sum_{n=2}^{\infty}\frac{1}{n(n-1)}$ which converges since $\sum_{n=2}^{\infty}\frac{1}{n(n-1)} = \sum_{n=2}^{\infty}\left[\frac{1}{n-1} - \frac{1}{n}\right]$, and

   $s_k = \left(1 - \frac{1}{2}\right) + \left(\frac{1}{2} - \frac{1}{3}\right) + \ldots + \left(\frac{1}{k-2} - \frac{1}{k-1}\right) + \left(\frac{1}{k-1} - \frac{1}{k}\right) = 1 - \frac{1}{k} \Rightarrow \lim_{k\to\infty} s_k = 1;$ for $n \geq 2$, $(n-2)! \geq 1$

   $\Rightarrow n(n-1)(n-2)! \geq n(n-1) \Rightarrow n! \geq n(n-1) \Rightarrow \frac{1}{n!} \leq \frac{1}{n(n-1)}$

45. diverges by the Limit Comparison Test (part 1) with $\frac{1}{n}$, the nth term of the divergent harmonic series:

   $\lim_{n\to\infty}\frac{\left(\sin\frac{1}{n}\right)}{\left(\frac{1}{n}\right)} = \lim_{x\to0}\frac{\sin x}{x} = 1$

47. converges by the Direct Comparison Test: $\frac{\tan^{-1}n}{n^{1.1}} < \frac{\frac{\pi}{2}}{n^{1.1}}$ and $\sum_{n=1}^{\infty}\frac{\frac{\pi}{2}}{n^{1.1}} = \frac{\pi}{2}\sum_{n=1}^{\infty}\frac{1}{n^{1.1}}$ is the product of a convergent

   p-series and a nonzero constant

49. converges by the Limit Comparison Test (part 1) with $\frac{1}{n^2}$: $\lim_{n\to\infty}\frac{\left(\frac{\coth n}{n^2}\right)}{\left(\frac{1}{n^2}\right)} = \lim_{n\to\infty}\coth n = \lim_{n\to\infty}\frac{e^n+e^{-n}}{e^n-e^{-n}}$

   $= \lim_{n\to\infty}\frac{1+e^{-2n}}{1-e^{-2n}} = 1$

51. diverges by the Limit Comparison Test (part 1) with $\frac{1}{n}$: $\lim_{n\to\infty}\frac{\left(\frac{1}{n\sqrt[n]{n}}\right)}{\left(\frac{1}{n}\right)} = \lim_{n\to\infty}\frac{1}{\sqrt[n]{n}} = 1$

53. $\dfrac{1}{1+2+3+\ldots+n}=\dfrac{1}{\left(\frac{n(n+1)}{2}\right)}=\dfrac{2}{n(n+1)}$. The series converges by the Limit Comparison Test (part 1) with $\dfrac{1}{n^2}$:

$$\lim_{n\to\infty}\frac{\left(\frac{2}{n(n+1)}\right)}{\left(\frac{1}{n^2}\right)}=\lim_{n\to\infty}\frac{2n^2}{n^2+n}=\lim_{n\to\infty}\frac{4n}{2n+1}=\lim_{n\to\infty}\frac{4}{2}=2.$$

55. (a) If $\lim\limits_{n\to\infty}\dfrac{a_n}{b_n}=0$, then there exists an integer N such that for all $n>N$, $\left|\dfrac{a_n}{b_n}-0\right|<1\Rightarrow -1<\dfrac{a_n}{b_n}<1$

$\Rightarrow a_n<b_n$. Thus, if Σb_n converges, then Σa_n converges by the Direct Comparison Test.

(b) If $\lim\limits_{n\to\infty}\dfrac{a_n}{b_n}=\infty$, then there exists an integer N such that for all $n>N$, $\dfrac{a_n}{b_n}>1\Rightarrow a_n>b_n$. Thus, if

Σb_n diverges, then Σa_n diverges by the Direct Comparison Test.

57. $\lim\limits_{n\to\infty}\dfrac{a_n}{b_n}=\infty\Rightarrow$ there exists an integer N such that for all $n>N$, $\dfrac{a_n}{b_n}>1\Rightarrow a_n>b_n$. If Σa_n converges,

then Σb_n converges by the Direct Comparison Test

59. Since $a_n>0$ and $\lim\limits_{n\to\infty}a_n=\infty\neq 0$, by n^{th} term test for divergence, Σa_n diverges.

61. Let $-\infty<q<\infty$ and $p>1$. If $q=0$, then $\sum\limits_{n=2}^{\infty}\dfrac{(\ln n)^q}{n^p}=\sum\limits_{n=2}^{\infty}\dfrac{1}{n^p}$, which is a convergent p-series. If $q\neq 0$,

compare with $\sum\limits_{n=2}^{\infty}\dfrac{1}{n^r}$ where $1<r<p$, then $\lim\limits_{n\to\infty}\dfrac{\frac{(\ln n)^q}{n^p}}{1/n^r}=\lim\limits_{n\to\infty}\dfrac{(\ln n)^q}{n^{p-r}}$, and $p-r>0$. If $q<0\Rightarrow -q>0$ and

$\lim\limits_{n\to\infty}\dfrac{(\ln n)^q}{n^{p-r}}=\lim\limits_{n\to\infty}\dfrac{1}{(\ln n)^{-q}n^{p-r}}=0$. If $q>0$, $\lim\limits_{n\to\infty}\dfrac{(\ln n)^q}{n^{p-r}}=\lim\limits_{n\to\infty}\dfrac{q(\ln n)^{q-1}\left(\frac{1}{n}\right)}{(p-r)n^{p-r-1}}=\lim\limits_{n\to\infty}\dfrac{q(\ln n)^{q-1}}{(p-r)n^{p-r}}$. If

$q-1\leq 0\Rightarrow 1-q\geq 0$ and $\lim\limits_{n\to\infty}\dfrac{q(\ln n)^{q-1}}{(p-r)n^{p-r}}=\lim\limits_{n\to\infty}\dfrac{q}{(p-r)n^{p-r}(\ln n)^{1-q}}=0$, otherwise, we apply L'Hopital's Rule

again. $\lim\limits_{n\to\infty}\dfrac{q(q-1)(\ln n)^{q-2}\left(\frac{1}{n}\right)}{(p-r)^2n^{p-r-1}}=\lim\limits_{n\to\infty}\dfrac{q(q-1)(\ln n)^{q-2}}{(p-r)^2n^{p-r}}$. If $q-2\leq 0\Rightarrow 2-q\geq 0$ and

$\lim\limits_{n\to\infty}\dfrac{q(q-1)(\ln n)^{q-2}}{(p-r)^2n^{p-r}}=\lim\limits_{n\to\infty}\dfrac{q(q-1)}{(p-r)^2n^{p-r}(\ln n)^{2-q}}=0$; otherwise, we apply L'Hopital's Rule again. Since q is finite,

there is a positive integer k such that $q-k\leq 0\Rightarrow k-q\geq 0$. Thus, after k applications of L'Hopital's Rule we

obtain $\lim\limits_{n\to\infty}\dfrac{q(q-1)\cdots(q-k+1)(\ln n)^{q-k}}{(p-r)^k n^{p-r}}=\lim\limits_{n\to\infty}\dfrac{q(q-1)\cdots(q-k+1)}{(p-r)^k n^{p-r}(\ln n)^{k-q}}=0$. Since the limit is 0 in every case, by Limit

Comparison Test, the series $\sum\limits_{n=1}^{\infty}\dfrac{(\ln n)^q}{n^p}$ converges.

63. Since $0\leq d_n\leq 9$ for all n and the geometric series $\sum\limits_{n=1}^{\infty}\dfrac{9}{10^n}$ converges to 1, $\sum\limits_{n=1}^{\infty}\dfrac{d_n}{10^n}$ converges.

65. Converges by Exercise 61 with $q=3$ and $p=4$.

67. Converges by Exercise 61 with $q=1000$ and $p=1.001$.

69. Converges by Exercise 61 with $q = -3$ and $p = 1.1$.

71. Example CAS commands:

Maple:

```
a := n -> 1./n^3/sin(n)^2;
s := k -> sum( a(n), n=1..k );                              # (a)]
limit( s(k), k=infinity );
pts := [seq( [k,s(k)], k=1..100 )]:                         # (b)
plot( pts, style=point, title="#71(b) (Section 10.4)" );
pts := [seq( [k,s(k)], k=1..200 )]:                         # (c)
plot( pts, style=point, title="#71(c) (Section 10.4)" );
pts := [seq( [k,s(k)], k=1..400 )]:                         # (d)
plot( pts, style=point, title="#71(d) (Section 10.4)" );
evalf( 355/113 );
```

Mathematica:

```
Clear[a, n, s, k, p]
```

$a[n\_] := 1 / (n^3 \, Sin[n]^2)$

```
s[k_]= Sum[ a[n], {n, 1, k}];
points[p_]:= Table[{k, N[s[k]]}, {k, 1, p}]
points[100]
ListPlot[points[100]]
points[200]
ListPlot[points[200]]
points[400]
ListPlot[points[400], PlotRange → All]
```

To investigate what is happening around k = 355, you could do the following.

```
N[355/113]
N[π − 355/113]
Sin[355]//N
a[355]//N
N[s[354]]
N[s[355]]
N[s[356]]
```

9.5   ABSOLUTE CONVERGENCE; THE RATIO AND ROOT TESTS

1. $\lim\limits_{n\to\infty}\left|\dfrac{\frac{2^{n+1}}{(n+1)!}}{\frac{2^n}{n!}}\right|=\lim\limits_{n\to\infty}\left(\dfrac{2^n\cdot 2}{(n+1)\cdot n!}\cdot\dfrac{n!}{2^n}\right)=\lim\limits_{n\to\infty}\left(\dfrac{2}{n+1}\right)=0<1\Rightarrow\sum\limits_{n=1}^{\infty}\dfrac{2^n}{n!}$ converges

3. $\lim\limits_{n\to\infty}\left|\dfrac{\frac{((n+1)-1)!}{((n+1)+1)^2}}{\frac{(n-1)!}{(n+1)^2}}\right|=\lim\limits_{n\to\infty}\left(\dfrac{n\cdot(n-1)!}{(n+2)^2}\cdot\dfrac{(n+1)^2}{(n-1)!}\right)=\lim\limits_{n\to\infty}\left(\dfrac{n^3+2n^2+n}{n^2+4n+4}\right)=\lim\limits_{n\to\infty}\left(\dfrac{3n^2+4n+1}{2n+4}\right)=\lim\limits_{n\to\infty}\left(\dfrac{6n+4}{2}\right)=\infty>1\Rightarrow\sum\limits_{n=1}^{\infty}\dfrac{(n-1)!}{(n+1)^2}$

   diverges

5. $\lim\limits_{n\to\infty}\left|\dfrac{\frac{(n+1)^4}{(-4)^{n+1}}}{\frac{n^4}{(-4)^n}}\right|=\lim\limits_{n\to\infty}\left(\dfrac{(n+1)^4}{4^n\cdot 4}\cdot\dfrac{4^n}{n^4}\right)=\lim\limits_{n\to\infty}\left(\dfrac{n^4+4n^3+6n^2+4n+1}{4n^4}\right)=\lim\limits_{n\to\infty}\left(\dfrac{1}{4}+\dfrac{1}{n}+\dfrac{3}{2n^2}+\dfrac{1}{n^3}+\dfrac{1}{4n^4}\right)=\dfrac{1}{4}<1\Rightarrow\sum\limits_{n=1}^{\infty}\dfrac{n^4}{(-4)^n}$

   converges

7. $\lim\limits_{n\to\infty}\left|\dfrac{(-1)^{n+1}\frac{(n+1)^2((n+1)+2)!}{(n+1)!3^{2(n+1)}}}{(-1)^n\frac{n^2(n+2)!}{n!3^{2n}}}\right|=\lim\limits_{n\to\infty}\left(\dfrac{(n+1)^2(n+3)(n+2)!}{(n+1)\cdot n!3^{2n}\cdot 3^2}\cdot\dfrac{n!3^{2n}}{n^2(n+2)!}\right)=\lim\limits_{n\to\infty}\left(\dfrac{n^3+5n^2+7n+3}{9n^3+9n^2}\right)$

   $=\lim\limits_{n\to\infty}\left(\dfrac{3n^2+15n+7}{27n^2+18n}\right)=\lim\limits_{n\to\infty}\left(\dfrac{6n+15}{54n+18}\right)=\lim\limits_{n\to\infty}\left(\dfrac{6}{54}\right)=\dfrac{1}{9}<1\Rightarrow\sum\limits_{n=1}^{\infty}(-1)^n\dfrac{n^2(n+2)!}{n!3^{2n}}$ converges

9. $\lim\limits_{n\to\infty}\sqrt[n]{\left|\dfrac{7}{(2n+5)^n}\right|}=\lim\limits_{n\to\infty}\left(\dfrac{\sqrt[n]{7}}{2n+5}\right)=0<1\Rightarrow\sum\limits_{n=1}^{\infty}\dfrac{7}{(2n+5)^n}$ converges

11. $\lim\limits_{n\to\infty}\sqrt[n]{\left|\left(\dfrac{4n+3}{3n-5}\right)^n\right|}=\lim\limits_{n\to\infty}\left(\dfrac{4n+3}{3n-5}\right)=\lim\limits_{n\to\infty}\left(\dfrac{4}{3}\right)=\dfrac{4}{3}>1\Rightarrow\sum\limits_{n=1}^{\infty}\left(\dfrac{4n+3}{3n-5}\right)^n$ diverges

13. $\lim\limits_{n\to\infty}\sqrt[n]{\left|\dfrac{-8}{\left(3+\frac{1}{n}\right)^{2n}}\right|}=\lim\limits_{n\to\infty}\left(\dfrac{\sqrt[n]{8}}{\left(3+\frac{1}{n}\right)^2}\right)=\dfrac{1}{9}<1\Rightarrow\sum\limits_{n=1}^{\infty}\dfrac{8}{\left(3+\frac{1}{n}\right)^{2n}}$ converges

15. $\lim\limits_{n\to\infty}\sqrt[n]{\left|(-1)^n\left(1-\dfrac{1}{n}\right)^{n^2}\right|}=\lim\limits_{n\to\infty}\left(1-\dfrac{1}{n}\right)^n=e^{-1}<1\Rightarrow\sum\limits_{n=1}^{\infty}\left(1-\dfrac{1}{n}\right)^{n^2}$ converges

17. converges by the Ratio Test: $\lim\limits_{n\to\infty}\left|\dfrac{a_{n+1}}{a_n}\right|=\lim\limits_{n\to\infty}\dfrac{\left[\frac{(n+1)\sqrt{2}}{2^{n+1}}\right]}{\left[\frac{n\sqrt{2}}{2^n}\right]}=\lim\limits_{n\to\infty}\dfrac{(n+1)^{\sqrt{2}}}{2^{n+1}}\cdot\dfrac{2^n}{n^{\sqrt{2}}}=\lim\limits_{n\to\infty}\left(1+\dfrac{1}{n}\right)^{\sqrt{2}}\left(\dfrac{1}{2}\right)=\dfrac{1}{2}<1$

19. diverges by the Ratio Test: $\lim\limits_{n\to\infty}\left|\dfrac{a_{n+1}}{a_n}\right|=\lim\limits_{n\to\infty}\dfrac{\left(\frac{(n+1)!}{e^{n+1}}\right)}{\left(\frac{n!}{e^n}\right)}=\lim\limits_{n\to\infty}\dfrac{(n+1)!}{e^{n+1}}\cdot\dfrac{e^n}{n!}=\lim\limits_{n\to\infty}\dfrac{n+1}{e}=\infty$

21. converges by the Ratio Test: $\lim\limits_{n\to\infty}\left|\dfrac{a_{n+1}}{a_n}\right| = \lim\limits_{n\to\infty}\dfrac{\left(\frac{(n+1)^{10}}{10^{n+1}}\right)}{\left(\frac{n^{10}}{10^n}\right)} = \lim\limits_{n\to\infty}\dfrac{(n+1)^{10}}{10^{n+1}}\cdot\dfrac{10^n}{n^{10}} = \lim\limits_{n\to\infty}\left(1+\frac{1}{n}\right)^{10}\left(\frac{1}{10}\right) = \frac{1}{10} < 1$

23. converges by the Direct Comparison Test: $\dfrac{2+(-1)^n}{(1.25)^n} = \left(\frac{4}{5}\right)^n\left[2+(-1)^n\right] \le \left(\frac{4}{5}\right)^n (3)$ which is the nth term of a convergent geometric series

25. diverges; $\lim\limits_{n\to\infty} a_n = \lim\limits_{n\to\infty}(-1)^n\left(1-\frac{3}{n}\right)^n = \lim\limits_{n\to\infty}(-1)^n\left(1+\frac{-3}{n}\right)^n$; $(-1)^n\left(1+\frac{-3}{n}\right)^n \Rightarrow e^{-3}$ for n even and $-e^{-3}$
    for n odd, so the limit does not exist

27. converges by the Direct Comparison Test: $\dfrac{\ln n}{n^3} < \dfrac{n}{n^3} = \dfrac{1}{n^2}$ for $n \ge 2$, the nth term of a convergent p-series

29. diverges by the Direct Comparison Test: $\dfrac{1}{n}-\dfrac{1}{n^2} = \dfrac{n-1}{n^2} > \dfrac{1}{2}\left(\frac{1}{n}\right)$ for $n > 2$ or by the Limit Comparison Test
    (part 1) with $\frac{1}{n}$.

31. diverges by the nth-Term Test: Any exponenetial with base > 1 grows faster than any fixed power, so
    $\lim\limits_{n\to\infty} a_n \ne 0$.

33. converges by the Ratio Test: $\lim\limits_{n\to\infty}\left|\dfrac{a_{n+1}}{a_n}\right| = \lim\limits_{n\to\infty}\dfrac{(n+2)(n+3)}{(n+1)!}\cdot\dfrac{n!}{(n+1)(n+2)} = 0 < 1$

35. converges by the Ratio Test: . $\lim\limits_{n\to\infty}\left|\dfrac{a_{n+1}}{a_n}\right| = \lim\limits_{n\to\infty}\dfrac{(n+4)!}{3!(n+1)!3^{n+1}}\cdot\dfrac{3!n!3^n}{(n+3)!} = \lim\limits_{n\to\infty}\dfrac{n+4}{3(n+1)} = \frac{1}{3} < 1$

37. converges by the Ratio Test: $\lim\limits_{n\to\infty}\left|\dfrac{a_{n+1}}{a_n}\right| = \lim\limits_{n\to\infty}\dfrac{(n+1)!}{(2n+3)!}\cdot\dfrac{(2n+1)!}{n!} = \lim\limits_{n\to\infty}\dfrac{n+1}{(2n+3)(2n+2)} = 0 < 1$

39. converges by the Root Test: $\lim\limits_{n\to\infty}\sqrt[n]{|a_n|} = \lim\limits_{n\to\infty}\sqrt[n]{\dfrac{n}{(\ln n)^n}} = \lim\limits_{n\to\infty}\dfrac{\sqrt[n]{n}}{\ln n} = \dfrac{\lim\limits_{n\to\infty}\sqrt[n]{n}}{\lim\limits_{n\to\infty}\ln n} = 0 < 1$

41. converges by the Direct Comparison Test: $\dfrac{n!\ln n}{n(n+2)!} = \dfrac{\ln n}{n(n+1)(n+2)} < \dfrac{n}{n(n+1)(n+2)} = \dfrac{1}{(n+1)(n+2)} < \dfrac{1}{n^2}$ which is the nth-
    term of a convergent p-series

43. converges by the Ratio Test: $\lim\limits_{n\to\infty}\left|\dfrac{a_{n+1}}{a_n}\right| = \lim\limits_{n\to\infty}\dfrac{[(n+1)!]^2}{[2(n+1)]!}\cdot\dfrac{(2n)!}{[n!]^2} = \lim\limits_{n\to\infty}\dfrac{(n+1)^2}{(2n+2)(2n+1)} = \lim\limits_{n\to\infty}\dfrac{n^2+2n+1}{4n^2+6n+2} = \frac{1}{4} < 1$

45. converges by the Ratio Test: $\lim\limits_{n\to\infty}\left|\dfrac{a_{n+1}}{a_n}\right| = \lim\limits_{n\to\infty}\dfrac{\left(\frac{1+\sin n}{n}\right)a_n}{a_n} = 0 < 1$

47. diverges by the Ratio Test: $\lim\limits_{n\to\infty}\left|\dfrac{a_{n+1}}{a_n}\right| = \lim\limits_{n\to\infty}\dfrac{\left(\frac{3n-1}{2n+5}\right)a_n}{a_n} = \lim\limits_{n\to\infty}\dfrac{3n-1}{2n+5} = \frac{3}{2} > 1$

49. converges by the Ratio Test: $\lim\limits_{n\to\infty}\left|\dfrac{a_{n+1}}{a_n}\right| = \lim\limits_{n\to\infty}\dfrac{\left(\frac{2}{n}\right)a_n}{a_n} = \lim\limits_{n\to\infty}\dfrac{2}{n} = 0 < 1$

51. converges by the Ratio Test: $\lim\limits_{n\to\infty}\left|\dfrac{a_{n+1}}{a_n}\right| = \lim\limits_{n\to\infty}\dfrac{\left(\frac{1+\ln n}{n}\right)a_n}{a_n} = \lim\limits_{n\to\infty}\dfrac{1+\ln n}{n} = \lim\limits_{n\to\infty}\dfrac{1}{n} = 0 < 1$

53. diverges by the nth-Term Test: $a_1 = \frac{1}{3}, a_2 = \sqrt[2]{\frac{1}{3}}, a_3 = \sqrt[3]{\sqrt[2]{\frac{1}{3}}} = \sqrt[6]{\frac{1}{3}}, a_4 = \sqrt[4]{\sqrt[3]{\sqrt[2]{\frac{1}{3}}}} = \sqrt[4]{\frac{1}{3}},\ldots, a_n = \sqrt[n]{\frac{1}{3}}$

$\Rightarrow \lim\limits_{n\to\infty} a_n = 1$ because $\left\{\sqrt[n]{\frac{1}{3}}\right\}$ is a subsequence of $\left\{\sqrt[n]{\frac{1}{3}}\right\}$ whose limit is 1 by Table 8.1

55. converges by the Ratio Test: $\lim\limits_{n\to\infty}\left|\dfrac{a_{n+1}}{a_n}\right| = \lim\limits_{n\to\infty}\dfrac{2^{n+1}(n+1)!(n+1)!}{(2n+2)!}\cdot\dfrac{(2n)!}{2^n n!n!} = \lim\limits_{n\to\infty}\dfrac{2(n+1)(n+1)}{(2n+2)(2n+1)} = \lim\limits_{n\to\infty}\dfrac{n+1}{2n+1} = \frac{1}{2} < 1$

57. diverges by the Root Test: $\lim\limits_{n\to\infty}\sqrt[n]{|a_n|} = \lim\limits_{n\to\infty}\sqrt[n]{\dfrac{(n!)^n}{\left(n^n\right)^2}} = \lim\limits_{n\to\infty}\dfrac{n!}{n^2} = \infty > 1$

59. converges by the Root Test: $\lim\limits_{n\to\infty}\sqrt[n]{|a_n|} = \lim\limits_{n\to\infty}\sqrt[n]{\dfrac{n^n}{2^{n^2}}} = \lim\limits_{n\to\infty}\dfrac{n}{2^n} = \lim\limits_{n\to\infty}\dfrac{1}{2^n\ln 2} = 0 < 1$

61. converges by the Ratio Test: $\lim\limits_{n\to\infty}\left|\dfrac{a_{n+1}}{a_n}\right| = \lim\limits_{n\to\infty}\dfrac{1\cdot3\cdots(2n-1)(2n+1)}{4^{n+1}2^{n+1}(n+1)!}\cdot\dfrac{4^n2^n n!}{1\cdot3\cdots(2n-1)} = \lim\limits_{n\to\infty}\dfrac{2n+1}{(4\cdot2)(n+1)} = \frac{1}{4} < 1$

63. Ratio: $\lim\limits_{n\to\infty}\left|\dfrac{a_{n+1}}{a_n}\right| = \lim\limits_{n\to\infty}\dfrac{1}{(n+1)^p}\cdot\dfrac{n^p}{1} = \lim\limits_{n\to\infty}\left(\dfrac{n}{n+1}\right)^p = 1^p = 1 \Rightarrow$ no conclusion

   Root: $\lim\limits_{n\to\infty}\sqrt[n]{|a_n|} = \lim\limits_{n\to\infty}\sqrt[n]{\dfrac{1}{n^p}} = \lim\limits_{n\to\infty}\dfrac{1}{\left(\sqrt[n]{n}\right)^p} = \dfrac{1}{(1)^p} = 1 \Rightarrow$ no conclusion

65. $a_n \leq \dfrac{n}{2^n}$ for every n and the series $\sum\limits_{n=1}^{\infty}\dfrac{n}{2^n}$ converges by the Ratio Test since $\lim\limits_{n\to\infty}\dfrac{(n+1)}{2^{n+1}}\cdot\dfrac{2^n}{n} = \frac{1}{2} < 1$

   $\Rightarrow \sum\limits_{n=1}^{\infty} a_n$ converges by the Direct Comparison Test

9.6 ALTERNATING SERIES AND CONDITIONAL CONVERGENCE

1. converges by the Alternating Convergence Test since: $u_n = \dfrac{1}{\sqrt{n}} > 0$ for all $n \geq 1$;

   $n \geq 1 \Rightarrow n+1 \geq n \Rightarrow \sqrt{n+1} \geq \sqrt{n} \Rightarrow \dfrac{1}{\sqrt{n+1}} \leq \dfrac{1}{\sqrt{n}} \Rightarrow u_{n+1} \leq u_n$; $\lim\limits_{n\to\infty} u_n = \lim\limits_{n\to\infty}\dfrac{1}{\sqrt{n}} = 0$.

3. converges $\Rightarrow$ converges by Alternating Series Test since: $u_n = \dfrac{1}{n3^n} > 0$ for all $n \geq 1$;

   $n \geq 1 \Rightarrow n+1 \geq n \Rightarrow 3^{n+1} \geq 3^n \Rightarrow (n+1)3^{n+1} \geq n3^n \Rightarrow \dfrac{1}{(n+1)3^{n+1}} \leq \dfrac{1}{n3^n} \Rightarrow u_{n+1} \leq u_n$; $\lim\limits_{n\to\infty} u_n = \lim\limits_{n\to\infty}\dfrac{1}{n3^n} = 0$.

5. converges $\Rightarrow$ converges by Alternating Series Test since: $u_n = \frac{n}{n^2+1} > 0$ for all $n \geq 1$;

$n \geq 1 \Rightarrow 2n^2 + 2n \geq n^2 + n + 1 \Rightarrow n^3 + 2n^2 + 2n \geq n^3 + n^2 + n + 1 \Rightarrow n\left(n^2 + 2n + 2\right) \geq n^3 + n^2 + n + 1$

$\Rightarrow n\left((n+1)^2 + 1\right) \geq \left(n^2 + 1\right)(n+1) \Rightarrow \frac{n}{n^2+1} \geq \frac{n+1}{(n+1)^2+1} \Rightarrow u_{n+1} \leq u_n$; $\lim\limits_{n \to \infty} u_n = \lim\limits_{n \to \infty} \frac{n}{n^2+1} = 0.$

7. diverges $\Rightarrow$ diverges by n^{th} Term Test for Divergence since: $\lim\limits_{n \to \infty} \frac{2^n}{n^2} = \infty \Rightarrow \lim\limits_{n \to \infty} (-1)^{n+1} \frac{2^n}{n^2} = $ does not exist

9. diverges by the nth-Term Test since for $n > 10 \Rightarrow \frac{n}{10} > 1 \Rightarrow \lim\limits_{n \to \infty} \left(\frac{n}{10}\right)^n \neq 0 \Rightarrow \sum\limits_{n=1}^{\infty} (-1)^{n+1} \left(\frac{n}{10}\right)^n$ diverges

11. converges by the Alternating Series Test since $f(x) = \frac{\ln x}{x} \Rightarrow f'(x) = \frac{1 - \ln x}{x^2} < 0$ when $x > e \Rightarrow f(x)$ is

decreasing $\Rightarrow u_n \geq u_{n+1}$; also $u_n \geq 0$ for $n \geq 1$ and $\lim\limits_{n \to \infty} u_n = \lim\limits_{n \to \infty} \frac{\ln n}{n} = \lim\limits_{n \to \infty} \frac{\left(\frac{1}{n}\right)}{1} = 0$

13. converges by the Alternating Series Test since $f(x) = \frac{\sqrt{x}+1}{x+1} \Rightarrow f'(x) = \frac{1 - x - 2\sqrt{x}}{2\sqrt{x}(x+1)^2} < 0 \Rightarrow f(x)$ is decreasing

$\Rightarrow u_n \geq u_{n+1}$; also $u_n \geq 0$ for $n \geq 1$ and $\lim\limits_{n \to \infty} u_n = \lim\limits_{n \to \infty} \frac{\sqrt{n}+1}{n+1} = 0$

15. converges absolutely since $\sum\limits_{n=1}^{\infty} |a_n| = \sum\limits_{n=1}^{\infty} \left(\frac{1}{10}\right)^n$ a convergent geometric series

17. converges conditionally since $\frac{1}{\sqrt{n}} > \frac{1}{\sqrt{n+1}} > 0$ and $\lim\limits_{n \to \infty} \frac{1}{\sqrt{n}} = 0 \Rightarrow$ convergence; but $\sum\limits_{n=1}^{\infty} |a_n| = \sum\limits_{n=1}^{\infty} \frac{1}{n^{1/2}}$

is a divergent p-series

19. converges absolutely since $\sum\limits_{n=1}^{\infty} |a_n| = \sum\limits_{n=1}^{\infty} \frac{n}{n^3+1}$ and $\frac{n}{n^3+1} < \frac{1}{n^2}$ which is the nth-term of a converging p-series

21. converges conditionally since $\frac{1}{n+3} > \frac{1}{(n+1)+3} > 0$ and $\lim\limits_{n \to \infty} \frac{1}{n+3} = 0 \Rightarrow$ convergence; but $\sum\limits_{n=1}^{\infty} |a_n| = \sum\limits_{n=1}^{\infty} \frac{1}{n+3}$

diverges because $\frac{1}{n+3} \geq \frac{1}{4n}$ and $\sum\limits_{n=1}^{\infty} \frac{1}{n}$ is a divergent series

23. diverges by the nth-Term Test since $\lim\limits_{n \to \infty} \frac{3+n}{5+n} = 1 \neq 0$

25. converges conditionally since $f(x) = \frac{1}{x^2} + \frac{1}{x} \Rightarrow f'(x) = -\left(\frac{2}{x^3} + \frac{1}{x^2}\right) < 0 \Rightarrow f(x)$ is decreasing and hence

$u_n > u_{n+1} > 0$ for $n \geq 1$ and $\lim\limits_{n \to \infty} \left(\frac{1}{n^2} + \frac{1}{n}\right) = 0 \Rightarrow$ convergence; but $\sum\limits_{n=1}^{\infty} |a_n| = \sum\limits_{n=1}^{\infty} \frac{1+n}{n^2} = \sum\limits_{n=1}^{\infty} \frac{1}{n^2} + \sum\limits_{n=1}^{\infty} \frac{1}{n}$ is the sum

of a convergent and divergent series, and hence diverges

27. converges absolutely by the Ratio Test: $\lim\limits_{n\to\infty}\left|\dfrac{u_{n+1}}{u_n}\right| = \lim\limits_{n\to\infty}\left[\dfrac{(n+1)^2\left(\frac{2}{3}\right)^{n+1}}{n^2\left(\frac{2}{3}\right)^n}\right] = \dfrac{2}{3} < 1$

29. converges absolutely by the Integral Test since $\displaystyle\int_1^\infty \left(\tan^{-1}x\right)\left(\dfrac{1}{1+x^2}\right)dx = \lim\limits_{b\to\infty}\left[\dfrac{\left(\tan^{-1}x\right)^2}{2}\right]_1^b$

$= \lim\limits_{b\to\infty}\left[\left(\tan^{-1}b\right)^2 - \left(\tan^{-1}1\right)^2\right] = \dfrac{1}{2}\left[\left(\dfrac{\pi}{2}\right)^2 - \left(\dfrac{\pi}{4}\right)^2\right] = \dfrac{3\pi^2}{32}$

31. diverges by the nth-Term Test since $\lim\limits_{n\to\infty}\dfrac{n}{n+1} = 1 \neq 0$

33. converges absolutely by the Ratio Test: $\lim\limits_{n\to\infty}\left|\dfrac{u_{n+1}}{u_n}\right| = \lim\limits_{n\to\infty}\dfrac{(100)^{n+1}}{(n+1)!}\cdot\dfrac{n!}{(100)^n} = \lim\limits_{n\to\infty}\dfrac{100}{n+1} = 0 < 1$

35. converges absolutely since $\displaystyle\sum_{n=1}^\infty |a_n| = \sum_{n=1}^\infty \left|\dfrac{(-1)^n}{n\sqrt{n}}\right| = \sum_{n=1}^\infty \dfrac{1}{n^{3/2}}$ is a convergent p-series

37. converges absolutely by the Root Test: $\lim\limits_{n\to\infty}\sqrt[n]{|a_n|} = \lim\limits_{n\to\infty}\left(\dfrac{(n+1)^n}{(2n)^n}\right)^{1/n} = \lim\limits_{n\to\infty}\dfrac{n+1}{2n} = \dfrac{1}{2} < 1$

39. diverges by the nth-Term Test since $\lim\limits_{n\to\infty}|a_n| = \lim\limits_{n\to\infty}\dfrac{(2n)!}{2^n\,n!n} = \lim\limits_{n\to\infty}\dfrac{(n+1)(n+2)\cdots(2n)}{2^n\,n} = \lim\limits_{n\to\infty}\dfrac{(n+1)(n+2)\cdots(n+(n-1))}{2^{n-1}}$

$> \lim\limits_{n\to\infty}\left(\dfrac{n+1}{2}\right)^{n-1} = \infty \neq 0$

41. converges conditionally since $\dfrac{\sqrt{n+1}-\sqrt{n}}{1}\cdot\dfrac{\sqrt{n+1}+\sqrt{n}}{\sqrt{n+1}+\sqrt{n}} = \dfrac{1}{\sqrt{n+1}+\sqrt{n}}$ and $\left\{\dfrac{1}{\sqrt{n+1}+\sqrt{n}}\right\}$ is a decreasing sequence of

positive terms which converges to $0 \Rightarrow \displaystyle\sum_{n=1}^\infty \dfrac{(-1)^n}{\sqrt{n+1}+\sqrt{n}}$ converges; but $\displaystyle\sum_{n=1}^\infty |a_n| = \sum_{n=1}^\infty \dfrac{1}{\sqrt{n+1}+\sqrt{n}}$ diverges by the

Limit Comparison Test (part 1) with $\dfrac{1}{\sqrt{n}}$; a divergent p-series $\lim\limits_{n\to\infty}\left(\dfrac{\frac{1}{\sqrt{n+1}+\sqrt{n}}}{\frac{1}{\sqrt{n}}}\right) = \lim\limits_{n\to\infty}\dfrac{\sqrt{n}}{\sqrt{n+1}+\sqrt{n}}$

$= \lim\limits_{n\to\infty}\dfrac{1}{\sqrt{1+\frac{1}{n}}+1} = \dfrac{1}{2}$

43. diverges by the nth-Term Test since $\lim\limits_{n\to\infty}\left(\sqrt{n+\sqrt{n}}-\sqrt{n}\right) = \lim\limits_{n\to\infty}\left[\left(\sqrt{n+\sqrt{n}}-\sqrt{n}\right)\left(\dfrac{\sqrt{n+\sqrt{n}}+\sqrt{n}}{\sqrt{n+\sqrt{n}}+\sqrt{n}}\right)\right]$

$= \lim\limits_{n\to\infty}\dfrac{\sqrt{n}}{\sqrt{n+\sqrt{n}}+\sqrt{n}} = \lim\limits_{n\to\infty}\dfrac{1}{\sqrt{1+\frac{1}{\sqrt{n}}}+1} = \dfrac{1}{2} \neq 0$

45. converges absolutely by the Direct Comparison Test since $\operatorname{sech}(n) = \dfrac{2}{e^n+e^{-n}} = \dfrac{2e^n}{e^{2n}+1} < \dfrac{2e^n}{e^{2n}} = \dfrac{2}{e^n}$ which is the nth

term of a convergent geometric series

47. $\frac{1}{4}-\frac{1}{6}+\frac{1}{8}-\frac{1}{10}+\frac{1}{12}-\frac{1}{14}+\ldots=\sum_{n=1}^{\infty}\frac{(-1)^{n+1}}{2(n+1)}$; converges by Alternating Series Test since: $u_n=\frac{1}{2(n+1)}>0$ for all

$n\geq1$; $n+2\geq n+1\Rightarrow 2(n+2)\geq 2(n+1)\Rightarrow\frac{1}{2((n+1)+1)}\leq\frac{1}{2(n+1)}\Rightarrow u_{n+1}\leq u_n$; $\lim_{n\to\infty}u_n=\lim_{n\to\infty}\frac{1}{2(n+1)}=0$.

49. $|\text{error}|<\left|(-1)^6\left(\frac{1}{5}\right)\right|=0.2$ 51. $|\text{error}|<\left|(-1)^6\frac{(0.01)^5}{5}\right|=2\times10^{-11}$

53. $|\text{error}|<0.001\Rightarrow u_{n+1}<0.001\Rightarrow\frac{1}{(n+1)^2+3}<0.001\Rightarrow(n+1)^2+3>1000\Rightarrow n>-1+\sqrt{997}\approx30.5753\Rightarrow n\geq31$

55. $|\text{error}|<0.001\Rightarrow u_{n+1}<0.001\Rightarrow\frac{1}{\left((n+1)+3\sqrt{n+1}\right)^3}<0.001\Rightarrow\left((n+1)+3\sqrt{n+1}\right)^3>1000$

$\Rightarrow\left(\sqrt{n+1}\right)^2+3\sqrt{n+1}-10>0\Rightarrow\sqrt{n+1}=\frac{-3+\sqrt{9+40}}{2}=2\Rightarrow n=3\Rightarrow n\geq4$

57. $\frac{1}{(2n)!}<\frac{5}{10^6}\Rightarrow(2n)!>\frac{10^6}{5}=200,000\Rightarrow n\geq5\Rightarrow1-\frac{1}{2!}+\frac{1}{4!}-\frac{1}{6!}+\frac{1}{8!}\approx0.54030$

59. (a) $a_n\geq a_{n+1}$ fails since $\frac{1}{3}<\frac{1}{2}$

(b) Since $\sum_{n=1}^{\infty}|a_n|=\sum_{n=1}^{\infty}\left[\left(\frac{1}{3}\right)^n+\left(\frac{1}{2}\right)^n\right]=\sum_{n=1}^{\infty}\left(\frac{1}{3}\right)^n+\sum_{n=1}^{\infty}\left(\frac{1}{2}\right)^n$ is the sum of two absolutely convergent series,

we can rearrange the terms of the original series to find its sum: $\left(\frac{1}{3}+\frac{1}{9}+\frac{1}{27}+\ldots\right)-\left(\frac{1}{2}+\frac{1}{4}+\frac{1}{8}+\ldots\right)$

$=\frac{\left(\frac{1}{3}\right)}{1-\left(\frac{1}{3}\right)}-\frac{\left(\frac{1}{2}\right)}{1-\left(\frac{1}{2}\right)}=\frac{1}{2}-1=-\frac{1}{2}$

61. The unused terms are $\sum_{j=n+1}^{\infty}(-1)^{j+1}a_j=(-1)^{n+1}\left(a_{n+1}-a_{n+2}\right)+(-1)^{n+3}\left(a_{n+3}-a_{n+4}\right)+\ldots$

$=(-1)^{n+1}\left[\left(a_{n+1}-a_{n+2}\right)+\left(a_{n+3}-a_{n+4}\right)+\ldots\right]$. Each grouped term is positive, so the remainder has the same

sign as $(-1)^{n+1}$, which is the sign of the first unused term.

63. Theorem 16 states that $\sum_{n=1}^{\infty}|a_n|$ converges $\Rightarrow\sum_{n=1}^{\infty}a_n$ converges. But this is equivalent to $\sum_{n=1}^{\infty}a_n$ diverges

$\Rightarrow\sum_{n=1}^{\infty}|a_n|$ diverges

65. (a) $\sum_{n=1}^{\infty}|a_n+b_n|$ converges by the Direct Comparison Test since $|a_n+b_n|\leq|a_n|+|b_n|$ and hence $\sum_{n=1}^{\infty}(a_n+b_n)$
converges absolutely

(b) $\sum_{n=1}^{\infty}|b_n|$ converges $\Rightarrow\sum_{n=1}^{\infty}-b_n$ converges absolutely; since $\sum_{n=1}^{\infty}a_n$ converges absolutely and $\sum_{n=1}^{\infty}-b_n$

converges absolutely, we have $\sum_{n=1}^{\infty}[a_n+(-b_n)]=\sum_{n=1}^{\infty}(a_n-b_n)$ converges absolutely by part (a)

(c) $\displaystyle\sum_{n=1}^{\infty}|a_n|$ converges $\Rightarrow |k|\displaystyle\sum_{n=1}^{\infty}|a_n| = \sum_{n=1}^{\infty}|ka_n|$ converges $\Rightarrow \displaystyle\sum_{n=1}^{\infty}ka_n$ converges absolutely

67. Since $\displaystyle\sum_{n=1}^{\infty}a_n$ converges, $a_n \to 0$ and for all n greater than some N, $|a_n| < 1$ and $(a_n)^2 < |a_n|$. Since

$\displaystyle\sum_{n=1}^{\infty}a_n$ is absolutely convergent, $\displaystyle\sum_{n=1}^{\infty}|a_n|$ converges and thus $\displaystyle\sum_{n=1}^{\infty}(a_n)^2$ converges by the Direct Comparison Test.

69. $s_1 = -\frac{1}{2}$, $s_2 = -\frac{1}{2} + 1 = \frac{1}{2}$,

$s_3 = -\frac{1}{2} + 1 - \frac{1}{4} - \frac{1}{6} - \frac{1}{8} - \frac{1}{10} - \frac{1}{12} - \frac{1}{14} - \frac{1}{16} - \frac{1}{18} - \frac{1}{20} - \frac{1}{22} \approx -0.5099$,

$s_4 = s_3 + \frac{1}{3} \approx -0.1766$,

$s_5 = s_4 - \frac{1}{24} - \frac{1}{26} - \frac{1}{28} - \frac{1}{30} - \frac{1}{32} - \frac{1}{34} - \frac{1}{36} - \frac{1}{38} - \frac{1}{40} - \frac{1}{42} - \frac{1}{44} \approx -0.512$,

$s_6 = s_5 + \frac{1}{5} \approx -0.312$,

$s_7 = s_6 - \frac{1}{46} - \frac{1}{48} - \frac{1}{50} - \frac{1}{52} - \frac{1}{54} - \frac{1}{56} - \frac{1}{58} - \frac{1}{60} - \frac{1}{62} - \frac{1}{64} - \frac{1}{66} \approx -0.51106$

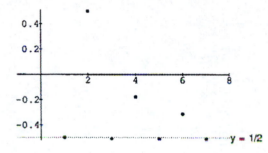

9.7   POWER SERIES

1. $\displaystyle\lim_{n\to\infty}\left|\frac{u_{n+1}}{u_n}\right| < 1 \Rightarrow \lim_{n\to\infty}\left|\frac{x^{n+1}}{x^n}\right| < 1 \Rightarrow |x| < 1 \Rightarrow -1 < x < 1$; when $x = -1$ we have $\displaystyle\sum_{n=1}^{\infty}(-1)^n$, a divergent series;

when $x = 1$ we have $\displaystyle\sum_{n=1}^{\infty}1$, a divergent series

(a) the radius is 1; the interval of convergence is $-1 < x < 1$

(b) the interval of absolute convergence is $-1 < x < 1$

(c) there are no values for which the series converges conditionally

3. $\displaystyle\lim_{n\to\infty}\left|\frac{u_{n+1}}{u_n}\right| < 1 \Rightarrow \lim_{n\to\infty}\left|\frac{(4x+1)^{n+1}}{(4x+1)^n}\right| < 1 \Rightarrow |4x+1| < 1 \Rightarrow -1 < 4x+1 < 1 \Rightarrow -\frac{1}{2} < x < 0$; when $x = -\frac{1}{2}$ we have

$\displaystyle\sum_{n=1}^{\infty}(-1)^n(-1)^n = \sum_{n=1}^{\infty}(-1)^{2n} = \sum_{n=1}^{\infty}1^n$, a divergent series; when $x = 0$ we have $\displaystyle\sum_{n=1}^{\infty}(-1)^n(1)^n = \sum_{n=1}^{\infty}(-1)^n$,

a divergent series

(a) the radius is $\frac{1}{4}$; the interval of convergence is $-\frac{1}{2} < x < 0$

(b) the interval of absolute convergence is $-\frac{1}{2} < x < 0$

(c) there are no values for which the series converges conditionally

5. $\lim\limits_{n\to\infty}\left|\dfrac{u_{n+1}}{u_n}\right|<1\Rightarrow\lim\limits_{n\to\infty}\left|\dfrac{(x-2)^{n+1}}{10^{n+1}}\cdot\dfrac{10^n}{(x-2)^n}\right|<1\Rightarrow\dfrac{|x-2|}{10}<1\Rightarrow|x-2|<10\Rightarrow-10<x-2<10\Rightarrow-8<x<12;$ when

$x=-8$ we have $\sum\limits_{n=1}^{\infty}(-1)^n$, a divergent series; when $x=12$ we have $\sum\limits_{n=1}^{\infty}1$, a divergent series

    (a) the radius is 10; the interval of convergence is $-8<x<12$

    (b) the interval of absolute convergence is $-8<x<12$

    (c) there are no values for which the series converges conditionally

7. $\lim\limits_{n\to\infty}\left|\dfrac{u_{n+1}}{u_n}\right|<1\Rightarrow\lim\limits_{n\to\infty}\left|\dfrac{(n+1)x^{n+1}}{(n+3)}\cdot\dfrac{(n+2)}{nx^n}\right|<1\Rightarrow|x|\lim\limits_{n\to\infty}\dfrac{(n+1)(n+2)}{(n+3)(n)}<1\Rightarrow|x|<1\Rightarrow-1<x<1;$ when $x=-1$ we have

$\sum\limits_{n=1}^{\infty}(-1)^n\dfrac{n}{n+2}$, a divergent series by the nth-term; Test; when $x=1$ we have $\sum\limits_{n=1}^{\infty}\dfrac{n}{n+2}$, a divergent series

    (a) the radius is 1; the interval of convergence is $-1<x<1$

    (b) the interval of absolute convergence is $-1<x<1$

    (c) there are no values for which the series converges conditionally

9. $\lim\limits_{n\to\infty}\left|\dfrac{u_{n+1}}{u_n}\right|<1\Rightarrow\lim\limits_{n\to\infty}\left|\dfrac{x^{n+1}}{(n+1)\sqrt{n+1}\,3^{n+1}}\cdot\dfrac{n\sqrt{n}\,3^n}{x^n}\right|<1\Rightarrow\dfrac{|x|}{3}\left(\lim\limits_{n\to\infty}\dfrac{n}{n+1}\right)\left(\sqrt{\lim\limits_{n\to\infty}\dfrac{n}{n+1}}\right)<1\Rightarrow\dfrac{|x|}{3}(1)(1)<1\Rightarrow|x|<3$

$\Rightarrow-3<x<3;$ when $x=-3$ we have $\sum\limits_{n=1}^{\infty}\dfrac{(-1)^n}{n^{3/2}}$, an absolutely convergent series; when $x=3$ we have

$\sum\limits_{n=1}^{\infty}\dfrac{1}{n^{3/2}}$, a convergent p-series

    (a) the radius is 3; the interval of convergence is $-3\le x\le3$

    (b) the interval of absolute convergence is $-3\le x\le3$

    (c) there are no values for which the series converges conditionally

11. $\lim\limits_{n\to\infty}\left|\dfrac{u_{n+1}}{u_n}\right|<1\Rightarrow\lim\limits_{n\to\infty}\left|\dfrac{x^{n+1}}{(n+1)!}\cdot\dfrac{n!}{x^n}\right|<1\Rightarrow|x|\lim\limits_{n\to\infty}\left(\dfrac{1}{n+1}\right)<1$ for all x

    (a) the radius is ∞; the series converges for all x

    (b) the series converges absolutely for all x

    (c) there are no values for which the series converges conditionally

13. $\lim\limits_{n\to\infty}\left|\dfrac{u_{n+1}}{u_n}\right|<1\Rightarrow\lim\limits_{n\to\infty}\left|\dfrac{4^{n+1}x^{2n+2}}{n+1}\cdot\dfrac{n}{4^n x^{2n}}\right|<1\Rightarrow x^2\lim\limits_{n\to\infty}\left(\dfrac{4n}{n+1}\right)=4x^2<1\Rightarrow x^2<\dfrac{1}{4}\Rightarrow-\dfrac{1}{2}<x<\dfrac{1}{2};$ when $x=-\dfrac{1}{2}$

we have $\sum\limits_{n=1}^{\infty}\dfrac{4^n}{n}\left(-\dfrac{1}{2}\right)^{2n}=\sum\limits_{n=1}^{\infty}\dfrac{1}{n}$, a divergent p-series when $x=\dfrac{1}{2}$ we have $\sum\limits_{n=1}^{\infty}\dfrac{4^n}{n}\left(\dfrac{1}{2}\right)^{2n}=\sum\limits_{n=1}^{\infty}\dfrac{1}{n}$, a divergent

p-series

    (a) the radius is $\dfrac{1}{2}$; the interval of convergence is $-\dfrac{1}{2}<x<\dfrac{1}{2}$

    (b) the interval of absolute convergence is $-\dfrac{1}{2}<x<\dfrac{1}{2}$

    (c) there are no values for which the series converges conditionally

15. $\lim\limits_{n\to\infty}\left|\dfrac{u_{n+1}}{u_n}\right|<1\Rightarrow\lim\limits_{n\to\infty}\left|\dfrac{x^{n+1}}{\sqrt{(n+1)^2+3}}\cdot\dfrac{\sqrt{n^2+3}}{x^n}\right|<1\Rightarrow|x|\sqrt{\lim\limits_{n\to\infty}\dfrac{n^2+3}{n^2+2n+4}}<1\Rightarrow|x|<1\Rightarrow-1<x<1;\ \text{when}\ x=-1$

we have $\displaystyle\sum_{n=1}^{\infty}\dfrac{(-1)^n}{\sqrt{n^2+3}}$, a conditionally convergent series; when $x=1$ we have $\displaystyle\sum_{n=1}^{\infty}\dfrac{1}{\sqrt{n^2+3}}$, a divergent series

(a) the radius is 1; the interval of convergence is $-1\le x<1$
(b) the interval of absolute convergence is $-1<x<1$
(c) the series converges conditionally at $x=-1$

17. $\lim\limits_{n\to\infty}\left|\dfrac{u_{n+1}}{u_n}\right|<1\Rightarrow\lim\limits_{n\to\infty}\left|\dfrac{(n+1)(x+3)^{n+1}}{5^{n+1}}\cdot\dfrac{5^n}{n(x+3)^n}\right|<1\Rightarrow\dfrac{|x+3|}{5}\lim\limits_{n\to\infty}\left(\dfrac{n+1}{n}\right)<1\Rightarrow\dfrac{|x+3|}{5}<1\Rightarrow|x+3|<5\Rightarrow-5<x+3<5$

$\Rightarrow-8<x<2;$ when $x=-8$ we have $\displaystyle\sum_{n=1}^{\infty}\dfrac{n(-5)^n}{5^n}=\sum_{n=1}^{\infty}(-1)^n\,n$, a divergent series; when $x=2$ we have

$\displaystyle\sum_{n=1}^{\infty}\dfrac{n5^n}{5^n}=\sum_{n=1}^{\infty}n$, a divergent series

(a) the radius is 5; the interval of convergence is $-8<x<2$
(b) the interval of absolute convergence is $-8<x<2$
(c) there are no values for which the series converges conditionally

19. $\lim\limits_{n\to\infty}\left|\dfrac{u_{n+1}}{u_n}\right|<1\Rightarrow\lim\limits_{n\to\infty}\left|\dfrac{\sqrt{n+1}\,x^{n+1}}{3^{n+1}}\cdot\dfrac{3^n}{\sqrt{n}\,x^n}\right|<1\Rightarrow\dfrac{|x|}{3}\sqrt{\lim\limits_{n\to\infty}\left(\dfrac{n+1}{n}\right)}<1\Rightarrow\dfrac{|x|}{3}<1\Rightarrow|x|<3\Rightarrow-3<x<3;\ \text{when}\ x=-3$

we have $\displaystyle\sum_{n=1}^{\infty}(-1)^n\sqrt{n}$, a divergent series; when $x=3$ we have $\displaystyle\sum_{n=1}^{\infty}\sqrt{n}$, a divergent series

(a) the radius is 3; the interval of convergence is $-3<x<3$
(b) the interval of absolute convergence is $-3<x<3$
(c) there are no values for which the series converges conditionally

21. First, rewrite the series as $\displaystyle\sum_{n=1}^{\infty}\left(2+(-1)^n\right)(x+1)^{n-1}=\sum_{n=1}^{\infty}2(x+1)^{n-1}+\sum_{n=1}^{\infty}(-1)^n(x+1)^{n-1}$.

For the series $\displaystyle\sum_{n=1}^{\infty}2(x+1)^{n-1}$: $\lim\limits_{n\to\infty}\left|\dfrac{u_{n+1}}{u_n}\right|<1\Rightarrow\lim\limits_{n\to\infty}\left|\dfrac{2(x+1)^n}{2(x+1)^{n-1}}\right|<1\Rightarrow|x+1|\lim\limits_{n\to\infty}1=|x+1|<1\Rightarrow-2<x<0;$

For the series $\displaystyle\sum_{n=1}^{\infty}(-1)^n(x+1)^{n-1}$: $\lim\limits_{n\to\infty}\left|\dfrac{u_{n+1}}{u_n}\right|<1\Rightarrow\lim\limits_{n\to\infty}\left|\dfrac{(-1)^{n+1}(x+1)^n}{(-1)^n(x+1)^{n-1}}\right|<1\Rightarrow|x+1|\lim\limits_{n\to\infty}1=|x+1|<1$

$\Rightarrow-2<x<0;$ when $x=-2$ we have $\displaystyle\sum_{n=1}^{\infty}\left(2+(-1)^n\right)(-1)^{n-1}$, a divergent series; when $x=0$ we have

$\displaystyle\sum_{n=1}^{\infty}\left(2+(-1)^n\right)$, a divergent series

(a) the radius is 1; the interval of convergence is $-2<x<0$
(b) the interval of absolute convergence is $-2<x<0$
(c) there are no values for which the series converges conditionally

23. $\lim\limits_{n\to\infty}\left|\dfrac{u_{n+1}}{u_n}\right|<1\Rightarrow\lim\limits_{n\to\infty}\left|\dfrac{\left(1+\frac{1}{n+1}\right)^{n+1}x^{n+1}}{\left(1+\frac{1}{n}\right)^n x^n}\right|<1\Rightarrow|x|\left(\dfrac{\lim\limits_{t\to\infty}\left(1+\frac{1}{t}\right)^t}{\lim\limits_{n\to\infty}\left(1+\frac{1}{n}\right)^n}\right)<1\Rightarrow|x|\left(\frac{e}{e}\right)<1\Rightarrow|x|<1\Rightarrow-1<x<1;$ when $x=-1$

we have $\sum\limits_{n=1}^{\infty}(-1)^n\left(1+\frac{1}{n}\right)^n$, a divergent series by the nth-Term Test since $\lim\limits_{n\to\infty}\left(1+\frac{1}{n}\right)^n=e\neq0$; when $x=1$

we have $\sum\limits_{n=1}^{\infty}\left(1+\frac{1}{n}\right)^n$, a divergent series

(a) the radius is 1; the interval of convergence is $-1<x<1$

(b) the interval of absolute convergence is $-1<x<1$

(c) there are no values for which the series converges conditionally

25. $\lim\limits_{n\to\infty}\left|\dfrac{u_{n+1}}{u_n}\right|<1\Rightarrow\lim\limits_{n\to\infty}\left|\dfrac{(n+1)^{n+1}x^{n+1}}{n^n x^n}\right|<1\Rightarrow|x|\left(\lim\limits_{n\to\infty}\left(1+\frac{1}{n}\right)^n\right)\left(\lim\limits_{n\to\infty}(n+1)\right)<1\Rightarrow e|x|\lim\limits_{n\to\infty}(n+1)<1\Rightarrow$ only

$x=0$ satisfies this inequality

(a) the radius is 0; the series converges only for $x=0$

(b) the series converges absolutely only for $x=0$

(c) there are no values for which the series converges conditionally

27. $\lim\limits_{n\to\infty}\left|\dfrac{u_{n+1}}{u_n}\right|<1\Rightarrow\lim\limits_{n\to\infty}\left|\dfrac{(x+2)^{n+1}}{(n+1)2^{n+1}}\cdot\dfrac{n2^n}{(x+2)^n}\right|<1\Rightarrow\dfrac{|x+2|}{2}\lim\limits_{n\to\infty}\left(\frac{n}{n+1}\right)<1\Rightarrow\dfrac{|x+2|}{2}<1\Rightarrow|x+2|<2\Rightarrow-2<x+2<2$

$\Rightarrow-4<x<0;$ when $x=-4$ we have $\sum\limits_{n=1}^{\infty}\frac{-1}{n}$, a divergent series; when $x=0$ we have $\sum\limits_{n=1}^{\infty}\frac{(-1)^{n+1}}{n}$, the

alternating harmonic series which converges conditionally

(a) the radius is 2; the interval of convergence is $-4<x\leq0$

(b) the interval of absolute convergence is $-4<x<0$

(c) the series converges conditionally at $x=0$

29. $\lim\limits_{n\to\infty}\left|\dfrac{u_{n+1}}{u_n}\right|<1\Rightarrow\lim\limits_{n\to\infty}\left|\dfrac{x^{n+1}}{(n+1)(\ln(n+1))^2}\cdot\dfrac{n(\ln n)^2}{x^n}\right|<1\Rightarrow|x|\left(\lim\limits_{n\to\infty}\frac{n}{n+1}\right)\left(\lim\limits_{n\to\infty}\frac{\ln n}{\ln(n+1)}\right)^2<1$

$\Rightarrow|x|(1)\left(\lim\limits_{n\to\infty}\dfrac{\left(\frac{1}{n}\right)}{\left(\frac{1}{n+1}\right)}\right)^2<1\Rightarrow|x|\left(\lim\limits_{n\to\infty}\frac{n+1}{n}\right)^2<1\Rightarrow|x|<1\Rightarrow-1<x<1;$ when $x=-1$ we have $\sum\limits_{n=1}^{\infty}\dfrac{(-1)^n}{n(\ln n)^2}$

which converges absolutely; when $x=1$ we have $\sum\limits_{n=1}^{\infty}\dfrac{1}{n(\ln n)^2}$ which converges

(a) the radius is 1; the interval of convergence is $-1\leq x\leq1$

(b) the interval of absolute convergence is $-1\leq x\leq1$

(c) there are no values for which the series converges conditionally

31. $\lim\limits_{n\to\infty}\left|\frac{u_{n+1}}{u_n}\right|<1 \Rightarrow \lim\limits_{n\to\infty}\left|\frac{(4x-5)^{2n+3}}{(n+1)^{3/2}}\cdot\frac{n^{3/2}}{(4x-5)^{2n+1}}\right|<1 \Rightarrow (4x-5)^2\left(\lim\limits_{n\to\infty}\frac{n}{n+1}\right)^{3/2}<1 \Rightarrow (4x-5)^2<1 \Rightarrow |4x-5|<1$

$\Rightarrow -1<4x-5<1 \Rightarrow 1<x<\frac{3}{2}$; when $x=1$ we have $\sum\limits_{n=1}^{\infty}\frac{(-1)^{2n+1}}{n^{3/2}}=\sum\limits_{n=1}^{\infty}\frac{-1}{n^{3/2}}$ which is absolutely convergent;

when $x=\frac{3}{2}$ we have $\sum\limits_{n=1}^{\infty}\frac{(1)^{2n+1}}{n^{3/2}}$, a convergent p-series

  (a) the radius is $\frac{1}{4}$; the interval of convergence is $1\le x\le\frac{3}{2}$

  (b) the interval of absolute convergence is $1\le x\le\frac{3}{2}$

  (c) there are no values for which the series converges conditionally

33. $\lim\limits_{n\to\infty}\left|\frac{u_{n+1}}{u_n}\right|<1 \Rightarrow \lim\limits_{n\to\infty}\left|\frac{x^{n+1}}{2\cdot4\cdot6\cdots(2n)(2(n+1))}\cdot\frac{2\cdot4\cdot6\cdots(2n)}{x^n}\right|<1 \Rightarrow |x|\lim\limits_{n\to\infty}\left(\frac{1}{2n+2}\right)<1$ for all x

  (a) the radius is ∞; the series converges for all x

  (b) the series converges absolutely for all x

  (c) there are no values for which the series converges conditionally

35. For the series $\sum\limits_{n=1}^{\infty}\frac{1+2+\cdots+n}{1^2+2^2+\cdots+n^2}x^n$, recall $1+2+\cdots+n=\frac{n(n+1)}{2}$ and $1^2+2^2+\cdots+n^2=\frac{n(n+1)(2n+1)}{6}$ so that we

can rewrite the series as $\sum\limits_{n=1}^{\infty}\left(\frac{\frac{n(n+1)}{2}}{\frac{n(n+1)(2n+1)}{6}}\right)x^n=\sum\limits_{n=1}^{\infty}\left(\frac{3}{2n+1}\right)x^n$; then $\lim\limits_{n\to\infty}\left|\frac{u_{n+1}}{u_n}\right|<1 \Rightarrow \lim\limits_{n\to\infty}\left|\frac{3x^{n+1}}{(2(n+1)+1)}\cdot\frac{(2n+1)}{3x^n}\right|<1$

$\Rightarrow |x|\lim\limits_{n\to\infty}\left|\frac{(2n+1)}{(2n+3)}\right|<1 \Rightarrow |x|<1 \Rightarrow -1<x<1$; when $x=-1$ we have $\sum\limits_{n=1}^{\infty}\left(\frac{3}{2n+1}\right)(-1)^n$, a conditionally

convergent series; when $x=1$ we have $\sum\limits_{n=1}^{\infty}\left(\frac{3}{2n+1}\right)$, a divergent series.

  (a) the radius is 1; the interval of convergence is $-1\le x<1$

  (b) the interval of absolute convergence is $-1<x<1$

  (c) the series converges conditionally at $x=-1$

37. $\lim\limits_{n\to\infty}\left|\frac{u_{n+1}}{u_n}\right|<1 \Rightarrow \lim\limits_{n\to\infty}\left|\frac{(n+1)!x^{n+1}}{3\cdot6\cdot9\cdots(3n)(3(n+1))}\cdot\frac{3\cdot6\cdot9\cdots(3n)}{n!x^n}\right|<1 \Rightarrow |x|\lim\limits_{n\to\infty}\left|\frac{(n+1)}{3(n+1)}\right|<1 \Rightarrow \frac{|x|}{3}<1 \Rightarrow |x|<3 \Rightarrow R=3$

39. $\lim\limits_{n\to\infty}\left|\frac{u_{n+1}}{u_n}\right|<1 \Rightarrow \lim\limits_{n\to\infty}\left|\frac{((n+1)!)^2 x^{n+1}}{2^{n+1}(2(n+1))!}\cdot\frac{2^n(2n)!}{(n!)^2 x^n}\right|<1 \Rightarrow |x|\lim\limits_{n\to\infty}\left|\frac{(n+1)^2}{2(2n+2)(2n+1)}\right|<1 \Rightarrow \frac{|x|}{8}<1 \Rightarrow |x|<8 \Rightarrow R=8$

41. $\lim\limits_{n\to\infty}\left|\frac{u_{n+1}}{u_n}\right|<1 \Rightarrow \lim\limits_{n\to\infty}\left|\frac{3^{n+1}x^{n+1}}{3^n x^n}\right|<1 \Rightarrow |x|\lim\limits_{n\to\infty}3<1 \Rightarrow |x|<\frac{1}{3} \Rightarrow -\frac{1}{3}<x<\frac{1}{3}$; at $x=-\frac{1}{3}$ we have

$\sum\limits_{n=0}^{\infty}3^n\left(-\frac{1}{3}\right)^n=\sum\limits_{n=0}^{\infty}(-1)^n$, which diverges; at $x=\frac{1}{3}$ we have $\sum\limits_{n=0}^{\infty}3^n\left(\frac{1}{3}\right)^n=\sum\limits_{n=0}^{\infty}1$, which diverges. The series

$\sum\limits_{n=0}^{\infty}3^n x^n=\sum\limits_{n=0}^{\infty}(3x)^n$, is a convergent geometric series when $-\frac{1}{3}<x<\frac{1}{3}$ and the sum is $\frac{1}{1-3x}$.

43. $\lim\limits_{n\to\infty}\left|\dfrac{u_{n+1}}{u_n}\right|<1\Rightarrow\lim\limits_{n\to\infty}\left|\dfrac{(x-1)^{2n+2}}{4^{n+1}}\cdot\dfrac{4^n}{(x-1)^{2n}}\right|<1\Rightarrow\dfrac{(x-1)^2}{4}\lim\limits_{n\to\infty}|1|<1\Rightarrow(x-1)^2<4\Rightarrow|x-1|<2\Rightarrow-2<x-1<2$

$\Rightarrow-1<x<3$; at $x=-1$ we have $\displaystyle\sum_{n=0}^{\infty}\dfrac{(-2)^{2n}}{4^n}=\sum_{n=0}^{\infty}\dfrac{4^n}{4^n}=\sum_{n=0}^{\infty}1$, which diverges; at $x=3$ we have $\displaystyle\sum_{n=0}^{\infty}\dfrac{2^{2n}}{4^n}$

$=\displaystyle\sum_{n=0}^{\infty}\dfrac{4^n}{4^n}=\sum_{n=0}^{\infty}1$, a divergent series; the interval of convergence is $-1<x<3$; the series $\displaystyle\sum_{n=0}^{\infty}\dfrac{(x-1)^{2n}}{4^n}$

$=\displaystyle\sum_{n=0}^{\infty}\left(\left(\dfrac{x-1}{2}\right)^2\right)^n$ is a convergent geometric series when $-1<x<3$ and sum is $\dfrac{1}{1-\left(\frac{x-1}{2}\right)^2}=\dfrac{1}{\left[\frac{4-(x-1)^2}{4}\right]}=\dfrac{4}{4-x^2+2x-1}$

$=\dfrac{4}{3+2x-x^2}$

45. $\lim\limits_{n\to\infty}\left|\dfrac{u_{n+1}}{u_n}\right|<1\Rightarrow\lim\limits_{n\to\infty}\left|\dfrac{(\sqrt{x}-2)^{n+1}}{2^{n+1}}\cdot\dfrac{2^n}{(\sqrt{x}-2)^n}\right|<1\Rightarrow\left|\sqrt{x}-2\right|<2\Rightarrow-2<\sqrt{x}-2<2\Rightarrow0<\sqrt{x}<4\Rightarrow0<x<16;$

when $x=0$ we have $\displaystyle\sum_{n=0}^{\infty}(-1)^n$, a divergent series; when $x=16$ we have $\displaystyle\sum_{n=0}^{\infty}(1)^n$, a divergent series; the

interval of convergence is $0<x<16$; the series $\displaystyle\sum_{n=0}^{\infty}\left(\dfrac{\sqrt{x}-2}{2}\right)^n$ is a convergent geometric series when $0<x<16$

and its sum is $\dfrac{1}{1-\left(\frac{\sqrt{x}-2}{2}\right)}=\dfrac{1}{\left(\frac{2-\sqrt{x}+2}{2}\right)}=\dfrac{2}{4-\sqrt{x}}$

47. $\lim\limits_{n\to\infty}\left|\dfrac{u_{n+1}}{u_n}\right|<1\Rightarrow\lim\limits_{n\to\infty}\left|\left(\dfrac{x^2+1}{3}\right)^{n+1}\cdot\left(\dfrac{3}{x^2+1}\right)^n\right|<1\Rightarrow\dfrac{(x^2+1)}{3}\lim\limits_{n\to\infty}|1|<1\Rightarrow\dfrac{x^2+1}{3}<1\Rightarrow x^2<2\Rightarrow|x|<\sqrt{2}$

$\Rightarrow-\sqrt{2}<x<\sqrt{2}$; at $x=\pm\sqrt{2}$ we have $\displaystyle\sum_{n=0}^{\infty}(1)^n$ which diverges; the interval of convergence is

$-\sqrt{2}<x<\sqrt{2}$; the series $\displaystyle\sum_{n=0}^{\infty}\left(\dfrac{x^2+1}{3}\right)^n$ is a convergent geometric series when $-\sqrt{2}<x<\sqrt{2}$ and its sum is

$\dfrac{1}{1-\left(\frac{x^2+1}{3}\right)}=\dfrac{1}{\left(\frac{3-x^2-1}{3}\right)}=\dfrac{3}{2-x^2}$

49. Writing $\dfrac{2}{x}$ as $\dfrac{2}{1-[-(x-1)]}$ we see that it can be written as the power

series $\displaystyle\sum_{n=0}^{\infty}2[-(x-1)]^n=\sum_{n=0}^{\infty}2(-1)^n(x-1)^n$. Since this is a geometric series with ratio $-(x-1)$ it will converge

for
$\left|-(x-1)\right|<1$ or $0<x<2$.

51. $g(x) = \frac{3}{x-2} = \frac{3}{3-[-(x-5)]} = \frac{1}{1 - -\frac{x-5}{3}} = \sum_{n=0}^{\infty} \left(-\frac{1}{3}\right)^n (x-5)^n$, which converges for

$\left|\frac{x-5}{3}\right| < 1$ or $2 < x < 8.$

53. $\lim_{n\to\infty} \left| \frac{(x-3)^{n+1}}{2^{n+1}} \cdot \frac{2^n}{(x-3)^n} \right| < 1 \Rightarrow |x-3| < 2 \Rightarrow 1 < x < 5$; when $x=1$ we have $\sum_{n=0}^{\infty} (1)^n$ which diverges; when $x=5$

we have $\sum_{n=1}^{\infty} (-1)^n$ which also diverges; the interval of convergence is $1 < x < 5$; the sum of this convergent

geometric series is $\frac{1}{1+\left(\frac{x-3}{2}\right)} = \frac{2}{x-1}$. If $f(x) = 1 - \frac{1}{2}(x-3) + \frac{1}{4}(x-3)^2 + \ldots + \left(-\frac{1}{2}\right)^n (x-3)^n + \ldots = \frac{2}{x-1}$ then

$f'(x) = -\frac{1}{2} + \frac{1}{2}(x-3) + \ldots + \left(-\frac{1}{2}\right)^n n(x-3)^{n-1} + \ldots$ is convergent when $1 < x < 5$, and diverges when $x=1$ or

5. The sum for $f'(x)$ is $\frac{-2}{(x-1)^2}$, the derivative of $\frac{2}{x-1}$.

55. (a) Differentiate the series for $\sin x$ to get $\cos x = 1 - \frac{3x^2}{3!} + \frac{5x^4}{5!} - \frac{7x^6}{7!} + \frac{9x^8}{9!} - \frac{11x^{10}}{11!} + \ldots$

$= 1 - \frac{x^2}{2!} + \frac{x^4}{4!} - \frac{x^6}{6!} + \frac{x^8}{8!} - \frac{x^{10}}{10!} + \ldots$. The series converges for all values of x since

$\lim_{n\to\infty} \left| \frac{x^{2n+2}}{(2n+2)!} \cdot \frac{(2n)!}{x^{2n}} \right| = x^2 \lim_{n\to\infty} \left(\frac{1}{(2n+1)(2n+2)} \right) = 0 < 1$ for all x.

(b) $\sin 2x = 2x - \frac{2^3 x^3}{3!} + \frac{2^5 x^5}{5!} - \frac{2^7 x^7}{7!} + \frac{2^9 x^9}{9!} - \frac{2^{11} x^{11}}{11!} + \ldots = 2x - \frac{8x^3}{3!} + \frac{32x^5}{5!} - \frac{128x^7}{7!} + \frac{512x^9}{9!} - \frac{2048x^{11}}{11!} + \ldots$

(c) $2\sin x \cos x = 2\Big[(0 \cdot 1) + (0 \cdot 0 + 1 \cdot 1)x + \left(0 \cdot \frac{-1}{2} + 1 \cdot 0 + 0 \cdot 1\right)x^2 + \left(0 \cdot 0 - 1 \cdot \frac{1}{2} + 0 \cdot 0 - 1 \cdot \frac{1}{3!}\right)x^3$

$+ \left(0 \cdot \frac{1}{4!} + 1 \cdot 0 - 0 \cdot \frac{1}{2} - 0 \cdot \frac{1}{3!} + 0 \cdot 1\right)x^4 + \left(0 \cdot 0 + 1 \cdot \frac{1}{4!} + 0 \cdot 0 + \frac{1}{2} \cdot \frac{1}{3!} + 0 \cdot 0 + 1 \cdot \frac{1}{5!}\right)x^5$

$+ \left(0 \cdot \frac{1}{6!} + 1 \cdot 0 + 0 \cdot \frac{1}{4!} + 0 \cdot \frac{1}{3!} + 0 \cdot \frac{1}{2} + 0 \cdot \frac{1}{5!} + 0 \cdot 1\right)x^6 + \ldots\Big] = 2\left[x - \frac{4x^3}{3!} + \frac{16x^5}{5!} - \ldots\right]$

$= 2x - \frac{2^3 x^3}{3!} + \frac{2^5 x^5}{5!} - \frac{2^7 x^7}{7!} + \frac{2^9 x^9}{9!} - \frac{2^{11} x^{11}}{11!} + \ldots$

57. (a) $\ln|\sec x| + C = \int \tan x \, dx = \int \left(x + \frac{x^3}{3} + \frac{2x^5}{15} + \frac{17x^7}{315} + \frac{62x^9}{2835} + \ldots\right) dx = \frac{x^2}{2} + \frac{x^4}{12} + \frac{x^6}{45} + \frac{17x^8}{2520} + \frac{31x^{10}}{14,175} + \ldots + C$;

$x = 0 \Rightarrow C = 0 \Rightarrow \ln|\sec x| = \frac{x^2}{2} + \frac{x^4}{12} + \frac{x^6}{45} + \frac{17x^8}{2520} + \frac{31x^{10}}{14,175} + \ldots$, converges when $-\frac{\pi}{2} < x < \frac{\pi}{2}$

(b) $\sec^2 x = \frac{d(\tan x)}{dx} = \frac{d}{dx}\left(x + \frac{x^3}{3} + \frac{2x^5}{15} + \frac{17x^7}{315} + \frac{62x^9}{2835} + \ldots\right) = 1 + x^2 + \frac{2x^4}{3} + \frac{17x^6}{45} + \frac{62x^8}{315} + \ldots$, converges

when $-\frac{\pi}{2} < x < \frac{\pi}{2}$

(c) $\sec^2 x = (\sec x)(\sec x) = \left(1 + \frac{x^2}{2} + \frac{5x^4}{24} + \frac{61x^6}{720} + \ldots\right)\left(1 + \frac{x^2}{2} + \frac{5x^4}{24} + \frac{61x^6}{720} + \ldots\right)$

$= 1 + \left(\frac{1}{2} + \frac{1}{2}\right)x^2 + \left(\frac{5}{24} + \frac{1}{4} + \frac{5}{24}\right)x^4 + \left(\frac{61}{720} + \frac{5}{48} + \frac{5}{48} + \frac{61}{720}\right)x^6 + \ldots = 1 + x^2 + \frac{2x^4}{3} + \frac{17x^6}{45} + \frac{62x^8}{315} + \ldots$,

$-\frac{\pi}{2} < x < \frac{\pi}{2}$

59. (a) If $f(x) = \sum\limits_{n=0}^{\infty} a_n x^n$, then $f^{(k)}(x) = \sum\limits_{n=k}^{\infty} n(n-1)(n-2)\cdots(n-(k-1))a_n x^{n-k}$ and $f^{(k)}(0) = k!a_k$

$\Rightarrow a_k = \dfrac{f^{(k)}(0)}{k!}$; likewise if $f(x) = \sum\limits_{n=0}^{\infty} b_n x^n$, then $b_k = \dfrac{f^{(k)}(0)}{k!} \Rightarrow a_k = b_k$ for every nonnegative integer k

(b) If $f(x) = \sum\limits_{n=0}^{\infty} a_n x^n = 0$ for all x, then $f^{(k)}(x) = 0$ for all $x \Rightarrow$ from part (a) that $a_k = 0$ for every nonnegative integer k

9.8   TAYLOR AND MACLAURIN SERIES

1. $f(x) = e^{2x},\ f'(x) = 2e^{2x},\ f''(x) = 4e^{2x},\ f'''(x) = 8e^{2x};\ f(0) = e^{2(0)} = 1,\ f'(0) = 2,\ f''(0) = 4,\ f'''(0) = 8$

$\Rightarrow P_0(x) = 1,\ P_1(x) = 1 + 2x,\ P_2(x) = 1 + 2x + 2x^2,\ P_3(x) = 1 + 2x + 2x^2 + \frac{4}{3}x^3$

3. $f(x) = \ln x,\ f'(x) = \frac{1}{x},\ f''(x) = -\frac{1}{x^2},\ f'''(x) = \frac{2}{x^3};\ f(1) = \ln 1 = 0,\ f'(1) = 1,\ f''(1) = -1,\ f'''(1) = 2$

$\Rightarrow P_0(x) = 0,\ P_1(x) = (x-1),\ P_2(x) = (x-1) - \frac{1}{2}(x-1)^2,\ P_3(x) = (x-1) - \frac{1}{2}(x-1)^2 + \frac{1}{3}(x-1)^3$

5. $f(x) = \frac{1}{x} = x^{-1},\ f'(x) = -x^{-2},\ f''(x) = 2x^{-3},\ f'''(x) = -6x^{-4};\ f(2) = \frac{1}{2},\ f'(2) = -\frac{1}{4},\ f''(2) = \frac{1}{4},\ f'''(x) = -\frac{3}{8}$

$\Rightarrow P_0(x) = \frac{1}{2},\ P_1(x) = \frac{1}{2} - \frac{1}{4}(x-2),\ P_2(x) = \frac{1}{2} - \frac{1}{4}(x-2) + \frac{1}{8}(x-2)^2,$

$\Rightarrow P_3(x) = \frac{1}{2} - \frac{1}{4}(x-2) + \frac{1}{8}(x-2)^2 - \frac{1}{16}(x-2)^3$

7. $f(x) = \sin x,\ f'(x) = \cos x,\ f''(x) = -\sin x,\ f'''(x) = -\cos x;\ f\left(\frac{\pi}{4}\right) = \sin\frac{\pi}{4} = \frac{\sqrt{2}}{2},\ f'\left(\frac{\pi}{4}\right) = \cos\frac{\pi}{4} = \frac{\sqrt{2}}{2},$

$f''\left(\frac{\pi}{4}\right) = -\sin\frac{\pi}{4} = -\frac{\sqrt{2}}{2},\ f'''\left(\frac{\pi}{4}\right) = -\cos\frac{\pi}{4} = -\frac{\sqrt{2}}{2} \Rightarrow P_0 = \frac{\sqrt{2}}{2},\ P_1(x) = \frac{\sqrt{2}}{2} + \frac{\sqrt{2}}{2}\left(x - \frac{\pi}{4}\right),$

$P_2(x) = \frac{\sqrt{2}}{2} + \frac{\sqrt{2}}{2}\left(x - \frac{\pi}{4}\right) - \frac{\sqrt{2}}{4}\left(x - \frac{\pi}{4}\right)^2,\ P_3(x) = \frac{\sqrt{2}}{2} + \frac{\sqrt{2}}{2}\left(x - \frac{\pi}{4}\right) - \frac{\sqrt{2}}{4}\left(x - \frac{\pi}{4}\right)^2 - \frac{\sqrt{2}}{12}\left(x - \frac{\pi}{4}\right)^3$

9. $f(x) = \sqrt{x} = x^{1/2},\ f'(x) = \left(\frac{1}{2}\right)x^{-1/2},\ f''(x) = \left(-\frac{1}{4}\right)x^{-3/2},\ f'''(x) = \left(\frac{3}{8}\right)x^{-5/2}; f(4) = \sqrt{4} = 2,$

$f'(4) = \left(\frac{1}{2}\right)4^{-1/2} = \frac{1}{4},\ f''(4) = \left(-\frac{1}{4}\right)4^{-3/2} = -\frac{1}{32},\ f'''(4) = \left(\frac{3}{8}\right)4^{-5/2} = \frac{3}{256} \Rightarrow P_0(x) = 2,\ P_1(x) = 2 + \frac{1}{4}(x-4),$

$P_2(x) = 2 + \frac{1}{4}(x-4) - \frac{1}{64}(x-4)^2,\ P_3(x) = 2 + \frac{1}{4}(x-4) - \frac{1}{64}(x-4)^2 + \frac{1}{512}(x-4)^3$

11. $f(x) = e^{-x},\ f'(x) = -e^{-x},\ f''(x) = e^{-x},\ f'''(x) = -e^{-x} \Rightarrow \ldots f^{(k)}(x) = (-1)^k e^{-x};\ f(0) = e^{-(0)} = 1,\ f'(0) = -1,$

$f''(0) = 1,\ f'''(0) = -1, \ldots, f^{(k)}(0) = (-1)^k \Rightarrow e^{-x} = 1 - x + \frac{1}{2}x^2 - \frac{1}{6}x^3 + \ldots = \sum\limits_{n=0}^{\infty} \frac{(-1)^n}{n!}x^n$

13. $f(x) = (1+x)^{-1} \Rightarrow f'(x) = -(1+x)^{-2},\ f''(x) = 2(1+x)^{-3},\ f'''(x) = -3!(1+x)^{-4}$

$\Rightarrow \ldots f^{(k)}(x) = (-1)^k k!(1+x)^{-k-1};\ f(0) = 1,\ f'(0) = -1,\ f''(0) = 2,\ f'''(0) = -3!, \ldots, f^{(k)}(0) = (-1)^k k!$

$\Rightarrow 1 - x + x^2 - x^3 + \ldots = \sum\limits_{n=0}^{\infty}(-x)^n = \sum\limits_{n=0}^{\infty}(-1)^n x^n$

15. $\sin x = \sum_{n=0}^{\infty} \frac{(-1)^n x^{2n+1}}{(2n+1)!} \Rightarrow \sin 3x = \sum_{n=0}^{\infty} \frac{(-1)^n (3x)^{2n+1}}{(2n+1)!} = \sum_{n=0}^{\infty} \frac{(-1)^n 3^{2n+1} x^{2n+1}}{(2n+1)!} = 3x - \frac{3^3 x^3}{3!} + \frac{3^5 x^5}{5!} - \cdots$

17. $7\cos(-x) = 7\cos x = 7\sum_{n=0}^{\infty} \frac{(-1)^n x^{2n}}{(2n)!} = 7 - \frac{7x^2}{2!} + \frac{7x^4}{4!} - \frac{7x^6}{6!} + \cdots$, since the cosine is an even function

19. $\cosh x = \frac{e^x + e^{-x}}{2} = \frac{1}{2}\left[\left(1 + x + \frac{x^2}{2!} + \frac{x^3}{3!} + \frac{x^4}{4!} + \cdots\right) + \left(1 - x + \frac{x^2}{2!} - \frac{x^3}{3!} + \frac{x^4}{4!} - \cdots\right)\right] = 1 + \frac{x^2}{2!} + \frac{x^4}{4!} + \frac{x^6}{6!} + \cdots = \sum_{n=0}^{\infty} \frac{x^{2n}}{(2n)!}$

21. $f(x) = x^4 - 2x^3 - 5x + 4 \Rightarrow f'(x) = 4x^3 - 6x^2 - 5, f''(x) = 12x^2 - 12x, f'''(x) = 24x - 12, f^{(4)}(x) = 24$
$\Rightarrow f^{(n)}(x) = 0$ if $n \geq 5$; $f(0) = 4, f'(0) = -5, f''(0) = 0, f'''(0) = -12, f^{(4)}(0) = 24, f^{(n)}(0) = 0$ if $n \geq 5$
$\Rightarrow x^4 - 2x^3 - 5x + 4 = 4 - 5x - \frac{12}{3!}x^3 + \frac{24}{4!}x^4 = x^4 - 2x^3 - 5x + 4$

23. $f(x) = x^3 - 2x + 4 \Rightarrow f'(x) = 3x^2 - 2, f''(x) = 6x, f'''(x) = 6 \Rightarrow f^{(n)}(x) = 0$ if $n \geq 4$;
$f(2) = 8, f'(2) = 10, f''(2) = 12, f'''(2) = 6, f^{(n)}(2) = 0$ if $n \geq 4$
$\Rightarrow x^3 - 2x + 4 = 8 + 10(x-2) + \frac{12}{2!}(x-2)^2 + \frac{6}{3!}(x-2)^3 = 8 + 10(x-2) + 6(x-2)^2 + (x-2)^3$

25. $f(x) = x^4 + x^2 + 1 \Rightarrow f'(x) = 4x^3 + 2x, f''(x) = 12x^2 + 2, f'''(x) = 24x, f^{(4)}(x) = 24, f^{(n)}(x) = 0$ if $n \geq 5$;
$f(-2) = 21, f'(-2) = -36, f''(-2) = 50, f'''(-2) = -48, f^{(4)}(-2) = 24, f^{(n)}(-2) = 0$ if $n \geq 5 \Rightarrow x^4 + x^2 + 1$
$= 21 - 36(x+2) + \frac{50}{2!}(x+2)^2 - \frac{48}{3!}(x+2)^3 + \frac{24}{4!}(x+2)^4 = 21 - 36(x+2) + 25(x+2)^2 - 8(x+2)^3 + (x+2)^4$

27. $f(x) = x^{-2} \Rightarrow f'(x) = -2x^{-3}, f''(x) = 3!x^{-4}, f'''(x) = -4!x^{-5} \Rightarrow f^{(n)}(x) = (-1)^n (n+1)! x^{-n-2}$;
$f(1) = 1, f'(1) = -2, f''(1) = 3!, f'''(1) = -4!, f^{(n)}(1) = (-1)^n (n+1)!$
$\Rightarrow \frac{1}{x^2} = 1 - 2(x-1) + 3(x-1)^2 - 4(x-1)^3 + \cdots = \sum_{n=0}^{\infty} (-1)^n (n+1)(x-1)^n$

29. $f(x) = e^x \Rightarrow f'(x) = e^x, f''(x) = e^x \Rightarrow f^{(n)}(x) = e^x$; $f(2) = e^2, f'(2) = e^2, \ldots f^{(n)}(2) = e^2$
$\Rightarrow e^x = e^2 + e^2(x-2) + \frac{e^2}{2}(x-2)^2 + \frac{e^3}{3!}(x-2)^3 + \cdots = \sum_{n=0}^{\infty} \frac{e^2}{n!}(x-2)^n$

31. $f(x) = \cos\left(2x + \frac{\pi}{2}\right), f'(x) = -2\sin\left(2x + \frac{\pi}{2}\right), f''(x) = -4\cos\left(2x + \frac{\pi}{2}\right), f'''(x) = 8\sin\left(2x + \frac{\pi}{2}\right)$,
$f^{(4)}(x) = 2^4 \cos\left(2x + \frac{\pi}{2}\right), f^{(5)}(x) = -2^5 \sin\left(2x + \frac{\pi}{2}\right), ..;$
$f\left(\frac{\pi}{4}\right) = -1, f'\left(\frac{\pi}{4}\right) = 0, f''\left(\frac{\pi}{4}\right) = 4, f'''\left(\frac{\pi}{4}\right) = 0, f^{(4)}\left(\frac{\pi}{4}\right) = 2^4, f^{(5)}\left(\frac{\pi}{4}\right) = 0, \ldots, f^{(2n)}\left(\frac{\pi}{4}\right) = (-1)^n 2^{2n}$
$\Rightarrow \cos\left(2x + \frac{\pi}{2}\right) = -1 + 2\left(x - \frac{\pi}{4}\right)^2 - \frac{2}{3}\left(x - \frac{\pi}{4}\right)^4 + \cdots = \sum_{n=0}^{\infty} \frac{(-1)^n 2^{2n}}{(2n)!}\left(x - \frac{\pi}{4}\right)^{2n}$

33. The Maclaurin series generated by $\cos x$ is $\displaystyle\sum_{n=0}^{\infty} \frac{(-1)^n}{(2n)!}x^{2n}$ which converges on $(-\infty, \infty)$ and the Maclaurin series generated by $\frac{2}{1-x}$ is $2\displaystyle\sum_{n=0}^{\infty} x^n$ which converges on $(-1, 1)$. Thus the Maclaurin series generated by $f(x) = \cos x - \frac{2}{1-x}$ is given by $\displaystyle\sum_{n=0}^{\infty} \frac{(-1)^n}{(2n)!}x^{2n} - 2\sum_{n=0}^{\infty} x^n = -1 - 2x - \frac{5}{2}x^2 - \dots$ which converges on the intersection of $(-\infty, \infty)$ and $(-1, 1)$, so the interval of convergence is $(-1, 1)$.

35. The Maclaurin series generated by $\sin x$ is $\displaystyle\sum_{n=0}^{\infty} \frac{(-1)^n}{(2n+1)!}x^{2n+1}$ which converges on $(-\infty, \infty)$ and the Maclaurin series generated by $\ln(1+x)$ is $\displaystyle\sum_{n=1}^{\infty} \frac{(-1)^{n-1}}{n}x^n$ which converges on $(-1, 1)$. Thus the Maclaurin series generated by $f(x) = \sin x \cdot \ln(1+x)$ is given by $\left(\displaystyle\sum_{n=0}^{\infty} \frac{(-1)^n}{(2n+1)!}x^{2n+1}\right)\left(\sum_{n=1}^{\infty} \frac{(-1)^{n-1}}{n}x^n\right) = x^2 - \frac{1}{2}x^3 + \frac{1}{6}x^4 - \dots$ which converges on the intersection of $(-\infty, \infty)$ and $(-1, 1)$, so the interval convergence is $(-1, 1)$.

37. If $e^x = \displaystyle\sum_{n=0}^{\infty} \frac{f^{(n)}(a)}{n!}(x-a)^n$ and $f(x) = e^x$, we have $f^{(n)}(a) = e^a$ for all $n = 0, 1, 2, 3, \dots$

$\Rightarrow e^x = e^a\left[\frac{(x-a)^0}{0!} + \frac{(x-a)^1}{1!} + \frac{(x-a)^2}{2!} + \dots\right] = e^a\left[1 + (x-a) + \frac{(x-a)^2}{2!} + \dots\right]$ at $x = a$

39. $f(x) = f(a) + f'(a)(x-a) + \frac{f''(a)}{2}(x-a)^2 + \frac{f'''(a)}{3!}(x-a)^3 + \dots$

$\Rightarrow f'(x) = f'(a) + f''(a)(x-a) + \frac{f'''(a)}{3!}3(x-a)^2 + \dots \Rightarrow f''(x) = f''(a) + f'''(a)(x-a) + \frac{f^{(4)}(a)}{4!}4\cdot 3(x-a)^2 + \dots$

$\Rightarrow f^{(n)}(x) = f^{(n)}(a) + f^{(n+1)}(a)(x-a) + \frac{f^{(n+2)}(a)}{2}(x-a)^2 + \dots$

$\Rightarrow f(a) = f(a) + 0, \; f'(a) = f'(a) + 0, \dots, f^{(n)}(a) = f^{(n)}(a) + 0$

41. $f(x) = \ln(\cos x) \Rightarrow f'(x) = -\tan x$ and $f''(x) = -\sec^2 x; \; f(0) = 0, \; f'(0) = 0, \; f''(0) = -1$

$\Rightarrow L(x) = 0$ and $Q(x) = -\frac{x^2}{2}$

43. $f(x) = \left(1-x^2\right)^{-1/2} \Rightarrow f'(x) = x\left(1-x^2\right)^{-3/2}$ and $f''(x) = \left(1-x^2\right)^{-3/2} + 3x^2\left(1-x^2\right)^{-5/2}$;

$f(0) = 1, \; f'(0) = 0, \; f''(0) = 1 \Rightarrow L(x) = 1$ and $Q(x) = 1 + \frac{x^2}{2}$

45. $f(x) = \sin x \Rightarrow f'(x) = \cos x$ and $f''(x) = -\sin x; \; f(0) = 0, \; f'(0) = 1, \; f''(0) = 0 \Rightarrow L(x) = x$ and $Q(x) = x$

9.9   CONVERGENCE OF TAYLOR SERIES

1. $e^x = 1 + x + \frac{x^2}{2!} + \dots = \sum_{n=0}^{\infty} \frac{x^n}{n!} \Rightarrow e^{-5x} = 1 + (-5x) + \frac{(-5x)^2}{2!} + \dots = 1 - 5x + \frac{5^2 x^2}{2!} - \frac{5^3 x^3}{3!} + \dots = \sum_{n=0}^{\infty} \frac{(-1)^n 5^n x^n}{n!}$

3. $\sin x = x - \frac{x^3}{3!} + \frac{x^5}{5!} - \dots = \sum_{n=0}^{\infty} \frac{(-1)^n x^{2n+1}}{(2n+1)!} \Rightarrow 5\sin(-x) = 5\left[(-x) - \frac{(-x)^3}{3!} + \frac{(-x)^5}{5!} - \dots \right] = \sum_{n=0}^{\infty} \frac{5(-1)^{n+1} x^{2n+1}}{(2n+1)!}$

5. $\cos x = \sum_{n=0}^{\infty} \frac{(-1)^n x^{2n}}{(2n)!} \Rightarrow \cos 5x^2 = \sum_{n=0}^{\infty} \frac{(-1)^n \left(5x^2\right)^{2n}}{(2n)!} = \sum_{n=0}^{\infty} \frac{(-1)^n 5^{2n} x^{4n}}{(2n)!} = 1 - \frac{25x^4}{2!} + \frac{625x^8}{4!} - \frac{15625x^{12}}{6!} + \dots$

7. $\ln(1+x) = \sum_{n=1}^{\infty} \frac{(-1)^{n-1} x^n}{n} \Rightarrow \ln\left(1+x^2\right) = \sum_{n=1}^{\infty} \frac{(-1)^{n-1}\left(x^2\right)^n}{n} = \sum_{n=1}^{\infty} \frac{(-1)^{n-1} x^{2n}}{n} = x^2 - \frac{x^4}{2} + \frac{x^6}{3} - \frac{x^8}{4} + \dots$

9. $\frac{1}{1+x} = \sum_{n=0}^{\infty} (-1)^n x^n \Rightarrow \frac{1}{1+\frac{3}{4}x^3} = \sum_{n=0}^{\infty} (-1)^n \left(\frac{3}{4}x^3\right)^n = \sum_{n=0}^{\infty} (-1)^n \left(\frac{3}{4}\right)^n x^{3n} = 1 - \frac{3}{4}x^3 + \frac{9}{16}x^6 - \frac{27}{64}x^9 + \dots$

11. $e^x = \sum_{n=0}^{\infty} \frac{x^n}{n!} \Rightarrow xe^x = x\left(\sum_{n=0}^{\infty} \frac{x^n}{n!} \right) = \sum_{n=0}^{\infty} \frac{x^{n+1}}{n!} = x + x^2 + \frac{x^3}{2!} + \frac{x^4}{3!} + \frac{x^5}{4!} + \dots$

13. $\cos x = \sum_{n=0}^{\infty} \frac{(-1)^n x^{2n}}{(2n)!} \Rightarrow \frac{x^2}{2} - 1 + \cos x = \frac{x^2}{2} - 1 + \sum_{n=0}^{\infty} \frac{(-1)^n x^{2n}}{(2n)!} = \frac{x^2}{2} - 1 + 1 - \frac{x^2}{2} + \frac{x^4}{4!} - \frac{x^6}{6!} + \frac{x^8}{8!} - \frac{x^{10}}{10!} + \dots$

$= \frac{x^4}{4!} - \frac{x^6}{6!} + \frac{x^8}{8!} - \frac{x^{10}}{10!} + \dots = \sum_{n=2}^{\infty} \frac{(-1)^n x^{2n}}{(2n)!}$

15. $\cos x = \sum_{n=0}^{\infty} \frac{(-1)^n x^{2n}}{(2n)!} \Rightarrow x\cos \pi x = x\sum_{n=0}^{\infty} \frac{(-1)^n (\pi x)^{2n}}{(2n)!} = \sum_{n=0}^{\infty} \frac{(-1)^n \pi^{2n} x^{2n+1}}{(2n)!} = x - \frac{\pi^2 x^3}{2!} + \frac{\pi^4 x^5}{4!} - \frac{\pi^6 x^7}{6!} + \dots$

17. $\cos^2 x = \frac{1}{2} + \frac{\cos 2x}{2} = \frac{1}{2} + \frac{1}{2}\sum_{n=0}^{\infty} \frac{(-1)^n (2x)^{2n}}{(2n)!} = \frac{1}{2} + \frac{1}{2}\left[1 - \frac{(2x)^2}{2!} + \frac{(2x)^4}{4!} - \frac{(2x)^6}{6!} + \frac{(2x)^8}{8!} - \dots \right]$

$= 1 - \frac{(2x)^2}{2\cdot 2!} + \frac{(2x)^4}{2\cdot 4!} - \frac{(2x)^6}{2\cdot 6!} + \frac{(2x)^8}{2\cdot 8!} - \dots = 1 + \sum_{n=1}^{\infty} \frac{(-1)^n (2x)^{2n}}{2\cdot (2n)!} = 1 + \sum_{n=1}^{\infty} \frac{(-1)^n 2^{2n-1} x^{2n}}{(2n)!}$

19. $\frac{x^2}{1-2x} = x^2\left(\frac{1}{1-2x} \right) = x^2 \sum_{n=0}^{\infty} (2x)^n = \sum_{n=0}^{\infty} 2^n x^{n+2} = x^2 + 2x^3 + 2^2 x^4 + 2^3 x^5 + \dots$

21. $\frac{1}{1-x} = \sum_{n=0}^{\infty} x^n = 1 + x + x^2 + x^3 + \dots \Rightarrow \frac{d}{dx}\left(\frac{1}{1-x} \right) = \frac{1}{(1-x)^2} = 1 + 2x + 3x^2 + \dots = \sum_{n=1}^{\infty} nx^{n-1} = \sum_{n=0}^{\infty} (n+1)x^n$

23. $\tan^{-1} x = x - \frac{1}{3}x^3 + \frac{1}{5}x^5 - \frac{1}{7}x^7 + \ldots \Rightarrow x\tan^{-1} x^2 = x\left(x^2 - \frac{1}{3}\left(x^2\right)^3 + \frac{1}{5}\left(x^2\right)^5 - \frac{1}{7}\left(x^2\right)^7 + \ldots\right)$

$= x^3 - \frac{1}{3}x^7 + \frac{1}{5}x^{11} - \frac{1}{7}x^{15} + \ldots = \sum_{n=1}^{\infty} \frac{(-1)^n x^{4n-1}}{2n-1}$

25. $e^x = 1 + x + \frac{x^2}{2!} + \frac{x^3}{3} + \ldots$ and $\frac{1}{1+x} = 1 - x + x^2 - x^3 + \ldots \Rightarrow e^x + \frac{1}{1+x} = \left(1 + x + \frac{x^2}{2!} + \frac{x^3}{3!} + \ldots\right) + \left(1 - x + x^2 - x^3 + \ldots\right)$

$= 2 + \frac{3}{2}x^2 - \frac{5}{6}x^3 + \frac{25}{24}x^4 + \ldots = \sum_{n=0}^{\infty} \left(\frac{1}{n!} + (-1)^n\right)x^n$

27. $\ln(1+x) = x - \frac{1}{2}x^2 + \frac{1}{3}x^3 - \frac{1}{4}x^4 + \ldots \Rightarrow \frac{x}{3}\ln\left(1+x^2\right) = \frac{x}{3}\left(x^2 - \frac{1}{2}\left(x^2\right)^2 + \frac{1}{3}\left(x^2\right)^3 - \frac{1}{4}\left(x^2\right)^4 + \ldots\right)$

$= \frac{1}{3}x^3 - \frac{1}{6}x^5 + \frac{1}{9}x^7 - \frac{1}{12}x^9 + \ldots = \sum_{n=1}^{\infty} \frac{(-1)^{n-1}}{3n}x^{2n+1}$

29. $e^x = 1 + x + \frac{x^2}{2!} + \frac{x^3}{3!} + \ldots$ and $\sin x = x - \frac{x^3}{3!} + \frac{x^5}{5!} - \frac{x^7}{7!} + \ldots$

$\Rightarrow e^x \cdot \sin x = \left(1 + x + \frac{x^2}{2!} + \frac{x^3}{3!} + \ldots\right)\left(x - \frac{x^3}{3!} + \frac{x^5}{5!} - \frac{x^7}{7!} + \ldots\right) = x + x^2 + \frac{1}{3}x^3 - \frac{1}{30}x^5 - \ldots$

31. $\tan^{-1} x = x - \frac{1}{3}x^3 + \frac{1}{5}x^5 - \frac{1}{7}x^7 + \ldots \Rightarrow \left(\tan^{-1} x\right)^2 = \left(\tan^{-1} x\right)\left(\tan^{-1} x\right)$

$= \left(x - \frac{1}{3}x^3 + \frac{1}{5}x^5 - \frac{1}{7}x^7 + \ldots\right)\left(x - \frac{1}{3}x^3 + \frac{1}{5}x^5 - \frac{1}{7}x^7 + \ldots\right) = x^2 - \frac{2}{3}x^4 - \frac{23}{45}x^6 - \frac{44}{105}x^8 + \ldots$

33. $\sin x = x - \frac{x^3}{3!} + \frac{x^5}{5!} - \frac{x^7}{7!} + \ldots$ and $e^x = 1 + x + \frac{x^2}{2!} + \frac{x^3}{3!} + \ldots$

$\Rightarrow e^{\sin x} = 1 + \left(x - \frac{x^3}{3!} + \frac{x^5}{5!} - \frac{x^7}{7!} + \ldots\right) + \frac{1}{2}\left(x - \frac{x^3}{3!} + \frac{x^5}{5!} - \frac{x^7}{7!} + \ldots\right)^2 + \frac{1}{6}\left(x - \frac{x^3}{3!} + \frac{x^5}{5!} - \frac{x^7}{7!} + \ldots\right)^3 + \ldots$

$= 1 + x + \frac{1}{2}x^2 - \frac{1}{8}x^4 + \ldots$

35. Since $n = 3$, then $f^{(4)}(x) = \sin x, \left|f^{(4)}(x)\right| \le M$ on $[0, 0.1] \Rightarrow |\sin x| \le 1$ on $[0, 0.1] \Rightarrow M = 1$.

Then $\left|R_3(0.1)\right| \le 1\frac{|0.1-0|^4}{4!} = 4.2 \times 10^{-6} \Rightarrow$ error $\le 4.2 \times 10^{-6}$

37. By the Alternating Series Estimation Theorem, the error is less than $\frac{|x|^5}{5!} \Rightarrow |x|^5 < (5!)\left(5 \times 10^{-4}\right)$

$\Rightarrow |x|^5 < 600 \times 10^{-4} \Rightarrow |x| < \sqrt[5]{6 \times 10^{-2}} \approx 0.56968$

39. If $\sin x = x$ and $|x| < 10^{-3}$, then the error is less than $\frac{\left(10^{-3}\right)^3}{3!} \approx 1.67 \times 10^{-10}$, by Alternating Series Estimation Theorem; The Alternating Series Estimation Theorem says $R_2(x)$ has the same sign as $-\frac{x^3}{3!}$. Moreover, $x < \sin x \Rightarrow 0 < \sin x - x = R_2(x) \Rightarrow x < 0 \Rightarrow -10^{-3} < x < 0$.

41. $|R_2(x)| = \left|\frac{e^c x^3}{3!}\right| < \frac{3^{(0.1)}(0.1)^3}{3!} < 1.87\times10^{-4}$, where c is between 0 and x

43. $\sin^2 x = \left(\frac{1-\cos 2x}{2}\right) = \frac{1}{2} - \frac{1}{2}\cos 2x = \frac{1}{2} - \frac{1}{2}\left(1 - \frac{(2x)^2}{2!} + \frac{(2x)^4}{4!} - \frac{(2x)^6}{6!} + \dots\right) = \frac{2x^2}{2!} - \frac{2^3 x^4}{4!} + \frac{2^5 x^6}{6!} - \dots$

$\Rightarrow \frac{d}{dx}\left(\sin^2 x\right) = \frac{d}{dx}\left(\frac{2x^2}{2!} - \frac{2^3 x^4}{4!} + \frac{2^5 x^6}{6!} - \dots\right) = 2x - \frac{(2x)^3}{3!} + \frac{(2x)^5}{5!} - \frac{(2x)^7}{7!} + \dots$

$\Rightarrow 2\sin x\cos x = 2x - \frac{(2x)^3}{3!} + \frac{(2x)^5}{5!} - \frac{(2x)^7}{7!} + \dots = \sin 2x$, which checks

45. A special case of Taylor's Theorem is $f(b) = f(a) + f'(c)(b-a)$, where c is between a and $b \Rightarrow f(b) - f(a) = f'(c)(b-a)$, the Mean Value Theorem.

47. (a) $f'' \le 0$, $f'(a) = 0$ and $x = a$ interior to the interval $I \Rightarrow f(x) - f(a) = \frac{f''(c_2)}{2}(x-a)^2 \le 0$ throughout I
$\Rightarrow f(x) \le f(a)$ throughout $I \Rightarrow f$ has a local maximum at $x = a$

(b) similar reasoning gives $f(x) - f(a) = \frac{f''(c_2)}{2}(x-a)^2 \ge 0$ throughout $I \Rightarrow f(x) \ge f(a)$ throughout I
$\Rightarrow f$ has a local minimum at $x = a$

49. (a) $f(x) = (1+x)^k \Rightarrow f'(x) = k(1+x)^{k-1} \Rightarrow f''(x) = k(k-1)(1+x)^{k-2}$; $f(0) = 1$, $f'(0) = k$, and
$f''(0) = k(k-1) \Rightarrow Q(x) = 1 + kx + \frac{k(k-1)}{2}x^2$

(b) $|R_2(x)| = \left|\frac{3\cdot2\cdot1}{3!}x^3\right| < \frac{1}{100} \Rightarrow |x^3| < \frac{1}{100} \Rightarrow 0 < x < \frac{1}{100^{1/3}}$ or $0 < x < .21544$

51. If $f(x) = \sum_{n=0}^{\infty} a_n x^n$, then $f^{(k)}(x) = \sum_{n=k}^{\infty} n(n-1)(n-2)\cdots(n-k+1)a_n x^{n-k}$ and $f^{(k)}(0) = k!a_k \Rightarrow a_k = \frac{f^{(k)}(0)}{k!}$
for k a nonnegative integer. Therefore, the coefficients of $f(x)$ are identical with the corresponding coefficients in the Maclaurin series of $f(x)$ and the statement follows.

53-57. Example CAS commands:

Maple:

```
f := x -> 1/sqrt(1+x);
x0 := -3/4;
x1 := 3/4;
# Step 1:
plot( f(x), x=x0..x1, title="Step 1: #53 (Section 10.9)" );
# Step 2:
P1 := unapply( TaylorApproximation(f(x), x = 0, order=1), x );
P2 := unapply( TaylorApproximation(f(x), x = 0, order=2), x );
P3 := unapply( TaylorApproximation(f(x), x = 0, order=3), x );
```

Step 3:

```
D2f := D(D(f));

D3f := D(D(D(f)));

D4f := D(D(D(D(f))));

plot( [D2f(x),D3f(x),D4f(x)], x=x0..x1, thickness=[0,2,4], color=[red,blue,green], title="Step 3: #57
   (Section 10.9)" );

c1 := x0;

M1 := abs( D2f(c1) );

c2 := x0;

M2 := abs( D3f(c2) );

c3 := x0;

M3 := abs( D4f(c3) );
```

Step 4:

```
R1 := unapply( abs(M1/2!*(x-0)^2), x );

R2 := unapply( abs(M2/3!*(x-0)^3), x );

R3 := unapply( abs(M3/4!*(x-0)^4), x );

plot( [R1(x),R2(x),R3(x)], x=x0..x1, thickness=[0,2,4], color=[red,blue,green], title="Step 4: #53
   (Section 10.9)" );
```

Step 5:

```
E1 := unapply( abs(f(x)-P1(x)), x );

E2 := unapply( abs(f(x)-P2(x)), x );

E3 := unapply( abs(f(x)-P3(x)), x );

plot( [E1(x),E2(x),E3(x),R1(x),R2(x),R3(x)], x=x0..x1, thickness=[0,2,4], color=[red,blue,green],
   linestyle=[1,1,1,3,3,3], title="Step 5: #53 (Section 10.9)" );
```

Step 6:

```
TaylorApproximation( f(x), view=[x0..x1,DEFAULT], x=0, output=animation, order=1..3 );

L1 := fsolve( abs(f(x)-P1(x))=0.01, x=x0/2 );                    # (a)

R1 := fsolve( abs(f(x)-P1(x))=0.01, x=x1/2 );

L2 := fsolve( abs(f(x)-P2(x))=0.01, x=x0/2 );

R2 := fsolve( abs(f(x)-P2(x))=0.01, x=x1/2 );

L3 := fsolve( abs(f(x)-P3(x))=0.01, x=x0/2 );

R3 := fsolve( abs(f(x)-P3(x))=0.01, x=x1/2 );

plot( [E1(x),E2(x),E3(x),0.01], x=min(L1,L2,L3)..max(R1,R2,R3), thickness=[0,2,4,0], linestyle=[0,0,0,2]
   color=[red,blue,green,black], view=[DEFAULT,0..0.01], title="#53(a) (Section 10.9)" );

abs(`f(x)`-`P`[1](x) ) <= evalf( E1 (x0) );                    # (b)

abs(`f(x)`-`P`[2](x) ) <= evalf( E2(x0) );

abs(`f(x)`-`P`[3](x) ) <= evalf( E3(x0) );
```

<u>Mathematica</u>: (assigned function and values for a, b, c, and n may vary)

        Clear[x, f, c]

        $f[x_] = (1 + x)^{3/2}$

        {a, b} = {−1/2, 2};

        pf = Plot[f[x], {x, a, b}];

        poly1[x_]=Series[f[x], {x,0,1}]//Normal

        poly2[x_]=Series[f[x], {x,0,2}]//Normal

        poly3[x_]=Series[f[x], {x,0,3}]//Normal

        Plot[{f[x], poly1[x], poly2[x], poly3[x]}, {x, a, b},

        PlotStyle → {RGBColor[1,0,0],RGBColor[0,1,0],RGBColor[0,0,1],RGBColor[0,.5,.5]}];

The above defines the approximations. The following analyzes the derivatives to determine their maximum values.

        f'[c]

        Plot[f'[x], {x, a, b}];

        f''[c]

        Plot[f''[x], {x, a, b}];

        f'''[c]

        [f'''[x], {x, a, b}];

Noting the upper bound for each of the above derivatives occurs at x = a, the upper bounds m1, m2, and m3 can be defined and bounds for remainders viewed as functions of x.

        m1=f'[a]

        m2=-f''[a]

        m3=f'''[a]

        $r1[x_]=m1\ x^2/2!$

        Plot[r1[x], {x, a, b}];

        $r2[x_]=m2\ x^3/3!$

        Plot[r2[x], {x, a, b}];

        $r3[x_]=m3\ x^4/4!$

        Plot[r3[x], {x, a, b}];

A three dimensional look at the error functions, allowing both c and x to vary can also be viewed. Recall that c must be a value between 0 and x. so some points on the surfaces where c is not in that interval are meaningless.

        Plot3D[f'[c] $x^2/2!$, {x, a, b}, {c, a, b}, PlotRange → All]

        Plot3D[f''[c] $x^3/3!$, {x, a, b}, {c, a, b}, PlotRange → All]

        Plot3D[f'''[c] $x^4/4!$, {x, a, b}, {c, a, b}, PlotRange → All]

9.10 THE BINOMIAL SERIES AND APPLICATIONS OF TAYLOR SERIES

1. $(1+x)^{1/2} = 1 + \frac{1}{2}x + \frac{\left(\frac{1}{2}\right)\left(-\frac{1}{2}\right)x^2}{2!} + \frac{\left(\frac{1}{2}\right)\left(-\frac{1}{2}\right)\left(-\frac{3}{2}\right)x^3}{3!} + \ldots = 1 + \frac{1}{2}x - \frac{1}{8}x^2 + \frac{1}{16}x^3 - \ldots$

3. $(1-x)^{-3} = 1 + (-3)(-x) + \frac{(-3)(-4)}{2!}(-x)^2 + \frac{(-3)(-4)(-5)}{3!}(-x)^3 + \cdots = 1 + 3x + 6x^2 + 10x^3 + \cdots$

5. $\left(1+\frac{x}{2}\right)^{-2} = 1 - 2\left(\frac{x}{2}\right) + \frac{(-2)(-3)\left(\frac{x}{2}\right)^2}{2!} + \frac{(-2)(-3)(-4)\left(\frac{x}{2}\right)^3}{3!} + \ldots = 1 - x + \frac{3}{4}x^2 - \frac{1}{2}x^3 + \ldots$

7. $\left(1+x^3\right)^{-1/2} = 1 - \frac{1}{2}x^3 + \frac{\left(-\frac{1}{2}\right)\left(-\frac{3}{2}\right)\left(x^3\right)^2}{2!} + \frac{\left(-\frac{1}{2}\right)\left(-\frac{3}{2}\right)\left(-\frac{5}{2}\right)\left(x^3\right)^3}{3!} + \ldots = 1 - \frac{1}{2}x^3 + \frac{3}{8}x^6 - \frac{5}{16}x^9 + \ldots$

9. $\left(1+\frac{1}{x}\right)^{1/2} = 1 + \frac{1}{2}\left(\frac{1}{x}\right) + \frac{\left(\frac{1}{2}\right)\left(-\frac{1}{2}\right)\left(\frac{1}{x}\right)^2}{2!} + \frac{\left(\frac{1}{2}\right)\left(-\frac{1}{2}\right)\left(-\frac{3}{2}\right)\left(\frac{1}{x}\right)^3}{3!} + \ldots = 1 + \frac{1}{2x} - \frac{1}{8x^2} + \frac{1}{16x^3} + \ldots$

11. $(1+x)^4 = 1 + 4x + \frac{(4)(3)x^2}{2!} + \frac{(4)(3)(2)x^3}{3!} + \frac{(4)(3)(2)x^4}{4!} = 1 + 4x + 6x^2 + 4x^3 + x^4$

13. $(1-2x)^3 = 1 + 3(-2x) + \frac{(3)(2)(-2x)^2}{2!} + \frac{(3)(2)(1)(-2x)^3}{3!} = 1 - 6x + 12x^2 - 8x^3$

15. Example 3 gives the indefinite integral as $C + \frac{x^3}{3} - \frac{x^7}{7\cdot3!} + \frac{x^{11}}{11\cdot5!} - \frac{x^{15}}{15\cdot7!} + \cdots$. Since the lower limit of integration is 0, the value of the definite integral will be the value of this series at the upper limit, with $C = 0$. Since $\frac{0.6^{11}}{11\cdot5!} \approx 2.75 \times 10^{-6}$ and the preceding term is greater than 10^{-5}, the first two terms should give the required accuracy, and the integral is approximated to within 10^{-5} by $\frac{0.6^3}{3} - \frac{0.6^7}{7\cdot3!} \approx 0.0713335$

17. Using a binomial series we find $\frac{1}{\sqrt{1+x^4}} = 1 - \frac{x^4}{2} + \frac{3x^8}{8} - \frac{5x^{12}}{16} + \cdots$. Integrating term by term and noting that the lower limit of integration is 0, the value of the definite integral from 0 to x is given by $x - \frac{x^5}{10} + \frac{x^9}{24} - \frac{5x^{13}}{13\cdot16} + \cdots$. Since $\frac{5\cdot0.5^{13}}{13\cdot16} \approx 2.93 \times 10^{-6}$ and the preceding term is greater than 10^{-5}, the first three terms should give the required accuracy, and the integral is approximated to within 10^{-5} by $0.5 - \frac{0.5^5}{10} + \frac{0.5^9}{24} \approx 0.4969564.$

19. $\int_0^{0.1} \frac{\sin x}{x} dx = \int_0^{0.1} \left(1 - \frac{x^2}{3!} + \frac{x^4}{5!} - \frac{x^6}{7!} + \ldots\right) dx = \left[x - \frac{x^3}{3\cdot3!} + \frac{x^5}{5\cdot5!} - \frac{x^7}{7\cdot7!} + \ldots\right]_0^{0.1} \approx \left[x - \frac{x^3}{3\cdot3!} + \frac{x^5}{5\cdot5!}\right]_0^{0.1}$

$\approx 0.0999444611, |E| \leq \frac{(0.1)^7}{7\cdot7!} \approx 2.8 \times 10^{-12}$

21. $\left(1+x^4\right)^{1/2} = (1)^{1/2} + \frac{\left(\frac{1}{2}\right)}{1}(1)^{-1/2}\left(x^4\right) + \frac{\left(\frac{1}{2}\right)\left(-\frac{1}{2}\right)}{2!}(1)^{-3/2}\left(x^4\right)^2 + \frac{\left(\frac{1}{2}\right)\left(-\frac{1}{2}\right)\left(-\frac{3}{2}\right)}{3!}(1)^{-5/2}\left(x^4\right)^3$

$\qquad + \frac{\left(\frac{1}{2}\right)\left(-\frac{1}{2}\right)\left(-\frac{3}{2}\right)\left(-\frac{5}{2}\right)}{4!}(1)^{-7/2}\left(x^4\right)^4 + \ldots = 1 + \frac{x^4}{2} - \frac{x^8}{8} + \frac{x^{12}}{16} - \frac{5x^{16}}{128} + \ldots$

$\Rightarrow \int_0^{0.1}\left(1 + \frac{x^4}{2} - \frac{x^8}{8} + \frac{x^{12}}{16} - \frac{5x^{16}}{128} + \ldots\right)dx \approx \left[x + \frac{x^5}{10}\right]_0^{0.1} \approx 0.100001, |E| \le \frac{(0.1)^9}{72} \approx 1.39 \times 10^{-11}$

23. $\int_0^1 \cos t^2\, dt = \int_0^1\left(1 - \frac{t^4}{2} + \frac{t^8}{4!} - \frac{t^{12}}{6!} + \ldots\right)dt = \left[t - \frac{t^5}{10} + \frac{t^9}{9\cdot 4!} - \frac{t^{13}}{13\cdot 6!} + \ldots\right]_0^1 \Rightarrow |\text{error}| < \frac{1}{13\cdot 6!} \approx .00011$

25. $F(x) = \int_0^x\left(t^2 - \frac{t^6}{3!} + \frac{t^{10}}{5!} - \frac{t^{14}}{7!} + \ldots\right)dt = \left[\frac{t^3}{3} - \frac{t^7}{7\cdot 3!} + \frac{t^{11}}{11\cdot 5!} - \frac{t^{15}}{15\cdot 7!} + \ldots\right]_0^x \approx \frac{x^3}{3} - \frac{x^7}{7\cdot 3!} + \frac{x^{11}}{11\cdot 5!}$

$\qquad \Rightarrow |\text{error}| < \frac{1}{15\cdot 7!} \approx 0.000013$

27. (a) $F(x) = \int_0^x\left(t - \frac{t^3}{3} + \frac{t^5}{5} - \frac{t^7}{7} + \ldots\right)dt = \left[\frac{t^2}{2} + \frac{t^4}{12} + \frac{t^6}{30} - \ldots\right]_0^x \approx \frac{x^2}{2} - \frac{x^4}{12} \Rightarrow |\text{error}| < \frac{(0.5)^6}{30} \approx .00052$

$\qquad$ (b) $|\text{error}| < \frac{1}{33.34} \approx .00089$ when $F(x) \approx \frac{x^2}{2} - \frac{x^4}{3\cdot 4} + \frac{x^6}{5\cdot 6} - \frac{x^8}{7\cdot 8} + \ldots + (-1)^{15}\frac{x^{32}}{31\cdot 22}$

29. $\frac{1}{x^2}\left(e^x - (1+x)\right) = \frac{1}{x^2}\left(\left(1 + x + \frac{x^2}{2} + \frac{x^3}{3!} + \ldots\right) - 1 - x\right) = \frac{1}{2} + \frac{x}{3!} + \frac{x^2}{4!} + \ldots \Rightarrow \lim_{x\to 0}\frac{e^x - (1+x)}{x^2} = \lim_{x\to 0}\left(\frac{1}{2} + \frac{x}{3!} + \frac{x^2}{4!} + \ldots\right)$

$\qquad = \frac{1}{2}$

31. $\frac{1}{t^4}\left(1 - \cos t - \frac{t^2}{2}\right) = \frac{1}{t^4}\left[1 - \frac{t^2}{2} - \left(1 - \frac{t^2}{2} + \frac{t^4}{4!} - \frac{t^6}{6!} + \ldots\right)\right] = -\frac{1}{4!} + \frac{t^2}{6!} - \frac{t^4}{8!} + \ldots \Rightarrow \lim_{t\to 0}\frac{1 - \cos t - \left(\frac{t^2}{2}\right)}{t^4}$

$\qquad = \lim_{t\to 0}\left(-\frac{1}{4!} + \frac{t^2}{6!} - \frac{t^4}{8!} + \ldots\right) = -\frac{1}{24}$

33. $\frac{1}{y^3}\left(y - \tan^{-1} y\right) = \frac{1}{y^3}\left[y - \left(y - \frac{y^3}{3} + \frac{y^5}{5} - \ldots\right)\right] = \frac{1}{3} - \frac{y^2}{5} + \frac{y^4}{7} - \ldots \Rightarrow \lim_{y\to 0}\frac{y - \tan^{-1} y}{y^3} = \lim_{y\to 0}\left(\frac{1}{3} - \frac{y^2}{5} + \frac{y^4}{7} - \ldots\right) = \frac{1}{3}$

35. $x^2\left(-1 + e^{-1/x^2}\right) = x^2\left(-1 + 1 - \frac{1}{x^2} + \frac{1}{2x^4} - \frac{1}{6x^6} + \ldots\right) = -1 + \frac{1}{2x^2} - \frac{1}{6x^4} + \ldots \Rightarrow \lim_{x\to\infty}x^2\left(e^{-1/x^2} - 1\right)$

$\qquad = \lim_{x\to\infty}\left(-1 + \frac{1}{2x^2} - \frac{1}{6x^4} + \ldots\right) = -1$

37. $\frac{\ln\left(1+x^2\right)}{1 - \cos x} = \frac{\left(x^2 - \frac{x^4}{2} + \frac{x^6}{3} - \ldots\right)}{1 - \left(1 - \frac{x^2}{2!} + \frac{x^4}{4!} - \ldots\right)} = \frac{\left(1 - \frac{x^2}{2} + \frac{x^4}{3} - \ldots\right)}{\left(\frac{1}{2!} - \frac{x^2}{4!} + \ldots\right)} \Rightarrow \lim_{x\to 0}\frac{\ln\left(1+x^2\right)}{1 - \cos x} = \lim_{x\to 0}\frac{\left(1 - \frac{x^2}{2} + \frac{x^4}{3} - \ldots\right)}{\left(\frac{1}{2!} - \frac{x^2}{4!} + \ldots\right)} = 2! = 2$

39. $\sin 3x^2 = 3x^2 - \frac{9}{2}x^6 + \frac{81}{40}x^{10} - \ldots$ and $1 - \cos 2x = 2x^2 - \frac{2}{3}x^4 + \frac{4}{45}x^6 - \ldots$

$\qquad \Rightarrow \lim_{x\to 0}\frac{\sin 3x^2}{1 - \cos 2x} = \lim_{x\to 0}\frac{3x^2 - \frac{9}{2}x^6 + \frac{81}{40}x^{10} - \ldots}{2x^2 - \frac{2}{3}x^4 + \frac{4}{45}x^6 - \ldots} = \lim_{x\to 0}\frac{3 - \frac{9}{2}x^4 + \frac{81}{40}x^8 - \ldots}{2 - \frac{2}{3}x^2 + \frac{4}{45}x^4 - \ldots} = \frac{3}{2}$

41. $1 + 1 + \frac{1}{2!} + \frac{1}{3!} + \frac{1}{4!} + \ldots = e^1 = e$

43. $1 - \frac{3^2}{4^2 2!} + \frac{3^4}{4^4 4!} - \frac{3^6}{4^6 6!} + \ldots = 1 - \frac{1}{2!}\left(\frac{3}{4}\right)^2 + \frac{1}{4!}\left(\frac{3}{4}\right)^4 - \frac{1}{6!}\left(\frac{3}{4}\right)^6 + \ldots = \cos\left(\frac{3}{4}\right)$

45. $\frac{\pi}{3} - \frac{\pi^3}{3^3 3!} + \frac{\pi^5}{3^5 5!} - \frac{\pi^7}{3^7 7!} + \ldots = \frac{\pi}{3} - \frac{1}{3!}\left(\frac{\pi}{3}\right)^3 + \frac{1}{5!}\left(\frac{\pi}{3}\right)^5 - \frac{1}{7!}\left(\frac{\pi}{3}\right)^7 + \ldots = \sin\left(\frac{\pi}{3}\right) = \frac{\sqrt{3}}{2}$

47. $x^3 + x^4 + x^5 + x^6 + \ldots = x^3\left(1 + x + x^2 + x^3 + \ldots\right) = x^3\left(\frac{1}{1-x}\right) = \frac{x^3}{1-x}$

49. $x^3 - x^5 + x^7 - x^9 + \ldots = x^3\left(1 - x^2 + \left(x^2\right)^2 - \left(x^2\right)^3 + \ldots\right) = x^3\left(\frac{1}{1+x^2}\right) = \frac{x^3}{1+x^2}$

51. $-1 + 2x - 3x^2 + 4x^3 - 5x^4 + \ldots = \frac{d}{dx}\left(1 - x + x^2 - x^3 + x^4 - x^5 + \ldots\right) = \frac{d}{dx}\left(\frac{1}{1+x}\right) = \frac{-1}{(1+x)^2}$

53. $\ln\left(\frac{1+x}{1-x}\right) = \ln(1+x) - \ln(1-x) = \left(x - \frac{x^2}{2} + \frac{x^3}{3} - \frac{x^4}{4} + \ldots\right) - \left(-x - \frac{x^2}{2} - \frac{x^3}{3} - \frac{x^4}{4} - \ldots\right) = 2\left(x + \frac{x^3}{3} + \frac{x^5}{5} + \ldots\right)$

55. $\tan^{-1} x = x - \frac{x^3}{3} + \frac{x^5}{5} - \frac{x^7}{7} + \frac{x^9}{9} - \ldots + \frac{(-1)^{n-1} x^{2n-1}}{2n-1} + \ldots \Rightarrow |\text{error}| = \left|\frac{(-1)^{n-1} x^{2n-1}}{2n-1}\right| = \frac{1}{2n-1}$ when $x = 1$;

$\frac{1}{2n-1} < \frac{1}{10^3} \Rightarrow n > \frac{1001}{2} = 500.5 \Rightarrow$ the first term not used is the $501^{st} \Rightarrow$ we must use 500 terms

57. $\tan^{-1} x = x - \frac{x^3}{3} + \frac{x^5}{5} - \frac{x^7}{7} + \frac{x^9}{9} - \ldots + \frac{(-1)^{n-1} x^{2n-1}}{2n-1} + \ldots$ and when the series representing $48\tan^{-1}\left(\frac{1}{18}\right)$ has an

error less than $\frac{1}{3} \cdot 10^{-6}$, then the series representing the sum $48\tan^{-1}\left(\frac{1}{18}\right) + 32\tan^{-1}\left(\frac{1}{57}\right) - 20\tan^{-1}\left(\frac{1}{239}\right)$ also

has an error of magnitude less than 10^{-6}; thus $|\text{error}| = 48\frac{\left(\frac{1}{18}\right)^{2n-1}}{2n-1} < \frac{1}{3 \cdot 10^6} \Rightarrow n \geq 4$ using a calculator $\Rightarrow 4$

terms

59. (a) $\left(1 - x^2\right)^{-1/2} \approx 1 + \frac{x^2}{2} + \frac{3x^4}{8} + \frac{5x^6}{16} \Rightarrow \sin^{-1} x \approx x + \frac{x^3}{6} + \frac{3x^5}{40} + \frac{5x^7}{112}$;

Using the Ratio Test: $\lim\limits_{n\to\infty}\left|\frac{1 \cdot 3 \cdot 5 \cdots (2n-1)(2n+1) x^{2n+3}}{2 \cdot 4 \cdot 6 \cdots (2n)(2n+2)(2n+3)} \cdot \frac{2 \cdot 4 \cdot 6 \cdots (2n)(2n+1)}{1 \cdot 3 \cdot 5 \cdots (2n-1) x^{2n+1}}\right| < 1 \Rightarrow x^2 \lim\limits_{n\to\infty}\left|\frac{(2n+1)(2n+1)}{(2n+1)(2n+3)}\right| < 1$

$\Rightarrow |x| < 1 \Rightarrow$ the radius convergence is 1. See Exercise 69.

(b) $\frac{d}{dx}\left(\cos^{-1} x\right) = -\left(1 - x^2\right)^{-1/2} \Rightarrow \cos^{-1} x = \frac{\pi}{2} - \sin^{-1} x \approx \frac{\pi}{2} - \left(x + \frac{x^3}{6} + \frac{3x^5}{40} + \frac{5x^7}{112}\right) \approx \frac{\pi}{2} - x - \frac{x^3}{6} - \frac{3x^5}{40} - \frac{5x^7}{112}$

61. $\frac{-1}{1+x} = -\frac{1}{1-(-x)} = -1 + x - x^2 + x^3 - \ldots \Rightarrow \frac{d}{dx}\left(\frac{-1}{1+x}\right) = \frac{1}{(1+x)^2} = \frac{d}{dx}\left(-1 + x - x^2 + x^3 - \ldots\right) = 1 - 2x + 3x^2 - 4x^3 + \ldots$

63. Wallis' formula gives the approximation $\pi \approx 4\left[\dfrac{2\cdot4\cdot4\cdot6\cdot6\cdot8\cdots(2n-2)\cdot(2n)}{3\cdot3\cdot5\cdot5\cdot7\cdot7\cdots(2n-1)\cdot(2n-1)}\right]$ to produce the table

n	$\sim \pi$
10	3.221088998
20	3.181104886
30	3.167880758
80	3.151425420
90	3.150331383
93	3.150049112
94	3.149959030
95	3.149870848
100	3.149456425

At $n = 1929$ we obtain the first approximation accurate to 3 decimals: 3.141999845. At $n = 30,000$ we still do not obtain accuracy to 4 decimals: 3.141617732, so the convergence to π is very slow. Here is a Maple CAS procedure to produce these approximations:

```
pie :=
    proc(n)
        local i, j;
        a(2) := evalf (8/9)
        for i from 3 to n do a(i) := evalf (2*(2*i − 2)*i/(2*i − 1)^2*a(i − 1))od;
        [[j, 4*a(j)] $ (j = n − 5 .. n)]
    end
```

65. $\left(1-x^2\right)^{-1/2} = \left(1+\left(-x^2\right)\right)^{-1/2} = (1)^{-1/2} + \left(-\frac{1}{2}\right)(1)^{-3/2}\left(-x^2\right) + \dfrac{\left(-\frac{1}{2}\right)\left(-\frac{3}{2}\right)(1)^{-5/2}\left(-x^2\right)^2}{2!} + \dfrac{\left(-\frac{1}{2}\right)\left(-\frac{3}{2}\right)\left(-\frac{5}{2}\right)(1)^{-7/2}\left(-x^2\right)^3}{3!} + \ldots$

$= 1 + \dfrac{x^2}{2} + \dfrac{1\cdot3 x^4}{2^2\cdot2!} + \dfrac{1\cdot3\cdot5 x^6}{2^3\cdot3!} + \ldots = 1 + \displaystyle\sum_{n=1}^{\infty} \dfrac{1\cdot3\cdot5\cdots(2n-1)x^{2n}}{2^n\cdot n!}$

$\Rightarrow \sin^{-1} x = \displaystyle\int_0^x \left(1-t^2\right)^{-1/2}\, dt = \int_0^x \left(1 + \sum_{n=1}^{\infty} \dfrac{1\cdot3\cdot5\cdots(2n-1)x^{2n}}{2^n\cdot n!}\right) dt = x + \sum_{n=1}^{\infty} \dfrac{1\cdot3\cdot5\cdots(2n-1)x^{2n+1}}{2\cdot4\cdots(2n)(2n+1)}$, where $|x| < 1$

67. (a) $e^{-i\pi} = \cos(-\pi) + i\sin(-\pi) = -1 + i(0) = -1$

(b) $e^{i\pi/4} = \cos\left(\frac{\pi}{4}\right) + i\sin\left(\frac{\pi}{4}\right) = \frac{1}{\sqrt{2}} + \frac{i}{\sqrt{2}} = \left(\frac{1}{\sqrt{2}}\right)(1+i)$

(c) $e^{-i\pi/2} = \cos\left(-\frac{\pi}{2}\right) + i\sin\left(-\frac{\pi}{2}\right) = 0 + i(-1) = -i$

69. $e^x = 1 + x + \frac{x^2}{2!} + \frac{x^3}{3!} + \frac{x^4}{4!} + \ldots \Rightarrow e^{i\theta} = 1 + i\theta + \frac{(i\theta)^2}{2!} + \frac{(i\theta)^3}{3!} + \frac{(i\theta)^4}{4!} + \ldots$ and

$e^{-i\theta} = 1 - i\theta + \frac{(-i\theta)^2}{2!} + \frac{(-i\theta)^3}{3!} + \frac{(-i\theta)^4}{4!} + \ldots = 1 - i\theta + \frac{(i\theta)^2}{2!} - \frac{(i\theta)^3}{3!} + \frac{(i\theta)^4}{4!} - \ldots$

$\Rightarrow \dfrac{e^{i\theta}+e^{-i\theta}}{2} = \dfrac{\left(1+i\theta+\frac{(i\theta)^2}{2!}+\frac{(i\theta)^3}{3!}+\frac{(i\theta)^4}{4!}+\ldots\right) + \left(1-i\theta+\frac{(i\theta)^2}{2!}-\frac{(i\theta)^3}{3!}+\frac{(i\theta)^4}{4!}\ldots\right)}{2} = 1 - \frac{\theta^2}{2!} + \frac{\theta^4}{4!} - \frac{\theta^6}{6!} + \ldots = \cos\theta;$

$\dfrac{e^{i\theta}-e^{-i\theta}}{2!} = \dfrac{\left(1+i\theta+\frac{(i\theta)^2}{2!}+\frac{(i\theta)^3}{3!}+\frac{(i\theta)^4}{4!}+\ldots\right) - \left(1-i\theta+\frac{(i\theta)^2}{2!}-\frac{(i\theta)^3}{3!}+\frac{(i\theta)^4}{4!}\ldots\right)}{2!} = \theta - \frac{\theta^3}{3!} + \frac{\theta^5}{5!} - \frac{\theta^7}{7!} + \ldots = \sin\theta$

71. $e^x \sin x = \left(1 + x + \frac{x^2}{2!} + \frac{x^3}{3!} + \frac{x^4}{4!} + \ldots\right)\left(x - \frac{x^3}{3!} + \frac{x^5}{5!} - \frac{x^7}{7!} + \ldots\right)$

$= (1)x + (1)x^2 + \left(-\frac{1}{6} + \frac{1}{2}\right)x^3 + \left(-\frac{1}{6} + \frac{1}{6}\right)x^4 + \left(\frac{1}{120} - \frac{1}{12} + \frac{1}{24}\right)x^5 + \ldots = x + x^2 + \frac{1}{3}x^3 - \frac{1}{30}x^5 + \ldots;$

$e^x \cdot e^{ix} = e^{(1+i)x} = e^x(\cos x + i \sin x) = e^x \cos x + i(e^x \sin x) \Rightarrow e^x \sin x$ is the series of the imaginary part

of $e^{(1+i)x}$ which we calculate next; $e^{(1+i)x} = \sum_{n=0}^{\infty} \frac{(x+ix)^n}{n!} = 1 + (x+ix) + \frac{(x+ix)^2}{2!} + \frac{(x+ix)^3}{3!} + \frac{(x+ix)^4}{4!} + \ldots$

$= 1 + x + ix + \frac{1}{2!}(2ix^2) + \frac{1}{3!}(2ix^3 - 2x^3) + \frac{1}{4!}(-4x^4) + \frac{1}{5!}(-4x^5 - 4ix^5) + \frac{1}{6!}(-8ix^6) + \ldots \Rightarrow$ the imaginary part of

$e^{(1+i)x}$ is $x + \frac{2}{2!}x^2 + \frac{2}{3!}x^3 - \frac{4}{5!}x^5 - \frac{8}{6!}x^6 + \ldots = x + x^2 + \frac{1}{3}x^3 - \frac{1}{30}x^5 - \frac{1}{90}x^6 + \ldots$ in agreement with our

product calculation. The series for $e^x \sin x$ converges for all values of x.

73. (a) $e^{i\theta_1}e^{i\theta_2} = (\cos\theta_1 + i\sin\theta_1)(\cos\theta_2 + i\sin\theta_2)$

$= (\cos\theta_1 \cos\theta_2 - \sin\theta_1 \sin\theta_2) + i(\sin\theta_1 \cos\theta_2 + \sin\theta_2 \cos\theta_1) = \cos(\theta_1 + \theta_2) + i\sin(\theta_1 + \theta_2) = e^{i(\theta_1+\theta_2)}$

(b) $e^{-i\theta} = \cos(-\theta) + i\sin(-\theta) = \cos\theta - i\sin\theta = (\cos\theta - i\sin\theta)\left(\frac{\cos\theta + i\sin\theta}{\cos\theta + i\sin\theta}\right) = \frac{1}{\cos\theta + i\sin\theta} = \frac{1}{e^{i\theta}}$

CHAPTER 9   PRACTICE EXERCISES

1. converges to 1, since $\lim_{n\to\infty} a_n = \lim_{n\to\infty}\left(1 + \frac{(-1)^n}{n}\right) = 1$

3. converges to -1, since $\lim_{n\to\infty} a_n = \lim_{n\to\infty}\left(\frac{1-2^n}{2^n}\right) = \lim_{n\to\infty}\left(\frac{1}{2^n} - 1\right) = -1$

5. diverges, since $\left\{\sin\frac{n\pi}{2}\right\} = \{0, 1, 0, -1, 0, 1, \ldots\}$

7. converges to 0, since $\lim_{n\to\infty} a_n = \lim_{n\to\infty}\frac{\ln n^2}{n} = 2\lim_{n\to\infty}\frac{\left(\frac{1}{n}\right)}{1} = 0$

9. converges to 1, since $\lim_{n\to\infty} a_n = \lim_{n\to\infty}\left(\frac{n+\ln n}{n}\right) = \lim_{n\to\infty}\frac{1+\left(\frac{1}{n}\right)}{1} = 1$

11. converges to e^{-5}, since $\lim_{n\to\infty} a_n = \lim_{n\to\infty}\left(\frac{n-5}{n}\right)^n = \lim_{n\to\infty}\left(1 + \frac{(-5)}{n}\right)^n = e^{-5}$ by Theorem 5

13. converges to 3, since $\lim_{n\to\infty} a_n = \lim_{n\to\infty}\left(\frac{3^n}{n}\right)^{1/n} = \lim_{n\to\infty}\frac{3}{n^{1/n}} = \frac{3}{1} = 3$ by Theorem 5

15. converges to $\ln 2$, since $\lim_{n\to\infty} a_n = \lim_{n\to\infty} n\left(2^{1/n} - 1\right) = \lim_{n\to\infty}\frac{2^{1/n} - 1}{\left(\frac{1}{n}\right)} = \lim_{n\to\infty}\frac{\left[\frac{(-2^{1/n}\ln 2)}{n^2}\right]}{\left(\frac{-1}{n^2}\right)} = \lim_{n\to\infty} 2^{1/n}\ln 2 = 2^0 \cdot \ln 2 = \ln 2$

17. diverges, since $\lim\limits_{n\to\infty} a_n = \lim\limits_{n\to\infty} \dfrac{(n+1)!}{n!} = \lim\limits_{n\to\infty} (n+1) = \infty$

19. $\dfrac{1}{(2n-3)(2n-1)} = \dfrac{\left(\frac{1}{2}\right)}{2n-3} - \dfrac{\left(\frac{1}{2}\right)}{2n-1} \Rightarrow s_n = \left[\dfrac{\left(\frac{1}{2}\right)}{3} - \dfrac{\left(\frac{1}{2}\right)}{5}\right] + \left[\dfrac{\left(\frac{1}{2}\right)}{5} - \dfrac{\left(\frac{1}{2}\right)}{7}\right] + \ldots + \left[\dfrac{\left(\frac{1}{2}\right)}{2n-3} - \dfrac{\left(\frac{1}{2}\right)}{2n-1}\right] = \dfrac{\left(\frac{1}{2}\right)}{3} - \dfrac{\left(\frac{1}{2}\right)}{2n-1}$

$\Rightarrow \lim\limits_{n\to\infty} s_n = \lim\limits_{n\to\infty} \left[\dfrac{1}{6} - \dfrac{\left(\frac{1}{2}\right)}{2n-1}\right] = \dfrac{1}{6}$

21. $\dfrac{9}{(3n-1)(3n+2)} = \dfrac{3}{3n-1} - \dfrac{3}{3n+2} \Rightarrow s_n = \left(\dfrac{3}{2}-\dfrac{3}{5}\right) + \left(\dfrac{3}{5}-\dfrac{3}{8}\right) + \left(\dfrac{3}{8}-\dfrac{3}{11}\right) + \ldots + \left(\dfrac{3}{3n-1} - \dfrac{3}{3n+2}\right) = \dfrac{3}{2} - \dfrac{3}{3n+2}$

$\Rightarrow \lim\limits_{n\to\infty} s_n = \lim\limits_{n\to\infty} \left(\dfrac{3}{2} - \dfrac{3}{3n+2}\right) = \dfrac{3}{2}$

23. $\displaystyle\sum_{n=0}^{\infty} e^{-n} = \sum_{n=0}^{\infty} \dfrac{1}{e^n}$, a convergent geometric series with $r = \dfrac{1}{e}$ and $a = 1 \Rightarrow$ the sum is $\dfrac{1}{1-\left(\frac{1}{e}\right)} = \dfrac{e}{e-1}$

25. diverges, a p-series with $p = \dfrac{1}{2}$

27. Since $f(x) = \dfrac{1}{x^{1/2}} \Rightarrow f'(x) = -\dfrac{1}{2x^{3/2}} < 0 \Rightarrow f(x)$ is decreasing $\Rightarrow a_{n+1} < a_n$, and $\lim\limits_{n\to\infty} a_n = \lim\limits_{n\to\infty} \dfrac{1}{\sqrt{n}} = 0$, the

series $\displaystyle\sum_{n=1}^{\infty} \dfrac{(-1)^n}{\sqrt{n}}$ converges by the Alternating Series Test. Since $\displaystyle\sum_{n=1}^{\infty} \dfrac{1}{\sqrt{n}}$ diverges, the given series converges

conditionally.

29. The given series does not converge absolutely by the Direct Comparison Test since $\dfrac{1}{\ln(n+1)} > \dfrac{1}{n+1}$, which is the

nth term of a divergent series. Since $f(x) = \dfrac{1}{\ln(x+1)} \Rightarrow f'(x) = -\dfrac{1}{(\ln(x+1))^2 (x+1)} < 0 \Rightarrow f(x)$ is decreasing

$\Rightarrow a_{n+1} < a_n$, and $\lim\limits_{n\to\infty} a_n = \lim\limits_{n\to\infty} \dfrac{1}{\ln(n+1)} = 0$, the given series converges conditionally by the Alternating

Series Test.

31. converges absolutely by the Direct Comparison Test since $\dfrac{\ln n}{n^3} < \dfrac{n}{n^3} = \dfrac{1}{n^2}$, the nth term of a convergent

p-series

33. $\lim\limits_{n\to\infty} \dfrac{\left(\dfrac{1}{n\sqrt{n^2+1}}\right)}{\left(\dfrac{1}{n^2}\right)} = \sqrt{\lim\limits_{n\to\infty} \dfrac{n^2}{n^2+1}} = \sqrt{1} = 1 \Rightarrow$ converges absolutely by the Limit Comparison Test

35. converges absolutely by the Ratio Test since $\lim\limits_{n\to\infty} \left[\dfrac{n+2}{(n+1)!} \cdot \dfrac{n!}{n+1}\right] = \lim\limits_{n\to\infty} \dfrac{n+2}{(n+1)^2} = 0 < 1$

37. converges absolutely by the Ratio Test since $\lim\limits_{n\to\infty} \left[\dfrac{3^{n+1}}{(n+1)!} \cdot \dfrac{n!}{3^n}\right] = \lim\limits_{n\to\infty} \dfrac{3}{n+1} = 0 < 1$

39. converges absolutely by the Limit Comparison Test since $\lim\limits_{n\to\infty}\dfrac{\left(\frac{1}{n^{3/2}}\right)}{\left(\frac{1}{\sqrt{n(n+1)(n+2)}}\right)}=\sqrt{\lim\limits_{n\to\infty}\dfrac{n(n+1)(n+2)}{n^3}}=1$

41. $\lim\limits_{n\to\infty}\left|\dfrac{u_{n+1}}{u_n}\right|<1\Rightarrow\lim\limits_{n\to\infty}\left|\dfrac{(x+4)^{n+1}}{(n+1)3^{n+1}}\cdot\dfrac{n3^n}{(x+4)^n}\right|<1\Rightarrow\dfrac{|x+4|}{3}\lim\limits_{n\to\infty}\left(\dfrac{n}{n+1}\right)<1\Rightarrow\dfrac{|x+4|}{3}<1\Rightarrow|x+4|<3\Rightarrow-3<x+4<3$

$\Rightarrow-7<x<-1$; at $x=-7$ we have $\sum\limits_{n=1}^{\infty}\dfrac{(-1)^n3^n}{n3^n}=\sum\limits_{n=1}^{\infty}\dfrac{(-1)^n}{n}$, the alternating harmonic series, which converges

conditionally; at $x=-1$ we have $\sum\limits_{n=1}^{\infty}\dfrac{3^n}{n3^n}=\sum\limits_{n=1}^{\infty}\dfrac{1}{n}$, the divergent harmonic series

(a) the radius is 3; the interval of convergence is $-7\le x<-1$
(b) the interval of absolute convergence is $-7<x<-1$
(c) the series converges conditionally at $x=-7$

43. $\lim\limits_{n\to\infty}\left|\dfrac{u_{n+1}}{u_n}\right|<1\Rightarrow\lim\limits_{n\to\infty}\left|\dfrac{(3x-1)^{n+1}}{(n+1)^2}\cdot\dfrac{n^2}{(3x-1)^n}\right|<1\Rightarrow|3x-1|\lim\limits_{n\to\infty}\dfrac{n^2}{(n+1)^2}<1\Rightarrow|3x-1|<1\Rightarrow-1<3x-1<1$

$\Rightarrow0<3x<2\Rightarrow0<x<\frac{2}{3}$; at $x=0$ we have $\sum\limits_{n=1}^{\infty}\dfrac{(-1)^{n-1}(-1)^n}{n^2}=\sum\limits_{n=1}^{\infty}\dfrac{(-1)^{2n-1}}{n^2}=-\sum\limits_{n=1}^{\infty}\dfrac{1}{n^2}$, a nonzero constant

multiple of a convergent p-series which is absolutely convergent; at $x=\frac{2}{3}$ we have $\sum\limits_{n=1}^{\infty}\dfrac{(-1)^{n-1}(1)^n}{n^2}=\sum\limits_{n=1}^{\infty}\dfrac{(-1)^{n-1}}{n^2}$,

which converges absolutely
(a) the radius is $\frac{1}{3}$; the interval of convergence is $0\le x\le\frac{2}{3}$

(b) the interval of absolute convergence is $0\le x\le\frac{2}{3}$

(c) there are no values for which the series converges conditionally

45. $\lim\limits_{n\to\infty}\left|\dfrac{u_{n+1}}{u_n}\right|<1\Rightarrow\lim\limits_{n\to\infty}\left|\dfrac{x^{n+1}}{(n+1)^{n+1}}\cdot\dfrac{n^n}{x^n}\right|<1\Rightarrow|x|\lim\limits_{n\to\infty}\left|\left(\dfrac{n}{n+1}\right)^n\left(\dfrac{1}{n+1}\right)\right|<1\Rightarrow\dfrac{|x|}{e}\lim\limits_{n\to\infty}\left(\dfrac{1}{n+1}\right)<1\Rightarrow\dfrac{|x|}{e}\cdot0<1$, which holds

for all x
(a) the radius is ∞; the series converges for all x
(b) the series converges absolutely for all x
(c) there are no values for which the series converges conditionally

47. $\lim\limits_{n\to\infty}\left|\dfrac{u_{n+1}}{u_n}\right|<1\Rightarrow\lim\limits_{n\to\infty}\left|\dfrac{(n+2)x^{2n+1}}{3^{n+1}}\cdot\dfrac{3^n}{(n+1)x^{2n-1}}\right|<1\Rightarrow\dfrac{x^2}{3}\lim\limits_{n\to\infty}\left(\dfrac{n+2}{n+1}\right)<1\Rightarrow-\sqrt{3}<x<\sqrt{3}$;

the series $\sum\limits_{n=1}^{\infty}-\dfrac{n+1}{\sqrt{3}}$ and $\sum\limits_{n=1}^{\infty}\dfrac{n+1}{\sqrt{3}}$, obtained with $x=\pm\sqrt{3}$, both diverge

(a) the radius is $\sqrt{3}$; the interval of convergence is $-\sqrt{3}<x<\sqrt{3}$
(b) the interval of absolute convergence is $-\sqrt{3}<x<\sqrt{3}$
(c) there are no values for which the series converges conditionally

49. $\lim\limits_{n\to\infty}\left|\dfrac{u_{n+1}}{u_n}\right|<1 \Rightarrow \lim\limits_{n\to\infty}\left|\dfrac{\operatorname{csch}(n+1)x^{n+1}}{\operatorname{csch}(n)x^n}\right|<1 \Rightarrow |x|\lim\limits_{n\to\infty}\left|\dfrac{\left(\frac{2}{e^{n+1}-e^{-n-1}}\right)}{\left(\frac{2}{e^n-e^{-n}}\right)}\right|<1 \Rightarrow |x|\lim\limits_{n\to\infty}\left|\dfrac{e^{-1}-e^{-2n-1}}{1-e^{-2n-2}}\right|<1 \Rightarrow \dfrac{|x|}{e}<1$

$\Rightarrow -e<x<e$; the series $\sum\limits_{n=1}^{\infty}(\pm e)^n\operatorname{csch}n$, obtained with $x=\pm e$, both diverge since $\lim\limits_{n\to\infty}(\pm e)^n\operatorname{csch}n\neq 0$

(a) the radius is e; the interval of convergence is $-e<x<e$

(b) the interval of absolute convergence is $-e<x<e$

(c) there are no values for which the series converges conditionally

51. The given series has the form $1-x+x^2-x^3+\ldots+(-x)^n+\ldots=\dfrac{1}{1+x}$, where $x=\frac{1}{4}$; the sum is $\dfrac{1}{1+\left(\frac{1}{4}\right)}=\dfrac{4}{5}$

53. The given series has the form $x-\dfrac{x^3}{3!}+\dfrac{x^5}{5!}-\ldots+(-1)^n\dfrac{x^{2n+1}}{(2n+1)!}+\ldots=\sin x$, where $x=\pi$; the sum is $\sin\pi=0$

55. The given series has the form $1+x+\dfrac{x^2}{2!}+\dfrac{x^2}{3!}+\ldots+\dfrac{x^n}{n!}+\ldots=e^x$, where $x=\ln 2$; the sum is $e^{\ln(2)}=2$

57. Consider $\dfrac{1}{1-2x}$ as the sum of a convergent geometric series with $a=1$ and $r=2x$

$\Rightarrow \dfrac{1}{1-2x}=1+(2x)+(2x)^2+(2x)^3+\ldots=\sum\limits_{n=0}^{\infty}(2x)^n=\sum\limits_{n=0}^{\infty}2^n x^n$ where $|2x|<1\Rightarrow |x|<\frac{1}{2}$

59. $\sin x=\sum\limits_{n=0}^{\infty}\dfrac{(-1)^n x^{2n+1}}{(2n+1)!}\Rightarrow \sin\pi x=\sum\limits_{n=0}^{\infty}\dfrac{(-1)^n(\pi x)^{2n+1}}{(2n+1)!}=\sum\limits_{n=0}^{\infty}\dfrac{(-1)^n\pi^{2n+1}x^{2n+1}}{(2n+1)!}$

61. $\cos x=\sum\limits_{n=0}^{\infty}\dfrac{(-1)^n x^{2n}}{(2n)!}\Rightarrow \cos\left(x^{5/3}\right)=\sum\limits_{n=0}^{\infty}\dfrac{(-1)^n\left(x^{5/3}\right)^{2n}}{(2n)!}=\sum\limits_{n=0}^{\infty}\dfrac{(-1)^n x^{10n/3}}{(2n)!}$

63. $e^x=\sum\limits_{n=0}^{\infty}\dfrac{x^n}{n!}\Rightarrow e^{(\pi x/2)}=\sum\limits_{n=0}^{\infty}\dfrac{\left(\frac{\pi x}{2}\right)^n}{n!}=\sum\limits_{n=0}^{\infty}\dfrac{\pi^n x^n}{2^n n!}$

65. $f(x)=\sqrt{3+x^2}=\left(3+x^2\right)^{1/2}\Rightarrow f'(x)=x\left(3+x^2\right)^{-1/2}\Rightarrow f''(x)=-x^2\left(3+x^2\right)^{-3/2}+\left(3+x^2\right)^{-1/2}$

$\Rightarrow f'''(x)=3x^3\left(3+x^2\right)^{-5/2}-3x\left(3+x^2\right)^{-3/2}$; $f(-1)=2, f'(-1)=-\frac{1}{2}, f''(-1)=-\frac{1}{8}+\frac{1}{2}=\frac{3}{8}$,

$f'''(-1)=-\frac{3}{32}+\frac{3}{8}=\frac{9}{32}\Rightarrow \sqrt{3+x^2}=2-\dfrac{(x+1)}{2\cdot 1!}+\dfrac{3(x+1)^2}{2^3\cdot 2!}+\dfrac{9(x+1)^3}{2^5\cdot 3!}+\ldots$

67. $f(x)=\dfrac{1}{x+1}=(x+1)^{-1}\Rightarrow f'(x)=-(x+1)^{-2}\Rightarrow f''(x)=2(x+1)^{-3}\Rightarrow f'''(x)=-6(x+1)^{-4}$;

$f(3)=\frac{1}{4}, f'(3)=-\frac{1}{4^2}, f''(3)=\frac{2}{4^3}, f'''(2)=\frac{-6}{4^4}\Rightarrow \dfrac{1}{x+1}=\frac{1}{4}-\frac{1}{4^2}(x-3)+\frac{1}{4^3}(x-3)^2-\frac{1}{4^4}(x-3)^3+\ldots$

69. $\int_0^{1/2} e^{-x^3}\,dx = \int_0^{1/2}\left(1 - x^3 + \frac{x^6}{2!} - \frac{x^9}{3!} + \frac{x^{12}}{4!} + \ldots\right)dx = \left[x - \frac{x^4}{4} + \frac{x^7}{7\cdot 2!} - \frac{x^{10}}{10\cdot 3!} + \frac{x^{13}}{13\cdot 4!} - \ldots\right]_0^{1/2}$

$\approx \frac{1}{2} - \frac{1}{2^4\cdot 4} + \frac{1}{2^7\cdot 7\cdot 2!} - \frac{1}{2^{10}\cdot 10\cdot 3!} + \frac{1}{2^{13}\cdot 13\cdot 4!} - \frac{1}{2^{16}\cdot 16\cdot 5!} \approx 0.484917143$

71. $\int_1^{1/2} \frac{\tan^{-1} x}{x}\,dx = \int_1^{1/2}\left(1 - \frac{x^2}{3} + \frac{x^4}{5} - \frac{x^6}{7} + \frac{x^8}{9} - \frac{x^{10}}{11} + \ldots\right)dx = \left[x - \frac{x^3}{9} + \frac{x^5}{25} - \frac{x^7}{49} + \frac{x^9}{81} - \frac{x^{11}}{121} + \ldots\right]_0^{1/2}$

$\approx \frac{1}{2} - \frac{1}{9\cdot 2^3} + \frac{1}{5^2\cdot 2^5} - \frac{1}{7^2\cdot 2^7} + \frac{1}{9^2\cdot 2^9} - \frac{1}{11^2\cdot 2^{11}} + \frac{1}{13^2\cdot 2^{13}} - \frac{1}{15^2\cdot 2^{15}} + \frac{1}{17^2\cdot 2^{17}} - \frac{1}{19^2\cdot 2^{19}} + \frac{1}{21^2\cdot 2^{21}} \approx 0.4872223583$

73. $\displaystyle\lim_{x\to 0} \frac{7\sin x}{e^{2x}-1} = \lim_{x\to 0} \frac{7\left(x - \frac{x^3}{3!} + \frac{x^5}{5!} - \ldots\right)}{\left(2x + \frac{2^2 x^2}{2!} + \frac{2^3 x^3}{3!} + \ldots\right)} = \lim_{x\to 0} \frac{7\left(1 - \frac{x^2}{3!} + \frac{x^4}{5!} - \ldots\right)}{\left(2 + \frac{2^2 x}{2!} + \frac{2^3 x^2}{3!} + \ldots\right)} = \frac{7}{2}$

75. $\displaystyle\lim_{t\to 0}\left(\frac{1}{2-2\cos t} - \frac{1}{t^2}\right) = \lim_{t\to 0} \frac{t^2 - 2 + 2\cos t}{2t^2(1-\cos t)} = \lim_{t\to 0} \frac{t^2 - 2 + 2\left(1 - \frac{t^2}{2} + \frac{t^4}{4!} - \ldots\right)}{2t^2\left(1 - 1 + \frac{t^2}{2} - \frac{t^4}{4!} + \ldots\right)} = \lim_{t\to 0} \frac{2\left(\frac{t^4}{4!} - \frac{t^6}{6!} + \ldots\right)}{\left(t^4 - \frac{2t^6}{4!} + \ldots\right)} = \lim_{t\to 0} \frac{2\left(\frac{1}{4!} - \frac{t^2}{6!} + \ldots\right)}{\left(1 - \frac{2t^2}{4!} + \ldots\right)} = \frac{1}{12}$

77. $\displaystyle\lim_{z\to 0} \frac{1-\cos^2 z}{\ln(1-z)+\sin z} = \lim_{z\to 0} \frac{1 - \left(1 - z^2 + \frac{z^4}{3} - \ldots\right)}{\left(-z - \frac{z^2}{2} + \frac{z^3}{3} - \ldots\right) + \left(z - \frac{z^3}{3!} + \frac{z^5}{5!} - \ldots\right)} = \lim_{z\to 0} \frac{\left(z^2 - \frac{z^4}{3} + \ldots\right)}{\left(-\frac{z^2}{2} - \frac{2z^3}{3} - \frac{z^4}{4} \ldots\right)} = \lim_{z\to 0} \frac{\left(1 - \frac{z^2}{3} + \ldots\right)}{\left(-\frac{1}{2} - \frac{2z}{3} - \frac{z^2}{4} \ldots\right)} = -2$

79. $\displaystyle\lim_{x\to 0}\left(\frac{\sin 3x}{x^3} + \frac{r}{x^2} + s\right) = \lim_{x\to 0}\left[\frac{\left(3x - \frac{(3x)^3}{6} + \frac{(3x)^5}{120} - \ldots\right)}{x^3} + \frac{r}{x^2} + s\right] = \lim_{x\to 0}\left(\frac{3}{x^2} - \frac{9}{2} + \frac{81x^2}{40} + \ldots + \frac{r}{x^2} + s\right) = 0$

$\Rightarrow \frac{r}{x^2} + \frac{3}{x^2} = 0$ and $s - \frac{9}{2} = 0 \Rightarrow r = -3$ and $s = \frac{9}{2}$

81. $\displaystyle\lim_{n\to\infty}\left|\frac{2\cdot 5\cdot 8\cdots(3n-1)(3n+2)x^{n+1}}{2\cdot 4\cdot 6\cdots(2n)(2n+2)} \cdot \frac{2\cdot 4\cdot 6\cdots(2n)}{2\cdot 5\cdot 8\cdots(3n-1)x^n}\right| < 1 \Rightarrow |x|\lim_{n\to\infty}\left|\frac{3n+2}{2n+2}\right| < 1 \Rightarrow |x| < \frac{2}{3} \Rightarrow$ the radius of convergence is $\frac{2}{3}$

83. $\displaystyle\sum_{k=2}^{n} \ln\left(1 - \frac{1}{k^2}\right) = \sum_{k=2}^{n}\left[\ln\left(1 + \frac{1}{k}\right) + \ln\left(1 - \frac{1}{k}\right)\right] = \sum_{k=2}^{n}\left[\ln(k+1) - \ln k + \ln(k-1) - \ln k\right]$

$= \left[\ln 3 - \ln 2 + \ln 1 - \ln 2\right] + \left[\ln 4 - \ln 3 + \ln 2 - \ln 3\right] + \left[\ln 5 - \ln 4 + \ln 3 - \ln 4\right] + \left[\ln 6 - \ln 5 + \ln 4 - \ln 5\right]$

$\quad + \ldots + \left[\ln(n+1) - \ln n + \ln(n-1) - \ln n\right] = \left[\ln 1 - \ln 2\right] + \left[\ln(n+1) - \ln n\right]$ after cancellation

$\Rightarrow \displaystyle\sum_{k=2}^{n} \ln\left(1 - \frac{1}{k^2}\right) = \ln\left(\frac{n+1}{2n}\right) \Rightarrow \sum_{k=2}^{\infty} \ln\left(1 - \frac{1}{k^2}\right) = \lim_{n\to\infty} \ln\left(\frac{n+1}{2n}\right) = \ln\frac{1}{2}$ is the sum

85. (a) $\displaystyle\lim_{n\to\infty}\left|\frac{1\cdot 4\cdot 7\cdots(3n-2)(3n+1)x^{3n+3}}{(3n+3)!} \cdot \frac{(3n)!}{1\cdot 4\cdot 7\cdots(3n-2)x^{3n}}\right| < 1 \Rightarrow \left|x^3\right|\lim_{n\to\infty} \frac{(3n+1)}{(3n+1)(3n+2)(3n+3)} = \left|x^3\right|\cdot 0 < 1$

$\Rightarrow$ the radius of convergence is ∞

(b) $y = 1 + \displaystyle\sum_{n=1}^{\infty} \frac{1\cdot 4\cdot 7\cdots(3n-2)}{(3n)!}x^{3n} \Rightarrow \frac{dy}{dx} = \sum_{n=1}^{\infty} \frac{1\cdot 4\cdot 7\cdots(3n-2)}{(3n-1)!}x^{3n-1} \Rightarrow \frac{d^2 y}{dx^2} = \sum_{n=1}^{\infty} \frac{1\cdot 4\cdot 7\cdots(3n-2)}{(3n-2)!}x^{3n-2}$

$= x + \displaystyle\sum_{n=2}^{\infty} \frac{1\cdot 4\cdot 7\cdots(3n-5)}{(3n-3)!}x^{3n-2} = x\left(1 + \sum_{n=1}^{\infty} \frac{1\cdot 4\cdot 7\cdots(3n-2)}{(3n)!}x^{3n}\right) = xy + 0 \Rightarrow a = 1$ and $b = 0$

87. Yes, the series $\sum_{n=1}^{\infty} a_n b_n$ converges as we now show. Since $\sum_{n=1}^{\infty} a_n$ converges it follows that $a_n \to 0 \Rightarrow a_n < 1$

for $n >$ some index $N \Rightarrow a_n b_n < b_n$ for $n > N \Rightarrow \sum_{n=1}^{\infty} a_n b_n$ converges by the Direct Comparison Test with

$\sum_{n=1}^{\infty} b_n$

89. $\sum_{n=1}^{\infty} (x_{n+1} - x_n) = \lim_{n \to \infty} \sum_{k=1}^{\infty} (x_{k+1} - x_k) = \lim_{n \to \infty} (x_{n+1} - x_1) = \lim_{n \to \infty} (x_{n+1}) - x_1 \Rightarrow$ both the series and sequence must
either converge or diverge.

91. $\sum_{n=1}^{\infty} \frac{a_n}{n} = a_1 + \frac{a_2}{2} + \frac{a_3}{3} + \frac{a_4}{4} + \ldots \geq a_1 + \left(\frac{1}{2}\right) a_2 + \left(\frac{1}{3} + \frac{1}{4}\right) a_4 + \left(\frac{1}{5} + \frac{1}{6} + \frac{1}{7} + \frac{1}{8}\right) a_8 + \left(\frac{1}{9} + \frac{1}{10} + \frac{1}{11} + \ldots + \frac{1}{16}\right) a_{16} + \ldots$

$\geq \frac{1}{2} (a_2 + a_4 + a_8 + a_{16} + \ldots)$ which is a divergent series

CHAPTER 9   ADDITIONAL AND ADVANCED EXERCISES

1. converges since $\frac{1}{(3n-2)^{(2n+1)/2}} < \frac{1}{(3n-2)^{3/2}}$ and $\sum_{n=1}^{\infty} \frac{1}{(3n-2)^{3/2}}$ converges by the Limit Comparison Test:

$\lim_{n \to \infty} \frac{\left(\frac{1}{n^{3/2}}\right)}{\left(\frac{1}{(3n-2)^{3/2}}\right)} = \lim_{n \to \infty} \left(\frac{3n-2}{n}\right)^{3/2} = 3^{3/2}$

3. diverges by the nth-Term Test since $\lim_{n \to \infty} a_n = \lim_{n \to \infty} (-1)^n \tanh n = \lim_{b \to \infty} (-1)^n \left(\frac{1 - e^{-2n}}{1 + e^{-2n}}\right) = \lim_{n \to \infty} (-1)^n$ does not
exist

5. converges by the Direct Comparison Test: $a_1 = 1 = \frac{12}{(1)(3)(2)^2}, a_2 = \frac{1 \cdot 2}{3 \cdot 4} = \frac{12}{(2)(4)(3)^2}, a_3 = \left(\frac{2 \cdot 3}{4 \cdot 5}\right)\left(\frac{1 \cdot 2}{3 \cdot 4}\right) = \frac{12}{(3)(5)(4)^2}$,

$a_4 = \left(\frac{3 \cdot 4}{5 \cdot 6}\right)\left(\frac{2 \cdot 3}{4 \cdot 5}\right)\left(\frac{1 \cdot 2}{3 \cdot 4}\right) = \frac{12}{(4)(6)(5)^2}, \ldots \Rightarrow 1 + \sum_{n=1}^{\infty} \frac{12}{(n+1)(n+3)(n+2)^2}$ represents the given series and

$\frac{12}{(n+1)(n+3)(n+2)^2} < \frac{12}{n^4}$, which is the nth-term of a convergent p-series

7. diverges by the nth-Term Test since if $a_n \to L$ as $n \to \infty$, then $L = \frac{1}{1+L} \Rightarrow L^2 + L - 1 = 0 \Rightarrow L = \frac{-1 \pm \sqrt{5}}{2} \neq 0$

9. $f(x) = \cos x$ with $a = \frac{\pi}{3} \Rightarrow f\left(\frac{\pi}{3}\right) = 0.5, f'\left(\frac{\pi}{3}\right) = -\frac{\sqrt{3}}{2}, f''\left(\frac{\pi}{3}\right) = -0.5, f'''\left(\frac{\pi}{3}\right) = \frac{\sqrt{3}}{2}, f^{(4)}\left(\frac{\pi}{3}\right) = 0.5;$

$\cos x = \frac{1}{2} - \frac{\sqrt{3}}{2}\left(x - \frac{\pi}{3}\right) - \frac{1}{4}\left(x - \frac{\pi}{3}\right)^2 + \frac{\sqrt{3}}{12}\left(x - \frac{\pi}{3}\right)^3 + \ldots$

11. $e^x = 1 + x + \frac{x^2}{2!} + \frac{x^3}{3!} + \ldots$ with $a = 0$

13. $f(x) = \cos x$ with $a = 22\pi \Rightarrow f(22\pi) = 1, \ f'(22\pi) = 0, \ f''(22\pi) = -1, \ f'''(22\pi) = 0, \ f^{(4)}(22\pi) = 1,$

$f^{(5)}(22\pi) = 0, \ f^{(6)}(22\pi) = -1; \ \cos x = 1 - \frac{1}{2}(x - 22\pi)^2 + \frac{1}{4!}(x - 22\pi)^4 - \frac{1}{6!}(x - 22\pi)^6 + \dots$

15. Yes, the sequence converges: $c_n = \left(a^n + b^n\right)^{1/n} \Rightarrow c_n = b\left(\left(\frac{a}{b}\right)^n + 1\right)^{1/n} \Rightarrow \lim_{n \to \infty} c_n = \ln b + \lim_{n \to \infty} \dfrac{\ln\left(\left(\frac{a}{b}\right)^n + 1\right)}{n}$

$= \ln b + \lim_{n \to \infty} \dfrac{\left(\frac{a}{b}\right)^n \ln\left(\frac{a}{b}\right)}{\left(\frac{a}{b}\right)^n + 1} = \ln b + \dfrac{0 \cdot \ln\left(\frac{a}{b}\right)}{0+1} = \ln b$ since $0 < a < b$. Thus, $\lim_{n \to \infty} c_n = e^{\ln b} = b.$

17. $s_n = \displaystyle\sum_{k=0}^{n-1} \int_k^{k+1} \frac{dx}{1+x^2} \Rightarrow s_n = \int_0^1 \frac{dx}{1+x^2} + \int_1^2 \frac{dx}{1+x^2} + \dots + \int_{n-1}^n \frac{dx}{1+x^2} \Rightarrow s_n = \int_0^n \frac{dx}{1+x^2}$

$\Rightarrow \lim_{n \to \infty} s_n = \lim_{n \to \infty} \left(\tan^{-1} n - \tan^{-1} 0\right) = \frac{\pi}{2}$

19. (a) No, the limit does not appear to depend on the value of the constant a
    (b) Yes, the limit depends on the value of b

   (c) $s = \left(1 - \dfrac{\cos\left(\frac{a}{n}\right)}{n}\right)^n \Rightarrow \ln s = \dfrac{\ln\left(1 - \dfrac{\cos\left(\frac{a}{n}\right)}{n}\right)}{\left(\frac{1}{n}\right)} \Rightarrow \lim_{n \to \infty} \ln s = \dfrac{\left(\dfrac{1}{1 - \frac{\cos\left(\frac{a}{n}\right)}{n}}\right)\left(\dfrac{-\frac{a}{n}\sin\left(\frac{a}{n}\right) + \cos\left(\frac{a}{n}\right)}{n^2}\right)}{\left(-\frac{1}{n^2}\right)} = \lim_{n \to \infty} \dfrac{\frac{a}{n}\sin\left(\frac{a}{n}\right) - \cos\left(\frac{a}{n}\right)}{1 - \frac{\cos\left(\frac{a}{n}\right)}{n}}$

   $= \frac{0-1}{1-0} = -1 \Rightarrow \lim_{n \to \infty} s = e^{-1} \approx 0.3678794412$; similarly, $\lim_{n \to \infty}\left(1 - \dfrac{\cos\left(\frac{a}{n}\right)}{bn}\right)^n = e^{-1/b}$

21. $\lim_{n \to \infty}\left|\dfrac{u_{n+1}}{u_n}\right| < 1 \Rightarrow \lim_{n \to \infty}\left|\dfrac{b^{n+1} x^{n+1}}{\ln(n+1)} \cdot \dfrac{\ln n}{b^n x^n}\right| < 1 \Rightarrow |bx| < 1 \Rightarrow -\frac{1}{b} < x < \frac{1}{b} = 5 \Rightarrow b = \pm\frac{1}{5}$

23. $\lim_{x \to 0} \dfrac{\sin(ax) - \sin x - x}{x^3} = \lim_{x \to 0} \dfrac{\left(ax - \frac{a^3 x^3}{3!} + \dots\right) - \left(x - \frac{x^3}{3!} + \dots\right) - x}{x^3} = \lim_{x \to 0}\left[\dfrac{a-2}{x^2} - \dfrac{a^3}{3!} + \dfrac{1}{3!} - \left(\dfrac{a^5}{5!} - \dfrac{1}{5!}\right)x^2 + \dots\right]$ is finite

   if $a - 2 = 0 \Rightarrow a = 2$; $\lim_{x \to 0} \dfrac{\sin 2x - \sin x - x}{x^3} = -\dfrac{2^3}{3!} + \dfrac{1}{3!} = -\dfrac{7}{6}$

25. (a) $\dfrac{u_n}{u_{n+1}} = \dfrac{(n+1)^2}{n^2} = 1 + \dfrac{2}{n} + \dfrac{1}{n^2} \Rightarrow C = 2 > 1$ and $\displaystyle\sum_{n=1}^{\infty} \dfrac{1}{n^2}$ converges

   (b) $\dfrac{u_n}{u_{n+1}} = \dfrac{n+1}{n} = 1 + \dfrac{1}{n} + \dfrac{0}{n^2} \Rightarrow C = 1 \leq 1$ and $\displaystyle\sum_{n=1}^{\infty} \dfrac{1}{n}$ diverges

27. (a) $\displaystyle\sum_{n=1}^{\infty} a_n = L \Rightarrow a_n^2 \leq a_n \sum_{n=1}^{\infty} a_n = a_n L \Rightarrow \sum_{n=1}^{\infty} a_n^2$ converges by the Direct Comparison Test

   (b) converges by the Limit Comparison Test: $\lim_{n \to \infty} \dfrac{\left(\frac{a_n}{1-a_n}\right)}{a_n} = \lim_{n \to \infty} \dfrac{1}{1-a_n} = 1$ since $\displaystyle\sum_{n=1}^{\infty} a_n$ converges and therefore

   $\lim_{n \to \infty} a_n = 0$

29. $(1-x)^{-1} = 1 + \sum_{n=1}^{\infty} x^n$ where $|x| < 1 \Rightarrow \frac{1}{(1-x)^2} = \frac{d}{dx}(1-x)^{-1} = \sum_{n=1}^{\infty} nx^{n-1}$ and when $x = \frac{1}{2}$ we have

$4 = 1 + 2\left(\frac{1}{2}\right) + 3\left(\frac{1}{2}\right)^2 + 4\left(\frac{1}{2}\right)^3 + \ldots + n\left(\frac{1}{2}\right)^{n-1} + \ldots$

31. (a) $\frac{1}{(1-x)^2} = \frac{d}{dx}\left(\frac{1}{1-x}\right) = \frac{d}{dx}\left(1 + x + x^2 + x^3 + \ldots\right) = 1 + 2x + 3x^2 + 4x^3 + \ldots = \sum_{n=1}^{\infty} nx^{n-1}$

(b) from part (a) we have $\sum_{n=1}^{\infty} n\left(\frac{5}{6}\right)^{n-1}\left(\frac{1}{6}\right) = \left(\frac{1}{6}\right)\left[\frac{1}{1-\left(\frac{5}{6}\right)}\right]^2 = 6$

(c) from part (a) we have $\sum_{n=1}^{\infty} np^{n-1}q = \frac{q}{(1-p)^2} = \frac{q}{q^2} = \frac{1}{q}$

33. (a) $R_n = C_0 e^{-kt_0} + C_0 e^{-2kt_0} + \ldots + C_0 e^{-nkt_0} = \frac{C_0 e^{-kt_0}\left(1 - e^{-nkt_0}\right)}{1 - e^{-kt_0}} \Rightarrow R = \lim_{n \to \infty} R_n = \frac{C_0 e^{-kt_0}}{1 - e^{-kt_0}} = \frac{C_0}{e^{kt_0} - 1}$

(b) $R_n = \frac{e^{-1}\left(1 - e^{-n}\right)}{1 - e^{-1}} \Rightarrow R_1 = e^{-1} \approx 0.36787944$ and $R_{10} = \frac{e^{-1}\left(1 - e^{-10}\right)}{1 - e^{-1}} \approx 0.58195028$; $R = \frac{1}{e-1} \approx 0.58197671$;

$R - R_{10} \approx 0.00002643 \Rightarrow \frac{R - R_{10}}{R} < 0.0001$

(c) $R_n = \frac{e^{-.1}\left(1 - e^{-.1n}\right)}{1 - e^{-.1}}$, $\frac{R}{2} = \frac{1}{2}\left(\frac{1}{e^{.1} - 1}\right) \approx 4.7541659$; $R_n > \frac{R}{2} \Rightarrow \frac{1 - e^{-.1n}}{e^{.1} - 1} > \left(\frac{1}{2}\right)\left(\frac{1}{e^{.1} - 1}\right) \Rightarrow 1 - e^{-n/10} > \frac{1}{2}$

$\Rightarrow e^{-n/10} < \frac{1}{2} \Rightarrow -\frac{n}{10} < \ln\left(\frac{1}{2}\right) \Rightarrow \frac{n}{10} > -\ln\left(\frac{1}{2}\right) \Rightarrow n > 6.93 \Rightarrow n = 7$

CHAPTER 10   PARAMETRIC EQUATIONS AND POLAR COORDINATES

10.1   PARAMETRIZATIONS OF PLANE CURVES

1. $x = 3t,\ y = 9t^2,\ -\infty < t < \infty \Rightarrow y = x^2$

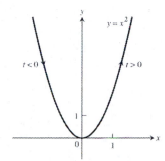

3. $x = 2t - 5,\ y = 4t - 7,\ -\infty < t < \infty$

$\Rightarrow x + 5 = 2t \Rightarrow 2(x+5) = 4t$

$\Rightarrow y = 2(x+5) - 7 \Rightarrow y = 2x + 3$

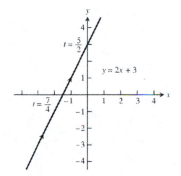

5. $x = \cos 2t,\ y = \sin 2t,\ 0 \le t \le \pi$

$\Rightarrow \cos^2 2t + \sin^2 2t = 1 \Rightarrow x^2 + y^2 = 1$

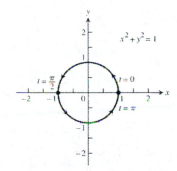

7. $x = 4 \cos t,\ y = 2 \sin t,\ 0 \le t \le 2\pi$

$\Rightarrow \dfrac{16 \cos^2 t}{16} + \dfrac{4 \sin^2 t}{4} = 1 \Rightarrow \dfrac{x^2}{16} + \dfrac{y^2}{4} = 1$

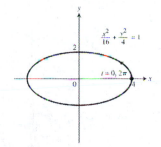

9. $x = \sin t,\ y = \cos 2t,\ -\dfrac{\pi}{2} \le t \le \dfrac{\pi}{2}$

$\Rightarrow y = \cos 2t = 1 - 2 \sin^2 t \Rightarrow y = 1 - 2x^2$

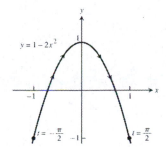

11. $x = t^2,\ y = t^6 - 2t^4,\ -\infty < t < \infty$

$\Rightarrow y = \left(t^2\right)^3 - 2\left(t^2\right)^2 \Rightarrow y = x^3 - 2x^2$

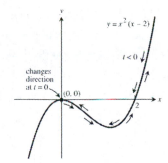

13. $x = t, y = \sqrt{1-t^2}, -1 \le t \le 0$

$\Rightarrow y = \sqrt{1-x^2}$

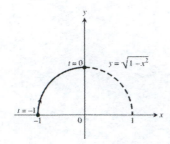

15. $x = \sec^2 t - 1, y = \tan t, -\frac{\pi}{2} < t < \frac{\pi}{2}$

$\Rightarrow \sec^2 t - 1 = \tan^2 t \Rightarrow x = y^2$

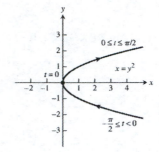

17. $x = -\cosh t, y = \sinh t, -\infty < t < \infty$

$\Rightarrow \cosh^2 t - \sinh^2 t = 1 \Rightarrow x^2 - y^2 = 1$

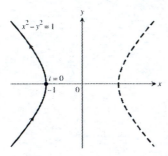

19. (a) $x = a \cos t, y = -a \sin t, 0 \le t \le 2\pi$

    (b) $x = a \cos t, y = a \sin t, 0 \le t \le 2\pi$

    (c) $x = a \cos t, y = -a \sin t, 0 \le t \le 4\pi$

    (d) $x = a \cos t, y = a \sin t, 0 \le t \le 4\pi$

21. Using $(-1, -3)$ we create the parametric equations $x = -1 + at$ and $y = -3 + bt$, representing a line which goes through $(-1, -3)$ at $t = 0$. We determine a and b so that the line goes through $(4, 1)$ when $t = 1$. Since $4 = -1 + a \Rightarrow a = 5$. Since $1 = -3 + b \Rightarrow b = 4$. Therefore, one possible parameterization is $x = -1 + 5t$, $y = -3 + 4t, 0 \le t \le 1$.

23. The lower half of the parabola is given by $x = y^2 + 1$ for $y \le 0$. Substituting t for y, we obtain one possible parameterization $x = t^2 + 1, y = t, t \le 0$.

25. For simplicity, we assume that x and y are linear functions of t and that the point (x, y) starts at $(2, 3)$ for $t = 0$ and passes through $(-1, -1)$ at $t = 1$. Then $x = f(t)$, where $f(0) = 2$ and $f(1) = -1$. Since slope $= \frac{\Delta x}{\Delta t} = \frac{-1-2}{1-0} = -3$, $x = f(t) = -3t + 2 = 2 - 3t$. Also, $y = g(t)$, where $g(0) = 3$ and $g(1) = -1$. Since slope $= \frac{\Delta y}{\Delta t} = \frac{-1-3}{1-0} = -4$. $y = g(t) = -4t + 3 = 3 - 4t$. One possible parameterization is: $x = 2 - 3t$, $y = 3 - 4t, t \ge 0$.

27. Since we only want the top half of a circle, $y \ge 0$, so let $x = 2\cos t, y = 2|\sin t|, 0 \le t \le 4\pi$

29. $x^2 + y^2 = a^2 \Rightarrow 2x + 2y\frac{dy}{dx} = 0 \Rightarrow \frac{dy}{dx} = -\frac{x}{y}$; let $t = \frac{dy}{dx} \Rightarrow -\frac{x}{y} = t \Rightarrow x = -yt.$ Substitution yields

$y^2t^2 + y^2 = a^2 \Rightarrow y = \frac{a}{\sqrt{1+t^2}}$ and $x = \frac{-at}{\sqrt{1+t}}, -\infty < t < \infty$

31. Drop a vertical line from the point (x, y) to the x-axis, then θ is an angle in a right triangle, and from

trigonometry we know that $\tan \theta = \frac{y}{x} \Rightarrow y = x \tan \theta.$ The equation of the line through $(0, 2)$ and $(4, 0)$

is given by $y = -\frac{1}{2}x + 2.$ Thus $x \tan \theta = -\frac{1}{2}x + 2 \Rightarrow x = \frac{4}{2\tan\theta+1}$ and $y = \frac{4\tan\theta}{2\tan\theta+1}$ where $0 \le \theta \le \frac{\pi}{2}.$

33. The equation of the circle is given by $(x-2)^2 + y^2 = 1.$ Drop a vertical line from the point (x, y) on the circle

to the x-axis, then θ is an angle in a right triangle. So that we can start at $(1, 0)$ and rotate in a clockwise

direction, let $x = 2 - \cos \theta, \ y = \sin \theta, \ 0 \le \theta \le 2\pi.$

35. Extend the vertical line through A to the x-axis and let C be the point of intersection. Then $OC = AQ = x$ and

$\tan t = \frac{2}{OC} = \frac{2}{x} \Rightarrow x = \frac{2}{\tan t} = 2 \cot t; \ \sin t = \frac{2}{OA} \Rightarrow OA = \frac{2}{\sin t};$ and $(AB)(OA) = (AQ)^2 \Rightarrow AB\left(\frac{2}{\sin t}\right) = x^2$

$\Rightarrow AB\left(\frac{2}{\sin t}\right) = \left(\frac{2}{\tan t}\right)^2 \Rightarrow AB = \frac{2\sin t}{\tan^2 t}.$ Next $y = 2 - AB \sin t \Rightarrow y = 2 - \left(\frac{2\sin t}{\tan^2 t}\right)\sin t = 2 - \frac{2\sin^2 t}{\tan^2 t}$

$= 2 - 2\cos^2 t = 2\sin^2 t.$ Therefore let $x = 2\cot t$ and $y = 2\sin^2 t, \ 0 < t < \pi.$

37. Draw line AM in the figure and note that $\angle AMO$ is a right angle since it is an inscribed angle which spans the diameter of a circle. Then $AN^2 = MN^2 + AM^2.$ Now, $OA = a, \ \frac{AN}{a} = \tan t,$ and

$\frac{AM}{a} = \sin t.$ Next $MN = OP$

$\Rightarrow OP^2 = AN^2 - AM^2 = a^2 \tan^2 t - a^2 \sin^2 t$

$\Rightarrow OP = \sqrt{a^2 \tan^2 t - a^2 \sin^2 t} = (a \sin t)\sqrt{\sec^2 t - 1} = \frac{a\sin^2 t}{\cos t}.$

In triangle $BPO, \ x = OP \sin t = \frac{a\sin^3 t}{\cos t} = a \sin^2 t \tan t$ and

$y = OP \cos t = a\sin^2 t \Rightarrow x = a\sin^2 t \tan t$ and $y = a\sin^2 t.$

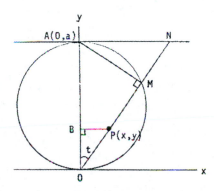

39. $D = \sqrt{(x-2)^2 + \left(y-\frac{1}{2}\right)^2} \Rightarrow D^2 = (x-2)^2 + \left(y-\frac{1}{2}\right)^2 = (t-2)^2 + \left(t^2 - \frac{1}{2}\right)^2 \Rightarrow D^2 = t^4 - 4t + \frac{17}{4}$

$\Rightarrow \frac{d(D^2)}{dt} = 4t^3 - 4 = 0 \Rightarrow t = 1.$ The second derivative is always positive for $t \ne 0 \Rightarrow t = 1$ gives a local

minimum for D^2 (and hence D) which is an absolute minimum since it is the only extremum $\Rightarrow$ the closest

point on the parabola is $(1, 1).$

10.2 CALCULUS WITH PARAMETRIC CURVES

1. $t = \frac{\pi}{4} \Rightarrow x = 2\cos\frac{\pi}{4} = \sqrt{2},\ \ y = 2\sin\frac{\pi}{4} = \sqrt{2};\ \ \frac{dx}{dt} = -2\sin t,\ \ \frac{dy}{dt} = 2\cos t \Rightarrow \frac{dy}{dx} = \frac{dy/dt}{dx/dt} = \frac{2\cos t}{-2\sin t} = -\cot t$

$\Rightarrow \left.\frac{dy}{dx}\right|_{t=\frac{\pi}{4}} = -\cot\frac{\pi}{4} = -1;$ tangent line is $y - \sqrt{2} = -1\left(x - \sqrt{2}\right)$ or $y = -x + 2\sqrt{2};\ \ \frac{dy'}{dt} = \csc^2 t$

$\Rightarrow \frac{d^2y}{dx^2} = \frac{dy'/dt}{dx/dt} = \frac{\csc^2 t}{-2\sin t} = -\frac{1}{2\sin^3 t} \Rightarrow \left.\frac{d^2y}{dx^2}\right|_{t=\frac{\pi}{4}} = -\sqrt{2}$

3. $t = \frac{\pi}{4} \Rightarrow x = 4\sin\frac{\pi}{4} = 2\sqrt{2},\ \ y = 2\cos\frac{\pi}{4} = \sqrt{2};\ \ \frac{dx}{dt} = 4\cos t,\ \ \frac{dy}{dt} = -2\sin t \Rightarrow \frac{dy}{dx} = \frac{dy/dt}{dx/dt} = \frac{-2\sin t}{4\cos t} = -\frac{1}{2}\tan t$

$\Rightarrow \left.\frac{dy}{dx}\right|_{t=\frac{\pi}{4}} = -\frac{1}{2}\tan\frac{\pi}{4} = -\frac{1}{2};$ tangent line is $y - \sqrt{2} = -\frac{1}{2}\left(x - 2\sqrt{2}\right)$ or $y = -\frac{1}{2}x + 2\sqrt{2};\ \ \frac{dy'}{dt} = -\frac{1}{2}\sec^2 t$

$\Rightarrow \frac{d^2y}{dx^2} = \frac{dy'/dt}{dx/dt} = \frac{-\frac{1}{2}\sec^2 t}{4\cos t} = -\frac{1}{8\cos^3 t} \Rightarrow \left.\frac{d^2y}{dx^2}\right|_{t=\frac{\pi}{4}} = -\frac{\sqrt{2}}{4}$

5. $t = \frac{1}{4} \Rightarrow x = \frac{1}{4},\ \ y = \frac{1}{2};\ \ \frac{dx}{dt} = 1,\ \ \frac{dy}{dt} = \frac{1}{2\sqrt{t}} \Rightarrow \frac{dy}{dx} = \frac{dy/dt}{dx/dt} = \frac{1}{2\sqrt{t}} \Rightarrow \left.\frac{dy}{dx}\right|_{t=\frac{1}{4}} = \frac{1}{2\sqrt{\frac{1}{4}}} = 1;$ tangent line is

$y - \frac{1}{2} = 1\cdot\left(x - \frac{1}{4}\right)$ or $y = x + \frac{1}{4};\ \ \frac{dy'}{dt} = -\frac{1}{4}t^{-3/2} \Rightarrow \frac{d^2y}{dx^2} = \frac{dy'/dt}{dx/dt} = -\frac{1}{4}t^{-3/2} \Rightarrow \left.\frac{d^2y}{dx^2}\right|_{t=\frac{1}{4}} = -2$

7. $t = \frac{\pi}{6} \Rightarrow x = \sec\frac{\pi}{6} = \frac{2}{\sqrt{3}},\ \ y = \tan\frac{\pi}{6} = \frac{1}{\sqrt{3}};\ \ \frac{dx}{dt} = \sec t\tan t,\ \ \frac{dy}{dt} = \sec^2 t \Rightarrow \frac{dy}{dx} = \frac{dy/dt}{dx/dt} = \frac{\sec^2 t}{\sec t\tan t} = \csc t$

$\Rightarrow \left.\frac{dy}{dx}\right|_{t=\frac{\pi}{6}} = \csc\frac{\pi}{6} = 2;$ tangent line is $y - \frac{1}{\sqrt{3}} = 2\left(x - \frac{2}{\sqrt{3}}\right)$ or $y = 2x - \sqrt{3};\ \ \frac{dy'}{dt} = -\csc t\cot t$

$\Rightarrow \frac{d^2y}{dx^2} = \frac{dy'/dt}{dx/dt} = \frac{-\csc t\cot t}{\sec t\tan t} = -\cot^3 t \Rightarrow \left.\frac{d^2y}{dx^2}\right|_{t=\frac{\pi}{6}} = -3\sqrt{3}$

9. $t = -1 \Rightarrow x = 5,\ \ y = 1;\ \ \frac{dx}{dt} = 4t,\ \ \frac{dy}{dt} = 4t^3 \Rightarrow \frac{dy}{dx} = \frac{dy/dt}{dx/dt} = \frac{4t^3}{4t} = t^2 \Rightarrow \left.\frac{dy}{dx}\right|_{t=-1} = (-1)^2 = 1;$ tangent line is

$y - 1 = 1\cdot(x - 5)$ or $y = x - 4;\ \ \frac{dy'}{dt} = 2t \Rightarrow \frac{d^2y}{dx^2} = \frac{dy'/dt}{dx/dt} = \frac{2t}{4t} = \frac{1}{2} \Rightarrow \left.\frac{d^2y}{dx^2}\right|_{t=-1} = \frac{1}{2}$

11. $t = \frac{\pi}{3} \Rightarrow x = \frac{\pi}{3} - \sin\frac{\pi}{3} = \frac{\pi}{3} - \frac{\sqrt{3}}{2},\ \ y = 1 - \cos\frac{\pi}{3} = 1 - \frac{1}{2} = \frac{1}{2};\ \ \frac{dx}{dt} = 1 - \cos t,\ \ \frac{dy}{dt} = \sin t \Rightarrow \frac{dy}{dx} = \frac{dy/dt}{dx/dt} = \frac{\sin t}{1-\cos t}$

$\Rightarrow \left.\frac{dy}{dx}\right|_{t=\frac{\pi}{3}} = \frac{\sin\left(\frac{\pi}{3}\right)}{1-\cos\left(\frac{\pi}{3}\right)} = \frac{\left(\frac{\sqrt{3}}{2}\right)}{\left(\frac{1}{2}\right)} = \sqrt{3};$ tangent line is $y - \frac{1}{2} = \sqrt{3}\left(x - \frac{\pi}{3} + \frac{\sqrt{3}}{2}\right) \Rightarrow y = \sqrt{3}x - \frac{\pi\sqrt{3}}{3} + 2;$

$\frac{dy'}{dt} = \frac{(1-\cos t)(\cos t) - (\sin t)(\sin t)}{(1-\cos t)^2} = \frac{-1}{1-\cos t} \Rightarrow \frac{d^2y}{dx^2} = \frac{dy'/dt}{dx/dt} = \frac{\left(\frac{-1}{1-\cos t}\right)}{1-\cos t} = \frac{-1}{(1-\cos t)^2} \Rightarrow \left.\frac{d^2y}{dx^2}\right|_{t=\frac{\pi}{3}} = -4$

13. $t = 2 \Rightarrow x = \frac{1}{2+1} = \frac{1}{3},\ \ y = \frac{2}{2-1} = 2;\ \ \frac{dx}{dt} = \frac{-1}{(t+1)^2},\ \ \frac{dy}{dt} = \frac{-1}{(t-1)^2} \Rightarrow \frac{dy}{dx} = \frac{(t+1)^2}{(t-1)^2} \Rightarrow \left.\frac{dy}{dx}\right|_{t=2} = \frac{(2+1)^2}{(2-1)^2} = 9;$ tangent line is

$y = 9x - 1;\ \ \frac{dy'}{dt} = -\frac{4(t+1)}{(t-1)^3} \Rightarrow \frac{d^2y}{dx^2} = \frac{4(t+1)^3}{(t-1)^3} \Rightarrow \left.\frac{d^2y}{dx^2}\right|_{t=2} = \frac{4(2+1)^3}{(2-1)^3} = 108$

15. $x^3 + 2t^2 = 9 \Rightarrow 3x^2 \frac{dx}{dt} + 4t = 0 \Rightarrow 3x^2 \frac{dx}{dt} = -4t \Rightarrow \frac{dx}{dt} = \frac{-4t}{3x^2}$; $2y^3 - 3t^2 = 4 \Rightarrow 6y^2 \frac{dy}{dt} - 6t = 0$

$\Rightarrow \frac{dy}{dx} = \frac{6t}{6y^2} = \frac{t}{y^2}$; thus $\frac{dy}{dx} = \frac{dy/dt}{dx/dt} = \frac{\left(\frac{t}{y^2}\right)}{\left(\frac{-4t}{3x^2}\right)} = \frac{t(3x^2)}{y^2(-4t)} = \frac{3x^2}{-4y^2}$; $t = 2 \Rightarrow x^3 + 2(2)^2 = 9 \Rightarrow x^3 + 8 = 9 \Rightarrow x^3 = 1$

$\Rightarrow x = 1$; $t = 2 \Rightarrow 2y^3 - 3(2)^2 = 4 \Rightarrow 2y^3 = 16 \Rightarrow y^3 = 8 \Rightarrow y = 2$; therefore $\left.\frac{dy}{dx}\right|_{t=2} = \frac{3(1)^2}{-4(2)^2} = -\frac{3}{16}$

17. $x + 2x^{3/2} = t^2 + t \Rightarrow \frac{dx}{dt} + 3x^{1/2}\frac{dx}{dt} = 2t + 1 \Rightarrow \left(1 + 3x^{1/2}\right)\frac{dx}{dt} = 2t + 1 \Rightarrow \frac{dx}{dt} = \frac{2t+1}{1+3x^{1/2}}$; $y\sqrt{t+1} + 2t\sqrt{y} = 4$

$\Rightarrow \frac{dy}{dt}\sqrt{t+1} + y\left(\frac{1}{2}\right)(t+1)^{-1/2} + 2\sqrt{y} + 2t\left(\frac{1}{2}y^{-1/2}\right)\frac{dy}{dt} = 0 \Rightarrow \frac{dy}{dt}\sqrt{t+1} + \frac{y}{2\sqrt{t+1}} + 2\sqrt{y} + \left(\frac{t}{\sqrt{y}}\right)\frac{dy}{dt} = 0$

$\Rightarrow \left(\sqrt{t+1} + \frac{t}{\sqrt{y}}\right)\frac{dy}{dt} = \frac{-y}{2\sqrt{t+1}} - 2\sqrt{y} \Rightarrow \frac{dy}{dt} = \frac{\left(\frac{-y}{2\sqrt{t+1}} - 2\sqrt{y}\right)}{\left(\sqrt{t+1} + \frac{t}{\sqrt{y}}\right)} = \frac{-y\sqrt{y} - 4y\sqrt{t+1}}{2\sqrt{y}(t+1) + 2t\sqrt{t+1}}$; thus $\frac{dy}{dx} = \frac{dy/dt}{dx/dt} = \frac{\left(\frac{-y\sqrt{y} - 4y\sqrt{t+1}}{2\sqrt{y}(t+1) + 2t\sqrt{t+1}}\right)}{\left(\frac{2t+1}{1+3x^{1/2}}\right)}$;

$t = 0 \Rightarrow x + 2x^{3/2} = 0 \Rightarrow x\left(1 + 2x^{1/2}\right) = 0 \Rightarrow x = 0$; $t = 0 \Rightarrow y\sqrt{0+1} + 2(0)\sqrt{y} = 4 \Rightarrow y = 4$;

therefore $\left.\frac{dy}{dx}\right|_{t=0} = \frac{\left(\frac{-4\sqrt{4} - 4(4)\sqrt{0+1}}{2\sqrt{4}(0+1) + 2(0)\sqrt{0+1}}\right)}{\left(\frac{2(0)+1}{1+3(0)^{1/2}}\right)} = -6$

19. $x = t^3 + t$, $y + 2t^3 = 2x + t^2 \Rightarrow \frac{dx}{dt} = 3t^2 + 1$, $\frac{dy}{dt} + 6t^2 = 2\frac{dx}{dt} + 2t \Rightarrow \frac{dy}{dt} = 2\left(3t^2 + 1\right) + 2t - 6t^2 = 2t + 2$

$\Rightarrow \frac{dy}{dx} = \frac{2t+2}{3t^2+1} \Rightarrow \left.\frac{dy}{dx}\right|_{t=1} = \frac{2(1)+2}{3(1)^2+1} = 1$

21. $A = \int_0^{2\pi} y\,dx = \int_0^{2\pi} a(1-\cos t)a(1-\cos t)\,dt = a^2\int_0^{2\pi}(1-\cos t)^2\,dt = a^2\int_0^{2\pi}\left(1 - 2\cos t + \cos^2 t\right)dt$

$= a^2\int_0^{2\pi}\left(1 - 2\cos t + \frac{1+\cos 2t}{2}\right)dt = a^2\int_0^{2\pi}\left(\frac{3}{2} - 2\cos t + \frac{1}{2}\cos 2t\right)dt = a^2\left[\frac{3}{2}t - 2\sin t + \frac{1}{4}\sin 2t\right]_0^{2\pi}$

$= a^2(3\pi - 0 + 0) - 0 = 3\pi a^2$

23. $A = 2\int_\pi^0 y\,dx = 2\int_\pi^0 (b\sin t)(-a\sin t)\,dt = 2ab\int_0^\pi \sin^2 t\,dt = 2ab\int_0^\pi \frac{1-\cos 2t}{2}\,dt = ab\int_0^\pi (1-\cos 2t)\,dt$

$= ab\left[t - \frac{1}{2}\sin 2t\right]_0^\pi = ab((\pi - 0) - 0) = \pi\,ab$

25. $\frac{dx}{dt} = -\sin t$ and $\frac{dy}{dt} = 1 + \cos t \Rightarrow \sqrt{\left(\frac{dx}{dt}\right)^2 + \left(\frac{dy}{dt}\right)^2} = \sqrt{(-\sin t)^2 + (1+\cos t)^2} = \sqrt{2 + 2\cos t}$

$\Rightarrow \text{Length} = \int_0^\pi \sqrt{2 + 2\cos t}\,dt = \sqrt{2}\int_0^\pi \sqrt{\left(\frac{1-\cos t}{1-\cos t}\right)(1+\cos t)}\,dt = \sqrt{2}\int_0^\pi \sqrt{\frac{\sin^2 t}{1-\cos t}}\,dt = \sqrt{2}\int_0^\pi \frac{\sin t}{\sqrt{1-\cos t}}\,dt$

(since $\sin t \geq 0$ on $[0, \pi]$); $\left[u = 1 - \cos t \Rightarrow du = \sin t\,dt; t = 0 \Rightarrow u = 0, t = \pi \Rightarrow u = 2\right]$

$\rightarrow \sqrt{2}\int_0^2 u^{-1/2}\,du = \sqrt{2}\left[2u^{1/2}\right]_0^2 = 4$

27. $\frac{dx}{dt} = t$ and $\frac{dy}{dt} = (2t+1)^{1/2} \Rightarrow \sqrt{\left(\frac{dx}{dt}\right)^2 + \left(\frac{dy}{dt}\right)^2} = \sqrt{t^2 + (2t+1)} = \sqrt{(t+1)^2} = |t+1| = t+1$ since $0 \le t \le 4$

$\Rightarrow$ Length $= \int_0^4 (t+1)\, dt = \left[\frac{t^2}{2} + t\right]_0^4 = (8+4) = 12$

29. $\frac{dx}{dt} = 8t \cos t$ and $\frac{dy}{dt} = 8t \sin t \Rightarrow \sqrt{\left(\frac{dx}{dt}\right)^2 + \left(\frac{dy}{dt}\right)^2} = \sqrt{(8t \cos t)^2 + (8t \sin t)^2} = \sqrt{64t^2 \cos^2 t + 64t^2 \sin^2 t}$

$= |8t| = 8t$ since $0 \le t \le \frac{\pi}{2} \Rightarrow$ Length $= \int_0^{\pi/2} 8t\, dt = \left[4t^2\right]_0^{\pi/2} = \pi^2$

31. $\frac{dx}{dt} = -\sin t$ and $\frac{dy}{dt} = \cos t \Rightarrow \sqrt{\left(\frac{dx}{dt}\right)^2 + \left(\frac{dy}{dt}\right)^2} = \sqrt{(-\sin t)^2 + (\cos t)^2} = 1$

$\Rightarrow$ Area $= \int 2\pi y\, ds = \int_0^{2\pi} 2\pi(2 + \sin t)(1)dt = 2\pi\left[2t - \cos t\right]_0^{2\pi} = 2\pi\left[(4\pi - 1) - (0-1)\right] = 8\pi^2$

33. $\frac{dx}{dt} = 1$ and $\frac{dy}{dt} = t + \sqrt{2} \Rightarrow \sqrt{\left(\frac{dx}{dt}\right)^2 + \left(\frac{dy}{dt}\right)^2} = \sqrt{1^2 + \left(t + \sqrt{2}\right)^2} = \sqrt{t^2 + 2\sqrt{2}t + 3}$

$\Rightarrow$ Area $= \int 2\pi x\, ds = \int_{-\sqrt{2}}^{\sqrt{2}} 2\pi\left(t + \sqrt{2}\right)\sqrt{t^2 + 2\sqrt{2}t + 3}\, dt;$

$\left[u = t^2 + 2\sqrt{2}t + 3 \Rightarrow du = \left(2t + 2\sqrt{2}\right)dt; t = -\sqrt{2} \Rightarrow u = 1, t = \sqrt{2} \Rightarrow u = 9\right]$

$\rightarrow \int_1^9 \pi\sqrt{u}\, du = \left[\frac{2}{3}\pi u^{3/2}\right]_1^9 = \frac{2\pi}{3}(27 - 1) = \frac{52\pi}{3}$

35. $\frac{dx}{dt} = 2$ and $\frac{dy}{dt} = 1 \Rightarrow \sqrt{\left(\frac{dx}{dt}\right)^2 + \left(\frac{dy}{dt}\right)^2} = \sqrt{2^2 + 1^2} = \sqrt{5} \Rightarrow$ Area $= \int 2\pi y\, ds = \int_0^1 2\pi(t+1)\sqrt{5}\, dt$

$= 2\pi\sqrt{5}\left[\frac{t^2}{2} + t\right]_0^1 = 3\pi\sqrt{5}$. Check: slant height is $\sqrt{5} \Rightarrow$ Area is $\pi(1+2)\sqrt{5} = 3\pi\sqrt{5}$.

37. Let the density be $\delta = 1$. Then $x = \cos t + t\sin t \Rightarrow \frac{dx}{dt} = t\cos t$, and $y = \sin t - t\cos t \Rightarrow \frac{dy}{dt} = t\sin t$

$\Rightarrow dm = 1 \cdot ds = \sqrt{\left(\frac{dx}{dt}\right)^2 + \left(\frac{dy}{dt}\right)^2}\, dt = \sqrt{(t\cos t)^2 + (t\sin t)^2} = |t|\, dt = t\, dt$ since $0 \le t \le \frac{\pi}{2}$. The curve's mass is

$M = \int dm = \int_0^{\pi/2} t\, dt = \frac{\pi^2}{8}$. Also $M_x = \int \tilde{y}\, dm = \int_0^{\pi/2} (\sin t - t\cos t)\, t\, dt = \int_0^{\pi/2} t\sin t\, dt - \int_0^{\pi/2} t^2 \cos t\, dt$

$= \left[\sin t - t\cos t\right]_0^{\pi/2} - \left[t^2 \sin t - 2\sin t + 2t\cos t\right]_0^{\pi/2} = 3 - \frac{\pi^2}{4}$, where we integrated by parts. Therefore,

$\bar{y} = \frac{M_x}{M} = \frac{\left(3 - \frac{\pi^2}{4}\right)}{\left(\frac{\pi^2}{8}\right)} = \frac{24}{\pi^2} - 2$. Next, $M_y = \int \tilde{x}\, dm = \int_0^{\pi/2} (\cos t + t\sin t)\, t\, dt = \int_0^{\pi/2} t\cos t\, dt + \int_0^{\pi/2} t^2 \sin t\, dt$

$= \left[\cos t + t\sin t\right]_0^{\pi/2} + \left[-t^2 \cos t + 2\cos t + 2t\sin t\right]_0^{\pi/2} = \frac{3\pi}{2} - 3$, again integrating by parts. Hence,

$\bar{x} = \frac{M_y}{M} = \frac{\left(\frac{3\pi}{2} - 3\right)}{\left(\frac{\pi^2}{8}\right)} = \frac{12}{\pi} - \frac{24}{\pi^2}$. Therefore $(\bar{x}, \bar{y}) = \left(\frac{12}{\pi} - \frac{24}{\pi^2}, \frac{24}{\pi^2} - 2\right)$.

39. Let the density be $\delta = 1$. Then $x = \cos t \Rightarrow \frac{dx}{dt} = -\sin t$, and $y = t + \sin t \Rightarrow \frac{dy}{dt} = 1 + \cos t$

$\Rightarrow dm = 1 \cdot ds = \sqrt{\left(\frac{dx}{dt}\right)^2 + \left(\frac{dy}{dt}\right)^2}\, dt = \sqrt{(-\sin t)^2 + (1 + \cos t)^2}\, dt = \sqrt{2 + 2\cos t}\, dt$. The curve's mass is

$M = \int dm = \int_0^\pi \sqrt{2 + 2\cos t}\, dt = \sqrt{2}\int_0^\pi \sqrt{1 + \cos t}\, dt = \sqrt{2}\int_0^\pi \sqrt{2\cos^2\left(\frac{t}{2}\right)}\, dt = 2\int_0^\pi \left|\cos\left(\frac{t}{2}\right)\right|\, dt = 2\int_0^\pi \cos\left(\frac{t}{2}\right)dt$

$\left(\text{since } 0 \le t \le \pi \Rightarrow 0 \le \frac{t}{2} \le \frac{\pi}{2}\right) = 2\left[2\sin\left(\frac{t}{2}\right)\right]_0^\pi = 4$. Also $M_x = \int \tilde{y}\, dm = \int_0^\pi (t + \sin t)\left(2\cos\frac{t}{2}\right)dt$

$= \int_0^\pi 2t\cos\left(\frac{t}{2}\right)dt + \int_0^\pi 2\sin t\cos\left(\frac{t}{2}\right)dt = 2\left[4\cos\left(\frac{t}{2}\right) + 2t\sin\left(\frac{t}{2}\right)\right]_0^\pi + 2\left[-\frac{1}{3}\cos\left(\frac{3}{2}t\right) - \cos\left(\frac{1}{2}t\right)\right]_0^\pi = 4\pi - \frac{16}{3}$

$\Rightarrow \bar{y} = \frac{M_x}{M} = \frac{\left(4\pi - \frac{16}{3}\right)}{4} = \pi - \frac{4}{3}$. Next $M_y = \int \tilde{x}\, dm = \int_0^\pi (\cos t)\left(2\cos\frac{t}{2}\right)dt = 2\int_0^\pi \cos t\cos\left(\frac{t}{2}\right)dt$

$= 2\left[\sin\left(\frac{t}{2}\right) + \frac{\sin\left(\frac{3}{2}t\right)}{3}\right]_0^\pi = 2 - \frac{2}{3} = \frac{4}{3} \Rightarrow \bar{x} = \frac{M_y}{M} = \frac{\left(\frac{4}{3}\right)}{4} = \frac{1}{3}$. Therefore $(\bar{x}, \bar{y}) = \left(\frac{1}{3}, \pi - \frac{4}{3}\right)$.

41. (a) $\frac{dx}{dt} = -2\sin 2t$ and $\frac{dy}{dt} = 2\cos 2t \Rightarrow \sqrt{\left(\frac{dx}{dy}\right)^2 + \left(\frac{dy}{dt}\right)^2} = \sqrt{(-2\sin 2t)^2 + (2\cos 2t)^2} = 2$

$\Rightarrow$ Length $= \int_0^{\pi/2} 2\, dt = [2t]_0^{\pi/2} = \pi$

(b) $\frac{dx}{dt} = \pi\cos \pi t$ and $\frac{dy}{dt} = -\pi\sin \pi t \Rightarrow \sqrt{\left(\frac{dx}{dy}\right)^2 + \left(\frac{dy}{dt}\right)^2} = \sqrt{(\pi\cos \pi t)^2 + (-\pi\sin \pi t)^2} = \pi$

$\Rightarrow$ Length $= \int_{-1/2}^{1/2} \pi\, dt = [\pi t]_{-1/2}^{1/2} = \pi$

43. $x = (1 + 2\sin\theta)\cos\theta,\ y = (1 + 2\sin\theta)\sin\theta \Rightarrow \frac{dx}{d\theta} = 2\cos^2\theta - \sin\theta(1 + 2\sin\theta)$,

$\frac{dy}{d\theta} = 2\cos\theta\sin\theta + \cos\theta(1 + 2\sin\theta) \Rightarrow \frac{dy}{dx} = \frac{2\cos\theta\sin\theta + \cos\theta(1 + 2\sin\theta)}{2\cos^2\theta - \sin\theta(1 + 2\sin\theta)} = \frac{4\cos\theta\sin\theta + \cos\theta}{2\cos^2\theta - 2\sin^2\theta - \sin\theta} = \frac{2\sin 2\theta + \cos\theta}{2\cos 2\theta - \sin\theta}$

(a) $x = (1 + 2\sin(0))\cos(0) = 1,\ y = (1 + 2\sin(0))\sin(0) = 0;\ \left.\frac{dy}{dx}\right|_{\theta=0} = \frac{2\sin(2(0)) + \cos(0)}{2\cos(2(0)) - \sin(0)} = \frac{0+1}{2-0} = \frac{1}{2}$

(b) $x = \left(1 + 2\sin\left(\frac{\pi}{2}\right)\right)\cos\left(\frac{\pi}{2}\right) = 0,\ y = \left(1 + 2\sin\left(\frac{\pi}{2}\right)\right)\sin\left(\frac{\pi}{2}\right) = 3;\ \left.\frac{dy}{dx}\right|_{\theta=\pi/2} = \frac{2\sin\left(2\left(\frac{\pi}{2}\right)\right) + \cos\left(\frac{\pi}{2}\right)}{2\cos\left(2\left(\frac{\pi}{2}\right)\right) - \sin\left(\frac{\pi}{2}\right)} = \frac{0+0}{-2-1} = 0$

(c) $x = \left(1 + 2\sin\left(\frac{4\pi}{3}\right)\right)\cos\left(\frac{4\pi}{3}\right) = \frac{\sqrt{3}-1}{2},\ y = \left(1 + 2\sin\left(\frac{4\pi}{3}\right)\right)\sin\left(\frac{4\pi}{3}\right) = \frac{3-\sqrt{3}}{2};$

$\left.\frac{dy}{dx}\right|_{\theta=4\pi/3} = \frac{2\sin\left(2\left(\frac{4\pi}{3}\right)\right) + \cos\left(\frac{4\pi}{3}\right)}{2\cos\left(2\left(\frac{4\pi}{3}\right)\right) - \sin\left(\frac{4\pi}{3}\right)} = \frac{\sqrt{3} - \frac{1}{2}}{-1 + \frac{\sqrt{3}}{2}} = \frac{2\sqrt{3}-1}{\sqrt{3}-2} = -\left(4 + 3\sqrt{3}\right)$

45. $\frac{dx}{dt} = \cos t$ and $\frac{dy}{dt} = 2\cos 2t \Rightarrow \frac{dy}{dx} = \frac{dy/dt}{dx/dt} = \frac{2\cos 2t}{\cos t} = \frac{2(2\cos^2 t - 1)}{\cos t};$ then $\frac{dy}{dx} = 0 \Rightarrow \frac{2(2\cos^2 t - 1)}{\cos t} = 0$

$\Rightarrow 2\cos^2 t - 1 = 0 \Rightarrow \cos t = \pm\frac{1}{\sqrt{2}} \Rightarrow t = \frac{\pi}{4}, \frac{3\pi}{4}, \frac{5\pi}{4}, \frac{7\pi}{4}$. In the 1st quadrant: $t = \frac{\pi}{4} \Rightarrow x = \sin\frac{\pi}{4} = \frac{\sqrt{2}}{2}$ and

$y = \sin 2\left(\frac{\pi}{4}\right) = 1 \Rightarrow \left(\frac{\sqrt{2}}{2}, 1\right)$ is the point where the tangent line is horizontal. At the origin: $x = 0$ and $y = 0$

$\Rightarrow \sin t = 0 \Rightarrow t = 0$ or $t = \pi$ and $\sin 2t = 0 \Rightarrow t = 0, \frac{\pi}{2}, \pi, \frac{3\pi}{2}$; thus $t = 0$ and $t = \pi$ give the tangent lines at the

origin. Tangents at origin: $\left.\frac{dy}{dx}\right|_{t=0} = 2 \Rightarrow y = 2x$ and $\left.\frac{dy}{dx}\right|_{t=\pi} = -2 \Rightarrow y = -2x$

47. (a) $x = a(t - \sin t), y = a(1 - \cos t), 0 \le t \le 2\pi \Rightarrow \frac{dx}{dt} = a(1 - \cos t), \ \frac{dy}{dt} = a \sin t$

$\Rightarrow$ Length $= \int_0^{2\pi} \sqrt{(a(1 - \cos t))^2 + (a \sin t)^2} \ dt = \int_0^{2\pi} \sqrt{a^2 - 2a^2 \cos t + a^2 \cos^2 t + a^2 \sin^2 t} \ dt$

$= a\sqrt{2} \int_0^{2\pi} \sqrt{1 - \cos t} \ dt = a\sqrt{2} \int_0^{2\pi} \sqrt{2 \sin^2 \left(\frac{t}{2}\right)} \ dt = 2a \int_0^{2\pi} \sin\left(\frac{t}{2}\right) dt = \left[-4a \cos\left(\frac{t}{2}\right)\right]_0^{2\pi}$

$= -4a \cos \pi + 4a \cos(0) = 8a$

(b) $a = 1 \Rightarrow x = t - \sin t, y = 1 - \cos t, 0 \le t \le 2\pi \Rightarrow \frac{dx}{dt} = 1 - \cos t, \frac{dy}{dt} = \sin t$

$\Rightarrow$ Surface area $= \int_0^{2\pi} 2\pi (1 - \cos t) \sqrt{(1 - \cos t)^2 + (\sin t)^2} \ dt$

$= \int_0^{2\pi} 2\pi (1 - \cos t) \sqrt{1 - 2\cos t + \cos^2 t + \sin^2 t} \ dt = 2\pi \int_0^{2\pi} (1 - \cos t) \sqrt{2 - 2\cos t} \ dt$

$= 2\sqrt{2}\pi \int_0^{2\pi} (1 - \cos t)^{3/2} dt = 2\sqrt{2}\pi \int_0^{2\pi} \left(1 - \cos\left(2 \cdot \frac{t}{2}\right)\right)^{3/2} dt = 2\sqrt{2}\pi \int_0^{2\pi} \left(2 \sin^2\left(\frac{t}{2}\right)\right)^{3/2} dt$

$= 8\pi \int_0^{2\pi} \sin^3 \left(\frac{t}{2}\right) dt \quad \left[u = \frac{t}{2} \Rightarrow du = \frac{1}{2} dt \Rightarrow dt = 2 \ du; t = 0 \Rightarrow u = 0, t = 2\pi \Rightarrow u = \pi \right]$

$= 16\pi \int_0^{\pi} \sin^3 u \ du = 16\pi \int_0^{\pi} \sin^2 u \sin u \ du = 16\pi \int_0^{\pi} \left(1 - \cos^2 u\right) \sin u \ du$

$= 16\pi \int_0^{\pi} \sin u \ du - 16\pi \int_0^{\pi} \cos^2 u \sin u \ du = \left[-16\pi \cos u + \frac{16\pi}{3} \cos^3 u\right]_0^{\pi}$

$= \left(16\pi - \frac{16\pi}{3}\right) - \left(-16\pi + \frac{16\pi}{3}\right) = \frac{64\pi}{3}$

49-51. Example CAS commands:

Maple:

```
with( plots );

with( student );

x := t -> t^3/3;

y := t -> t^2/2;

a := 0;

b := 1;

N := [2, 4, 8 ];

for n in N do

tt := [seq( a+i*(b-a)/n, i=0..n )];

pts := [seq([x(t),y(t)],t=tt)];

L := simplify(add( student[distance](pts[i+1],pts[i], i=1..n ));     # (b)

T := sprintf("#49(a) (Section 11.2)\nn=%3d L=%8.5f\n", n, L );

P[n] := plot( [[x(t),y(t),t=a..b],pts], title=T ):     # (a)

end do:

display( [seq(P[n],n=N)], insequence=true );

ds := t ->sqrt( simplify(D(x)(t)^2 + D(y)(t)^2) );     # (c)

L := Int( ds(t), t=a..b ):

L = evalf(L);
```

10.3 POLAR COORDINATES

1. $a, e; b, g; c, h; d, f$

3. (a) $\left(2, \frac{\pi}{2}+2n\pi\right)$ and $\left(-2, \frac{\pi}{2}+(2n+1)\pi\right)$, n an integer

    (b) $(2, 2n\pi)$ and $(-2, (2n+1)\pi)$, n an integer

    (c) $\left(2, \frac{3\pi}{2}+2n\pi\right)$ and $\left(-2, \frac{3\pi}{2}+(2n+1)\pi\right)$, n an integer

    (d) $(2, (2n+1)\pi)$ and $(-2, 2n\pi)$, n an integer

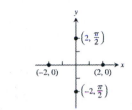

5. (a) $x = r\cos\theta = 3\cos 0 = 3$, $y = r\sin\theta = 3\sin 0 = 0 \Rightarrow$ Cartesian coordinates are $(3, 0)$

    (b) $x = r\cos\theta = -3\cos 0 = -3$, $y = r\sin\theta = -3\sin 0 = 0 \Rightarrow$ Cartesian coordinates are $(-3, 0)$

    (c) $x = r\cos\theta = 2\cos\frac{2\pi}{3} = -1$, $y = r\sin\theta = 2\sin\frac{2\pi}{3} = \sqrt{3} \Rightarrow$ Cartesian coordinates are $\left(-1, \sqrt{3}\right)$

    (d) $x = r\cos\theta = 2\cos\frac{7\pi}{3} = 1$, $y = r\sin\theta = 2\sin\frac{7\pi}{3} = \sqrt{3} \Rightarrow$ Cartesian coordinates are $\left(1, \sqrt{3}\right)$

    (e) $x = r\cos\theta = -3\cos\pi = 3$, $y = r\sin\theta = -3\sin\pi = 0 \Rightarrow$ Cartesian coordinates are $(3, 0)$

    (f) $x = r\cos\theta = 2\cos\frac{\pi}{3} = 1$, $y = r\sin\theta = 2\sin\frac{\pi}{3} = \sqrt{3} \Rightarrow$ Cartesian coordinates are $\left(1, \sqrt{3}\right)$

    (g) $x = r\cos\theta = -3\cos 2\pi = -3$, $y = r\sin\theta = -3\sin 2\pi = 0 \Rightarrow$ Cartesian coordinates are $(-3, 0)$

    (h) $x = r\cos\theta = -2\cos\left(-\frac{\pi}{3}\right) = -1$, $y = r\sin\theta = -2\sin\left(-\frac{\pi}{3}\right) = \sqrt{3} \Rightarrow$ Cartesian coordinates are $\left(-1, \sqrt{3}\right)$

7. (a) $(1, 1) \Rightarrow r = \sqrt{1^2 + 1^2} = \sqrt{2}$, $\sin\theta = \frac{1}{\sqrt{2}}$ and $\cos\theta = \frac{1}{\sqrt{2}} \Rightarrow \theta = \frac{\pi}{4} \Rightarrow$ Polar coordinates are $\left(\sqrt{2}, \frac{\pi}{4}\right)$

    (b) $(-3, 0) \Rightarrow r = \sqrt{(-3)^2 + 0^2} = 3$, $\sin\theta = 0$ and $\cos\theta = -1 \Rightarrow \theta = \pi \Rightarrow$ Polar coordinates are $(3, \pi)$

    (c) $\left(\sqrt{3}, -1\right) \Rightarrow r = \sqrt{\left(\sqrt{3}\right)^2 + (-1)^2} = 2$, $\sin\theta = -\frac{1}{2}$ and $\cos\theta = \frac{\sqrt{3}}{2} \Rightarrow \theta = \frac{11\pi}{6} \Rightarrow$ Polar coordinates are $\left(2, \frac{11\pi}{6}\right)$

    (d) $(-3, 4) \Rightarrow r = \sqrt{(-3)^2 + 4^2} = 5$, $\sin\theta = \frac{4}{5}$ and $\cos\theta = -\frac{3}{5} \Rightarrow \theta = \pi - \arctan\left(\frac{4}{3}\right) \Rightarrow$ Polar coordinates are $\left(5, \pi - \arctan\left(\frac{4}{3}\right)\right)$

9. (a) $(3, 3) \Rightarrow r = -\sqrt{3^2 + 3^2} = -3\sqrt{2}$, $\sin\theta = -\frac{1}{\sqrt{2}}$ and $\cos\theta = -\frac{1}{\sqrt{2}} \Rightarrow \theta = \frac{5\pi}{4} \Rightarrow$ Polar coordinates are $\left(-3\sqrt{2}, \frac{5\pi}{4}\right)$

    (b) $(-1, 0) \Rightarrow r = -\sqrt{(-1)^2 + 0^2} = -1$, $\sin\theta = 0$ and $\cos\theta = 1 \Rightarrow \theta = 0 \Rightarrow$ Polar coordinates are $(-1, 0)$

    (c) $\left(-1, \sqrt{3}\right) \Rightarrow r = -\sqrt{(-1)^2 + \left(\sqrt{3}\right)^2} = -2$, $\sin\theta = -\frac{\sqrt{3}}{2}$ and $\cos\theta = \frac{1}{2} \Rightarrow \theta = \frac{5\pi}{3} \Rightarrow$ Polar coordinates are $\left(-2, \frac{5\pi}{3}\right)$

    (d) $(4, -3) \Rightarrow r = -\sqrt{4^2 + (-3)^2} = -5$, $\sin\theta = \frac{3}{5}$ and $\cos\theta = -\frac{4}{5} \Rightarrow \theta = \pi - \arctan\left(\frac{3}{4}\right) \Rightarrow$ Polar coordinates are $\left(-5, \pi - \arctan\left(\frac{4}{3}\right)\right)$

11.

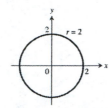

13.

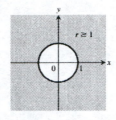

15.

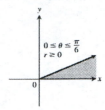

17.

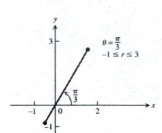

19.

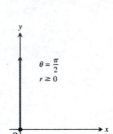

21.

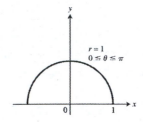

23.

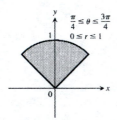

25.

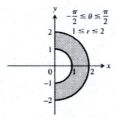

27. $r \cos \theta = 2 \Rightarrow x = 2$, vertical line through $(2, 0)$

29. $r \sin \theta = 0 \Rightarrow y = 0$, the x-axis

31. $r = 4 \csc \theta \Rightarrow r = \frac{4}{\sin \theta} \Rightarrow r \sin \theta = 4 \Rightarrow y = 4$, a horizontal line through $(0, 4)$

33. $r \cos \theta + r \sin \theta = 1 \Rightarrow x + y = 1$, line with slope $m = -1$ and intercept $b = 1$

35. $r^2 = 1 \Rightarrow x^2 + y^2 = 1$, circle with center $C = (0, 0)$ and radius 1

37. $r = \frac{5}{\sin \theta - 2\cos \theta} \Rightarrow r \sin \theta - 2r \cos \theta = 5 \Rightarrow y - 2x = 5$, line with slope $m = 2$ and intercept $b = 5$

39. $r = \cot \theta \csc \theta = \left(\frac{\cos \theta}{\sin \theta}\right)\left(\frac{1}{\sin \theta}\right) \Rightarrow r \sin^2 \theta = \cos \theta \Rightarrow r^2 \sin^2 \theta = r \cos \theta \Rightarrow y^2 = x$, parabola with vertex $(0, 0)$
   which opens to the right

41. $r = (\csc \theta)e^{r\cos\theta} \Rightarrow r\sin\theta = e^{r\cos\theta} \Rightarrow y = e^x$, graph of the natural exponential function

43. $r^2 + 2r^2 \cos\theta \sin\theta = 1 \Rightarrow x^2 + y^2 + 2xy = 1 \Rightarrow x^2 + 2xy + y^2 = 1 \Rightarrow (x+y)^2 = 1 \Rightarrow x+y = \pm 1$, two parallel straight lines of slope -1 and y-intercepts $b = \pm 1$

45. $r^2 = -4r\cos\theta \Rightarrow x^2 + y^2 = -4x \Rightarrow x^2 + 4x + y^2 = 0 \Rightarrow x^2 + 4x + 4 + y^2 = 4 \Rightarrow (x+2)^2 + y^2 = 4$, a circle with center $C(-2, 0)$ and radius 2

47. $r = 8\sin\theta \Rightarrow r^2 = 8r\sin\theta \Rightarrow x^2 + y^2 = 8y \Rightarrow x^2 + y^2 - 8y = 0 \Rightarrow x^2 + y^2 - 8y + 16 = 16 \Rightarrow x^2 + (y-4)^2 = 16$, a circle with center $C(0, 4)$ and radius 4

49. $r = 2\cos\theta + 2\sin\theta \Rightarrow r^2 = 2r\cos\theta + 2r\sin\theta \Rightarrow x^2 + y^2 = 2x + 2y \Rightarrow x^2 - 2x + y^2 - 2y = 0$
     $\Rightarrow (x-1)^2 + (y-1)^2 = 2$, a circle with center $C(1, 1)$ and radius $\sqrt{2}$

51. $r\sin\left(\theta + \frac{\pi}{6}\right) = 2 \Rightarrow r\left(\sin\theta\cos\frac{\pi}{6} + \cos\theta\sin\frac{\pi}{6}\right) = 2 \Rightarrow \frac{\sqrt{3}}{2}r\sin\theta + \frac{1}{2}r\cos\theta = 2 \Rightarrow \frac{\sqrt{3}}{2}y + \frac{1}{2}x = 2$
     $\Rightarrow \sqrt{3}\,y + x = 4$, line with slope $m = -\frac{1}{\sqrt{3}}$ and intercept $b = \frac{4}{\sqrt{3}}$

53. $x = 7 \Rightarrow r\cos\theta = 7$ 55. $x = y \Rightarrow r\cos\theta = r\sin\theta \Rightarrow \theta = \frac{\pi}{4}$

57. $x^2 + y^2 = 4 \Rightarrow r^2 = 4 \Rightarrow r = 2$ or $r = -2$

59. $\frac{x^2}{9} + \frac{y^2}{4} = 1 \Rightarrow 4x^2 + 9y^2 = 36 \Rightarrow 4r^2\cos^2\theta + 9r^2\sin^2\theta = 36$

61. $y^2 = 4x \Rightarrow r^2\sin^2\theta = 4r\cos\theta \Rightarrow r\sin^2\theta = 4\cos\theta$

63. $x^2 + (y-2)^2 = 4 \Rightarrow x^2 + y^2 - 4y + 4 = 4 \Rightarrow x^2 + y^2 = 4y \Rightarrow r^2 = 4r\sin\theta \Rightarrow r = 4\sin\theta$

65. $(x-3)^2 + (y+1)^2 = 4 \Rightarrow x^2 - 6x + 9 + y^2 + 2y + 1 = 4 \Rightarrow x^2 + y^2 = 6x - 2y - 6 \Rightarrow r^2 = 6r\cos\theta - 2r\sin\theta - 6$

67. $(0, \theta)$ where θ is any angle

10.4 GRAPHING POLAR COORDINATE EQUATIONS

1. $1 + \cos(-\theta) = 1 + \cos\theta = r \Rightarrow$ symmetric about the x-axis;
    $1 + \cos(-\theta) \neq -r$ and $1 + \cos(\pi - \theta) = 1 - \cos\theta \neq r \Rightarrow$ not symmetric
    about the y-axis; therefore not symmetric about the origin

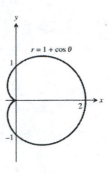

3. $1 - \sin(-\theta) = 1 + \sin\theta \neq r$ and $1 - \sin(\pi - \theta) = 1 - \sin\theta \neq -r$
    $\Rightarrow$ not symmetric about the x-axis; $1 - \sin(\pi - \theta) = 1 - \sin\theta = r$
    $\Rightarrow$ symmetric about the y-axis; therefore not symmetric
    about the origin

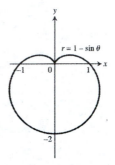

5. $2 + \sin(-\theta) = 2 - \sin\theta \neq r$ and $2 + \sin(\pi - \theta) = 2 + \sin\theta \neq -r \Rightarrow$ not
    symmetric about the x-axis; $2 + \sin(\pi - \theta) = 2 + \sin\theta = r \Rightarrow$ symmetric
    about the y-axis; therefore not symmetric about the origin

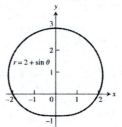

7. $\sin\left(-\frac{\theta}{2}\right) = -\sin\left(\frac{\theta}{2}\right) = -r \Rightarrow$ symmetric about the y-axis;
    $\sin\left(\frac{2\pi - \theta}{2}\right) = \sin\left(\frac{\theta}{2}\right)$, so the graph is symmetric about the x-axis, and
    hence the origin.

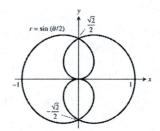

9. $\cos(-\theta) = \cos\theta = r^2 \Rightarrow (r, -\theta)$ and $(-r, -\theta)$ are on the graph when
    (r, θ) is on the graph $\Rightarrow$ symmetric about the x-axis and y-axis;
    therefore symmetric about the origin

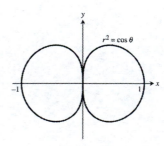

11. $-\sin(\pi-\theta)=-\sin\theta=r^2 \Rightarrow (r,\pi-\theta)$ and $(-r,\pi-\theta)$ are on the graph
     when (r,θ) is on the graph $\Rightarrow$ symmetric about the y-axis and the
     x-axis; therefore symmetric about the origin

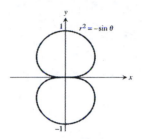

13. Since $(\pm r,-\theta)$ are on the graph when (r,θ) is on the graph,
     $(\pm r)^2 = 4\cos 2(-\theta) \Rightarrow r^2 = 4\cos 2\theta$, the graph is symmetric about
     the x-axis and the y-axis $\Rightarrow$ the graph is symmetric about the origin

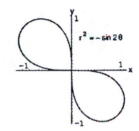

15. Since (r,θ) on the graph $\Rightarrow (-r,\theta)$ is on the graph,
     $(\pm r)^2 = -\sin 2\theta \Rightarrow r^2 = -\sin 2\theta$, the graph is symmetric
     about the origin. But $-\sin 2(-\theta) = -(-\sin 2\theta) = \sin 2\theta \neq r^2$ and
     $-\sin 2(\pi-\theta) = -\sin(2\pi-2\theta) = -\sin(-2\theta) = -(-\sin 2\theta) = \sin 2\theta \neq r^2$
     $\Rightarrow$ the graph is not symmetric about the x-axis; therefore the graph is
     not symmetric about the y-axis

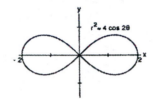

17. $\theta = \frac{\pi}{2} \Rightarrow r = -1 \Rightarrow \left(-1,\frac{\pi}{2}\right)$, and $\theta = -\frac{\pi}{2} \Rightarrow r = -1 \Rightarrow \left(-1,-\frac{\pi}{2}\right)$;

     $r' = \frac{dr}{d\theta} = -\sin\theta$; Slope $= \frac{r'\sin\theta + r\cos\theta}{r'\cos\theta - r\sin\theta} = \frac{-\sin^2\theta + r\cos\theta}{-\sin\theta\cos\theta - r\sin\theta}$

     $\Rightarrow$ Slope at $\left(-1,\frac{\pi}{2}\right)$ is $\dfrac{-\sin^2\left(\frac{\pi}{2}\right) + (-1)\cos\frac{\pi}{2}}{-\sin\frac{\pi}{2}\cos\frac{\pi}{2} - (-1)\sin\frac{\pi}{2}} = -1;$

     Slope at $\left(-1,-\frac{\pi}{2}\right)$ is $\dfrac{-\sin^2\left(-\frac{\pi}{2}\right) + (-1)\cos\left(-\frac{\pi}{2}\right)}{-\sin\left(-\frac{\pi}{2}\right)\cos\left(-\frac{\pi}{2}\right) - (-1)\sin\left(-\frac{\pi}{2}\right)} = 1$

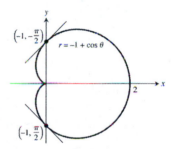

19. $\theta = \frac{\pi}{4} \Rightarrow r = 1 \Rightarrow \left(1,\frac{\pi}{4}\right)$; $\theta = -\frac{\pi}{4} \Rightarrow r = -1 \Rightarrow \left(-1,-\frac{\pi}{4}\right)$;

     $\theta = \frac{3\pi}{4} \Rightarrow r = -1 \Rightarrow \left(-1,\frac{3\pi}{4}\right)$; $\theta = -\frac{3\pi}{4} \Rightarrow r = 1 \Rightarrow \left(1,-\frac{3\pi}{4}\right)$;

     $r' = \frac{dr}{d\theta} = 2\cos 2\theta$; Slope $= \frac{r'\sin\theta + r\cos\theta}{r'\cos\theta - r\sin\theta} = \frac{2\cos 2\theta\sin\theta + r\cos\theta}{2\cos 2\theta\cos\theta - r\sin\theta}$

     $\Rightarrow$ Slope at $\left(1,\frac{\pi}{4}\right)$ is $\dfrac{2\cos\left(\frac{\pi}{2}\right)\sin\left(\frac{\pi}{4}\right) + (1)\cos\left(\frac{\pi}{4}\right)}{2\cos\left(\frac{\pi}{2}\right)\cos\left(\frac{\pi}{4}\right) - (1)\sin\left(\frac{\pi}{4}\right)} = -1;$

     Slope at $\left(-1,-\frac{\pi}{4}\right)$ is $\dfrac{2\cos\left(-\frac{\pi}{2}\right)\sin\left(-\frac{\pi}{4}\right) + (-1)\cos\left(-\frac{\pi}{4}\right)}{2\cos\left(-\frac{\pi}{2}\right)\cos\left(-\frac{\pi}{4}\right) - (-1)\sin\left(-\frac{\pi}{4}\right)} = 1;$

     Slope at $\left(-1,\frac{3\pi}{4}\right)$ is $\dfrac{2\cos\left(\frac{3\pi}{2}\right)\sin\left(\frac{3\pi}{4}\right) + (-1)\cos\left(\frac{3\pi}{4}\right)}{2\cos\left(\frac{3\pi}{2}\right)\cos\left(\frac{3\pi}{4}\right) - (-1)\sin\left(\frac{3\pi}{4}\right)} = 1;$

     Slope at $\left(1,-\frac{3\pi}{4}\right)$ is $\dfrac{2\cos\left(-\frac{3\pi}{2}\right)\sin\left(-\frac{3\pi}{4}\right) + (1)\cos\left(-\frac{3\pi}{4}\right)}{2\cos\left(-\frac{3\pi}{2}\right)\cos\left(-\frac{3\pi}{4}\right) - (1)\sin\left(-\frac{3\pi}{4}\right)} = -1;$

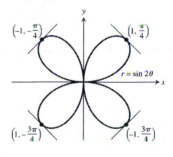

21. (a)

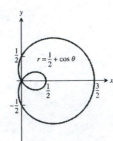

(b)

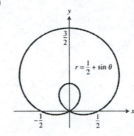

23. (a)

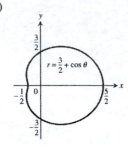

(b)

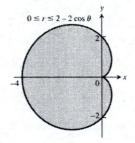

25.

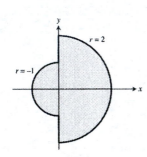

27.

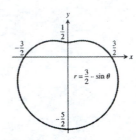

29. Note that (r, θ) and $(-r, \theta + \pi)$ describe the same point in the plane. Then $r = 1 - \cos\theta \Leftrightarrow -1 - \cos(\theta + \pi)$
$= -1 - (\cos\theta\cos\pi - \sin\theta\sin\pi) = -1 + \cos\theta = -(1 - \cos\theta) = -r$; therefore (r, θ) is on the graph of
$r = 1 - \cos\theta \Leftrightarrow (-r, \theta + \pi)$ is on the graph of $r = -1 - \cos\theta \Rightarrow$ the answer is (a).

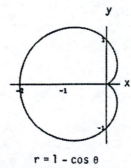

r = 1 – cos θ

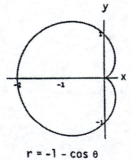

r = –1 – cos θ

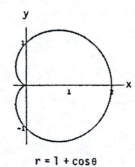

r = 1 + cos θ

31.

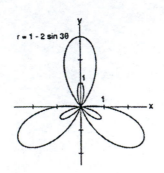

33. (a)

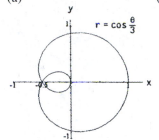

(b)

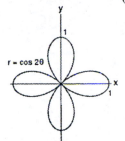

(c)

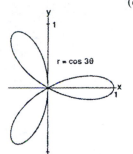

(d)

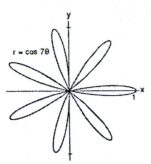

35.

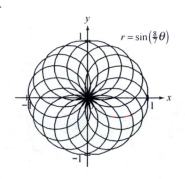

10.5 AREAS AND LENGTHS IN POLAR COORDINATES

1. $A = \int_0^\pi \frac{1}{2}\theta^2 \, d\theta = \left[\frac{1}{6}\theta^3\right]_0^\pi = \frac{\pi^3}{6}$

3. $A = \int_0^{2\pi} \frac{1}{2}(4+2\cos\theta)^2 \, d\theta = \int_0^{2\pi} \frac{1}{2}\left(16+16\cos\theta+4\cos^2\theta\right) d\theta = \int_0^{2\pi}\left[8+8\cos\theta+2\left(\frac{1+\cos 2\theta}{2}\right)\right]d\theta$

   $= \int_0^{2\pi}(9+8\cos\theta+\cos 2\theta)\,d\theta = \left[9\theta+8\sin\theta+\frac{1}{2}\sin 2\theta\right]_0^{2\pi} = 18\pi$

5. $A = 2\int_0^{\pi/4}\frac{1}{2}\cos^2 2\theta \, d\theta = \int_0^{\pi/4}\frac{1+\cos 4\theta}{2} \, d\theta = \frac{1}{2}\left[\theta+\frac{\sin 4\theta}{4}\right]_0^{\pi/4} = \frac{\pi}{8}$

7. $A = \int_0^{\pi/2}\frac{1}{2}(4\sin 2\theta)\,d\theta = \int_0^{\pi/2} 2\sin 2\theta \, d\theta = \left[-\cos 2\theta\right]_0^{\pi/2} = 2$

9. $r = 2\cos\theta$ and $r = 2\sin\theta \Rightarrow 2\cos\theta = 2\sin\theta$

   $\Rightarrow \cos\theta = \sin\theta \Rightarrow \theta = \frac{\pi}{4}$; therefore

   $A = 2\int_0^{\pi/4}\frac{1}{2}(2\sin\theta)^2 d\theta = \int_0^{\pi/4} 4\sin^2\theta \, d\theta$

   $= \int_0^{\pi/4} 4\left(\frac{1-\cos 2\theta}{2}\right) d\theta = \int_0^{\pi/4}(2-2\cos 2\theta)\,d\theta$

   $= \left[2\theta-\sin 2\theta\right]_0^{\pi/4} = \frac{\pi}{2}-1$

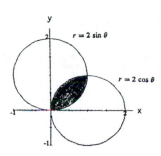

11. $r = 2$ and $r = 2(1 - \cos\theta) \Rightarrow 2 = 2(1 - \cos\theta) \Rightarrow \cos\theta = 0$

$\Rightarrow \theta = \pm\frac{\pi}{2}$; therefore

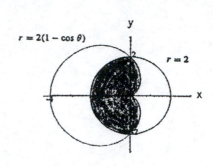

$A = 2\int_0^{\pi/2} \frac{1}{2}[2(1 - \cos\theta)]^2 \, d\theta + \frac{1}{2}$ area of the circle

$= \int_0^{\pi/2} 4\left(1 - 2\cos\theta + \cos^2\theta\right) d\theta + \left(\frac{1}{2}\pi\right)(2)^2$

$= \int_0^{\pi/2} 4\left(1 - 2\cos\theta + \frac{1 + \cos 2\theta}{2}\right) d\theta + 2\pi$

$= \int_0^{\pi/2} \left(4 - 8\cos\theta + 2 + 2\cos 2\theta\right) d\theta + 2\pi$

$= \left[6\theta - 8\sin\theta + \sin 2\theta\right]_0^{\pi/2} + 2\pi = 5\pi - 8$

13. $r = \sqrt{3}$ and $r^2 = 6\cos 2\theta \Rightarrow 3 = 6\cos 2\theta \Rightarrow \cos 2\theta = \frac{1}{2}$

$\Rightarrow \theta = \frac{\pi}{6}$ (in the 1st quadrant); we use symmetry of the graph

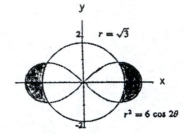

to find the area, so $A = 4\int_0^{\pi/6} \left[\frac{1}{2}(6\cos 2\theta) - \frac{1}{2}\left(\sqrt{3}\right)^2\right] d\theta$

$= 2\int_0^{\pi/6} (6\cos 2\theta - 3) \, d\theta = 2\left[3\sin 2\theta - 3\theta\right]_0^{\pi/6} = 3\sqrt{3} - \pi$

15. $r = 1$ and $r = -2\cos\theta \Rightarrow 1 = -2\cos\theta \Rightarrow \cos\theta = -\frac{1}{2} \Rightarrow \theta = \frac{2\pi}{3}$ in

quadrant II; therefore

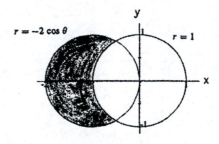

$A = 2\int_{2\pi/3}^{\pi} \frac{1}{2}\left[(-2\cos\theta)^2 - 1^2\right] d\theta = \int_{2\pi/3}^{\pi} \left(4\cos^2\theta - 1\right) d\theta$

$= \int_{2\pi/3}^{\pi} \left[2(1 + \cos 2\theta) - 1\right] d\theta = \int_{2\pi/3}^{\pi} (1 + 2\cos 2\theta) \, d\theta$

$= \left[\theta + \sin 2\theta\right]_{2\pi/3}^{\pi} = \frac{\pi}{3} + \frac{\sqrt{3}}{2}$

17. $r = \sec\theta$ and $r = 4\cos\theta \Rightarrow 4\cos\theta = \sec\theta \Rightarrow \cos^2\theta = \frac{1}{4}$

$\Rightarrow \theta = \frac{\pi}{3}, \frac{2\pi}{3}, \frac{4\pi}{3}$, or $\frac{5\pi}{3}$; therefore

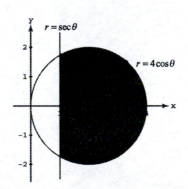

$A = 2\int_0^{\pi/3} \frac{1}{2}\left(16\cos^2\theta - \sec^2\theta\right) d\theta$

$= \int_0^{\pi/3} \left(8 + 8\cos 2\theta - \sec^2\theta\right) d\theta = \left[8\theta + 4\sin 2\theta - \tan\theta\right]_0^{\pi/3}$

$= \left(\frac{8\pi}{3} + 2\sqrt{3} - \sqrt{3}\right) - (0 + 0 - 0) = \frac{8\pi}{3} + \sqrt{3}$

19. (a) $r = \tan\theta$ and $r = \left(\frac{\sqrt{2}}{2}\right)\csc\theta \Rightarrow \tan\theta = \left(\frac{\sqrt{2}}{2}\right)\csc\theta$

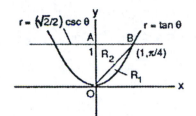

$\Rightarrow \sin^2\theta = \left(\frac{\sqrt{2}}{2}\right)\cos\theta \Rightarrow 1 - \cos^2\theta = \left(\frac{\sqrt{2}}{2}\right)\cos\theta$

$\Rightarrow \cos^2\theta + \left(\frac{\sqrt{2}}{2}\right)\cos\theta - 1 = 0 \Rightarrow \cos\theta = -\sqrt{2}$ or $\frac{\sqrt{2}}{2}$

(use the quadratic formula) $\Rightarrow \theta = \frac{\pi}{4}$ (the solution in the

first quadrant); therefore the area of R_1 is

$A_1 = \int_0^{\pi/4} \frac{1}{2}\tan^2\theta\, d\theta = \frac{1}{2}\int_0^{\pi/4}\left(\sec^2\theta - 1\right)d\theta = \frac{1}{2}\left[\tan\theta - \theta\right]_0^{\pi/4} = \frac{1}{2}\left(\tan\frac{\pi}{4} - \frac{\pi}{4}\right) = \frac{1}{2} - \frac{\pi}{8};$

$AO = \left(\frac{\sqrt{2}}{2}\right)\csc\frac{\pi}{2} = \frac{\sqrt{2}}{2}$ and $OB = \left(\frac{\sqrt{2}}{2}\right)\csc\frac{\pi}{4} = 1 \Rightarrow AB = \sqrt{1^2 - \left(\frac{\sqrt{2}}{2}\right)^2} = \frac{\sqrt{2}}{2} \Rightarrow$ the area of R_2 is

$A_2 = \frac{1}{2}\left(\frac{\sqrt{2}}{2}\right)\left(\frac{\sqrt{2}}{2}\right) = \frac{1}{4};$ therefore the area of the region shaded in the text is $2\left(\frac{1}{2} - \frac{\pi}{8} + \frac{1}{4}\right) = \frac{3}{2} - \frac{\pi}{4}.$ Note:

The area must be found this way since no common interval generates the region. For example, the interval

$0 \le \theta \le \frac{\pi}{4}$ generates the arc OB of $r = \tan\theta$ but does not generate the segment AB of the liner $= \frac{\sqrt{2}}{2}\csc\theta.$

Instead the interval generates the half-line from B to $+\infty$ on the line $r = \frac{\sqrt{2}}{2}\csc\theta.$

(b) $\lim\limits_{\theta\to\frac{\pi}{2}^-}\tan\theta = \infty$ and the line $x = 1$ is $r = \sec\theta$ in polar coordinates; then

$\lim\limits_{\theta\to\frac{\pi}{2}^-}\left(\tan\theta - \sec\theta\right) = \lim\limits_{\theta\to\frac{\pi}{2}^-}\left(\frac{\sin\theta}{\cos\theta} - \frac{1}{\cos\theta}\right) = \lim\limits_{\theta\to\frac{\pi}{2}^-}\left(\frac{\sin\theta - 1}{\cos\theta}\right) = \lim\limits_{\theta\to\frac{\pi}{2}^-}\left(\frac{\cos\theta}{-\sin\theta}\right) = 0 \Rightarrow r = \tan\theta$ approaches

$r = \sec\theta$ as $\theta \to \left(\frac{\pi}{2}\right)^- \Rightarrow r = \sec\theta$(or $x = 1$) is a vertical asymptote of $r = \tan\theta.$ Similarly,

$r = -\sec\theta$(or $x = -1$) is a vertical asymptote of $r = \tan\theta.$

21. $r = \theta^2, 0 \le \theta \le \sqrt{5} \Rightarrow \frac{dr}{d\theta} = 2\theta;$ therefore Length $= \int_0^{\sqrt{5}}\sqrt{\left(\theta^2\right)^2 + (2\theta)^2}\, d\theta = \int_0^{\sqrt{5}}\sqrt{\theta^4 + 4\theta^2}\, d\theta$

$= \int_0^{\sqrt{5}}|\theta|\sqrt{\theta^2 + 4}\, d\theta = \int_0^{\sqrt{5}}\theta\sqrt{\theta^2 + 4}\, d\theta;$ (since $\theta \ge 0$)

$\left[u = \theta^2 + 4 \Rightarrow \frac{1}{2}du = \theta\, d\theta; \theta = 0 \Rightarrow u = 4, \theta = \sqrt{5} \Rightarrow u = 9\right] \to \int_4^9 \frac{1}{2}\sqrt{u}\, du = \frac{1}{2}\left[\frac{2}{3}u^{3/2}\right]_4^9 = \frac{19}{3}$

23. $r = 1 + \cos\theta \Rightarrow \frac{dr}{d\theta} = -\sin\theta;$ therefore Length $= \int_0^{2\pi}\sqrt{(1 + \cos\theta)^2 + (-\sin\theta)^2}\, d\theta$

$= 2\int_0^{\pi}\sqrt{2 + 2\cos\theta}\, d\theta = 2\int_0^{\pi}\sqrt{\frac{4(1+\cos\theta)}{2}}\, d\theta = 4\int_0^{\pi}\sqrt{\frac{1+\cos\theta}{2}}\, d\theta = 4\int_0^{\pi}\cos\left(\frac{\theta}{2}\right)d\theta = 4\left[2\sin\frac{\theta}{2}\right]_0^{\pi} = 8$

25. $r = \frac{6}{1+\cos\theta}, 0 \le \theta \le \frac{\pi}{2} \Rightarrow \frac{dr}{d\theta} = \frac{6\sin\theta}{(1+\cos\theta)^2}$; therefore Length $= \int_0^{\pi/2} \sqrt{\left(\frac{6}{1+\cos\theta}\right)^2 + \left(\frac{6\sin\theta}{(1+\cos\theta)^2}\right)^2}\, d\theta$

$= \int_0^{\pi/2} \sqrt{\frac{36}{(1+\cos\theta)^2} + \frac{36\sin^2\theta}{(1+\cos\theta)^4}}\, d\theta = 6\int_0^{\pi/2} \left|\frac{1}{1+\cos\theta}\right| \sqrt{1 + \frac{\sin^2\theta}{(1+\cos\theta)^2}}\, d\theta$

$= 6\int_0^{\pi/2} \left(\frac{1}{1+\cos\theta}\right) \sqrt{\frac{1+2\cos\theta+\cos^2\theta+\sin^2\theta}{(1+\cos\theta)^2}}\, d\theta \quad \left(\text{since } \frac{1}{1+\cos\theta} > 0 \text{ on } 0 \le \theta \le \frac{\pi}{2}\right)$

$= 6\int_0^{\pi/2} \left(\frac{1}{1+\cos\theta}\right) \sqrt{\frac{2+2\cos\theta}{(1+\cos\theta)^2}}\, d\theta = 6\sqrt{2}\int_0^{\pi/2} \frac{d\theta}{(1+\cos\theta)^{3/2}} = 6\sqrt{2}\int_0^{\pi/2} \frac{d\theta}{\left(2\cos^2\frac{\theta}{2}\right)^{3/2}} = 3\int_0^{\pi/2} \left|\sec^3\frac{\theta}{2}\right| d\theta$

$= 3\int_0^{\pi/2} \sec^3\frac{\theta}{2}\, d\theta = 6\int_0^{\pi/4} \sec^3 u\, du = 6\left(\left[\frac{\sec u\tan u}{2}\right]_0^{\pi/4} + \frac{1}{2}\int_0^{\pi/4} \sec u\, du\right) \quad \text{(use tables)}$

$= 6\left(\frac{1}{\sqrt{2}} + \left[\frac{1}{2}\ln|\sec u + \tan u|\right]_0^{\pi/4}\right) = 3\left[\sqrt{2} + \ln\left(1+\sqrt{2}\right)\right]$

27. $r = \cos^3\frac{\theta}{3} \Rightarrow \frac{dr}{d\theta} = -\sin\frac{\theta}{3}\cos^2\frac{\theta}{3}$; therefore Length $= \int_0^{\pi/4} \sqrt{\left(\cos^3\frac{\theta}{3}\right)^2 + \left(-\sin\frac{\theta}{3}\cos^2\frac{\theta}{3}\right)^2}\, d\theta$

$= \int_0^{\pi/4} \sqrt{\cos^6\left(\frac{\theta}{3}\right) + \sin^2\left(\frac{\theta}{3}\right)\cos^4\left(\frac{\theta}{3}\right)}\, d\theta = \int_0^{\pi/4} \left(\cos^2\frac{\theta}{3}\right)\sqrt{\cos^2\left(\frac{\theta}{3}\right) + \sin^2\left(\frac{\theta}{3}\right)}\, d\theta = \int_0^{\pi/4} \cos^2\left(\frac{\theta}{3}\right) d\theta$

$= \int_0^{\pi/4} \frac{1+\cos\left(\frac{2\theta}{3}\right)}{2}\, d\theta = \frac{1}{2}\left[\theta + \frac{3}{2}\sin\frac{2\theta}{3}\right]_0^{\pi/4} = \frac{\pi}{8} + \frac{3}{8}$

29. Let $r = f(\theta)$. Then $x = f(\theta)\cos\theta \Rightarrow \frac{dx}{d\theta} = f'(\theta)\cos\theta - f(\theta)\sin\theta \Rightarrow \left(\frac{dx}{d\theta}\right)^2 = [f'(\theta)\cos\theta - f(\theta)\sin\theta]^2$

$= [f'(\theta)]^2\cos^2\theta - 2f'(\theta)f(\theta)\sin\theta\cos\theta + [f(\theta)]^2\sin^2\theta; \ y = f(\theta)\sin\theta \Rightarrow \frac{dy}{d\theta} = f'(\theta)\sin\theta + f(\theta)\cos\theta$

$\Rightarrow \left(\frac{dy}{d\theta}\right)^2 = [f'(\theta)\sin\theta + f(\theta)\cos\theta]^2 = [f'(\theta)]^2\sin^2\theta + 2f'(\theta)f(\theta)\sin\theta\cos\theta + [f(\theta)]^2\cos^2\theta.$ Therefore

$\left(\frac{dx}{d\theta}\right)^2 + \left(\frac{dy}{d\theta}\right)^2 = [f'(\theta)]^2\left(\cos^2\theta + \sin^2\theta\right) + [f(\theta)]^2\left(\cos^2\theta + \sin^2\theta\right) = [f'(\theta)]^2 + [f(\theta)]^2 = r^2 + \left(\frac{dr}{d\theta}\right)^2.$

Thus, $L = \int_\alpha^\beta \sqrt{\left(\frac{dx}{d\theta}\right)^2 + \left(\frac{dy}{d\theta}\right)^2}\, d\theta = \int_\alpha^\beta \sqrt{r^2 + \left(\frac{dr}{d\theta}\right)^2}\, d\theta.$

31. (a) $r_{av} = \frac{1}{2\pi - 0} \int_0^{2\pi} a(1 - \cos\theta)\, d\theta = \frac{a}{2\pi}[\theta - \sin\theta]_0^{2\pi} = a$

(b) $r_{av} = \frac{1}{2\pi - 0} \int_0^{2\pi} a\, d\theta = \frac{1}{2\pi}[a\theta]_0^{2\pi} = a$

(c) $r_{av} = \frac{1}{\left(\frac{\pi}{2}\right) - \left(-\frac{\pi}{2}\right)} \int_{-\pi/2}^{\pi/2} a\cos\theta\, d\theta = \frac{1}{\pi}[a\sin\theta]_{-\pi/2}^{\pi/2} = \frac{2a}{\pi}$

10.6 CONICS IN POLAR COORDINATES

1. $16x^2 + 25y^2 = 400 \Rightarrow \frac{x^2}{25} + \frac{y^2}{16} = 1 \Rightarrow c = \sqrt{a^2 - b^2}$

$= \sqrt{25 - 16} = 3 \Rightarrow e = \frac{c}{a} = \frac{3}{5};\ F(\pm 3, 0);$ directrices are

$x = 0 \pm \frac{a}{e} = \pm \frac{5}{\left(\frac{3}{5}\right)} = \pm \frac{25}{3}$

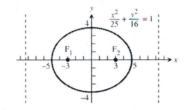

3. $2x^2 + y^2 = 2 \Rightarrow x^2 + \frac{y^2}{2} = 1 \Rightarrow c = \sqrt{a^2 - b^2}$

$= \sqrt{2 - 1} = 1 \Rightarrow e = \frac{c}{a} = \frac{1}{\sqrt{2}};\ F(0, \pm 1);$

directrices are $y = 0 \pm \frac{a}{e} = \pm \frac{\sqrt{2}}{\left(\frac{1}{\sqrt{2}}\right)} = \pm 2$

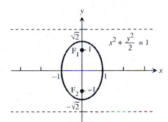

5. $3x^2 + 2y^2 = 6 \Rightarrow \frac{x^2}{2} + \frac{y^2}{3} = 1 \Rightarrow c = \sqrt{a^2 - b^2}$

$= \sqrt{3 - 2} = 1 \Rightarrow e = \frac{c}{a} = \frac{1}{\sqrt{3}};\ F(0, \pm 1);$

directrices are $y = 0 \pm \frac{a}{e} = \pm \frac{\sqrt{3}}{\left(\frac{1}{\sqrt{3}}\right)} = \pm 3$

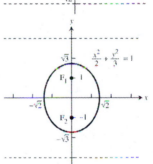

7. $6x^2 + 9y^2 = 54 \Rightarrow \frac{x^2}{9} + \frac{y^2}{6} = 1 \Rightarrow c = \sqrt{a^2 - b^2}$

$= \sqrt{9 - 6} = \sqrt{3} \Rightarrow e = \frac{c}{a} = \frac{\sqrt{3}}{3};\ F\left(\pm\sqrt{3}, 0\right);$

directrices are $x = 0 \pm \frac{a}{e} = \pm \frac{3}{\left(\frac{\sqrt{3}}{3}\right)} = \pm 3\sqrt{3}$

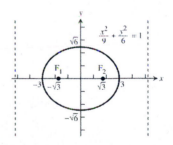

9. Foci: $(0, \pm 3),\ e = 0.5 \Rightarrow c = 3$ and $a = \frac{c}{e} = \frac{3}{0.5} = 6 \Rightarrow b^2 = 36 - 9 = 27 \Rightarrow \frac{x^2}{27} + \frac{y^2}{36} = 1$

11. Vertices: $(0, \pm 70),\ e = 0.1 \Rightarrow a = 70$ and $c = ae = 70(0.1) = 7 \Rightarrow b^2 = 4900 - 49 = 4851 \Rightarrow \frac{x^2}{4851} + \frac{y^2}{4900} = 1$

13. Focus: $\left(\sqrt{5}, 0\right)$, Directrix: $x = \frac{9}{\sqrt{5}} \Rightarrow c = ae = \sqrt{5}$ and $\frac{a}{e} = \frac{9}{\sqrt{5}} \Rightarrow \frac{ae}{e^2} = \frac{9}{\sqrt{5}} \Rightarrow \frac{\sqrt{5}}{e^2} = \frac{9}{\sqrt{5}} \Rightarrow e^2 = \frac{5}{9} \Rightarrow e = \frac{\sqrt{5}}{3}$.

Then $PF = \frac{\sqrt{5}}{3} PD \Rightarrow \sqrt{\left(x - \sqrt{5}\right)^2 + (y-0)^2} = \frac{\sqrt{5}}{3}\left|x - \frac{9}{\sqrt{5}}\right| \Rightarrow \left(x - \sqrt{5}\right)^2 + y^2 = \frac{5}{9}\left(x - \frac{9}{\sqrt{5}}\right)^2$

$\Rightarrow x^2 - 2\sqrt{5}x + 5 + y^2 = \frac{5}{9}\left(x^2 - \frac{18}{\sqrt{5}}x + \frac{81}{5}\right) \Rightarrow \frac{4}{9}x^2 + y^2 = 4 \Rightarrow \frac{x^2}{9} + \frac{y^2}{4} = 1$

15. Focus: $(-4, 0)$, Directrix: $x = -16 \Rightarrow c = ae = 4$ and $\frac{a}{e} = 16 \Rightarrow \frac{ae}{e^2} = 16 \Rightarrow \frac{4}{e^2} = 16 \Rightarrow e^2 = \frac{1}{4} \Rightarrow e = \frac{1}{2}$.

Then $PF = \frac{1}{2} PD \Rightarrow \sqrt{(x+4)^2 + (y-0)^2} = \frac{1}{2}|x+16| \Rightarrow (x+4)^2 + y^2 = \frac{1}{4}(x+16)^2$

$\Rightarrow x^2 + 8x + 16 + y^2 = \frac{1}{4}\left(x^2 + 32x + 256\right) \Rightarrow \frac{3}{4}x^2 + y^2 = 48 \Rightarrow \frac{x^2}{64} + \frac{y^2}{48} = 1$

17. $x^2 - y^2 = 1 \Rightarrow c = \sqrt{a^2 + b^2} = \sqrt{1+1} = \sqrt{2} \Rightarrow e = \frac{c}{a}$

$= \frac{\sqrt{2}}{1} = \sqrt{2}$; asymptotes are $y = \pm x$; $F\left(\pm\sqrt{2}, 0\right)$;

directrices are $x = 0 \pm \frac{a}{e} = \pm\frac{1}{\sqrt{2}}$

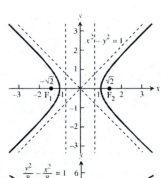

19. $y^2 - x^2 = 8 \Rightarrow \frac{y^2}{8} - \frac{x^2}{8} = 1 \Rightarrow c = \sqrt{a^2 + b^2} = \sqrt{8+8} = 4$

$\Rightarrow e = \frac{c}{a} = \frac{4}{\sqrt{8}} = \sqrt{2}$; asymptotes are $y = \pm x$; $F(0, \pm 4)$;

directrices are $y = 0 \pm \frac{a}{e} = \pm\frac{\sqrt{8}}{\sqrt{2}} = \pm 2$

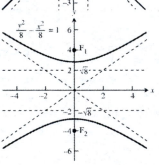

21. $8x^2 - 2y^2 = 16 \Rightarrow \frac{x^2}{2} - \frac{y^2}{8} = 1 \Rightarrow c = \sqrt{a^2 + b^2}$

$= \sqrt{2+8} = \sqrt{10} \Rightarrow e = \frac{c}{a} = \frac{\sqrt{10}}{\sqrt{2}} = \sqrt{5}$;

asymptotes are $y = \pm 2x$; $F\left(\pm\sqrt{10}, 0\right)$;

directrices are $x = 0 \pm \frac{a}{e} = \pm\frac{\sqrt{2}}{\sqrt{5}} = \pm\frac{2}{\sqrt{10}}$

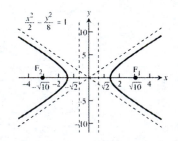

23. $8y^2 - 2x^2 = 16 \Rightarrow \frac{y^2}{2} - \frac{x^2}{8} = 1 \Rightarrow c = \sqrt{a^2 + b^2}$

$= \sqrt{2+8} = \sqrt{10} \Rightarrow e = \frac{c}{a} = \frac{\sqrt{10}}{\sqrt{2}} = \sqrt{5}$; asymptotes are $y = \pm\frac{x}{2}$;

$F\left(0, \pm\sqrt{10}\right)$; directrices are $y = 0 \pm \frac{a}{e} = \pm\frac{\sqrt{2}}{\sqrt{5}} = \pm\frac{2}{\sqrt{10}}$

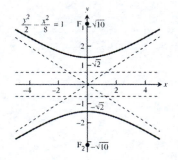

25. Vertices $(0, \pm 1)$ and $e = 3 \Rightarrow a = 1$ and $e = \frac{c}{a} = 3 \Rightarrow c = 3a = 3 \Rightarrow b^2 = c^2 - a^2 = 9 - 1 = 8 \Rightarrow y^2 - \frac{x^2}{8} = 1$

27. Foci $(\pm 3, 0)$ and $e = 3 \Rightarrow c = 3$ and $e = \frac{c}{a} = 3 \Rightarrow c = 3a \Rightarrow a = 1 \Rightarrow b^2 = c^2 - a^2 = 9 - 1 = 8 \Rightarrow x^2 - \frac{y^2}{8} = 1$

29. $e = 1, x = 2 \Rightarrow k = 2 \Rightarrow r = \frac{2(1)}{1 + (1)\cos\theta} = \frac{2}{1 + \cos\theta}$

31. $e = 5, y = -6 \Rightarrow k = 6 \Rightarrow r = \frac{6(5)}{1 - 5\sin\theta} = \frac{30}{1 - 5\sin\theta}$

33. $e = \frac{1}{2}, x = 1 \Rightarrow k = 1 \Rightarrow r = \frac{\left(\frac{1}{2}\right)(1)}{1 + \left(\frac{1}{2}\right)\cos\theta} = \frac{1}{2 + \cos\theta}$

35. $e = \frac{1}{5}, y = -10 \Rightarrow k = 10 \Rightarrow r = \frac{\left(\frac{1}{5}\right)(10)}{1 - \left(\frac{1}{5}\right)\sin\theta} = \frac{10}{5 - \sin\theta}$

37. $r = \frac{1}{1 + \cos\theta} \Rightarrow e = 1, k = 1 \Rightarrow x = 1$

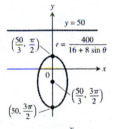

39. $r = \frac{25}{10 - 5\cos\theta} \Rightarrow r = \frac{\left(\frac{25}{10}\right)}{1 - \left(\frac{5}{10}\right)\cos\theta} = \frac{\left(\frac{5}{2}\right)}{1 - \left(\frac{1}{2}\right)\cos\theta}$

$\Rightarrow e = \frac{1}{2}, k = 5 \Rightarrow x = -5; \quad a\left(1 - e^2\right) = ke$

$\Rightarrow a\left[1 - \left(\frac{1}{2}\right)^2\right] = \frac{5}{2} \Rightarrow \frac{3}{4}a = \frac{5}{2} \Rightarrow a = \frac{10}{3} \Rightarrow ea = \frac{5}{3}$

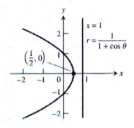

41. $r = \frac{400}{16 + 8\sin\theta} \Rightarrow r = \frac{\left(\frac{400}{16}\right)}{1 + \left(\frac{8}{16}\right)\sin\theta} \Rightarrow r = \frac{25}{1 + \left(\frac{1}{2}\right)\sin\theta}$

$e = \frac{1}{2}, k = 50 \Rightarrow y = 50; \quad a\left(1 - e^2\right) = ke$

$\Rightarrow a\left[1 - \left(\frac{1}{2}\right)^2\right] = 25 \Rightarrow \frac{3}{4}a = 25 \Rightarrow a = \frac{100}{3} \Rightarrow ea = \frac{50}{3}$

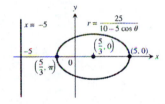

43. $r = \frac{8}{2 - 2\sin\theta} \Rightarrow r = \frac{4}{1 - \sin\theta} \Rightarrow e = 1, k = 4 \Rightarrow y = -4$

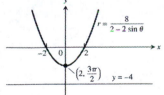

45. $r\cos\left(\theta - \frac{\pi}{4}\right) = \sqrt{2} \Rightarrow r\left(\cos\theta\cos\frac{\pi}{4} + \sin\theta\sin\frac{\pi}{4}\right) = \sqrt{2}$

$\Rightarrow \frac{1}{\sqrt{2}}r\cos\theta + \frac{1}{\sqrt{2}}r\sin\theta = \sqrt{2} \Rightarrow \frac{1}{\sqrt{2}}x + \frac{1}{\sqrt{2}}y = \sqrt{2}$

$\Rightarrow x + y = 2 \Rightarrow y = 2 - x$

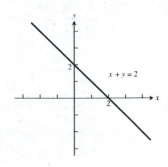

47. $r\cos\left(\theta - \frac{2\pi}{3}\right) = 3 \Rightarrow r\left(\cos\theta\cos\frac{2\pi}{3} + \sin\theta\sin\frac{2\pi}{3}\right) = 3$

$\Rightarrow -\frac{1}{2}r\cos\theta + \frac{\sqrt{3}}{2}r\sin\theta = 3 \Rightarrow -\frac{1}{2}x + \frac{\sqrt{3}}{2}y = 3$

$\Rightarrow -x + \sqrt{3}y = 6 \Rightarrow y = \frac{\sqrt{3}}{3}x + 2\sqrt{3}$

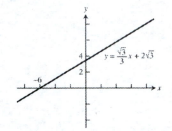

49. $\sqrt{2}x + \sqrt{2}y = 6 \Rightarrow \sqrt{2}\,r\cos\theta + \sqrt{2}\,r\sin\theta = 6 \Rightarrow r\left(\frac{\sqrt{2}}{2}\cos\theta + \frac{\sqrt{2}}{2}\sin\theta\right) = 3 \Rightarrow r\left(\cos\frac{\pi}{4}\cos\theta + \sin\frac{\pi}{4}\sin\theta\right) = 3$

$\Rightarrow r\cos\left(\theta - \frac{\pi}{4}\right) = 3$

51. $y = -5 \Rightarrow r\sin\theta = -5 \Rightarrow -r\sin\theta = 5 \Rightarrow r\sin(-\theta) = 5 \Rightarrow r\cos\left(\frac{\pi}{2} - (-\theta)\right) = 5 \Rightarrow r\cos\left(\theta + \frac{\pi}{2}\right) = 5$

53.

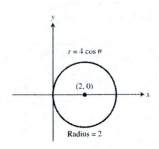

55.

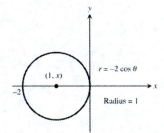

57. $(x - 6)^2 + y^2 = 36 \Rightarrow C = (6, 0), a = 6$

$\Rightarrow r = 12\cos\theta$ is the polar equation

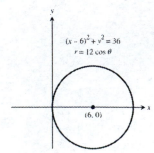

59. $x^2 + (y - 5)^2 = 25 \Rightarrow C = (0, 5), a = 5 \Rightarrow r = 10\sin\theta$

is the polar equation

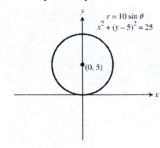

61. $x^2 + 2x + y^2 = 0 \Rightarrow (x+1)^2 + y^2 = 1$

$\Rightarrow C = (-1, 0), \quad a = 1 \Rightarrow r = -2\cos\theta$

is the polar equation

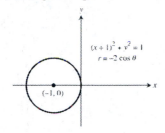

63. $x^2 + y^2 + y = 0 \Rightarrow x^2 + \left(y + \frac{1}{2}\right)^2 = \frac{1}{4}$

$\Rightarrow C = \left(0, -\frac{1}{2}\right), \quad a = \frac{1}{2} \Rightarrow r = -\sin\theta$

is the polar equation

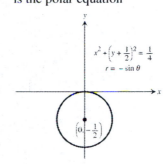

65.

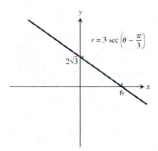

67.

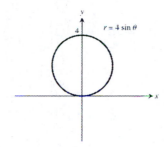

69.

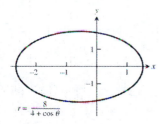

71.

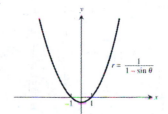

73.

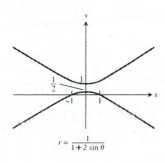

75. (a) Perihelion $= a - ae = a(1-e)$, Aphelion $= ea + a = a(1+e)$

(b)

Planet	Perihelion	Aphelion
Mercury	0.3075 AU	0.4667 AU
Venus	0.7184 AU	0.7282 AU
Earth	0.9833 AU	1.0167 AU
Mars	1.3817 AU	1.6663 AU
Jupiter	4.9512 AU	5.4548 AU
Saturn	9.0210 AU	10.0570 AU
Uranus	18.2977 AU	20.0623 AU
Neptune	29.8135 AU	30.3065 AU

CHAPTER 10 PRACTICE EXERCISES

1. $x = \frac{t}{2}$ and $y = t + 1 \Rightarrow 2x = t \Rightarrow y = 2x + 1$

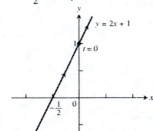

3. $x = \frac{1}{2}\tan t$ and $y = \frac{1}{2}\sec t \Rightarrow x^2 = \frac{1}{4}\tan^2 t$ and

$y^2 = \frac{1}{4}\sec^2 t \Rightarrow 4x^2 = \tan^2 t$ and $4y^2 = \sec^2 t$

$\Rightarrow 4x^2 + 1 = 4y^2 \Rightarrow 4y^2 - 4x^2 = 1$

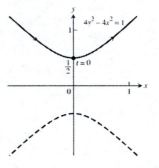

5. $x = -\cos t$ and $y = \cos^2 t \Rightarrow y = (-x)^2 = x^2$

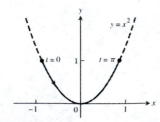

7. $16x^2 + 9y^2 = 144 \Rightarrow \frac{x^2}{9} + \frac{y^2}{16} = 1 \Rightarrow a = 3$ and $b = 4 \Rightarrow x = 3\cos t$ and $y = 4\sin t, 0 \le t \le 2\pi$

9. $x = \frac{1}{2}\tan t, y = \frac{1}{2}\sec t \Rightarrow \frac{dy}{dx} = \frac{dy/dt}{dx/dt} = \frac{\frac{1}{2}\sec t \tan t}{\frac{1}{2}\sec^2 t} = \frac{\tan t}{\sec t} = \sin t \Rightarrow \frac{dy}{dx}\Big|_{t=\pi/3} = \sin\frac{\pi}{3} = \frac{\sqrt{3}}{2}$;

$t = \frac{\pi}{3} \Rightarrow x = \frac{1}{2}\tan\frac{\pi}{3} = \frac{\sqrt{3}}{2}$ and $y = \frac{1}{2}\sec\frac{\pi}{3} = 1 \Rightarrow y = \frac{\sqrt{3}}{2}x + \frac{1}{4}; \frac{d^2y}{dx^2} = \frac{dy'/dt}{dx/dt} = \frac{\cos t}{\frac{1}{2}\sec^2 t} = 2\cos^3 t$

$\Rightarrow \frac{d^2y}{dx^2}\Big|_{t=\pi/3} = 2\cos^3\left(\frac{\pi}{3}\right) = \frac{1}{4}$

11. (a) $x = 4t^2,\ y = t^3 - 1 \Rightarrow t = \pm\frac{\sqrt{x}}{2} \Rightarrow y = \left(\pm\frac{\sqrt{x}}{2}\right)^3 - 1 = \pm\frac{x^{3/2}}{8} - 1$

(b) $x = \cos t,\ y = \tan t \Rightarrow \sec t = \frac{1}{x} \Rightarrow \tan^2 t + 1 = \sec^2 t \Rightarrow y^2 = \frac{1}{x^2} - 1 = \frac{1-x^2}{x^2} \Rightarrow y = \pm\frac{\sqrt{1-x^2}}{x}$

13. $y = x^{1/2} - \frac{x^{3/2}}{3} \Rightarrow \frac{dy}{dx} = \frac{1}{2}x^{-1/2} - \frac{1}{2}x^{1/2} \Rightarrow \left(\frac{dy}{dx}\right)^2 = \frac{1}{4}\left(\frac{1}{x} - 2 + x\right) \Rightarrow L = \int_1^4 \sqrt{1 + \frac{1}{4}\left(\frac{1}{x} - 2 + x\right)}\ dx$

$\Rightarrow L = \int_1^4 \sqrt{\frac{1}{4}\left(\frac{1}{x} + 2 + x\right)}\ dx = \int_1^4 \sqrt{\frac{1}{4}\left(x^{-1/2} + x^{1/2}\right)^2}\ dx = \int_1^4 \frac{1}{2}\left(x^{-1/2} + x^{1/2}\right)dx = \frac{1}{2}\left[2x^{1/2} + \frac{2}{3}x^{3/2}\right]_1^4$

$= \frac{1}{2}\left[\left(4 + \frac{2}{3}\cdot 8\right) - \left(2 + \frac{2}{3}\right)\right] = \frac{1}{2}\left(2 + \frac{14}{3}\right) = \frac{10}{3}$

15. $y = \frac{5}{12}x^{6/5} - \frac{5}{8}x^{4/5} \Rightarrow \frac{dy}{dx} = \frac{1}{2}x^{1/5} - \frac{1}{2}x^{-1/5} \Rightarrow \left(\frac{dy}{dx}\right)^2 = \frac{1}{4}\left(x^{2/5} - 2 + x^{-2/5}\right)$

$\Rightarrow L = \int_1^{32}\sqrt{1 + \frac{1}{4}\left(x^{2/5} - 2 + x^{-2/5}\right)}\ dx \Rightarrow L = \int_1^{32}\sqrt{\frac{1}{4}\left(x^{2/5} + 2 + x^{-2/5}\right)}\ dx = \int_1^{32}\sqrt{\frac{1}{4}\left(x^{1/5} + x^{-1/5}\right)^2}\ dx$

$= \int_1^{32}\frac{1}{2}\left(x^{1/5} + x^{-1/5}\right)dx = \frac{1}{2}\left[\frac{5}{6}x^{6/5} + \frac{5}{4}x^{4/5}\right]_1^{32} = \frac{1}{2}\left[\left(\frac{5}{6}\cdot 2^6 + \frac{5}{4}\cdot 2^4\right) - \left(\frac{5}{6} + \frac{5}{4}\right)\right] = \frac{1}{2}\left(\frac{315}{6} + \frac{75}{4}\right)$

$= \frac{1}{48}(1260 + 450) = \frac{1710}{48} = \frac{285}{8}$

17. $\frac{dx}{dt} = -5\sin t + 5\sin 5t$ and $\frac{dy}{dt} = 5\cos t - 5\cos 5t \Rightarrow \sqrt{\left(\frac{dx}{dt}\right)^2 + \left(\frac{dy}{dt}\right)^2}$

$= \sqrt{(-5\sin t + 5\sin 5t)^2 + (5\cos t - 5\cos 5t)^2} = 5\sqrt{\sin^2 5t - 2\sin t \sin 5t + \sin^2 t + \cos^2 t - 2\cos t \cos 5t + \cos^2 5t}$

$= 5\sqrt{2 - 2(\sin t \sin 5t + \cos t \cos 5t)} = 5\sqrt{2(1 - \cos 4t)} = 5\sqrt{4\left(\frac{1}{2}\right)(1 - \cos 4t)} = 10\sqrt{\sin^2 2t} = 10|\sin 2t|$

$= 10\sin 2t\ \left(\text{since } 0 \le t \le \frac{\pi}{2}\right) \Rightarrow \text{Length} = \int_0^{\pi/2}10\sin 2t\ dt = \left[-5\cos 2t\right]_0^{\pi/2} = (-5)(-1) - (-5)(1) = 10$

19. $\frac{dx}{d\theta} = -3\sin\theta$ and $\frac{dy}{d\theta} = 3\cos\theta \Rightarrow \sqrt{\left(\frac{dx}{d\theta}\right)^2 + \left(\frac{dy}{d\theta}\right)^2} = \sqrt{(-3\sin\theta)^2 + (3\cos\theta)^2} = \sqrt{3(\sin^2\theta + \cos^2\theta)} = 3$

$\Rightarrow \text{Length} = \int_0^{3\pi/2}3\ d\theta = 3\int_0^{3\pi/2}d\theta = 3\left(\frac{3\pi}{2} - 0\right) = \frac{9\pi}{2}$

21. $x = \frac{t^2}{2}$ and $y = 2t,\ 0 \le t \le \sqrt{5} \Rightarrow \frac{dx}{dt} = t$ and $\frac{dy}{dt} = 2 \Rightarrow \text{Surface Area} = \int_0^{\sqrt{5}}2\pi(2t)\sqrt{t^2 + 4}\ dt$

$\left[u = t^2 + 4 \Rightarrow du = 2t\ dt;\ t = 0 \Rightarrow u = 4, t = \sqrt{5} \Rightarrow u = 9\right] \to \int_4^9 2\pi u^{1/2}du = 2\pi\left[\frac{2}{3}u^{3/2}\right]_4^9 = \frac{76\pi}{3}$

23. $r\cos\left(\theta + \frac{\pi}{3}\right) = 2\sqrt{3} \Rightarrow r\left(\cos\theta\cos\frac{\pi}{3} - \sin\theta\sin\frac{\pi}{3}\right) = 2\sqrt{3}$

$\Rightarrow \frac{1}{2}r\cos\theta - \frac{\sqrt{3}}{2}r\sin\theta = 2\sqrt{3}$

$\Rightarrow r\cos\theta - \sqrt{3}\ r\sin\theta = 4\sqrt{3} \Rightarrow x - \sqrt{3}y = 4\sqrt{3}$

$\Rightarrow y = \frac{\sqrt{3}}{3}x - 4$

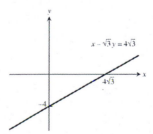

25. $r = 2\sec\theta \Rightarrow r = \frac{2}{\cos\theta} \Rightarrow r\cos\theta = 2 \Rightarrow x = 2$

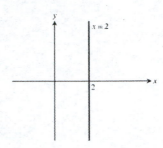

27. $r = -\frac{3}{2}\csc\theta \Rightarrow r\sin\theta = -\frac{3}{2} \Rightarrow y = -\frac{3}{2}$

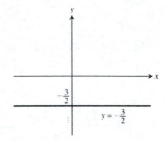

29. $r = -4\sin\theta \Rightarrow r^2 = -4r\sin\theta \Rightarrow x^2 + y^2 + 4y = 0$
$\Rightarrow x^2 + (y+2)^2 = 4;$
circle with center $(0, -2)$ and radius 2.

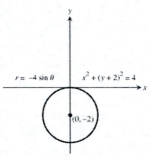

31. $r = 2\sqrt{2}\cos\theta \Rightarrow r^2 = 2\sqrt{2}\, r\cos\theta$
$\Rightarrow x^2 + y^2 - 2\sqrt{2}x = 0 \Rightarrow \left(x - \sqrt{2}\right)^2 + y^2 = 2;$
circle with center $\left(\sqrt{2}, 0\right)$ and radius $\sqrt{2}$

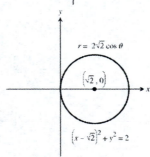

33. $x^2 + y^2 + 5y = 0 \Rightarrow x^2 + \left(y + \frac{5}{2}\right)^2 = \frac{25}{4}$
$\Rightarrow C = \left(0, -\frac{5}{2}\right)$ and $a = \frac{5}{2};$
$r^2 + 5r\sin\theta = 0 \Rightarrow r = -5\sin\theta$

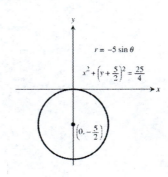

35. $x^2 + y^2 - 3x = 0 \Rightarrow \left(x - \frac{3}{2}\right)^2 + y^2 = \frac{9}{4}$

$\Rightarrow C = \left(\frac{3}{2}, 0\right)$ and $a = \frac{3}{2}$;

$r^2 - 3r\cos\theta = 0 \Rightarrow r = 3\cos\theta$

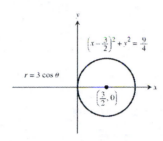

37.

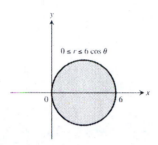

39. d **41.** l **43.** k **45.** i

47. $A = 2\int_0^\pi \frac{1}{2} r^2 d\theta = \int_0^\pi (2 - \cos\theta)^2\, d\theta = \int_0^\pi \left(4 - 4\cos\theta + \cos^2\theta\right)d\theta = \int_0^\pi \left(4 - 4\cos\theta + \frac{1 + \cos 2\theta}{2}\right)d\theta$

$= \int_0^\pi \left(\frac{9}{2} - 4\cos\theta + \frac{\cos 2\theta}{2}\right)d\theta = \left[\frac{9}{2}\theta - 4\sin\theta + \frac{\sin 2\theta}{4}\right]_0^\pi = \frac{9}{2}\pi$

49. $r = 1 + \cos 2\theta$ and $r = 1 \Rightarrow 1 = 1 + \cos 2\theta \Rightarrow 0 = \cos 2\theta \Rightarrow 2\theta = \frac{\pi}{2} \Rightarrow \theta = \frac{\pi}{4}$;

$A = 4\int_0^{\pi/4} \frac{1}{2}\left[(1 + \cos 2\theta)^2 - 1^2\right]d\theta = 2\int_0^{\pi/4}\left(1 + 2\cos 2\theta + \cos^2 2\theta - 1\right)d\theta$

$= 2\int_0^{\pi/4}\left(2\cos 2\theta + \frac{1}{2} + \frac{\cos 4\theta}{2}\right)d\theta = 2\left[\sin 2\theta + \frac{1}{2}\theta + \frac{\sin 4\theta}{8}\right]_0^{\pi/4} = 2\left(1 + \frac{\pi}{8} + 0\right) = 2 + \frac{\pi}{4}$

51. $r = -1 + \cos\theta \Rightarrow \frac{dr}{d\theta} = -\sin\theta$; Length $= \int_0^{2\pi}\sqrt{(-1 + \cos\theta)^2 + (-\sin\theta)^2}\,d\theta = \int_0^{2\pi}\sqrt{2 - 2\cos\theta}\,d\theta$

$= \int_0^{2\pi}\sqrt{\frac{4(1 - \cos\theta)}{2}}\,d\theta = \int_0^{2\pi} 2\sin\frac{\theta}{2}\,d\theta = \left[-4\cos\frac{\theta}{2}\right]_0^{2\pi} = (-4)(-1) - (-4)(1) = 8$

53. $r = 8\sin^3\left(\frac{\theta}{3}\right), 0 \le \theta \le \frac{\pi}{4} \Rightarrow \frac{dr}{d\theta} = 8\sin^2\left(\frac{\theta}{3}\right)\cos\left(\frac{\theta}{3}\right); r^2 + \left(\frac{dr}{d\theta}\right)^2 = \left[8\sin^3\left(\frac{\theta}{3}\right)\right]^2 + \left[8\sin^2\left(\frac{\theta}{3}\right)\cos\left(\frac{\theta}{3}\right)\right]^2$

$= 64\sin^4\left(\frac{\theta}{3}\right) \Rightarrow L = \int_0^{\pi/4}\sqrt{64\sin^4\left(\frac{\theta}{3}\right)}\,d\theta = \int_0^{\pi/4} 8\sin^2\left(\frac{\theta}{3}\right)d\theta = \int_0^{\pi/4} 8\left[\frac{1 - \cos\left(\frac{2\theta}{3}\right)}{2}\right]d\theta$

$= \int_0^{\pi/4}\left[4 - 4\cos\left(\frac{2\theta}{3}\right)\right]d\theta = \left[4\theta - 6\sin\left(\frac{2\theta}{3}\right)\right]_0^{\pi/4} = 4\left(\frac{\pi}{4}\right) - 6\sin\left(\frac{\pi}{6}\right) - 0 = \pi - 3$

55. $r = \frac{2}{1+\cos\theta} \Rightarrow e = 1 \Rightarrow$ parabola with vertex at $(1, 0)$

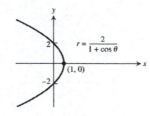

57. $r = \frac{6}{1-2\cos\theta} \Rightarrow e = 2 \Rightarrow$ hyperbola; $ke = 6 \Rightarrow 2k = 6$
    $\Rightarrow k = 3 \Rightarrow$ vertices are $(2, \pi)$ and $(6, \pi)$

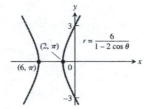

59. $e = 2$ and $r\cos\theta = 2 \Rightarrow x = 2$ is directrix $\Rightarrow k = 2$; the conic is a hyperbola; $r = \frac{ke}{1+e\cos\theta} \Rightarrow r = \frac{(2)(2)}{1+2\cos\theta}$

    $\Rightarrow r = \frac{4}{1+2\cos\theta}$

61. $e = \frac{1}{2}$ and $r\sin\theta = 2 \Rightarrow y = 2$ is directrix $\Rightarrow k = 2$; the conic is an ellipse; $r = \frac{ke}{1+e\sin\theta} \Rightarrow r = \frac{(2)\left(\frac{1}{2}\right)}{1+\left(\frac{1}{2}\right)\sin\theta}$

    $\Rightarrow r = \frac{2}{2+\sin\theta}$

CHAPTER 10 ADDITIONAL AND ADVANCED EXERCISES

1. (a) $x = e^{2t}\cos t$ and $y = e^{2t}\sin t \Rightarrow x^2 + y^2 = e^{4t}\cos^2 t + e^{4t}\sin^2 t = e^{4t}$. Also $\frac{y}{x} = \frac{e^{2t}\sin t}{e^{2t}\cos t} = \tan t$

    $\Rightarrow t = \tan^{-1}\left(\frac{y}{x}\right) \Rightarrow x^2 + y^2 = e^{4\tan^{-1}(y/x)}$ is the Cartesian equation. Since $r^2 = x^2 + y^2$ and

    $\theta = \tan^{-1}\left(\frac{y}{x}\right)$, the polar equation is $r^2 = e^{4\theta}$ or $r = e^{2\theta}$ for $r > 0$

   (b) $ds^2 = r^2 d\theta^2 + dr^2; r = e^{2\theta} \Rightarrow dr = 2e^{2\theta} d\theta$

    $\Rightarrow ds^2 = r^2 d\theta^2 + \left(2e^{2\theta} d\theta\right)^2 = \left(e^{2\theta}\right)^2 d\theta^2 + 4e^{4\theta} d\theta^2$

    $= 5e^{4\theta} d\theta^2 \Rightarrow ds = \sqrt{5}e^{2\theta} d\theta$

    $\Rightarrow L = \int_0^{2\pi} \sqrt{5}e^{2\theta} d\theta = \left[\frac{\sqrt{5}e^{2\theta}}{2}\right]_0^{2\pi} = \frac{\sqrt{5}}{2}\left(e^{4\pi} - 1\right)$

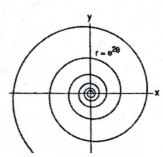

3. $e = 2$ and $r\cos\theta = 2 \Rightarrow x = 2$ is the directrix $\Rightarrow k = 2$; the conic is a hyperbola with $r = \frac{ke}{1+e\cos\theta}$

    $\Rightarrow r = \frac{(2)(2)}{1+2\cos\theta} = \frac{4}{1+2\cos\theta}$

5. $e = \frac{1}{2}$ and $r\sin\theta = 2 \Rightarrow y = 2$ is the directrix $\Rightarrow k = 2$; the conic is an ellipse with $r = \frac{ke}{1+e\sin\theta}$

    $\Rightarrow r = \frac{2\left(\frac{1}{2}\right)}{1+\left(\frac{1}{2}\right)\sin\theta} = \frac{2}{2+\sin\theta}$

7. Arc PF = Arc AF since each is the distance rolled;

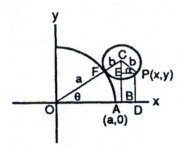

$\angle PCF = \frac{\text{Arc } PF}{b} \Rightarrow$ Arc $PF = b(\angle PCF)$;

$\theta = \frac{\text{Arc } AF}{a} \Rightarrow$ Arc $AF = a\theta \Rightarrow a\theta = b(\angle PCF)$

$\Rightarrow \angle PCF = \left(\frac{a}{b}\right)\theta$; $\quad \angle OCB = \frac{\pi}{2} - \theta$ and

$\angle OCB = \angle PCF - \angle PCE = \angle PCF - \left(\frac{\pi}{2} - \alpha\right)$

$= \left(\frac{a}{b}\right)\theta - \left(\frac{\pi}{2} - \alpha\right) \Rightarrow \frac{\pi}{2} - \theta = \left(\frac{a}{b}\right)\theta - \left(\frac{\pi}{2} - \alpha\right)$

$\Rightarrow \frac{\pi}{2} - \theta = \left(\frac{a}{b}\right)\theta - \frac{\pi}{2} + \alpha \Rightarrow \alpha = \pi - \theta - \left(\frac{a}{b}\right)\theta$

$\Rightarrow \alpha = \pi - \left(\frac{a+b}{b}\right)\theta.$

Now $x = OB + BD = OB + EP = (a+b)\cos\theta + b\cos\alpha = (a+b)\cos\theta + b\cos\left(\pi - \left(\frac{a+b}{b}\right)\theta\right)$

$= (a+b)\cos\theta + b\cos\pi\cos\left(\left(\frac{a+b}{b}\right)\theta\right) + b\sin\pi\sin\left(\left(\frac{a+b}{b}\right)\theta\right) = (a+b)\cos\theta - b\cos\left(\left(\frac{a+b}{b}\right)\theta\right)$ and

$y = PD = CB - CE = (a+b)\sin\theta - b\sin\alpha = (a+b)\sin\theta - b\sin\left(\left(\frac{a+b}{b}\right)\theta\right)$

$= (a+b)\sin\theta - b\sin\pi\cos\left(\left(\frac{a+b}{b}\right)\theta\right) + b\cos\pi\sin\left(\left(\frac{a+b}{b}\right)\theta\right) = (a+b)\sin\theta - b\sin\left(\left(\frac{a+b}{b}\right)\theta\right)$;

therefore $x = (a+b)\cos\theta - b\cos\left(\left(\frac{a+b}{b}\right)\theta\right)$ and $y = (a+b)\sin\theta - b\sin\left(\left(\frac{a+b}{b}\right)\theta\right)$

11. $r = 2a\sin 3\theta \Rightarrow \frac{dr}{d\theta} = 6a\cos 3\theta \Rightarrow \tan\psi = \frac{r}{\left(\frac{dr}{d\theta}\right)} = \frac{2a\sin 3\theta}{6a\cos 3\theta} = \frac{1}{3}\tan 3\theta$; when $\theta = \frac{\pi}{6}$, $\tan\psi = \frac{1}{3}\tan\frac{\pi}{2} \Rightarrow \psi = \frac{\pi}{2}$

CHAPTER 11   VECTORS AND THE GEOMETRY OF SPACE

11.1 THREE-DIMENSIONAL COORDINATE SYSTEMS

1. The line through the point $(2, 3, 0)$ parallel to the z-axis

3. The x-axis

5. The circle $x^2 + y^2 = 4$ in the xy-plane

7. The circle $x^2 + z^2 = 4$ in the xz-plane

9. The circle $y^2 + z^2 = 1$ in the yz-plane

11. The circle $x^2 + y^2 = 16$ in the xy-plane

13. The ellipse formed by the intersection of the cylinder $x^2 + y^2 = 4$ and the plane $z = y$.

15. The parabola $y = x^2$ in the xy-plane.

17. (a) The first quadrant of the xy-plane   (b) The fourth quadrant of the xy-plane

19. (a) The solid ball of radius 1 centered at the origin
     (b) The exterior of the sphere of radius 1 centered at the origin

21. (a) The solid enclosed between the sphere of radius 1 and radius 2 centered at the origin
     (b) The solid upper hemisphere of radius 1 centered at the origin

23. (a) The region on or inside the parabola $y = x^2$ in the xy-plane and all points above this region.

     (b) The region on or to the left of the parabola $x = y^2$ in the xy-plane and all points above it that are 2 units
          or less away from the xy-plane.

25. (a) $x = 3$ (b) $y = -1$ (c) $z = -2$

27. (a) $z = 1$ (b) $x = 3$ (c) $y = -1$

29. (a) $x^2 + (y-2)^2 = 4, z = 0$ (b) $(y-2)^2 + z^2 = 4, x = 0$ (c) $x^2 + z^2 = 4, y = 2$

31. (a) $y = 3, z = -1$ (b) $x = 1, z = -1$ (c) $x = 1, y = 3$

33. $x^2 + y^2 + z^2 = 25, z = 3 \Rightarrow x^2 + y^2 = 16$ in the plane $z = 3$

35. $0 \le z \le 1$

37. $z \le 0$

39. (a) $(x-1)^2 + (y-1)^2 + (z-1)^2 < 1$ (b) $(x-1)^2 + (y-1)^2 + (z-1)^2 > 1$

41. $|P_1 P_2| = \sqrt{(3-1)^2 + (3-1)^2 + (0-1)^2} = \sqrt{9} = 3$

43. $|P_1 P_2| = \sqrt{(4-1)^2 + (-2-4)^2 + (7-5)^2} = \sqrt{49} = 7$

45. $|P_1 P_2| = \sqrt{(2-0)^2 + (-2-0)^2 + (-2-0)^2} = \sqrt{3 \cdot 4} = 2\sqrt{3}$

47. center $(-2, 0, 2)$, radius $2\sqrt{2}$ **49.** center $\left(\sqrt{2}, \sqrt{2}, -\sqrt{2}\right)$, radius $\sqrt{2}$

51. $(x-1)^2 + (y-2)^2 + (z-3)^2 = 14$ **53.** $(x+1)^2 + \left(y-\frac{1}{2}\right)^2 + \left(z+\frac{2}{3}\right)^2 = \frac{16}{81}$

55. $x^2 + y^2 + z^2 + 4x - 4z = 0 \Rightarrow \left(x^2 + 4x + 4\right) + y^2 + \left(z^2 - 4z + 4\right) = 4 + 4 \Rightarrow (x+2)^2 + (y-0)^2 + (z-2)^2 = \left(\sqrt{8}\right)^2$

$\Rightarrow$ the center is at $(-2, 0, 2)$ and the radius is $\sqrt{8}$

57. $2x^2 + 2y^2 + 2z^2 + x + y + z = 9 \Rightarrow x^2 + \frac{1}{2}x + y^2 + \frac{1}{2}y + z^2 + \frac{1}{2}z = \frac{9}{2}$

$\Rightarrow \left(x^2 + \frac{1}{2}x + \frac{1}{16}\right) + \left(y^2 + \frac{1}{2}y + \frac{1}{16}\right) + \left(z^2 + \frac{1}{2}z + \frac{1}{16}\right) = \frac{9}{2} + \frac{3}{16} \Rightarrow \left(x+\frac{1}{4}\right)^2 + \left(y+\frac{1}{4}\right)^2 + \left(z+\frac{1}{4}\right)^2 = \left(\frac{5\sqrt{3}}{4}\right)^2$

$\Rightarrow$ the center is at $\left(-\frac{1}{4}, -\frac{1}{4}, -\frac{1}{4}\right)$ and the radius is $\frac{5\sqrt{3}}{4}$

59. (a) the distance between (x, y, z) and $(x, 0, 0)$ is $\sqrt{y^2 + z^2}$

     (b) the distance between (x, y, z) and $(0, y, 0)$ is $\sqrt{x^2 + z^2}$

     (c) the distance between (x, y, z) and $(0, 0, z)$ is $\sqrt{x^2 + y^2}$

61. $|AB| = \sqrt{(1-(-1))^2 + (-1-2)^2 + (3-1)^2} = \sqrt{4+9+4} = \sqrt{17}$

$|BC| = \sqrt{(3-1)^2 + (4-(-1))^2 + (5-3)^2} = \sqrt{4+25+4} = \sqrt{33}$

$|CA| = \sqrt{(-1-3)^2 + (2-4)^2 + (1-5)^2} = \sqrt{16+4+16} = \sqrt{36} = 6$

Thus the perimeter of triangle ABC is $\sqrt{17} + \sqrt{33} + 6$.

63. $\sqrt{(x-x)^2 + (y-(-1))^2 + (z-z)^2} = \sqrt{(x-x)^2 + (y-3)^2 + (z-z)^2} \Rightarrow (y+1)^2 = (y-3)^2 \Rightarrow 2y + 1 = -6y + 9$

$\Rightarrow y = 1$

65. (a) Since the entire sphere is below the xy-plane, the point on the sphere closest to the xy-plane is the point

     at the top of the sphere, which occurs when $x = 0$ and $y = 3 \Rightarrow 0^2 + (3-3)^2 + (z+5)^2 = 4 \Rightarrow z = -5 \pm 2$

     $\Rightarrow z = -3 \Rightarrow (0, 3, -3)$.

(b) Both the center $(0, 3, -5)$ and the point $(0, 7, -5)$ lie in the plane $z = -5$, so the point on the sphere

   closest to $(0, 7, -5)$ should also be in the same plane. In fact it should lie on the line segment between

   $(0, 3, -5)$ and $(0, 7, -5)$, thus the point occurs when $x = 0$ and $z = -5 \Rightarrow 0^2 + (y-3)^2 + (-5+5)^2 = 4$

   $\Rightarrow y = 3 \pm 2 \Rightarrow y = 5 \Rightarrow (0, 5, -5)$.

11.2 VECTORS

1. (a) $\langle 3(3), 3(-2) \rangle = \langle 9, -6 \rangle$

   (b) $\sqrt{9^2 + (-6)^2} = \sqrt{117} = 3\sqrt{13}$

3. (a) $\langle 3+(-2), -2+5 \rangle = \langle 1, 3 \rangle$

   (b) $\sqrt{1^2 + 3^2} = \sqrt{10}$

5. (a) $2\mathbf{u} = \langle 2(3), 2(-2) \rangle = \langle 6, -4 \rangle$

   $3\mathbf{v} = \langle 3(-2), 3(5) \rangle = \langle -6, 15 \rangle$

   $2\mathbf{u} - 3\mathbf{v} = \langle 6-(-6), -4-15 \rangle = \langle 12, -19 \rangle$

   (b) $\sqrt{12^2 + (-19)^2} = \sqrt{505}$

7. (a) $\frac{3}{5}\mathbf{u} = \langle \frac{3}{5}(3), \frac{3}{5}(-2) \rangle = \langle \frac{9}{5}, -\frac{6}{5} \rangle$

   $\frac{4}{5}\mathbf{v} = \langle \frac{4}{5}(-2), \frac{4}{5}(5) \rangle = \langle -\frac{8}{5}, 4 \rangle$

   $\frac{3}{5}\mathbf{u} + \frac{4}{5}\mathbf{v} = \langle \frac{9}{5} + (-\frac{8}{5}), -\frac{6}{5} + 4 \rangle = \langle \frac{1}{5}, \frac{14}{5} \rangle$

   (b) $\sqrt{\left(\frac{1}{5}\right)^2 + \left(\frac{14}{5}\right)^2} = \frac{\sqrt{197}}{5}$

9. $\langle 2-1, -1-3 \rangle = \langle 1, -4 \rangle$

11. $\langle 0-2, 0-3 \rangle = \langle -2, -3 \rangle$

13. $\langle \cos\frac{2\pi}{3}, \sin\frac{2\pi}{3} \rangle = \langle -\frac{1}{2}, \frac{\sqrt{3}}{2} \rangle$

15. This is the unit vector which makes an angle of $120° + 90° = 210°$ with the positive x-axis;

   $\langle \cos 210°, \sin 210° \rangle = \langle -\frac{\sqrt{3}}{2}, -\frac{1}{2} \rangle$

17. $\overrightarrow{P_1 P_2} = (2-5)\mathbf{i} + (9-7)\mathbf{j} + (-2-(-1))\mathbf{k} = -3\mathbf{i} + 2\mathbf{j} - \mathbf{k}$

19. $\overrightarrow{AB} = (-10-(-7))\mathbf{i} + (8-(-8))\mathbf{j} + (1-1)\mathbf{k} = -3\mathbf{i} + 16\mathbf{j}$

21. $5\mathbf{u} - \mathbf{v} = 5\langle 1, 1, -1 \rangle - \langle 2, 0, 3 \rangle = \langle 5, 5, -5 \rangle - \langle 2, 0, 3 \rangle = \langle 5-2, 5-0, -5-3 \rangle = \langle 3, 5, -8 \rangle = 3\mathbf{i} + 5\mathbf{j} - 8\mathbf{k}$

23. The vector $\mathbf{v}$ is horizontal and 1 in. long. The vectors $\mathbf{u}$ and $\mathbf{w}$ are $\frac{11}{16}$ in. long. $\mathbf{w}$ is vertical and $\mathbf{u}$ makes

   a $45°$ angle with the horizontal. All vectors must be drawn to scale.

(a)

(b)

(c)

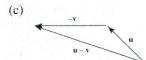

(d)

25. length $=|2\mathbf{i}+\mathbf{j}-2\mathbf{k}|=\sqrt{2^2+1^2+(-2)^2}=3,$ the direction is $\frac{2}{3}\mathbf{i}+\frac{1}{3}\mathbf{j}-\frac{2}{3}\mathbf{k}\Rightarrow 2\mathbf{i}+\mathbf{j}-2\mathbf{k}=3\left(\frac{2}{3}\mathbf{i}+\frac{1}{3}\mathbf{j}-\frac{2}{3}\mathbf{k}\right)$

27. length $=|5\mathbf{k}|=\sqrt{25}=5,$ the direction is $\mathbf{k}\Rightarrow 5\mathbf{k}=5(\mathbf{k})$

29. length $=\left|\frac{1}{\sqrt{6}}\mathbf{i}-\frac{1}{\sqrt{6}}\mathbf{j}-\frac{1}{\sqrt{6}}\mathbf{k}\right|=\sqrt{3\left(\frac{1}{\sqrt{6}}\right)^2}=\sqrt{\frac{1}{2}},$ the direction is $\frac{1}{\sqrt{3}}\mathbf{i}-\frac{1}{\sqrt{3}}\mathbf{j}-\frac{1}{\sqrt{3}}\mathbf{k}$

$\Rightarrow\frac{1}{\sqrt{6}}\mathbf{i}-\frac{1}{\sqrt{6}}\mathbf{j}-\frac{1}{\sqrt{6}}\mathbf{k}=\sqrt{\frac{1}{2}}\left(\frac{1}{\sqrt{3}}\mathbf{i}-\frac{1}{\sqrt{3}}\mathbf{j}-\frac{1}{\sqrt{3}}\mathbf{k}\right)$

31. (a) $2\mathbf{i}$ (b) $-\sqrt{3}\mathbf{k}$ (c) $\frac{3}{10}\mathbf{j}+\frac{2}{5}\mathbf{k}$ (d) $6\mathbf{i}-2\mathbf{j}+3\mathbf{k}$

33. $|\mathbf{v}|=\sqrt{12^2+5^2}=\sqrt{169}=13;\ \frac{\mathbf{v}}{|\mathbf{v}|}=\frac{1}{13}\mathbf{v}=\frac{1}{13}(12\mathbf{i}-5\mathbf{k})\Rightarrow$ the desired vector is $\frac{7}{13}(12\mathbf{i}-5\mathbf{k})$

35. (a) $3\mathbf{i}+4\mathbf{j}-5\mathbf{k}=5\sqrt{2}\left(\frac{3}{5\sqrt{2}}\mathbf{i}+\frac{4}{5\sqrt{2}}\mathbf{j}-\frac{1}{\sqrt{2}}\mathbf{k}\right)\Rightarrow$ the direction is $\frac{3}{5\sqrt{2}}\mathbf{i}+\frac{4}{5\sqrt{2}}\mathbf{j}-\frac{1}{\sqrt{2}}\mathbf{k}$

(b) the midpoint is $\left(\frac{1}{2},3,\frac{5}{2}\right)$

37. (a) $-\mathbf{i}-\mathbf{j}-\mathbf{k}=\sqrt{3}\left(-\frac{1}{\sqrt{3}}\mathbf{i}-\frac{1}{\sqrt{3}}\mathbf{j}-\frac{1}{\sqrt{3}}\mathbf{k}\right)\Rightarrow$ the direction is $-\frac{1}{\sqrt{3}}\mathbf{i}-\frac{1}{\sqrt{3}}\mathbf{j}-\frac{1}{\sqrt{3}}\mathbf{k}$

(b) the midpoint is $\left(\frac{5}{2},\frac{7}{2},\frac{9}{2}\right)$

39. $\overrightarrow{AB}=(5-a)\mathbf{i}+(1-b)\mathbf{j}+(3-c)\mathbf{k}=\mathbf{i}+4\mathbf{j}-2\mathbf{k}\Rightarrow 5-a=1,1-b=4,$ and $3-c=-2\Rightarrow a=4,b=-3,$ and

$c=15\Rightarrow A$ is the point $(4,-3,5)$

41. $2\mathbf{i}+\mathbf{j}=a(\mathbf{i}+\mathbf{j})+b(\mathbf{i}-\mathbf{j})=(a+b)\mathbf{i}+(a-b)\mathbf{j}\Rightarrow a+b=2$ and $a-b=1\Rightarrow 2a=3\Rightarrow a=\frac{3}{2}$ and $b=a-1=\frac{1}{2}$

43. $25°$ west of north is $90°+25°=115°$ north of east. $800\langle\cos 115°,\sin 115°\rangle\approx\langle -338.095,725.046\rangle$

45. $\mathbf{F}_1=\langle -|\mathbf{F}_1|\cos 30°,|\mathbf{F}_1|\sin 30°\rangle=\left\langle -\frac{\sqrt{3}}{2}|\mathbf{F}_1|,\frac{1}{2}|\mathbf{F}_1|\right\rangle,\ \mathbf{F}_2=\langle|\mathbf{F}_2|\cos 45°,|\mathbf{F}_2|\sin 45°\rangle=\left\langle\frac{1}{\sqrt{2}}|\mathbf{F}_2|,\frac{1}{\sqrt{2}}|\mathbf{F}_2|\right\rangle,$ and

$\mathbf{w}=\langle 0,-100\rangle.$ Since $\mathbf{F}_1+\mathbf{F}_2=\langle 0,100\rangle\Rightarrow\left\langle -\frac{\sqrt{3}}{2}|\mathbf{F}_1|+\frac{1}{\sqrt{2}}|\mathbf{F}_2|,\frac{1}{2}|\mathbf{F}_1|+\frac{1}{\sqrt{2}}|\mathbf{F}_2|\right\rangle=\langle 0,100\rangle$

$\Rightarrow -\frac{\sqrt{3}}{2}|\mathbf{F}_1|+\frac{1}{\sqrt{2}}|\mathbf{F}_2|=0$ and $\frac{1}{2}|\mathbf{F}_1|+\frac{1}{\sqrt{2}}|\mathbf{F}_2|=100.$ Solving the first equation for $|\mathbf{F}_2|$ results in:

$|\mathbf{F}_2|=\frac{\sqrt{6}}{2}|\mathbf{F}_1|.$ Substituting this result into the second equation gives us: $\frac{1}{2}|\mathbf{F}_1|+\frac{1}{\sqrt{2}}\left(\frac{\sqrt{6}}{2}|\mathbf{F}_1|\right)=100$

$\Rightarrow|\mathbf{F}_1|=\frac{200}{1+\sqrt{3}}\approx 73.205$ N $\Rightarrow|\mathbf{F}_2|=\frac{100\sqrt{6}}{1+\sqrt{3}}\approx 89.658$ N $\Rightarrow\mathbf{F}_1\approx\langle -63.397,36.603\rangle$ and $\mathbf{F}_2=\approx\langle 63.397,63.397\rangle$

47. $\mathbf{F}_1=\langle -|\mathbf{F}_1|\cos 40°,|\mathbf{F}_1|\sin 40°\rangle,\ \mathbf{F}_2=\langle 100\cos 35°,100\sin 35°\rangle,$ and $\mathbf{w}=\langle 0,-w\rangle.$ Since $\mathbf{F}_1+\mathbf{F}_2=\langle 0,w\rangle$

$\Rightarrow\langle -|\mathbf{F}_1|\cos 40°+100\cos 35°,|\mathbf{F}_1|\sin 40°+100\sin 35°\rangle=\langle 0,w\rangle\Rightarrow -|\mathbf{F}_1|\cos 40°+100\cos 35°=0$ and

$|\mathbf{F}_1|\sin 40°+100\sin 35°=w.$ Solving the first equation for $|\mathbf{F}_1|$ results in: $|\mathbf{F}_1|=\frac{100\cos 35°}{\cos 40°}\approx 106.933$ N.

Substituting this result into the second equation gives us: $w\approx 126.093$ N.

49. (a) The tree is located at the tip of the vector $\overrightarrow{OP} = \left(5\cos 60°\right)\mathbf{i} + \left(5\sin 60°\right)\mathbf{j} = \frac{5}{2}\mathbf{i} + \frac{5\sqrt{3}}{2}\mathbf{j} \Rightarrow P = \left(\frac{5}{2}, \frac{5\sqrt{3}}{2}\right)$

   (b) The telephone pole is located at the point Q, which is the tip of the vector $\overrightarrow{OP} + \overrightarrow{PQ}$

   $= \left(\frac{5}{2}\mathbf{i} + \frac{5\sqrt{3}}{2}\mathbf{j}\right) + \left(10\cos 315°\right)\mathbf{i} + \left(10\sin 315°\right)\mathbf{j} = \left(\frac{5}{2} + \frac{10\sqrt{2}}{2}\right)\mathbf{i} + \left(\frac{5\sqrt{3}}{2} - \frac{10\sqrt{2}}{2}\right)\mathbf{j} \Rightarrow Q = \left(\frac{5+10\sqrt{2}}{2}, \frac{5\sqrt{3}-10\sqrt{2}}{2}\right)$

51. (a) the midpoint of AB is $M\left(\frac{5}{2}, \frac{5}{2}, 0\right)$ and $\overrightarrow{CM} = \left(\frac{5}{2}-1\right)\mathbf{i} + \left(\frac{5}{2}-1\right)\mathbf{j} + (0-3)\mathbf{k} = \frac{3}{2}\mathbf{i} + \frac{3}{2}\mathbf{j} - 3\mathbf{k}$

   (b) the desired vector is $\left(\frac{2}{3}\right)\overrightarrow{CM} = \frac{2}{3}\left(\frac{3}{2}\mathbf{i} + \frac{3}{2}\mathbf{j} - 3\mathbf{k}\right) = \mathbf{i} + \mathbf{j} - 2\mathbf{k}$

   (c) the vector whose sum is the vector from the origin to C and the result of part (b) will terminate at the
       center of mass $\Rightarrow$ the terminal point of $(\mathbf{i} + \mathbf{j} + 3\mathbf{k}) + (\mathbf{i} + \mathbf{j} - 2\mathbf{k}) = 2\mathbf{i} + 2\mathbf{j} + \mathbf{k}$ is the point $(2, 2, 1)$, which
       is the location of the center of mass

53. Without loss of generality we identify the vertices of the quadrilateral such that $A(0, 0, 0)$, $B(x_b, 0, 0)$,

   $C(x_c, y_c, 0)$ and $D(x_d, y_d, z_d) \Rightarrow$ the midpoint of AB is $M_{AB}\left(\frac{x_b}{2}, 0, 0\right)$, the midpoint of BC is

   $M_{BC}\left(\frac{x_b+x_c}{2}, \frac{y_c}{2}, 0\right)$, the midpoint of CD is $M_{CD}\left(\frac{x_c+x_d}{2}, \frac{y_c+y_d}{2}, \frac{z_d}{2}\right)$ and the midpoint of AD is

   $M_{AD}\left(\frac{x_d}{2}, \frac{y_d}{2}, \frac{z_d}{2}\right) \Rightarrow$ the midpoint of $M_{AB}M_{CD}$ is $\left(\frac{\frac{x_b}{2} + \frac{x_c+x_d}{2}}{2}, \frac{y_c+y_d}{4}, \frac{z_d}{4}\right)$ which is the same as the

   midpoint of $M_{AD}M_{BC} = \left(\frac{\frac{x_b+x_c}{2} + \frac{x_d}{2}}{2}, \frac{y_c+y_d}{4}, \frac{z_d}{4}\right)$.

55. Without loss of generality we can coordinatize the vertices of the triangle such that $A(0, 0)$, $B(b, 0)$ and

   $C(x_c, y_c) \Rightarrow a$ is located at $\left(\frac{b+x_c}{2}, \frac{y_c}{2}\right)$, b is at $\left(\frac{x_c}{2}, \frac{y_c}{2}\right)$ and c is at $\left(\frac{b}{2}, 0\right)$. Therefore, $\overrightarrow{Aa} = \left(\frac{b}{2} + \frac{x_c}{2}\right)\mathbf{i} + \left(\frac{y_c}{2}\right)\mathbf{j}$,

   $\overrightarrow{Bb} = \left(\frac{x_c}{2} - b\right)\mathbf{i} + \left(\frac{y_c}{2}\right)\mathbf{j}$, and $\overrightarrow{Cc} = \left(\frac{b}{2} - x_c\right)\mathbf{i} + \left(-y_c\right)\mathbf{j} \Rightarrow \overrightarrow{Aa} + \overrightarrow{Bb} + \overrightarrow{Cc} = \mathbf{0}$.

11.3 THE DOT PRODUCT

NOTE: In Exercises 1-8 below we calculate $\text{proj}_\mathbf{v}\,\mathbf{u}$ as the vector $\left(\frac{|\mathbf{u}|\cos\theta}{|\mathbf{v}|}\right)\mathbf{v}$, so the scalar multiplier of $\mathbf{v}$ is the
number in column 5 divided by the number in column 2.

	$\mathbf{v}\cdot\mathbf{u}$	$|\mathbf{v}|$	$|\mathbf{u}|$	$\cos\theta$	$|\mathbf{u}|\cos\theta$	$\text{proj}_\mathbf{v}\,\mathbf{u}$
1.	-25	5	5	-1	-5	$-2\mathbf{i} + 4\mathbf{j} - \sqrt{5}\mathbf{k}$
2.	3	1	13	$\frac{3}{13}$	3	$3\left(\frac{3}{5}\mathbf{i} + \frac{4}{5}\mathbf{k}\right)$
3.	25	15	5	$\frac{1}{3}$	$\frac{5}{3}$	$\frac{1}{9}(10\mathbf{i} + 11\mathbf{j} - 2\mathbf{k})$
4.	13	15	3	$\frac{13}{45}$	$\frac{13}{15}$	$\frac{13}{225}(2\mathbf{i} + 10\mathbf{j} - 11\mathbf{k})$
5.	2	$\sqrt{34}$	$\sqrt{3}$	$\frac{2}{\sqrt{3}\sqrt{34}}$	$\frac{2}{\sqrt{34}}$	$\frac{1}{17}(5\mathbf{j} - 3\mathbf{k})$
6.	$\sqrt{3}-\sqrt{2}$	$\sqrt{2}$	3	$\frac{\sqrt{3}-\sqrt{2}}{3\sqrt{2}}$	$\frac{\sqrt{3}-\sqrt{2}}{\sqrt{2}}$	$\frac{\sqrt{3}-\sqrt{2}}{2}(-\mathbf{i} + \mathbf{j})$
7.	$10+\sqrt{17}$	$\sqrt{26}$	$\sqrt{21}$	$\frac{10+\sqrt{17}}{\sqrt{546}}$	$\frac{10+\sqrt{17}}{\sqrt{26}}$	$\frac{10+\sqrt{17}}{26}(5\mathbf{i} + \mathbf{j})$

9. $\theta = \cos^{-1}\left(\dfrac{\mathbf{u}\cdot\mathbf{v}}{|\mathbf{u}||\mathbf{v}|}\right) = \cos^{-1}\left(\dfrac{(2)(1)+(1)(2)+(0)(-1)}{\sqrt{2^2+1^2+0^2}\sqrt{1^2+2^2+(-1)^2}}\right) = \cos^{-1}\left(\dfrac{4}{\sqrt{5}\sqrt{6}}\right) = \cos^{-1}\left(\dfrac{4}{\sqrt{30}}\right) \approx 0.75$ rad

11. $\theta = \cos^{-1}\left(\dfrac{\mathbf{u}\cdot\mathbf{v}}{|\mathbf{u}||\mathbf{v}|}\right) = \cos^{-1}\left(\dfrac{(\sqrt{3})(\sqrt{3})+(-7)(1)+(0)(-2)}{\sqrt{(\sqrt{3})^2+(-7)^2+0^2}\sqrt{(\sqrt{3})^2+(1)^2+(-2)^2}}\right) = \cos^{-1}\left(\dfrac{3-7}{\sqrt{52}\sqrt{8}}\right) = \cos^{-1}\left(\dfrac{-1}{\sqrt{26}}\right) \approx 1.77$ rad

13. $\overrightarrow{AB} = \langle 3, 1\rangle$, $\overrightarrow{BC} = \langle -1, -3\rangle$, and $\overrightarrow{AC} = \langle 2, -2\rangle$. $\overrightarrow{BA} = \langle -3, -1\rangle$, $\overrightarrow{CB} = \langle 1, 3\rangle$, $\overrightarrow{CA} = \langle -2, 2\rangle$.

$|\overrightarrow{AB}| = |\overrightarrow{BA}| = \sqrt{10}, |\overrightarrow{BC}| = |\overrightarrow{CB}| = \sqrt{10}, |\overrightarrow{AC}| = |\overrightarrow{CA}| = 2\sqrt{2}$,

Angle at $A = \cos^{-1}\left(\dfrac{\overrightarrow{AB}\cdot\overrightarrow{AC}}{|\overrightarrow{AB}||\overrightarrow{AC}|}\right) = \cos^{-1}\left(\dfrac{3(2)+1(-2)}{(\sqrt{10})(2\sqrt{2})}\right) = \cos^{-1}\left(\dfrac{1}{\sqrt{5}}\right) \approx 63.435°$

Angle at $B = \cos^{-1}\left(\dfrac{\overrightarrow{BC}\cdot\overrightarrow{BA}}{|\overrightarrow{BC}||\overrightarrow{BA}|}\right) = \cos^{-1}\left(\dfrac{(-1)(-3)+(-3)(-1)}{(\sqrt{10})(\sqrt{10})}\right) = \cos^{-1}\left(\dfrac{3}{5}\right) \approx 53.130°$, and

Angle at $C = \cos^{-1}\left(\dfrac{\overrightarrow{CB}\cdot\overrightarrow{CA}}{|\overrightarrow{CB}||\overrightarrow{CA}|}\right) = \cos^{-1}\left(\dfrac{1(-2)+3(2)}{(\sqrt{10})(2\sqrt{2})}\right) = \cos^{-1}\left(\dfrac{1}{\sqrt{5}}\right) \approx 63.435°$

15. (a) $\cos\alpha = \dfrac{\mathbf{i}\cdot\mathbf{v}}{|\mathbf{i}||\mathbf{v}|} = \dfrac{a}{|\mathbf{v}|}$, $\cos\beta = \dfrac{\mathbf{j}\cdot\mathbf{v}}{|\mathbf{j}||\mathbf{v}|} = \dfrac{b}{|\mathbf{v}|}$, $\cos\gamma = \dfrac{\mathbf{k}\cdot\mathbf{v}}{|\mathbf{k}||\mathbf{v}|} = \dfrac{c}{|\mathbf{v}|}$ and

$\cos^2\alpha + \cos^2\beta + \cos^2\gamma = \left(\dfrac{a}{|\mathbf{v}|}\right)^2 + \left(\dfrac{b}{|\mathbf{v}|}\right)^2 + \left(\dfrac{c}{|\mathbf{v}|}\right)^2 = \dfrac{a^2+b^2+c^2}{|\mathbf{v}||\mathbf{v}|} = \dfrac{|\mathbf{v}||\mathbf{v}|}{|\mathbf{v}||\mathbf{v}|} = 1$

(b) $|\mathbf{v}| = 1 \Rightarrow \cos\alpha = \dfrac{a}{|\mathbf{v}|} = a, \cos\beta = \dfrac{b}{|\mathbf{v}|} = b$ and $\cos\gamma = \dfrac{c}{|\mathbf{v}|} = c$ are the direction cosines of $\mathbf{v}$

17. The sum of two vectors of equal length is *always* orthogonal to their difference, as we can see from the equation $(\mathbf{v}_1 + \mathbf{v}_2)\cdot(\mathbf{v}_1 - \mathbf{v}_2) = \mathbf{v}_1\cdot\mathbf{v}_1 + \mathbf{v}_2\cdot\mathbf{v}_1 - \mathbf{v}_1\cdot\mathbf{v}_2 - \mathbf{v}_2\cdot\mathbf{v}_2 = |\mathbf{v}_1|^2 - |\mathbf{v}_2|^2 = 0$

19. Let $\mathbf{u}$ and $\mathbf{v}$ be the sides of a rhombus $\Rightarrow$ the diagonals are $\mathbf{d}_1 = \mathbf{u}+\mathbf{v}$ and $\mathbf{d}_2 = -\mathbf{u}+\mathbf{v}$

$\Rightarrow \mathbf{d}_1\cdot\mathbf{d}_2 = (\mathbf{u}+\mathbf{v})\cdot(-\mathbf{u}+\mathbf{v}) = -\mathbf{u}\cdot\mathbf{u}+\mathbf{u}\cdot\mathbf{v}-\mathbf{v}\cdot\mathbf{u}+\mathbf{v}\cdot\mathbf{v} = |\mathbf{v}|^2 - |\mathbf{u}|^2 = 0$ because $|\mathbf{u}| = |\mathbf{v}|$, since a rhombus has equal sides.

21. Clearly the diagonals of a rectangle are equal in length. What is not as obvious is the statement that equal diagonals happen only in a rectangle. We show this is true by letting the adjacent sides of a parallelogram be the vectors $(v_1\mathbf{i}+v_2\mathbf{j})$ and $(u_1\mathbf{i}+u_2\mathbf{j})$. The equal diagonals of the parallelogram are

$\mathbf{d}_1 = (v_1\mathbf{i}+v_2\mathbf{j})+(u_1\mathbf{i}+u_2\mathbf{j})$ and $\mathbf{d}_2 = (v_1\mathbf{i}+v_2\mathbf{j})-(u_1\mathbf{i}+u_2\mathbf{j})$. Hence $|\mathbf{d}_1| = |\mathbf{d}_2| = |(v_1\mathbf{i}+v_2\mathbf{j})+(u_1\mathbf{i}+u_2\mathbf{j})|$

$= |(v_1\mathbf{i}+v_2\mathbf{j})-(u_1\mathbf{i}+u_2\mathbf{j})| \Rightarrow |(v_1+u_1)\mathbf{i}+(v_2+u_2)\mathbf{j}| = |(v_1-u_1)\mathbf{i}+(v_2-u_2)\mathbf{j}| \Rightarrow \sqrt{(v_1+u_1)^2+(v_2+u_2)^2}$

$= \sqrt{(v_1-u_1)^2+(v_2-u_2)^2} \Rightarrow v_1^2+2v_1u_1+u_1^2+v_2^2+2v_2u_2+u_2^2 = v_1^2-2v_1u_1+u_1^2+v_2^2-2v_2u_2+u_2^2$

$\Rightarrow 2(v_1u_1+v_2u_2) = -2(v_1u_1+v_2u_2) \Rightarrow v_1u_1+v_2u_2 = 0 \Rightarrow (v_1\mathbf{i}+v_2\mathbf{j})\cdot(u_1\mathbf{i}+u_2\mathbf{j}) = 0 \Rightarrow$ the vectors $(v_1\mathbf{i}+v_2\mathbf{j})$ and $(u_1\mathbf{i}+u_2\mathbf{j})$ are perpendicular and the parallelogram must be a rectangle.

23. horizontal component: $1200\cos(8°) \approx 1188$ ft/s; vertical component: $1200\sin(8°) \approx 167$ ft/s

25. (a) Since $\left|\cos\theta\right|\le 1$, we have $\left|\mathbf{u}\cdot\mathbf{v}\right|=\left|\mathbf{u}\right|\left|\mathbf{v}\right|\left|\cos\theta\right|\le\left|\mathbf{u}\right|\left|\mathbf{v}\right|(1)=\left|\mathbf{u}\right|\left|\mathbf{v}\right|$.

    (b) We have equality precisely when $\left|\cos\theta\right|=1$ or when one or both of $\mathbf{u}$ and $\mathbf{v}$ is $\mathbf{0}$. In the case of nonzero vectors, we have equality when $\theta=0$ or π, i.e., when the vectors are parallel.

27. $\mathbf{v}\cdot\mathbf{u}_1=\left(a\mathbf{u}_1+b\mathbf{u}_2\right)\cdot\mathbf{u}_1=a\mathbf{u}_1\cdot\mathbf{u}_1+b\mathbf{u}_2\cdot\mathbf{u}_1=a\left|\mathbf{u}_1\right|^2+b\left(\mathbf{u}_2\cdot\mathbf{u}_1\right)=a(1)^2+b(0)=a$

29. $\mathrm{proj}_{\mathbf{v}}\mathbf{u}=\dfrac{\mathbf{u}\cdot\mathbf{v}}{\left|\mathbf{v}\right|^2}\mathbf{v}\Rightarrow\left(\mathbf{u}-\dfrac{\mathbf{u}\cdot\mathbf{v}}{\left|\mathbf{v}\right|^2}\mathbf{v}\right)\cdot\left(\dfrac{\mathbf{u}\cdot\mathbf{v}}{\left|\mathbf{v}\right|^2}\mathbf{v}\right)=\mathbf{u}\cdot\left(\dfrac{\mathbf{u}\cdot\mathbf{v}}{\left|\mathbf{v}\right|^2}\mathbf{v}\right)-\left(\dfrac{\mathbf{u}\cdot\mathbf{v}}{\left|\mathbf{v}\right|^2}\mathbf{v}\right)\cdot\left(\dfrac{\mathbf{u}\cdot\mathbf{v}}{\left|\mathbf{v}\right|^2}\mathbf{v}\right)=\left(\dfrac{\mathbf{u}\cdot\mathbf{v}}{\left|\mathbf{v}\right|^2}\right)(\mathbf{u}\cdot\mathbf{v})-\left(\dfrac{\mathbf{u}\cdot\mathbf{v}}{\left|\mathbf{v}\right|^2}\right)^2(\mathbf{v}\cdot\mathbf{v})$

    $=\dfrac{(\mathbf{u}\cdot\mathbf{v})^2}{\left|\mathbf{v}\right|^2}-\dfrac{(\mathbf{u}\cdot\mathbf{v})^2}{\left|\mathbf{v}\right|^4}\left|\mathbf{v}\right|^2=0$

31. $P(x_1,y_1)=P\left(x_1,\dfrac{c}{b}-\dfrac{a}{b}x_1\right)$ and $Q(x_2,y_2)=Q\left(x_2,\dfrac{c}{b}-\dfrac{a}{b}x_2\right)$ and any two points P and Q on the line with

    $b\ne 0\Rightarrow\overrightarrow{PQ}=\left(x_2-x_1\right)\mathbf{i}+\dfrac{a}{b}\left(x_1-x_2\right)\mathbf{j}\Rightarrow\overrightarrow{PQ}\cdot\mathbf{v}=\left[\left(x_2-x_1\right)\mathbf{i}+\dfrac{a}{b}\left(x_1-x_2\right)\mathbf{j}\right]\cdot(a\mathbf{i}+b\mathbf{j})$

    $=a\left(x_2-x_1\right)+b\left(\dfrac{a}{b}\right)\left(x_1-x_2\right)=0\Rightarrow\mathbf{v}$ is perpendicular to $\overrightarrow{PQ}$ for $b\ne 0$. If $b=0$, then $\mathbf{v}=a\mathbf{i}$ is

    perpendicular to the vertical line $ax=c$. Alternatively, the slope of $\mathbf{v}$ is $\dfrac{b}{a}$ and the slope of the line $ax+by=c$

    is $-\dfrac{a}{b}$, so the slopes are negative reciprocals $\Rightarrow$ the vector $\mathbf{v}$ and the line are perpendicular.

33. $\mathbf{v}=\mathbf{i}+2\mathbf{j}$ is perpendicular to the line $x+2y=c$;

    $P(2,1)$ on the line $\Rightarrow 2+2=c\Rightarrow x+2y=4$

35. $\mathbf{v}=-2\mathbf{i}+\mathbf{j}$ is perpendicular to the line $-2x+y=c$;

    $P(-2,-7)$ on the line $\Rightarrow (-2)(-2)-7=c$

    $\Rightarrow -2x+y=-3$

37. $\mathbf{v}=\mathbf{i}-\mathbf{j}$ is parallel to the line $-x-y=c$;

    $P(-2,1)$ on the line $\Rightarrow -(-2)-1=c\Rightarrow -x-y=1$

    or $x+y=-1$.

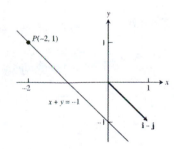

39. $\mathbf{v} = -\mathbf{i} - 2\mathbf{j}$ is parallel to the line $-2x + y = c$;
     $P(1,2)$ on the line $\Rightarrow -2(1) + 2 = c \Rightarrow -2x - y = 0$
     or $2x - y = 0$.

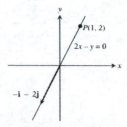

41. $P(0,0),\ Q(1,1)$ and $\mathbf{F} = 5\mathbf{j} \Rightarrow \overrightarrow{PQ} = \mathbf{i} + \mathbf{j}$ and $W = \mathbf{F} \cdot \overrightarrow{PQ} = (5\mathbf{j}) \cdot (\mathbf{i} + \mathbf{j}) = 5 \text{ N} \cdot \text{m} = 5 \text{ J}$

43. $W = |\mathbf{F}|\left|\overrightarrow{PQ}\right| \cos \theta = (200)(20)(\cos 30°) = 2000\sqrt{3} = 3464.10 \text{ N} \cdot \text{m} = 3464.10 \text{ J}$

45. $\mathbf{n}_1 = 3\mathbf{i} + \mathbf{j}$ and $\mathbf{n}_2 = 2\mathbf{i} - \mathbf{j} \Rightarrow \theta = \cos^{-1}\left(\frac{\mathbf{n}_1 \cdot \mathbf{n}_2}{|\mathbf{n}_1||\mathbf{n}_2|}\right) = \cos^{-1}\left(\frac{6-1}{\sqrt{10}\sqrt{5}}\right) = \cos^{-1}\left(\frac{1}{\sqrt{2}}\right) = \frac{\pi}{4}$

47. $\mathbf{n}_1 = \sqrt{3}\mathbf{i} - \mathbf{j}$ and $\mathbf{n}_2 = \mathbf{i} - \sqrt{3}\mathbf{j} \Rightarrow \theta = \cos^{-1}\left(\frac{\mathbf{n}_1 \cdot \mathbf{n}_2}{|\mathbf{n}_1||\mathbf{n}_2|}\right) = \cos^{-1}\left(\frac{\sqrt{3}+\sqrt{3}}{\sqrt{4}\sqrt{4}}\right) = \cos^{-1}\left(\frac{\sqrt{3}}{2}\right) = \frac{\pi}{6}$

49. $\mathbf{n}_1 = 3\mathbf{i} - 4\mathbf{j}$ and $\mathbf{n}_2 = \mathbf{i} - \mathbf{j} \Rightarrow \theta = \cos^{-1}\left(\frac{\mathbf{n}_1 \cdot \mathbf{n}_2}{|\mathbf{n}_1||\mathbf{n}_2|}\right) = \cos^{-1}\left(\frac{3+4}{\sqrt{25}\sqrt{2}}\right) = \cos^{-1}\left(\frac{7}{5\sqrt{2}}\right) \approx 0.14 \text{ rad}$

11.4 THE CROSS PRODUCT

1. $\mathbf{u} \times \mathbf{v} = \begin{vmatrix} \mathbf{i} & \mathbf{j} & \mathbf{k} \\ 2 & -2 & -1 \\ 1 & 0 & -1 \end{vmatrix} = 3\left(\frac{2}{3}\mathbf{i} + \frac{1}{3}\mathbf{j} + \frac{2}{3}\mathbf{k}\right) \Rightarrow$ length $= 3$ and the direction is $\frac{2}{3}\mathbf{i} + \frac{1}{3}\mathbf{j} + \frac{2}{3}\mathbf{k}$;

    $\mathbf{v} \times \mathbf{u} = -(\mathbf{u} \times \mathbf{v}) = -3\left(\frac{2}{3}\mathbf{i} + \frac{1}{3}\mathbf{j} + \frac{2}{3}\mathbf{k}\right) \Rightarrow$ length $= 3$ and the direction is $-\frac{2}{3}\mathbf{i} - \frac{1}{3}\mathbf{j} - \frac{2}{3}\mathbf{k}$

3. $\mathbf{u} \times \mathbf{v} = \begin{vmatrix} \mathbf{i} & \mathbf{j} & \mathbf{k} \\ 2 & -2 & 4 \\ -1 & 1 & -2 \end{vmatrix} = \mathbf{0} \Rightarrow$ length $= 0$ and has no direction

    $\mathbf{v} \times \mathbf{u} = -(\mathbf{u} \times \mathbf{v}) = \mathbf{0} \Rightarrow$ length $= 0$ and has no direction

5. $\mathbf{u} \times \mathbf{v} = \begin{vmatrix} \mathbf{i} & \mathbf{j} & \mathbf{k} \\ 2 & 0 & 0 \\ 0 & -3 & 0 \end{vmatrix} = -6(\mathbf{k}) \Rightarrow$ length $= 6$ and the direction is $-\mathbf{k}$

    $\mathbf{v} \times \mathbf{u} = -(\mathbf{u} \times \mathbf{v}) = 6(\mathbf{k}) \Rightarrow$ length $= 6$ and the direction is $\mathbf{k}$

7. $\mathbf{u} \times \mathbf{v} = \begin{vmatrix} \mathbf{i} & \mathbf{j} & \mathbf{k} \\ -8 & -2 & -4 \\ 2 & 2 & 1 \end{vmatrix} = 6\mathbf{i} - 12\mathbf{k} \Rightarrow$ length $= 6\sqrt{5}$ and the direction is $\frac{1}{\sqrt{5}}\mathbf{i} - \frac{2}{\sqrt{5}}\mathbf{k}$

    $\mathbf{v} \times \mathbf{u} = -(\mathbf{u} \times \mathbf{v}) = -(6\mathbf{i} - 12\mathbf{k}) \Rightarrow$ length $= 6\sqrt{5}$ and the direction is $-\frac{1}{\sqrt{5}}\mathbf{i} + \frac{2}{\sqrt{5}}\mathbf{k}$

9. $\mathbf{u} \times \mathbf{v} = \begin{vmatrix} \mathbf{i} & \mathbf{j} & \mathbf{k} \\ 1 & 0 & 0 \\ 0 & 1 & 0 \end{vmatrix} = \mathbf{k}$

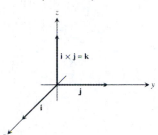

11. $\mathbf{u} \times \mathbf{v} = \begin{vmatrix} \mathbf{i} & \mathbf{j} & \mathbf{k} \\ 1 & 0 & -1 \\ 0 & 1 & 1 \end{vmatrix} = \mathbf{i} - \mathbf{j} + \mathbf{k}$

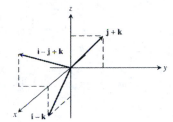

13. $\mathbf{u} \times \mathbf{v} = \begin{vmatrix} \mathbf{i} & \mathbf{j} & \mathbf{k} \\ 1 & 1 & 0 \\ 1 & -1 & 0 \end{vmatrix} = -2\mathbf{k}$

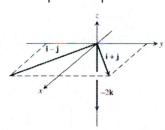

15. (a) $\overrightarrow{PQ} \times \overrightarrow{PR} = \begin{vmatrix} \mathbf{i} & \mathbf{j} & \mathbf{k} \\ 1 & 1 & -3 \\ -1 & 3 & -1 \end{vmatrix} = 8\mathbf{i} + 4\mathbf{j} + 4\mathbf{k} \Rightarrow$ Area $= \frac{1}{2}\left|\overrightarrow{PQ} \times \overrightarrow{PR}\right| = \frac{1}{2}\sqrt{64 + 16 + 16} = 2\sqrt{6}$

   (b) $\mathbf{u} = \dfrac{\overrightarrow{PQ} \times \overrightarrow{PR}}{\left|\overrightarrow{PQ} \times \overrightarrow{PR}\right|} = \dfrac{1}{\sqrt{6}}(2\mathbf{i} + \mathbf{j} + \mathbf{k})$

17. (a) $\overrightarrow{PQ} \times \overrightarrow{PR} = \begin{vmatrix} \mathbf{i} & \mathbf{j} & \mathbf{k} \\ 1 & 1 & 1 \\ 1 & 1 & 0 \end{vmatrix} = -\mathbf{i} + \mathbf{j} \Rightarrow$ Area $\frac{1}{2}\left|\overrightarrow{PQ} \times \overrightarrow{PR}\right| = \frac{1}{2}\sqrt{1 + 1} = \frac{\sqrt{2}}{2}$

   (b) $\mathbf{u} = \dfrac{\overrightarrow{PQ} \times \overrightarrow{PR}}{\left|\overrightarrow{PQ} \times \overrightarrow{PR}\right|} = \dfrac{1}{\sqrt{2}}(-\mathbf{i} + \mathbf{j}) \Rightarrow -\dfrac{1}{\sqrt{2}}(\mathbf{i} - \mathbf{j})$

19. If $\mathbf{u} = a_1\mathbf{i} + a_2\mathbf{j} + a_3\mathbf{k}$, $\mathbf{v} = b_1\mathbf{i} + b_2\mathbf{j} + b_3\mathbf{k}$, and $\mathbf{w} = c_1\mathbf{i} + c_2\mathbf{j} + c_3\mathbf{k}$, then $(\mathbf{u} \times \mathbf{v}) \cdot \mathbf{w} = \begin{vmatrix} a_1 & a_2 & a_3 \\ b_1 & b_2 & b_3 \\ c_1 & c_2 & c_3 \end{vmatrix}$,

   $(\mathbf{v} \times \mathbf{w}) \cdot \mathbf{u} = \begin{vmatrix} b_1 & b_2 & b_3 \\ c_1 & c_2 & c_3 \\ a_1 & a_2 & a_3 \end{vmatrix}$ and $(\mathbf{w} \times \mathbf{u}) \cdot \mathbf{v} = \begin{vmatrix} c_1 & c_2 & c_3 \\ a_1 & a_2 & a_3 \\ b_1 & b_2 & b_3 \end{vmatrix}$ which all have the same absolute value,

   since interchanging two rows in a determinant does not change its absolute value

   $\Rightarrow$ the volume is $|(\mathbf{u} \times \mathbf{v}) \cdot \mathbf{w}| = abs\begin{vmatrix} 2 & 0 & 0 \\ 0 & 2 & 0 \\ 0 & 0 & 2 \end{vmatrix} = 8$

21. $|(\mathbf{u} \times \mathbf{v}) \cdot \mathbf{w}| = abs\begin{vmatrix} 2 & 1 & 0 \\ 2 & -1 & 1 \\ 1 & 0 & 2 \end{vmatrix} = |-7| = 7$ (for details about verification, see Exercise 19)

23. (a) $\mathbf{u} \cdot \mathbf{v} = -6, \mathbf{u} \cdot \mathbf{w} = -81, \mathbf{v} \cdot \mathbf{w} = 18 \Rightarrow$ none are perpendicular

   (b) $\mathbf{u} \times \mathbf{v} = \begin{vmatrix} \mathbf{i} & \mathbf{j} & \mathbf{k} \\ 5 & -1 & 1 \\ 0 & 1 & -5 \end{vmatrix} \neq \mathbf{0}, \mathbf{u} \times \mathbf{w} = \begin{vmatrix} \mathbf{i} & \mathbf{j} & \mathbf{k} \\ 5 & -1 & 1 \\ -15 & 3 & -3 \end{vmatrix} = \mathbf{0}, \mathbf{v} \times \mathbf{w} = \begin{vmatrix} \mathbf{i} & \mathbf{j} & \mathbf{k} \\ 0 & 1 & -5 \\ -15 & 3 & -3 \end{vmatrix} \neq \mathbf{0} \Rightarrow \mathbf{u}$ and $\mathbf{w}$ are parallel

25. $\left|\overrightarrow{PQ} \times \mathbf{F}\right| = \left|\overrightarrow{PQ}\right| |\mathbf{F}| \sin(60°) = \frac{2}{3} \cdot 30 \cdot \frac{\sqrt{3}}{2} \text{ ft} \cdot \text{lb} = 10\sqrt{3} \text{ ft} \cdot \text{lb}$

27. (a) true, $|\mathbf{u}| = \sqrt{a_1^2 + a_2^2 + a_3^2} = \sqrt{\mathbf{u} \cdot \mathbf{u}}$

   (b) not always true, $\mathbf{u} \cdot \mathbf{u} = |\mathbf{u}|^2$

   (c) true, $\mathbf{u} \times \mathbf{0} = \begin{vmatrix} \mathbf{i} & \mathbf{j} & \mathbf{k} \\ u_1 & u_2 & u_3 \\ 0 & 0 & 0 \end{vmatrix} = 0\mathbf{i} + 0\mathbf{j} + 0\mathbf{k} = \mathbf{0}$ and $\mathbf{0} \times \mathbf{u} = \begin{vmatrix} \mathbf{i} & \mathbf{j} & \mathbf{k} \\ 0 & 0 & 0 \\ u_1 & u_2 & u_3 \end{vmatrix} = 0\mathbf{i} + 0\mathbf{j} + 0\mathbf{k} = \mathbf{0}$

   (d) true, $\mathbf{u} \times (-\mathbf{u}) = \begin{vmatrix} \mathbf{i} & \mathbf{j} & \mathbf{k} \\ u_1 & u_2 & u_3 \\ -u_1 & -u_2 & -u_3 \end{vmatrix} = (-u_2 u_3 + u_2 u_3)\mathbf{i} - (-u_1 u_3 + u_1 u_3)\mathbf{j} + (-u_1 u_2 + u_1 u_2)\mathbf{k} = \mathbf{0}$

   (e) not always true, $\mathbf{i} \times \mathbf{j} = \mathbf{k} \neq -\mathbf{k} = \mathbf{j} \times \mathbf{i}$ for example

   (f) true, distributive property of the cross product

   (g) true, $(\mathbf{u} \times \mathbf{v}) \cdot \mathbf{v} = \mathbf{u} \cdot (\mathbf{v} \times \mathbf{v}) = \mathbf{u} \cdot \mathbf{0} = 0$

   (h) true, the volume of a parallelepiped with $\mathbf{u}, \mathbf{v}$, and $\mathbf{w}$ along the three edges is the same whether the plane containing $\mathbf{u}$ and $\mathbf{v}$ or the plane containing $\mathbf{v}$ and $\mathbf{w}$ is used as the base plane, and the dot product is commutative.

29. (a) $\text{proj}_\mathbf{v} \mathbf{u} = \left(\frac{\mathbf{u} \cdot \mathbf{v}}{|\mathbf{v}||\mathbf{v}|}\right)\mathbf{v}$ (b) $(\mathbf{u} \times \mathbf{v})$ (c) $((\mathbf{u} \times \mathbf{v}) \times \mathbf{w})$ (d) $|(\mathbf{u} \times \mathbf{v}) \cdot \mathbf{w}|$

   (e) $(\mathbf{u} \times \mathbf{v}) \times (\mathbf{u} \times \mathbf{w})$ (f) $|\mathbf{u}| \frac{\mathbf{v}}{|\mathbf{v}|}$

31. (a) yes, $\mathbf{u} \times \mathbf{v}$ and $\mathbf{w}$ are both vectors   (b) no, $\mathbf{u}$ is a vector but $\mathbf{v} \cdot \mathbf{w}$ is a scalar
    (c) yes, $\mathbf{u}$ and $\mathbf{u} \times \mathbf{w}$ are both vectors   (d) no, $\mathbf{u}$ is a vector but $\mathbf{v} \cdot \mathbf{w}$ is a scalar

33. No, $\mathbf{v}$ need not equal $\mathbf{w}$. For example, $\mathbf{i} + \mathbf{j} \neq -\mathbf{i} + \mathbf{j}$, but $\mathbf{i} \times (\mathbf{i} + \mathbf{j}) = \mathbf{i} \times \mathbf{i} + \mathbf{i} \times \mathbf{j} = \mathbf{0} + \mathbf{k} = \mathbf{k}$ and $\mathbf{i} \times (-\mathbf{i} + \mathbf{j}) = \mathbf{i} \times (-\mathbf{i}) + \mathbf{i} \times \mathbf{j} = \mathbf{0} + \mathbf{k} = \mathbf{k}$.

35. $\overrightarrow{AB} = -\mathbf{i} + \mathbf{j}$ and $\overrightarrow{AD} = -\mathbf{i} - \mathbf{j} \Rightarrow \overrightarrow{AB} \times \overrightarrow{AD} = \begin{vmatrix} \mathbf{i} & \mathbf{j} & \mathbf{k} \\ -1 & 1 & 0 \\ -1 & -1 & 0 \end{vmatrix} = 2\mathbf{k} \Rightarrow$ area $= \left|\overrightarrow{AB} \times \overrightarrow{AD}\right| = 2$

37. $\overrightarrow{AB} = 3\mathbf{i} - 2\mathbf{j}$ and $\overrightarrow{AD} = 5\mathbf{i} + \mathbf{j} \Rightarrow \overrightarrow{AB} \times \overrightarrow{AD} = \begin{vmatrix} \mathbf{i} & \mathbf{j} & \mathbf{k} \\ 3 & -2 & 0 \\ 5 & 1 & 0 \end{vmatrix} = 13\mathbf{k} \Rightarrow$ area $= \left|\overrightarrow{AB} \times \overrightarrow{AD}\right| = 13$

39. $\overrightarrow{AB} = 3\mathbf{i} + 2\mathbf{j} + 4\mathbf{k}$ and $\overrightarrow{DC} = 3\mathbf{i} + 2\mathbf{j} + 4\mathbf{k} \Rightarrow \overrightarrow{AB}$ is parallel to $\overrightarrow{DC}; \overrightarrow{BC} = 2\mathbf{i} - \mathbf{j}$ and $\overrightarrow{AD} = 2\mathbf{i} - \mathbf{j} \Rightarrow \overrightarrow{BC}$ is

parallel to $\overrightarrow{AD}$. $\overrightarrow{AB} \times \overrightarrow{BC} = \begin{vmatrix} \mathbf{i} & \mathbf{j} & \mathbf{k} \\ 3 & 2 & 4 \\ 2 & -1 & 0 \end{vmatrix} = 4\mathbf{i} + 8\mathbf{j} - 7\mathbf{k} \Rightarrow$ area $= \left|\overrightarrow{AB} \times \overrightarrow{BC}\right| = \sqrt{129}$

41. $\overrightarrow{AB} = -2\mathbf{i} + 3\mathbf{j}$ and $\overrightarrow{AC} = 3\mathbf{i} + \mathbf{j} \Rightarrow \overrightarrow{AB} \times \overrightarrow{AC} = \begin{vmatrix} \mathbf{i} & \mathbf{j} & \mathbf{k} \\ -2 & 3 & 0 \\ 3 & 1 & 0 \end{vmatrix} = -11\mathbf{k} \Rightarrow$ area $= \frac{1}{2}\left|\overrightarrow{AB} \times \overrightarrow{AC}\right| = \frac{11}{2}$

43. $\overrightarrow{AB} = 6\mathbf{i} - 5\mathbf{j}$ and $\overrightarrow{AC} = 11\mathbf{i} - 5\mathbf{j} \Rightarrow \overrightarrow{AB} \times \overrightarrow{AC} = \begin{vmatrix} \mathbf{i} & \mathbf{j} & \mathbf{k} \\ 6 & -5 & 0 \\ 11 & -5 & 0 \end{vmatrix} = 25\mathbf{k} \Rightarrow$ area $= \frac{1}{2}\left|\overrightarrow{AB} \times \overrightarrow{AC}\right| = \frac{25}{2}$

45. $\overrightarrow{AB} = -\mathbf{i} + 2\mathbf{j}$ and $\overrightarrow{AC} = -\mathbf{i} - \mathbf{k} \Rightarrow \overrightarrow{AB} \times \overrightarrow{AC} = \begin{vmatrix} \mathbf{i} & \mathbf{j} & \mathbf{k} \\ -1 & 2 & 0 \\ -1 & 0 & -1 \end{vmatrix} = -2\mathbf{i} - \mathbf{j} + 2\mathbf{k} \Rightarrow$ area $= \frac{1}{2}\left|\overrightarrow{AB} \times \overrightarrow{AC}\right| = \frac{3}{2}$

47. $\overrightarrow{AB} = -\mathbf{i} + 2\mathbf{j}$ and $\overrightarrow{AC} = \mathbf{j} - 2\mathbf{k} \Rightarrow \overrightarrow{AB} \times \overrightarrow{AC} = \begin{vmatrix} \mathbf{i} & \mathbf{j} & \mathbf{k} \\ -1 & 2 & 0 \\ 0 & 1 & -2 \end{vmatrix} = -4\mathbf{i} - 2\mathbf{j} - \mathbf{k} \Rightarrow$ area $= \frac{1}{2}\left|\overrightarrow{AB} \times \overrightarrow{AC}\right| = \frac{\sqrt{21}}{2}$

49. If $\mathbf{A} = a_1\mathbf{i} + a_2\mathbf{j}$ and $\mathbf{B} = b_1\mathbf{i} + b_2\mathbf{j}$, then $\mathbf{A} \times \mathbf{B} = \begin{vmatrix} \mathbf{i} & \mathbf{j} & \mathbf{k} \\ a_1 & a_2 & 0 \\ b_1 & b_2 & 0 \end{vmatrix} = \begin{vmatrix} a_1 & a_2 \\ b_1 & b_2 \end{vmatrix}\mathbf{k}$ and the triangle's area is

$\frac{1}{2}|\mathbf{A} \times \mathbf{B}| = \pm\frac{1}{2}\begin{vmatrix} a_1 & a_2 \\ b_1 & b_2 \end{vmatrix}$. The applicable sign is (+) if the acute angle from $\mathbf{A}$ to $\mathbf{B}$ runs counterclockwise in

the xy-plane, and (−) if it runs clockwise, because the area must be a nonnegative number.

11.5 LINES AND PLANES IN SPACE

1. The direction $\mathbf{i} + \mathbf{j} + \mathbf{k}$ and $P(3, -4, -1) \Rightarrow x = 3 + t, y = -4 + t, z = -1 + t$

3. The direction $\overrightarrow{PQ} = 5\mathbf{i} + 5\mathbf{j} - 5\mathbf{k}$ and $P(-2, 0, 3) \Rightarrow x = -2 + 5t, y = 5t, z = 3 - 5t$

5. The direction $2\mathbf{j} + \mathbf{k}$ and $P(0, 0, 0) \Rightarrow x = 0, y = 2t, z = t$

7. The direction $\mathbf{k}$ and $P(1, 1, 1) \Rightarrow x = 1, y = 1, z = 1 + t$

9. The direction $\mathbf{i} + 2\mathbf{j} + 2\mathbf{k}$ and $P(0, -7, 0) \Rightarrow x = t, y = -7 + 2t, z = 2t$

11. The direction $\mathbf{i}$ and $P(0, 0, 0) \Rightarrow x = t, y = 0, z = 0$

13. The direction $\overrightarrow{PQ} = \mathbf{i} + \mathbf{j} + \frac{3}{2}\mathbf{k}$ and $P(0,0,0)$

$\Rightarrow x = t,\, y = t,\, z = \frac{3}{2}t,$ where $0 \le t \le 1$

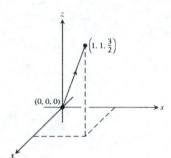

15. The direction $\overrightarrow{PQ} = \mathbf{j}$ and $P(1,1,0)$

$\Rightarrow x = 1,\, y = 1 + t,\, z = 0,$ where $-1 \le t \le 0$

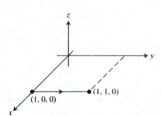

17. The direction $\overrightarrow{PQ} = -2\mathbf{j}$ and $P(0,1,1)$

$\Rightarrow x = 0,\, y = 1 - 2t,\, z = 1,$ where $0 \le t \le 1$

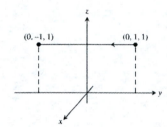

19. The direction $\overrightarrow{PQ} = -2\mathbf{i} + 2\mathbf{j} - 2\mathbf{k}$ and $P(2,0,2)$

$\Rightarrow x = 2 - 2t,\, y = 2t,\, z = 2 - 2t,$ where $0 \le t \le 1$

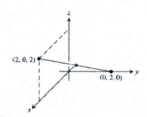

21. $3(x - 0) + (-2)(y - 2) + (-1)(z + 1) = 0 \Rightarrow 3x - 2y - z = -3$

23. $\overrightarrow{PQ} = \mathbf{i} - \mathbf{j} + 3\mathbf{k},\ \overrightarrow{PS} = -\mathbf{i} - 3\mathbf{j} + 2\mathbf{k} \Rightarrow \overrightarrow{PQ} \times \overrightarrow{PS} = \begin{vmatrix} \mathbf{i} & \mathbf{j} & \mathbf{k} \\ 1 & -1 & 3 \\ -1 & -3 & 2 \end{vmatrix} = 7\mathbf{i} - 5\mathbf{j} - 4\mathbf{k}$ is normal to the plane

$\Rightarrow 7(x - 2) + (-5)(y - 0) + (-4)(z - 2) = 0 \Rightarrow 7x - 5y - 4z = 6$

25. $\mathbf{n} = \mathbf{i} + 3\mathbf{j} + 4\mathbf{k},\ P(2,4,5) \Rightarrow (1)(x - 2) + (3)(y - 4) + (4)(z - 5) = 0 \Rightarrow x + 3y + 4z = 34$

27. $\begin{cases} x = 2t+1 = s+2 \\ y = 3t+2 = 2s+4 \end{cases} \Rightarrow \begin{cases} 2t-s=1 \\ 3t-2s=2 \end{cases} \Rightarrow \begin{cases} 4t-2s=2 \\ 3t-2s=2 \end{cases} \Rightarrow t=0 \text{ and } s=-1;$ then $z = 4t+3 = -4s-1$

$\Rightarrow 4(0)+3 = (-4)(-1)-1$ is satisfied $\Rightarrow$ the lines intersect when $t=0$ and $s=-1 \Rightarrow$ the point of intersection

is $x=1$, $y=2$, and $z=3$ or $P(1,2,3)$. A vector normal to the plane determined by these lines is

$\mathbf{n_1} \times \mathbf{n_2} = \begin{vmatrix} \mathbf{i} & \mathbf{j} & \mathbf{k} \\ 2 & 3 & 4 \\ 1 & 2 & -4 \end{vmatrix} = -20\mathbf{i}+12\mathbf{j}+\mathbf{k}$, where $\mathbf{n_1}$ and $\mathbf{n_2}$ are directions of the lines $\Rightarrow$ the plane containing the

lines is represented by $(-20)(x-1)+(12)(y-2)+(1)(z-3)=0 \Rightarrow -20x+12y+z=7$.

29. The cross product of $\mathbf{i}+\mathbf{j}-\mathbf{k}$ and $-4\mathbf{i}+2\mathbf{j}-2\mathbf{k}$ has the same direction as the normal to the plane

$\Rightarrow \mathbf{n} = \begin{vmatrix} \mathbf{i} & \mathbf{j} & \mathbf{k} \\ 1 & 1 & -1 \\ -4 & 2 & -2 \end{vmatrix} = 6\mathbf{j}+6\mathbf{k}$. Select a point on either line, such as $P(-1,2,1)$. Since the lines are given to

intersect, the desired plane is $0(x+1)+6(y-2)+6(z-1)=0 \Rightarrow 6y+6z=18 \Rightarrow y+z=3$.

31. $\mathbf{n_1} \times \mathbf{n_2} = \begin{vmatrix} \mathbf{i} & \mathbf{j} & \mathbf{k} \\ 2 & 1 & -1 \\ 1 & 2 & 1 \end{vmatrix} = 3\mathbf{i}-3\mathbf{j}+3\mathbf{k}$ is a vector in the direction of the line of intersection of the planes

$\Rightarrow 3(x-2)+(-3)(y-1)+3(z+1)=0 \Rightarrow 3x-3y+3z=0 \Rightarrow x-y+z=0$ is the desired plane containing
$P_0(2,1,-1)$

33. $S(0,0,12)$, $P(0,0,0)$ and $\mathbf{v}=4\mathbf{i}-2\mathbf{j}+2\mathbf{k} \Rightarrow \overrightarrow{PS} \times \mathbf{v} = \begin{vmatrix} \mathbf{i} & \mathbf{j} & \mathbf{k} \\ 0 & 0 & 12 \\ 4 & -2 & 2 \end{vmatrix} = 24\mathbf{i}+48\mathbf{j} = 24(\mathbf{i}+2\mathbf{j})$

$\Rightarrow d = \dfrac{|\overrightarrow{PS} \times \mathbf{v}|}{|\mathbf{v}|} = \dfrac{24\sqrt{1+4}}{\sqrt{16+4+4}} = \dfrac{24\sqrt{5}}{\sqrt{24}} = \sqrt{5 \cdot 24} = 2\sqrt{30}$ is the distance from S to the line

35. $S(2,1,3)$, $P(2,1,3)$ and $\mathbf{v}=2\mathbf{i}+6\mathbf{j} \Rightarrow \overrightarrow{PS} \times \mathbf{v} = \mathbf{0} \Rightarrow d = \dfrac{|\overrightarrow{PS} \times \mathbf{v}|}{|\mathbf{v}|} = \dfrac{0}{\sqrt{40}} = 0$ is the distance from S to the line

(i.e., the point S lies on the line)

37. $S(3,-1,4)$, $P(4,3,-5)$ and $\mathbf{v}=-\mathbf{i}+2\mathbf{j}+3\mathbf{k} \Rightarrow \overrightarrow{PS} \times \mathbf{v} = \begin{vmatrix} \mathbf{i} & \mathbf{j} & \mathbf{k} \\ -1 & -4 & 9 \\ -1 & 2 & 3 \end{vmatrix} = -30\mathbf{i}-6\mathbf{j}-6\mathbf{k}$

$\Rightarrow d = \dfrac{|\overrightarrow{PS} \times \mathbf{v}|}{|\mathbf{v}|} = \dfrac{\sqrt{900+36+36}}{\sqrt{1+4+9}} = \dfrac{\sqrt{972}}{\sqrt{14}} = \dfrac{\sqrt{486}}{\sqrt{7}} = \dfrac{\sqrt{81\cdot6}}{\sqrt{7}} = \dfrac{9\sqrt{42}}{7}$ is the distance from S to the line

39. $S(2,-3,4)$, $x+2y+2z=13$ and $P(13,0,0)$ is on the plane $\Rightarrow \overrightarrow{PS} = -11\mathbf{i}-3\mathbf{j}+4\mathbf{k}$ and $\mathbf{n}=\mathbf{i}+2\mathbf{j}+2\mathbf{k}$

$\Rightarrow d = \left| \overrightarrow{PS} \cdot \dfrac{\mathbf{n}}{|\mathbf{n}|} \right| = \left| \dfrac{-11-6+8}{\sqrt{1+4+4}} \right| = \left| \dfrac{-9}{\sqrt{9}} \right| = 3$

41. $S(0,1,1)$, $4y+3z=-12$ and $P(0,-3,0)$ is on the plane $\Rightarrow \overrightarrow{PS} = 4\mathbf{j}+\mathbf{k}$ and $\mathbf{n}=4\mathbf{j}+3\mathbf{k}$

$\Rightarrow d = \left| \overrightarrow{PS} \cdot \dfrac{\mathbf{n}}{|\mathbf{n}|} \right| = \left| \dfrac{16+3}{\sqrt{16+9}} \right| = \dfrac{19}{5}$

43. $S(0, -1, 0), 2x + y + 2z = 4$ and $P(2, 0, 0)$ is on the plane $\Rightarrow \overrightarrow{PS} = -2i - j$ and $n = 2i + j + 2k$

$\Rightarrow d = \left| \overrightarrow{PS} \cdot \frac{n}{|n|} \right| = \left| \frac{-4-1+0}{\sqrt{4+1+4}} \right| = \frac{5}{3}$

45. The point $P(1, 0, 0)$ is on the first plane and $S(10, 0, 0)$ is a point on the second plane $\Rightarrow \overrightarrow{PS} = 9i$,

and $n = i + 2j + 6k$ is normal to the first plane $\Rightarrow$ the distance from S to the first plane is

$d = \left| \overrightarrow{PS} \cdot \frac{n}{|n|} \right| = \left| \frac{9}{\sqrt{1+4+36}} \right| = \frac{9}{\sqrt{41}}$, which is also the distance between the planes.

47. $n_1 = i + j$ and $n_2 = 2i + j - 2k \Rightarrow \theta = \cos^{-1}\left(\frac{n_1 \cdot n_2}{|n_1||n_2|} \right) = \cos^{-1}\left(\frac{2+1}{\sqrt{2}\sqrt{9}} \right) = \cos^{-1}\left(\frac{1}{\sqrt{2}} \right) = \frac{\pi}{4}$

49. $n_1 = 2i + 2j + 2k$ and $n_2 = 2i - 2j - k \Rightarrow \theta = \cos^{-1}\left(\frac{n_1 \cdot n_2}{|n_1||n_2|} \right) = \cos^{-1}\left(\frac{4-4-2}{\sqrt{12}\sqrt{9}} \right) = \cos^{-1}\left(\frac{-1}{3\sqrt{3}} \right) \approx 1.76$ rad

51. $n_1 = 2i + 2j - k$ and $n_2 = i + 2j + k \Rightarrow \theta = \cos^{-1}\left(\frac{n_1 \cdot n_2}{|n_1||n_2|} \right) = \cos^{-1}\left(\frac{2+4-1}{\sqrt{9}\sqrt{6}} \right) = \cos^{-1}\left(\frac{5}{3\sqrt{6}} \right) \approx 0.82$ rad

53. $2x - y + 3z = 6 \Rightarrow 2(1-t) - (3t) + 3(1+t) = 6 \Rightarrow -2t + 5 = 6 \Rightarrow t = -\frac{1}{2} \Rightarrow x = \frac{3}{2}, y = -\frac{3}{2}$ and $z = \frac{1}{2}$

$\Rightarrow \left(\frac{3}{2}, -\frac{3}{2}, \frac{1}{2} \right)$ is the point

55. $x + y + z = 2 \Rightarrow (1+2t) + (1+5t) + (3t) = 2 \Rightarrow 10t + 2 = 2 \Rightarrow t = 0 \Rightarrow x = 1, y = 1$ and $z = 0 \Rightarrow (1, 1, 0)$ is the point

57. $n_1 = i + j + k$ and $n_2 = i + j \Rightarrow n_1 \times n_2 = \begin{vmatrix} i & j & k \\ 1 & 1 & 1 \\ 1 & 1 & 0 \end{vmatrix} = -i + j$, the direction of the desired line;

$(1, 1, -1)$ is on both planes $\Rightarrow$ the desired line is $x = 1 - t, y = 1 + t, z = -1$

59. $n_1 = i - 2j + 4k$ and $n_2 = i + j - 2k \Rightarrow n_1 \times n_2 = \begin{vmatrix} i & j & k \\ 1 & -2 & 4 \\ 1 & 1 & -2 \end{vmatrix} = 6j + 3k$, the direction of the desired line;

$(4, 3, 1)$ is on both planes $\Rightarrow$ the desired line is $x = 4, y = 3 + 6t, z = 1 + 3t$

61. <u>L1 & L2</u>: $\begin{cases} x = 3 + 2t = 1 + 4s \\ y = -1 + 4t = 1 + 2s \end{cases} \Rightarrow \begin{cases} 2t - 4s = -2 \\ 4t - 2s = 2 \end{cases} \Rightarrow \begin{cases} 2t - 4s = -2 \\ 2t - s = 1 \end{cases} \Rightarrow -3s = -3 \Rightarrow s = 1$ and $t = 1$

$\Rightarrow$ on L1, $z = 1$ and on L2, $z = 1 \Rightarrow L1$ and $L2$ intersect at $(5, 3, 1)$.

<u>L2 & L3</u>: The direction of L2 is $\frac{1}{6}(4i + 2j + 4k) = \frac{1}{3}(2i + j + 2k)$ which is the same as the direction

$\frac{1}{3}(2i + j + 2k)$ of L3; hence L2 and L3 are parallel.

<u>L1 & L3</u>: $\begin{cases} x = 3 + 2t = 3 + 2r \\ y = -1 + 4t = 2 + r \end{cases} \Rightarrow \begin{cases} 2t - 2r = 0 \\ 4t - r = 3 \end{cases} \Rightarrow \begin{cases} t - r = 0 \\ 4t - r = 3 \end{cases} \Rightarrow 3t = 3 \Rightarrow t = 1$ and $r = 1 \Rightarrow$ on L1, $z = 2$

while on L3, $z = 0 \Rightarrow L1$ and $L2$ do not intersect. The direction of L1 is $\frac{1}{\sqrt{21}}(2i + 4j - k)$ while the direction of

L3 is $\frac{1}{3}(2i + j + 2k)$ and neither is a multiple of the other; hence L1 and L3 are skew.

63. $x = 2 + 2t, \ y = -4 - t, \ z = 7 + 3t; \ x = -2 - t, \ y = -2 + \frac{1}{2}t, \ z = 1 - \frac{3}{2}t$

65. $x = 0 \Rightarrow t = -\frac{1}{2}, \ y = -\frac{1}{2}, \ z = -\frac{3}{2} \Rightarrow \left(0, -\frac{1}{2}, -\frac{3}{2}\right); \ \ y = 0 \Rightarrow t = -1, \ x = -1, \ z = -3 \Rightarrow (-1, 0, -3);$

      $z = 0 \Rightarrow t = 0, \ x = 1, \ y = -1 \Rightarrow (1, -1, 0)$

67.   With substitution of the line into the plane we have $2(1 - 2t) + (2 + 5t) - (-3t) = 8 \Rightarrow 2 - 4t + 2 + 5t + 3t = 8$
      $\Rightarrow 4t + 4 = 8 \Rightarrow t = 1 \Rightarrow$ the point $(-1, 7, -3)$ is contained in both the line and plane, so they are not parallel.

69.   There are many possible answers. One is found as follows: eliminate t to get $t = x - 1 = 2 - y = \frac{z-3}{2}$
      $\Rightarrow x - 1 = 2 - y$ and $2 - y = \frac{z-3}{2} \Rightarrow x + y = 3$ and $2y + z = 7$ are two such planes.

71.   The points $(a, 0, 0), \ (0, b, 0)$ and $(0, 0, c)$ are the x, y, and z intercepts of the plane. Since a, b, and c are all
      nonzero, the plane must intersect all three coordinate axes and cannot pass through the origin. Thus,
      $\frac{x}{a} + \frac{y}{b} + \frac{z}{c} = 1$ describes all planes <u>except</u> those through the origin or parallel to a coordinate axis.

73.   (a) $\overrightarrow{EP} = c\overrightarrow{EP_1} \Rightarrow -x_0\mathbf{i} + y\mathbf{j} + z\mathbf{k} = c\left[(x_1 - x_0)\mathbf{i} + y_1\mathbf{j} + z_1\mathbf{k}\right] \Rightarrow -x_0 = c(x_1 - x_0), \ y = cy_1$ and $z = cz_1$, where c
            is a positive real number

      (b)   At $x_1 = 0 \Rightarrow c = 1 \Rightarrow y = y_1$ and $z = z_1$; at $x_1 = x_0 \Rightarrow x_0 = 0, \ y = 0, \ z = 0; \ \lim_{x_0 \to \infty} c = \lim_{x_0 \to \infty} \frac{-x_0}{x_1 - x_0}$

            $= \lim_{x_0 \to \infty} \frac{-1}{-1} = 1 \Rightarrow c \to 1$ so that $y \to y_1$ and $z \to z_1$

11.6 CYLINDERS AND QUADRIC SURFACES

1. d, ellipsoid

3. a, cylinder

5. l, hyperbolic paraboloid

7. b, cylinder

9. k, hyperbolic paraboloid

11. h, cone

13. $x^2 + y^2 = 4$

15. $x^2 + 4z^2 = 16$

17. $9x^2 + y^2 + z^2 = 9$

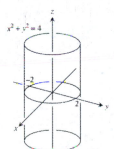

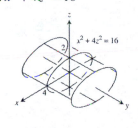

19. $4x^2 + 9y^2 + 4z^2 = 36$

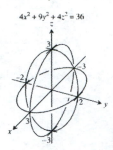

21. $x^2 + 4y^2 = z$

23. $x = 4 - 4y^2 - z^2$

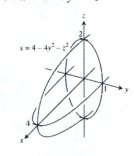

25. $x^2 + y^2 = z^2$

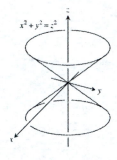

27. $x^2 + y^2 - z^2 = 1$

29. $z^2 - x^2 - y^2 = 1$

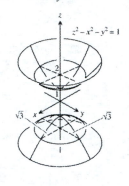

31. $y^2 - x^2 = z$

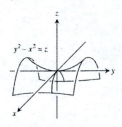

33. $z = 1 + y^2 - x^2$

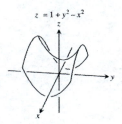

35. $y = -\left(x^2 + z^2\right)$

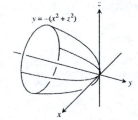

37. $x^2 + y^2 - z^2 = 4$

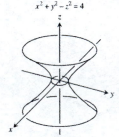

39. $x^2 + z^2 = 1$

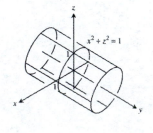

41. $z = -\left(x^2 + y^2\right)$

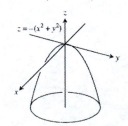

43. $4y^2 + z^2 - 4x^2 = 4$

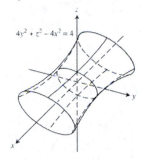
$4y^2 + z^2 - 4x^2 = 4$

45. (a) If $x^2 + \dfrac{y^2}{4} + \dfrac{z^2}{9} = 1$ and $z = c$, then $x^2 + \dfrac{y^2}{4} = \dfrac{9-c^2}{9} \Rightarrow \dfrac{x^2}{\left(\frac{9-c^2}{9}\right)} + \dfrac{y^2}{\left[\frac{4(9-c^2)}{9}\right]} = 1 \Rightarrow A = ab\pi$

$= \pi\left(\dfrac{\sqrt{9-c^2}}{3}\right)\left(\dfrac{2\sqrt{9-c^2}}{3}\right) = \dfrac{2\pi(9-c^2)}{9}$

   (b) From part (a), each slice has the area $\dfrac{2\pi(9-z^2)}{9}$, where $-3 \le z \le 3$. Thus $V = 2\displaystyle\int_0^3 \dfrac{2\pi}{9}\left(9 - z^2\right) dz$

$= \dfrac{4\pi}{9}\displaystyle\int_0^3 \left(9 - z^2\right) dz = \dfrac{4\pi}{9}\left[9z - \dfrac{z^3}{3}\right]_0^3 = \dfrac{4\pi}{9}(27 - 9) = 8\pi$

   (c) $\dfrac{x^2}{a^2} + \dfrac{y^2}{b^2} + \dfrac{z^2}{c^2} = 1 \Rightarrow \dfrac{x^2}{\left[\frac{a^2(c^2-z^2)}{c^2}\right]} + \dfrac{y^2}{\left[\frac{b^2(c^2-z^2)}{c^2}\right]} = 1 \Rightarrow A = \pi\left(\dfrac{a\sqrt{c^2-z^2}}{c}\right)\left(\dfrac{b\sqrt{c^2-z^2}}{c}\right)$

$\Rightarrow V = 2\displaystyle\int_0^c \dfrac{\pi ab}{c^2}\left(c^2 - z^2\right) dz = \dfrac{2\pi ab}{c^2}\left[c^2 z - \dfrac{z^3}{3}\right]_0^c = \dfrac{2\pi ab}{c^2}\left(\dfrac{2}{3}c^3\right) = \dfrac{4\pi abc}{3}$. Note that if $r = a = b = c$,

then $V = \dfrac{4\pi r^3}{3}$, which is the volume of a sphere.

47. We calculate the volume by the slicing method, taking slices parallel to the xy-plane. For fixed z, $\dfrac{x^2}{a^2} + \dfrac{y^2}{b^2} = \dfrac{z}{c}$

gives the ellipse $\dfrac{x^2}{\left(\frac{za^2}{c}\right)} + \dfrac{y^2}{\left(\frac{zb^2}{c}\right)} = 1$. The area of this ellipse is $\pi\left(a\sqrt{\dfrac{z}{c}}\right)\left(b\sqrt{\dfrac{z}{c}}\right) = \dfrac{\pi abz}{c}$ (see Exercise 45a). Hence

the volume is given by $V = \displaystyle\int_0^h \dfrac{\pi abz}{c} dz = \left[\dfrac{\pi abz^2}{2c}\right]_0^h = \dfrac{\pi abh^2}{c}$. Now the area of the elliptic base when $z = h$ is

$A = \dfrac{\pi abh}{c}$, as determined previously. Thus, $V = \dfrac{\pi abh^2}{c} = \dfrac{1}{2}\left(\dfrac{\pi abh}{c}\right)h = \dfrac{1}{2}$ (base)(altitude), as claimed.

49. $z = y^2$ 51. $z = x^2 + y^2$

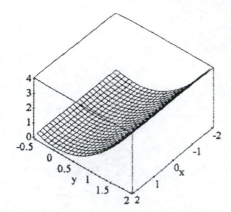   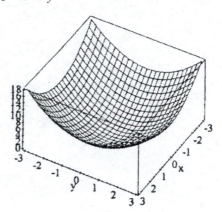

53-58.   Example CAS commands:

Maple:

with(plots);

eq := x^2/9 + y^2/36 = 1 - z^2/25;

implicitplot3d(eq, x=-3..3, y=-6..6, z=-5..5, scaling=constrained,

shading=zhue, axes=boxed, title="#53 (Section 12.6)");

Mathematica: (functions and domains may vary):

In the following chapter, you will consider contours or level curves for surfaces in three dimensions. For the purposes of plotting the functions of two variables expressed implicitly in this section, we will call upon the function **ContourPlot3D**. To insert the stated function, write all terms on the same side of the equal sign and the default contour equating that expression to zero will be plotted.

This built-in function requires the loading of a special graphics package.

<<Graphics`ContourPlot3D`

Clear[x, y, z]

ContourPlot3D[$x^2/9 - y^2/16 - z^2/2 - 1$, {x, −9, 9}, {y, −12, 12}, {z, −5, 5},

Axes → True, AxesLabel → {x, y, z}, Boxed → False,

PlotLabel → "Elliptic Hyperboloid of Two Sheets"]

Your identification of the plot may or may not be able to be done without considering the graph.

CHAPTER 11 PRACTICE EXERCISES

1. (a) $3\langle-3, 4\rangle - 4\langle 2, -5\rangle = \langle -9-8, 12+20\rangle = \langle -17, 32\rangle$

   (b) $\sqrt{17^2 + 32^2} = \sqrt{1313}$

4. (a) $\langle 5(2), 5(-5)\rangle = \langle 10, -25\rangle$

   (b) $\sqrt{10^2 + (-25)^2} = \sqrt{725} = 5\sqrt{29}$

6. $\left\langle \frac{\sqrt{3}}{2}, \frac{1}{2}\right\rangle$

7. $2\left(\frac{1}{\sqrt{4^2+1^2}}\right)(4\mathbf{i} - \mathbf{j}) = \left(\frac{8}{\sqrt{17}}\mathbf{i} - \frac{2}{\sqrt{17}}\mathbf{j}\right)$

9. length $= \left|\sqrt{2}\mathbf{i} + \sqrt{2}\mathbf{j}\right| = \sqrt{2+2} = 2$, $\sqrt{2}\mathbf{i} + \sqrt{2}\mathbf{j} = 2\left(\frac{1}{\sqrt{2}}\mathbf{i} + \frac{1}{\sqrt{2}}\mathbf{j}\right) \Rightarrow$ the direction is $\frac{1}{\sqrt{2}}\mathbf{i} + \frac{1}{\sqrt{2}}\mathbf{j}$

11. $t = \frac{\pi}{2} \Rightarrow \mathbf{v} = \left(-2\sin\frac{\pi}{2}\right)\mathbf{i} + \left(2\cos\frac{\pi}{2}\right)\mathbf{j} = -2\mathbf{i}$; length $= |-2\mathbf{i}| = \sqrt{4+0} = 2$; $-2\mathbf{i} = 2(-\mathbf{i}) \Rightarrow$ the direction is $-\mathbf{i}$

13. length $= |2\mathbf{i} - 3\mathbf{j} + 6\mathbf{k}| = \sqrt{4+9+36} = 7$, $2\mathbf{i} - 3\mathbf{j} + 6\mathbf{k} = 7\left(\frac{2}{7}\mathbf{i} - \frac{3}{7}\mathbf{j} + \frac{6}{7}\mathbf{k}\right) \Rightarrow$ the direction is $\frac{2}{7}\mathbf{i} - \frac{3}{7}\mathbf{j} + \frac{6}{7}\mathbf{k}$

15. $2\frac{\mathbf{v}}{|\mathbf{v}|} = 2\cdot\frac{4\mathbf{i} - \mathbf{j} + 4\mathbf{k}}{\sqrt{4^2 + (-1)^2 + 4^2}} = 2\cdot\frac{4\mathbf{i} - \mathbf{j} + 4\mathbf{k}}{\sqrt{33}} = \frac{8}{\sqrt{33}}\mathbf{i} - \frac{2}{\sqrt{33}}\mathbf{j} + \frac{8}{\sqrt{33}}\mathbf{k}$

17. $|\mathbf{v}| = \sqrt{1+1} = \sqrt{2}$, $|\mathbf{u}| = \sqrt{4+1+4} = 3$, $\mathbf{v}\cdot\mathbf{u} = 3$, $\mathbf{u}\cdot\mathbf{v} = 3$, $\mathbf{v}\times\mathbf{u} = \begin{vmatrix} \mathbf{i} & \mathbf{j} & \mathbf{k} \\ 1 & 1 & 0 \\ 2 & 1 & -2 \end{vmatrix} = -2\mathbf{i} + 2\mathbf{j} - \mathbf{k}$,

    $\mathbf{u}\times\mathbf{v} = -(\mathbf{v}\times\mathbf{u}) = 2\mathbf{i} - 2\mathbf{j} + \mathbf{k}$, $|\mathbf{v}\times\mathbf{u}| = \sqrt{4+4+1} = 3$, $\theta = \cos^{-1}\left(\frac{\mathbf{v}\cdot\mathbf{u}}{|\mathbf{v}||\mathbf{u}|}\right) = \cos^{-1}\left(\frac{1}{\sqrt{2}}\right) = \frac{\pi}{4}$,

    $|\mathbf{u}|\cos\theta = \frac{3}{\sqrt{2}}$, $\text{proj}_{\mathbf{v}}\mathbf{u} = \left(\frac{\mathbf{v}\cdot\mathbf{u}}{|\mathbf{v}||\mathbf{v}|}\right)\mathbf{v} = \frac{3}{2}(\mathbf{i}+\mathbf{j})$

19. $\text{proj}_{\mathbf{v}}\mathbf{u} = \left(\frac{\mathbf{v}\cdot\mathbf{u}}{|\mathbf{v}||\mathbf{v}|}\right)\mathbf{v} = \frac{4}{3}(2\mathbf{i}+\mathbf{j}+\mathbf{k})$ where $\mathbf{v}\cdot\mathbf{u} = 8$ and $\mathbf{v}\cdot\mathbf{v} = 6$

21. $\mathbf{u}\times\mathbf{v} = \begin{vmatrix} \mathbf{i} & \mathbf{j} & \mathbf{k} \\ 1 & 0 & 0 \\ 1 & 1 & 0 \end{vmatrix} = \mathbf{k}$

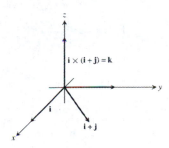

23. Let $\mathbf{v} = v_1\mathbf{i} + v_2\mathbf{j} + v_3\mathbf{k}$ and $\mathbf{w} = w_1\mathbf{i} + w_2\mathbf{j} + w_3\mathbf{k}$. Then $|\mathbf{v} - 2\mathbf{w}|^2 = |(v_1\mathbf{i} + v_2\mathbf{j} + v_3\mathbf{k}) - 2(w_1\mathbf{i} + w_2\mathbf{j} + w_3\mathbf{k})|^2$

$$= |(v_1 - 2w_1)\mathbf{i} + (v_2 - 2w_2)\mathbf{j} + (v_3 - 2w_3)\mathbf{k}|^2 = \left(\sqrt{(v_1 - 2w_1)^2 + (v_2 - 2w_2)^2 + (v_3 - 2w_3)^2}\right)^2$$

$$= (v_1^2 + v_2^2 + v_3^2) - 4(v_1 w_1 + v_2 w_2 + v_3 w_3) + 4(w_1^2 + w_2^2 + w_3^2) = |\mathbf{v}|^2 - 4\mathbf{v} \cdot \mathbf{w} + 4|\mathbf{w}|^2$$

$$|\mathbf{v}|^2 - 4|\mathbf{v}||\mathbf{w}|\cos\theta + 4|\mathbf{w}|^2 = 4 - 4(2)(3)\left(\cos\frac{\pi}{3}\right) + 36 = 40 - 24\left(\frac{1}{2}\right) = 40 - 12 = 28 \Rightarrow |\mathbf{v} - 2\mathbf{w}| = \sqrt{28} = 2\sqrt{7}$$

25. (a) area $= |\mathbf{u} \times \mathbf{v}| = abs\begin{vmatrix} \mathbf{i} & \mathbf{j} & \mathbf{k} \\ 1 & 1 & -1 \\ 2 & 1 & 1 \end{vmatrix} = |2\mathbf{i} - 3\mathbf{j} - \mathbf{k}| = \sqrt{4 + 9 + 1} = \sqrt{14}$

   (b) volume $= (\mathbf{u} \times \mathbf{v}) \cdot \mathbf{w} = \begin{vmatrix} 1 & 1 & -1 \\ 2 & 1 & 1 \\ -1 & -2 & 3 \end{vmatrix} = 1(3 + 2) - 1(6 - (-1)) - 1(-4 + 1) = 1$

27. The desired vector is $\mathbf{n} \times \mathbf{v}$ or $\mathbf{v} \times \mathbf{n}$ since $\mathbf{n} \times \mathbf{v}$ is perpendicular to both $\mathbf{n}$ and $\mathbf{v}$ and, therefore, also parallel to the plane.

29. The line L passes through the point $P(0, 0, -1)$ parallel to $\mathbf{v} = -\mathbf{i} + \mathbf{j} + \mathbf{k}$. With $\overrightarrow{PS} = 2\mathbf{i} + 2\mathbf{j} + \mathbf{k}$ and

   $\overrightarrow{PS} \times \mathbf{v} = \begin{vmatrix} \mathbf{i} & \mathbf{j} & \mathbf{k} \\ 2 & 2 & 1 \\ -1 & 1 & 1 \end{vmatrix} = (2 - 1)\mathbf{i} - (2 + 1)\mathbf{j} + (2 + 2)\mathbf{k} = \mathbf{i} - 3\mathbf{j} + 4\mathbf{k}$, we find the distance $d = \frac{|\overrightarrow{PS} \times \mathbf{v}|}{|\mathbf{v}|} = \frac{\sqrt{1 + 9 + 16}}{\sqrt{1 + 1 + 1}}$

   $= \frac{\sqrt{26}}{\sqrt{3}} = \frac{\sqrt{78}}{3}$.

31. Parametric equations for the line are $x = 1 - 3t$, $y = 2$, $z = 3 + 7t$.

33. The point $P(4, 0, 0)$ lies on the plane $x - y = 4$, and $\overrightarrow{PS} = (6 - 4)\mathbf{i} + 0\mathbf{j} + (-6 + 0)\mathbf{k} = 2\mathbf{i} - 6\mathbf{k}$ with $\mathbf{n} = \mathbf{i} - \mathbf{j}$

   $\Rightarrow d = \frac{|\mathbf{n} \cdot \overrightarrow{PS}|}{|\mathbf{n}|} = \left|\frac{2 + 0 + 0}{\sqrt{1 + 1 + 0}}\right| = \frac{2}{\sqrt{2}} = \sqrt{2}$.

35. $P(3, -2, 1)$ and $\mathbf{n} = 2\mathbf{i} + \mathbf{j} + \mathbf{k} \Rightarrow (2)(x - 3) + (1)(y - (-2)) + (1)(z - 1) = 0 \Rightarrow 2x + y + z = 5$

37. $P(1, -1, 2)$, $Q(2, 1, 3)$ and $R(-1, 2, -1) \Rightarrow \overrightarrow{PQ} = \mathbf{i} + 2\mathbf{j} + \mathbf{k}$, $\overrightarrow{PR} = -2\mathbf{i} + 3\mathbf{j} - 3\mathbf{k}$ and

   $\overrightarrow{PQ} \times \overrightarrow{PR} = \begin{vmatrix} \mathbf{i} & \mathbf{j} & \mathbf{K} \\ 1 & 2 & 1 \\ -2 & 3 & -3 \end{vmatrix} = -9\mathbf{i} + \mathbf{j} + 7\mathbf{k}$ is normal to the plane $\Rightarrow (-9)(x - 1) + (1)(y + 1) + (7)(z - 2) = 0$

   $\Rightarrow -9x + y + 7z = 4$

39. $\left(0, -\frac{1}{2}, -\frac{3}{2}\right)$, since $t = -\frac{1}{2}$, $y = -\frac{1}{2}$ and $z = -\frac{3}{2}$ when $x = 0$; $(-1, 0, -3)$, since $t = -1$, $x = -1$ and $z = -3$ when $y = 0$; $(1, -1, 0)$, since $t = 0$, $x = 1$ and $y = -1$ when $z = 0$

41. $\mathbf{n}_1 = \mathbf{i}$ and $\mathbf{n}_2 = \mathbf{i} + \mathbf{j} + \sqrt{2}\mathbf{k} \Rightarrow$ the desired angle is $\cos^{-1}\left(\frac{\mathbf{n}_1 \cdot \mathbf{n}_2}{|\mathbf{n}_1||\mathbf{n}_2|}\right) = \cos^{-1}\left(\frac{1}{2}\right) = \frac{\pi}{3}$

43. The direction of the line is $\mathbf{n}_1 \times \mathbf{n}_2 = \begin{vmatrix} \mathbf{i} & \mathbf{j} & \mathbf{k} \\ 1 & 2 & 1 \\ 1 & -1 & 2 \end{vmatrix} = 5\mathbf{i} - \mathbf{j} - 3\mathbf{k}$. Since the point $(-5, 3, 0)$ is on both planes, the

desired line is $x = -5 + 5t,\ y = 3 - t,\ z = -3t$.

45. (a) The corresponding normals are $\mathbf{n}_1 = 3\mathbf{i} + 6\mathbf{k}$ and $\mathbf{n}_2 = 2\mathbf{i} + 2\mathbf{j} - \mathbf{k}$ and since

$\mathbf{n}_1 \cdot \mathbf{n}_2 = (3)(2) + (0)(2) + (6)(-1) = 6 + 0 - 6 = 0$, we have that the plans are orthogonal

(b) The line of intersection is parallel to $\mathbf{n}_1 \times \mathbf{n}_2 = \begin{vmatrix} \mathbf{i} & \mathbf{j} & \mathbf{k} \\ 3 & 0 & 6 \\ 2 & 2 & -1 \end{vmatrix} = -12\mathbf{i} + 15\mathbf{j} + 6\mathbf{k}$. Now to find a point in the

intersection, solve $\begin{cases} 3x + 6z = 1 \\ 2x + 2y - z = 3 \end{cases} \Rightarrow \begin{cases} 3x + 6z = 1 \\ 12x + 12y - 6z = 18 \end{cases} \Rightarrow 15x + 12y = 19 \Rightarrow x = 0$ and $y = \frac{19}{12}$

$\Rightarrow \left(0, \frac{19}{12}, \frac{1}{6}\right)$ is a point on the line we seek. Therefore, the line is $x = -12t,\ y = \frac{19}{12} + 15t$ and $z = \frac{1}{6} + 6t$.

47. Yes; $\mathbf{v} \cdot \mathbf{n} = (2\mathbf{i} - 4\mathbf{j} + \mathbf{k}) \cdot (2\mathbf{i} + \mathbf{j} + 0\mathbf{k}) = 2 \cdot 2 - 4 \cdot 1 + 1 \cdot 0 = 0 \Rightarrow$ the vector is orthogonal to the plane's normal
$\Rightarrow \mathbf{v}$ is parallel to the plane

49. A normal to the plane is $\mathbf{n} = \overline{AB} \times \overline{AC} = \begin{vmatrix} \mathbf{i} & \mathbf{j} & \mathbf{k} \\ 2 & 0 & -1 \\ 2 & -1 & 0 \end{vmatrix} = -\mathbf{i} - 2\mathbf{j} - 2\mathbf{k} \Rightarrow$ the distance is

$d = \left| \frac{\overline{AP} \cdot \mathbf{n}}{\mathbf{n}} \right| = \left| \frac{(\mathbf{i} + 4\mathbf{j}) \cdot (-\mathbf{i} - 2\mathbf{j} - 2\mathbf{k})}{\sqrt{1 + 4 + 4}} \right| = \left| \frac{-1 - 8 + 0}{3} \right| = 3$

51. $\mathbf{n} = 2\mathbf{i} - \mathbf{j} - \mathbf{k}$ is normal to the plane $\Rightarrow \mathbf{n} \times \mathbf{v} = \begin{vmatrix} \mathbf{i} & \mathbf{j} & \mathbf{k} \\ 2 & -1 & -1 \\ 1 & 1 & 1 \end{vmatrix} = 0\mathbf{i} - 3\mathbf{j} + 3\mathbf{k} = -3\mathbf{j} + 3\mathbf{k}$ is orthogonal to $\mathbf{v}$ and

parallel to the plane

53. A vector parallel to the line of intersection is $\mathbf{v} = \mathbf{n}_1 \times \mathbf{n}_2 = \begin{vmatrix} \mathbf{i} & \mathbf{j} & \mathbf{k} \\ 1 & 2 & 1 \\ 1 & -1 & 2 \end{vmatrix} = 5\mathbf{i} - \mathbf{j} - 3\mathbf{k} \Rightarrow |\mathbf{v}| = \sqrt{25 + 1 + 9} = \sqrt{35}$

$\Rightarrow 2\left(\frac{\mathbf{v}}{|\mathbf{v}|}\right) = \frac{2}{\sqrt{35}}(5\mathbf{i} - \mathbf{j} - 3\mathbf{k})$ is the desired vector.

55. The line is represented by $x = 3 + 2t,\ y = 2 - t$, and $z = 1 + 2t$. It meets the plane $2x - y + 2z = -2$ when
$2(3 + 2t) - (2 - t) + 2(1 + 2t) = -2 \Rightarrow t = -\frac{8}{9} \Rightarrow$ the point is $\left(\frac{11}{9}, \frac{26}{9}, -\frac{7}{9}\right)$.

57. The intersection occurs when $(3 + 2t) + 3(2t) - t = -4 \Rightarrow t = -1 \Rightarrow$ the point is $(1, -2, -1)$. The required line
must be perpendicular to both the given line and to the normal, and hence is parallel to
$\begin{vmatrix} \mathbf{i} & \mathbf{j} & \mathbf{k} \\ 2 & 2 & 1 \\ 1 & 3 & -1 \end{vmatrix} = -5\mathbf{i} + 3\mathbf{j} + 4\mathbf{k} \Rightarrow$ the line is represented by $x = 1 - 5t,\ y = -2 + 3t$, and $z = -1 + 4t$.

59. The vector $\overrightarrow{AB} \times \overrightarrow{CD} = \begin{vmatrix} \mathbf{i} & \mathbf{j} & \mathbf{k} \\ 3 & -2 & 4 \\ \frac{26}{5} & 0 & -\frac{26}{5} \end{vmatrix} = \frac{26}{5}(2\mathbf{i} + 7\mathbf{j} + 2\mathbf{k})$ is normal to the plane and $A(-2, 0, -3)$ lies on the

plane $\Rightarrow 2(x+2) + 7(y-0) + 2(z-(-3)) = 0 \Rightarrow 2x + 7y + 2z + 10 = 0$ is an equation of the plane.

61. The vector $\overrightarrow{PQ} \times \overrightarrow{PR} = \begin{vmatrix} \mathbf{i} & \mathbf{j} & \mathbf{k} \\ 2 & -1 & 3 \\ -3 & 0 & 1 \end{vmatrix} = -\mathbf{i} - 11\mathbf{j} - 3\mathbf{k}$ is normal to the plane.

(a) No, the plane is not orthogonal to $\overrightarrow{PQ} \times \overrightarrow{PR}$.

(b) No, these equations represent a line, not a plane.

(c) No, the plane $(x+2) + 11(y-1) - 3z = 0$ has normal $\mathbf{i} + 11\mathbf{j} - 3\mathbf{k}$ which is not parallel to $\overrightarrow{PQ} \times \overrightarrow{PR}$.

(d) No, this vector equation is equivalent to the equations $3y + 3z = 3, 3x - 2z = -6,$ and $3x + 2y = -4$

$\Rightarrow x = -\frac{4}{3} - \frac{2}{3}t, y = t, z = 1 - t,$ which represents a line, not a plane.

(e) yes, this is a plane containing the point $R(-2, 1, 0)$ with normal $\overrightarrow{PQ} \times \overrightarrow{PR}$.

63. $\overrightarrow{AB} = -2\mathbf{i} + \mathbf{j} + \mathbf{k}, \overrightarrow{CD} = \mathbf{i} + 4\mathbf{j} - \mathbf{k},$ and $\overrightarrow{AC} = 2\mathbf{i} + \mathbf{j} \Rightarrow \mathbf{n} = \begin{vmatrix} \mathbf{i} & \mathbf{j} & \mathbf{k} \\ -2 & 1 & 1 \\ 1 & 4 & -1 \end{vmatrix} = -5\mathbf{i} - \mathbf{j} - 9\mathbf{k} \Rightarrow$ the distance is

$d = \left| \dfrac{(2\mathbf{i}+\mathbf{j})\cdot(-5\mathbf{i}-\mathbf{j}-9\mathbf{k})}{\sqrt{25+1+81}} \right| = \dfrac{11}{\sqrt{107}}$

65. $x^2 + y^2 + z^2 = 4$

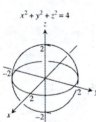

67. $4x^2 + 4y^2 + z^2 = 4$

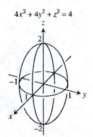

69. $z = -(x^2 + y^2)$

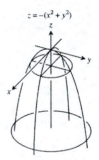

71. $x^2 + y^2 = z^2$

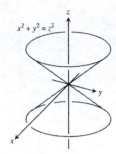

73. $x^2 - y^2 - z^2 = 4$

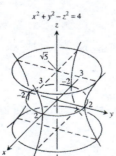

75. $y^2 - x^2 - z^2 = 1$

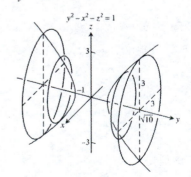

CHAPTER 11 ADDITIONAL AND ADVANCED EXERCISES

1. Information from ship A indicates the submarine is now on the line $L_1\colon x = 4 + 2t,\ y = 3t,\ z = -\tfrac{1}{3}t$; information from ship B indicates the submarine is now on the line $L_2\colon x = 18s,\ y = 5 - 6s,\ z = -s$. The current position of the sub is $\left(6, 3, -\tfrac{1}{3}\right)$ and occurs when the lines intersect at $t = 1$ and $s = \tfrac{1}{3}$. The straight line path of the submarine contains both point $P\left(2, -1, -\tfrac{1}{3}\right)$ and $Q\left(6, 3, -\tfrac{1}{3}\right)$; the line representing this path is $L\colon x = 2 + 4t,\ y = -1 + 4t,\ z = -\tfrac{1}{3}$. The submarine traveled the distance between P and Q in 4 minutes $\Rightarrow$ a speed of $\dfrac{|\overrightarrow{PQ}|}{4} = \dfrac{\sqrt{32}}{4} = \sqrt{2}$ thousand ft/min. In 20 minutes the submarine will move $20\sqrt{2}$ thousand ft from Q along the line $L \Rightarrow 20\sqrt{2} = \sqrt{(2 + 4t - 6)^2 + (-1 + 4t - 3)^2 + 0^2} \Rightarrow 800 = 16(t-1)^2 + 16(t-1)^2 = 32(t-1)^2$ $\Rightarrow (t-1)^2 = \tfrac{800}{32} = 25 \Rightarrow t = 6 \Rightarrow$ the submarine will be located at $\left(26, 23, -\tfrac{1}{3}\right)$ in 20 minutes.

3. Torque $= \left|\overrightarrow{PQ} \times \mathbf{F}\right| \Rightarrow 15\,\text{ft-lb} = \left|\overrightarrow{PQ}\right| |\mathbf{F}| \sin\tfrac{\pi}{2} = \tfrac{3}{4}\,\text{ft} \cdot |\mathbf{F}| \Rightarrow |\mathbf{F}| = 20\,\text{lb}$

5. (a) By the Law of Cosines we have $\cos\alpha = \dfrac{3^2 + 5^2 - 4^2}{2(3)(5)} = \tfrac{3}{5}$ and $\cos\beta = \dfrac{4^2 + 5^2 - 3^2}{2(4)(5)} = \tfrac{4}{5} \Rightarrow \sin\alpha = \tfrac{4}{5}$ and $\sin\beta = \tfrac{3}{5}$

$\Rightarrow \mathbf{F}_1 = \langle -|\mathbf{F}_1|\cos\alpha, |\mathbf{F}_1|\sin\alpha \rangle = \left\langle -\tfrac{3}{5}|\mathbf{F}_1|, \tfrac{4}{5}|\mathbf{F}_1| \right\rangle$, $\mathbf{F}_2 = \langle |\mathbf{F}_2|\cos\beta, |\mathbf{F}_2|\sin\beta \rangle = \left\langle \tfrac{4}{5}|\mathbf{F}_2|, \tfrac{3}{5}|\mathbf{F}_2| \right\rangle$, and

$\mathbf{w} = \langle 0, -100 \rangle$. Since $\mathbf{F}_1 + \mathbf{F}_2 = \langle 0, 100 \rangle \Rightarrow \left\langle -\tfrac{3}{5}|\mathbf{F}_1| + \tfrac{4}{5}|\mathbf{F}_2|, \tfrac{4}{5}|\mathbf{F}_1| + \tfrac{3}{5}|\mathbf{F}_2| \right\rangle = \langle 0, 100 \rangle \Rightarrow -\tfrac{3}{5}|\mathbf{F}_1| + \tfrac{4}{5}|\mathbf{F}_2| = 0$

and $\tfrac{4}{5}|\mathbf{F}_1| + \tfrac{3}{5}|\mathbf{F}_2| = 100$. Solving the first equation for $|\mathbf{F}_2|$ results in: $|\mathbf{F}_2| = \tfrac{3}{4}|\mathbf{F}_1|$. Substituting this result into the second equation gives us: $\tfrac{4}{5}|\mathbf{F}_1| + \tfrac{9}{20}|\mathbf{F}_1| = 100 \Rightarrow |\mathbf{F}_1| = 80\,\text{lb}. \Rightarrow |\mathbf{F}_2| = 60\,\text{lb}. \Rightarrow \mathbf{F}_1 = \langle -48, 64 \rangle$ and

$\mathbf{F}_2 = \langle 48, 36 \rangle$, and $\alpha = \tan^{-1}\left(\tfrac{4}{3}\right)$ and $\beta = \tan^{-1}\left(\tfrac{3}{4}\right)$

(b) By the Law of Cosines we have $\cos\alpha = \dfrac{5^2 + 13^2 - 12^2}{2(5)(13)} = \tfrac{5}{13}$ and $\cos\beta = \dfrac{12^2 + 13^2 - 5^2}{2(12)(13)} = \tfrac{12}{13} \Rightarrow \sin\alpha = \tfrac{12}{13}$ and

$\sin\beta = \tfrac{5}{13} \Rightarrow \mathbf{F}_1 = \langle -|\mathbf{F}_1|\cos\alpha, |\mathbf{F}_1|\sin\alpha \rangle = \left\langle -\tfrac{5}{13}|\mathbf{F}_1|, \tfrac{12}{13}|\mathbf{F}_1| \right\rangle$, $\mathbf{F}_2 = \langle |\mathbf{F}_2|\cos\beta, |\mathbf{F}_2|\sin\beta \rangle$

$= \left\langle \tfrac{12}{13}|\mathbf{F}_2|, \tfrac{5}{13}|\mathbf{F}_2| \right\rangle$, and $\mathbf{w} = \langle 0, -200 \rangle$. Since $\mathbf{F}_1 + \mathbf{F}_2 = \langle 0, 200 \rangle \Rightarrow \left\langle -\tfrac{5}{13}|\mathbf{F}_1| + \tfrac{12}{13}|\mathbf{F}_2|, \tfrac{12}{13}|\mathbf{F}_1| + \tfrac{5}{13}|\mathbf{F}_2| \right\rangle$

$= \langle 0, 200 \rangle \Rightarrow -\tfrac{5}{13}|\mathbf{F}_1| + \tfrac{12}{13}|\mathbf{F}_2| = 0$ and $\tfrac{12}{13}|\mathbf{F}_1| + \tfrac{5}{13}|\mathbf{F}_2| = 200$. Solving the first equation for $|\mathbf{F}_2|$ results in:

$|\mathbf{F}_2| = \tfrac{5}{12}|\mathbf{F}_1|$. Substituting this result into the second equation gives us: $\tfrac{12}{13}|\mathbf{F}_1| + \tfrac{25}{156}|\mathbf{F}_1| = 200$

$\Rightarrow |\mathbf{F}_1| = \tfrac{2400}{13} \approx 184.615\,\text{lb}. \Rightarrow |\mathbf{F}_2| = \tfrac{1000}{13} \approx 76.923\,\text{lb}. \Rightarrow \mathbf{F}_1 = \left\langle -\tfrac{12000}{1169}, \tfrac{28800}{1169} \right\rangle \approx \langle -71.006, 170.414 \rangle$ and

$\mathbf{F}_2 = \left\langle \tfrac{12000}{1169}, \tfrac{5000}{1169} \right\rangle \approx \langle 71.006, 29.586 \rangle$.

7. (a) If $P(x, y, z)$ is a point in the plane determined by the three points $P_1(x_1, y_1, z_1)$, $P_2(x_2, y_2, z_2)$ and

$P_3(x_3, y_3, z_3)$, then the vectors $\overrightarrow{PP_1}, \overrightarrow{PP_2}$ and $\overrightarrow{PP_3}$ all lie in the plane. Thus $\overrightarrow{PP_1} \cdot \left(\overrightarrow{PP_2} \times \overrightarrow{PP_3}\right) = 0$

$\Rightarrow \begin{vmatrix} x_1 - x & y_1 - y & z_1 - z \\ x_2 - x & y_2 - y & z_2 - z \\ x_3 - x & y_3 - y & z_3 - z \end{vmatrix} = 0$ by the determinant formula for the triple scalar product in Section 12.4.

(b) Subtract row 1 from rows 2, 3, and 4 and evaluate the resulting determinant (which has the same value as the given determinant) by cofactor expansion about column 4. This expansion is exactly the determinant in part (a) so we have all points $P(x, y, z)$ in the plane determined by $P_1(x_1, y_1, z_1)$, $P_2(x_2, y_2, z_2)$, and $P_3(x_3, y_3, z_3)$.

9. If $Q(x, y)$ is a point on the line $ax + by = c$, then $\overrightarrow{P_1 Q} = (x - x_1)\mathbf{i} + (y - y_1)\mathbf{j}$, and $\mathbf{n} = a\mathbf{i} + b\mathbf{j}$ is normal to the line. The distance is $\left|\text{proj}_{\mathbf{n}} \overrightarrow{P_1 Q}\right| = \left|\dfrac{[(x-x_1)\mathbf{i}+(y-y_1)\mathbf{j}]\cdot(a\mathbf{i}+b\mathbf{j})}{\sqrt{a^2+b^2}}\right| = \dfrac{|a(x-x_1)+b(y-y_1)|}{\sqrt{a^2+b^2}} = \dfrac{|ax_1+by_1-c|}{\sqrt{a^2+b^2}}$, since $c = ax + by$.

11. (a) If (x_1, y_1, z_1) is on the plane $Ax + By + Cz = D_1$, then the distance d between the planes is
$$d = \frac{|Ax_1 + By_1 + Cz_1 - D_2|}{\sqrt{A^2 + B^2 + C^2}} = \frac{|D_1 - D_2|}{|A\mathbf{i} + B\mathbf{j} + C\mathbf{k}|},\ \text{ since } Ax_1 + By_1 + Cz_1 = D_1,\ \text{ by Exercise 12(a).}$$

(b) $d = \dfrac{|12 - 6|}{\sqrt{4 + 9 + 1}} = \dfrac{6}{\sqrt{14}}$

(c) $\dfrac{|2(3)+(-1)(2)+2(-1)+4|}{\sqrt{14}} = \dfrac{|2(3)+(-1)(2)+2(-1)-D|}{\sqrt{14}} \Rightarrow D = 8 \text{ or } -4 \Rightarrow$ the desired plane is $2x - y + 2x = 8$

(d) Choose the point $(2, 0, 1)$ on the plane. Then $\dfrac{|3-D|}{\sqrt{6}} = 5 \Rightarrow D = 3 \pm 5\sqrt{6} \Rightarrow$ the desired planes are
$$x - 2y + z = 3 + 5\sqrt{6} \ \text{ and } \ x - 2y + z = 3 - 5\sqrt{6}.$$

13. (a) $\mathbf{u} \times \mathbf{v} = 2\mathbf{i} \times 2\mathbf{j} = 4\mathbf{k} \Rightarrow (\mathbf{u} \times \mathbf{v}) \times \mathbf{w} = \mathbf{0}; (\mathbf{u} \cdot \mathbf{w})\mathbf{v} - (\mathbf{v} \cdot \mathbf{w})\mathbf{u} = 0\mathbf{v} - 0\mathbf{u} = \mathbf{0}; \mathbf{v} \times \mathbf{w} = 4\mathbf{i} \Rightarrow \mathbf{u} \times (\mathbf{v} \times \mathbf{w}) = \mathbf{0};$
$(\mathbf{u} \cdot \mathbf{w})\mathbf{v} - (\mathbf{u} \cdot \mathbf{v})\mathbf{w} = 0\mathbf{v} - 0\mathbf{w} = \mathbf{0}$

(b) $\mathbf{u} \times \mathbf{v} = \begin{vmatrix} \mathbf{i} & \mathbf{j} & \mathbf{k} \\ 1 & -1 & 1 \\ 2 & 1 & -2 \end{vmatrix} = \mathbf{i} + 4\mathbf{j} + 3\mathbf{k} \Rightarrow (\mathbf{u} \times \mathbf{v}) \times \mathbf{w} = \begin{vmatrix} \mathbf{i} & \mathbf{j} & \mathbf{k} \\ 1 & 4 & 3 \\ -1 & 2 & -1 \end{vmatrix} = -10\mathbf{i} - 2\mathbf{j} + 6\mathbf{k};$

$(\mathbf{u} \cdot \mathbf{w})\mathbf{v} - (\mathbf{v} \cdot \mathbf{w})\mathbf{u} = -4(2\mathbf{i} + \mathbf{j} - 2\mathbf{k}) - 2(\mathbf{i} - \mathbf{j} + \mathbf{k}) = -10\mathbf{i} - 2\mathbf{j} + 6\mathbf{k};$

$\mathbf{v} \times \mathbf{w} = \begin{vmatrix} \mathbf{i} & \mathbf{j} & \mathbf{k} \\ 2 & 1 & -2 \\ -1 & 2 & -1 \end{vmatrix} = 3\mathbf{i} + 4\mathbf{j} + 5\mathbf{k} \Rightarrow \mathbf{u} \times (\mathbf{v} \times \mathbf{w}) = \begin{vmatrix} \mathbf{i} & \mathbf{j} & \mathbf{k} \\ 1 & -1 & 1 \\ 3 & 4 & 5 \end{vmatrix} = -9\mathbf{i} - 2\mathbf{j} + 7\mathbf{k};$

$(\mathbf{u} \cdot \mathbf{w})\mathbf{v} - (\mathbf{u} \cdot \mathbf{v})\mathbf{w} = -4(2\mathbf{i} + \mathbf{j} - 2\mathbf{k}) - (-1)(-\mathbf{i} + 2\mathbf{j} - \mathbf{k}) = -9\mathbf{i} - 2\mathbf{j} + 7\mathbf{k}$

(c) $\mathbf{u} \times \mathbf{v} = \begin{vmatrix} \mathbf{i} & \mathbf{j} & \mathbf{k} \\ 2 & 1 & 0 \\ 2 & -1 & 1 \end{vmatrix} = \mathbf{i} - 2\mathbf{j} - 4\mathbf{k} \Rightarrow (\mathbf{u} \times \mathbf{v}) \times \mathbf{w} = \begin{vmatrix} \mathbf{i} & \mathbf{j} & \mathbf{k} \\ 1 & -2 & -4 \\ 1 & 0 & 2 \end{vmatrix} = -4\mathbf{i} - 6\mathbf{j} + 2\mathbf{k};$

$(\mathbf{u} \cdot \mathbf{w})\mathbf{v} - (\mathbf{v} \cdot \mathbf{w})\mathbf{u} = 2(2\mathbf{i} - \mathbf{j} + \mathbf{k}) - 4(2\mathbf{i} + \mathbf{j}) = -4\mathbf{i} - 6\mathbf{j} + 2\mathbf{k};$

$\mathbf{v} \times \mathbf{w} = \begin{vmatrix} \mathbf{i} & \mathbf{j} & \mathbf{k} \\ 2 & -1 & 1 \\ 1 & 0 & 2 \end{vmatrix} = -2\mathbf{i} - 3\mathbf{j} + \mathbf{k} \Rightarrow \mathbf{u} \times (\mathbf{v} \times \mathbf{w}) = \begin{vmatrix} \mathbf{i} & \mathbf{j} & \mathbf{k} \\ 2 & 1 & 0 \\ -2 & -3 & 1 \end{vmatrix} = \mathbf{i} - 2\mathbf{j} - 4\mathbf{k};$

$(\mathbf{u} \cdot \mathbf{w})\mathbf{v} - (\mathbf{u} \cdot \mathbf{v})\mathbf{w} = 2(2\mathbf{i} - \mathbf{j} + \mathbf{k}) - 3(\mathbf{i} + 2\mathbf{k}) = \mathbf{i} - 2\mathbf{j} - 4\mathbf{k}$

(d) $\mathbf{u} \times \mathbf{v} = \begin{vmatrix} \mathbf{i} & \mathbf{j} & \mathbf{k} \\ 1 & 1 & -2 \\ -1 & 0 & -1 \end{vmatrix} = -\mathbf{i} + 3\mathbf{j} + \mathbf{k} \Rightarrow (\mathbf{u} \times \mathbf{v}) \times \mathbf{w} = \begin{vmatrix} \mathbf{i} & \mathbf{j} & \mathbf{k} \\ -1 & 3 & 1 \\ 2 & 4 & -2 \end{vmatrix} = -10\mathbf{i} - 10\mathbf{k};$

$(\mathbf{u} \cdot \mathbf{w})\mathbf{v} - (\mathbf{v} \cdot \mathbf{w})\mathbf{u} = 10(-\mathbf{i} - \mathbf{k}) - 0(\mathbf{i} + \mathbf{j} - 2\mathbf{k}) = -10\mathbf{i} - 10\mathbf{k};$

$\mathbf{v} \times \mathbf{w} = \begin{vmatrix} \mathbf{i} & \mathbf{j} & \mathbf{k} \\ -1 & 0 & -1 \\ 2 & 4 & -2 \end{vmatrix} = 4\mathbf{i} - 4\mathbf{j} - 4\mathbf{k} \Rightarrow \mathbf{u} \times (\mathbf{v} \times \mathbf{w}) = \begin{vmatrix} \mathbf{i} & \mathbf{j} & \mathbf{k} \\ 1 & 1 & -2 \\ 4 & -4 & -4 \end{vmatrix} = -12\mathbf{i} - 4\mathbf{j} - 8\mathbf{k};$

$(\mathbf{u} \cdot \mathbf{w})\mathbf{v} - (\mathbf{u} \cdot \mathbf{v})\mathbf{w} = 10(-\mathbf{i} - \mathbf{k}) - 1(2\mathbf{i} + 4\mathbf{j} - 2\mathbf{k}) = -12\mathbf{i} - 4\mathbf{j} - 8\mathbf{k}$

15. The formula is always true; $\mathbf{u} \times [\mathbf{u} \times (\mathbf{u} \times \mathbf{v})] \cdot \mathbf{w} = \mathbf{u} \times [(\mathbf{u} \cdot \mathbf{v})\mathbf{u} - (\mathbf{u} \cdot \mathbf{u})\mathbf{v}] \cdot \mathbf{w} = [(\mathbf{u} \cdot \mathbf{v})\mathbf{u} \times \mathbf{u} - (\mathbf{u} \cdot \mathbf{u})\mathbf{u} \times \mathbf{v}] \cdot \mathbf{w}$

$= -|\mathbf{u}|^2 \, \mathbf{u} \times \mathbf{v} \cdot \mathbf{w} = -|\mathbf{u}|^2 \, \mathbf{u} \cdot \mathbf{v} \times \mathbf{w}$

17. If $\mathbf{u} = a\mathbf{i} + b\mathbf{j}$ and $\mathbf{v} = c\mathbf{i} + d\mathbf{j}$, then $\mathbf{u} \cdot \mathbf{v} = |\mathbf{u}||\mathbf{v}|\cos\theta \Rightarrow ac + bd = \sqrt{a^2 + b^2}\sqrt{c^2 + d^2}\cos\theta$

$\Rightarrow (ac + bd)^2 = (a^2 + b^2)(c^2 + d^2)\cos^2\theta \Rightarrow (ac + bd)^2 \le (a^2 + b^2)(c^2 + d^2),$ since $\cos^2\theta \le 1.$

CHAPTER 12   VECTOR-VALUED FUNCTIONS AND MOTION IN SPACE

12.1 CURVES IN SPACE AND THEIR TANGENTS

1. $x = t+1$ and $y = t^2 - 1 \Rightarrow y = (x-1)^2 - 1 = x^2 - 2x$; $\mathbf{v} = \frac{d\mathbf{r}}{dt} = \mathbf{i} + 2t\mathbf{j} \Rightarrow \mathbf{a} = \frac{d\mathbf{v}}{dt} = 2\mathbf{j} \Rightarrow \mathbf{v} = \mathbf{i} + 2\mathbf{j}$ and $\mathbf{a} = 2\mathbf{j}$ at $t = 1$

3. $x = e^t$ and $y = \frac{2}{9}e^{2t} \Rightarrow y = \frac{2}{9}x^2$; $\mathbf{v} = \frac{d\mathbf{r}}{dt} = e^t\mathbf{i} + \frac{4}{9}e^{2t}\mathbf{j} \Rightarrow \mathbf{a} = e^t\mathbf{i} + \frac{8}{9}e^{2t}\mathbf{j} \Rightarrow \mathbf{v} = 3\mathbf{i} + 4\mathbf{j}$ and $\mathbf{a} = 3\mathbf{i} + 8\mathbf{j}$ at $t = \ln 3$

5. $\mathbf{v} = \frac{d\mathbf{r}}{dt} = (\cos t)\mathbf{i} - (\sin t)\mathbf{j}$ and $\mathbf{a} = \frac{d\mathbf{v}}{dt} = -(\sin t)\mathbf{i} - (\cos t)\mathbf{j}$

$\Rightarrow$ for $t = \frac{\pi}{4}$, $\mathbf{v}\left(\frac{\pi}{4}\right) = \frac{\sqrt{2}}{2}\mathbf{i} - \frac{\sqrt{2}}{2}\mathbf{j}$ and $\mathbf{a}\left(\frac{\pi}{4}\right) = -\frac{\sqrt{2}}{2}\mathbf{i} - \frac{\sqrt{2}}{2}\mathbf{j}$;

for $t = \frac{\pi}{2}$, $\mathbf{v}\left(\frac{\pi}{2}\right) = -\mathbf{j}$ and $\mathbf{a}\left(\frac{\pi}{2}\right) = -\mathbf{i}$

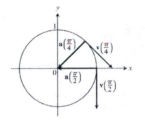

7. $\mathbf{v} = \frac{d\mathbf{r}}{dt} = (1 - \cos t)\mathbf{i} + (\sin t)\mathbf{j}$ and $\mathbf{a} = \frac{d\mathbf{v}}{dt} = (\sin t)\mathbf{i} + (\cos t)\mathbf{j}$

$\Rightarrow$ for $t = \pi$, $\mathbf{v}(\pi) = 2\mathbf{i}$ and $\mathbf{a}(\pi) = -\mathbf{j}$;

for $t = \frac{3\pi}{2}$, $\mathbf{v}\left(\frac{3\pi}{2}\right) = \mathbf{i} - \mathbf{j}$ and $\mathbf{a}\left(\frac{3\pi}{2}\right) = -\mathbf{i}$

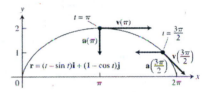

9. $\mathbf{r} = (t+1)\mathbf{i} + \left(t^2 - 1\right)\mathbf{j} + 2t\mathbf{k} \Rightarrow \mathbf{v} = \frac{d\mathbf{r}}{dt} = \mathbf{i} + 2t\mathbf{j} + 2\mathbf{k} \Rightarrow \mathbf{a} = \frac{d^2\mathbf{r}}{dt^2} = 2\mathbf{j}$; Speed: $|\mathbf{v}(1)| = \sqrt{1^2 + (2(1))^2 + 2^2} = 3$;

Direction: $\frac{\mathbf{v}(1)}{|\mathbf{v}(1)|} = \frac{\mathbf{i} + 2(1)\mathbf{j} + 2\mathbf{k}}{3} = \frac{1}{3}\mathbf{i} + \frac{2}{3}\mathbf{j} + \frac{2}{3}\mathbf{k} \Rightarrow \mathbf{v}(1) = 3\left(\frac{1}{3}\mathbf{i} + \frac{2}{3}\mathbf{j} + \frac{2}{3}\mathbf{k}\right)$

11. $\mathbf{r} = (2\cos t)\mathbf{i} + (3\sin t)\mathbf{j} + 4t\mathbf{k} \Rightarrow \mathbf{v} = \frac{d\mathbf{r}}{dt} = (-2\sin t)\mathbf{i} + (3\cos t)\mathbf{j} + 4\mathbf{k} \Rightarrow \mathbf{a} = \frac{d^2\mathbf{r}}{dt^2} = (-2\cos t)\mathbf{i} - (3\sin t)\mathbf{j}$;

Speed: $\left|\mathbf{v}\left(\frac{\pi}{2}\right)\right| = \sqrt{\left(-2\sin \frac{\pi}{2}\right)^2 + \left(3\cos \frac{\pi}{2}\right)^2 + 4^2} = 2\sqrt{5}$;

Direction: $\frac{\mathbf{v}\left(\frac{\pi}{2}\right)}{\left|\mathbf{v}\left(\frac{\pi}{2}\right)\right|} = \left(-\frac{2}{2\sqrt{5}}\sin \frac{\pi}{2}\right)\mathbf{i} + \left(\frac{3}{2\sqrt{5}}\cos \frac{\pi}{2}\right)\mathbf{j} + \frac{4}{2\sqrt{5}}\mathbf{k} = -\frac{1}{\sqrt{5}}\mathbf{i} + \frac{2}{\sqrt{5}}\mathbf{k} \Rightarrow \mathbf{v}\left(\frac{\pi}{2}\right) = 2\sqrt{5}\left(-\frac{1}{\sqrt{5}}\mathbf{i} + \frac{2}{\sqrt{5}}\mathbf{k}\right)$

13. $\mathbf{r} = (2\ln(t+1))\mathbf{i} + t^2\mathbf{j} + \frac{t^2}{2}\mathbf{k} \Rightarrow \mathbf{v} = \frac{d\mathbf{r}}{dt} = \left(\frac{2}{t+1}\right)\mathbf{i} + 2t\mathbf{j} + t\mathbf{k} \Rightarrow \mathbf{a} = \frac{d^2\mathbf{r}}{dt^2} = \left[\frac{-2}{(t+1)^2}\right]\mathbf{i} + 2\mathbf{j} + \mathbf{k}$;

Speed: $|\mathbf{v}(1)| = \sqrt{\left(\frac{2}{1+1}\right)^2 + (2(1))^2 + 1^2} = \sqrt{6}$;

Direction: $\frac{\mathbf{v}(1)}{|\mathbf{v}(1)|} = \frac{\left(\frac{2}{1+1}\right)\mathbf{i} + 2(1)\mathbf{j} + (1)\mathbf{k}}{\sqrt{6}} = \frac{1}{\sqrt{6}}\mathbf{i} + \frac{2}{\sqrt{6}}\mathbf{j} + \frac{1}{\sqrt{6}}\mathbf{k} \Rightarrow \mathbf{v}(1) = \sqrt{6}\left(\frac{1}{\sqrt{6}}\mathbf{i} + \frac{2}{\sqrt{6}}\mathbf{j} + \frac{1}{\sqrt{6}}\mathbf{k}\right)$

15. $\mathbf{v} = 3\mathbf{i} + \sqrt{3}\mathbf{j} + 2t\mathbf{k}$ and $\mathbf{a} = 2\mathbf{k} \Rightarrow \mathbf{v}(0) = 3\mathbf{i} + \sqrt{3}\mathbf{j}$ and $\mathbf{a}(0) = 2\mathbf{k} \Rightarrow |\mathbf{v}(0)| = \sqrt{3^2 + \left(\sqrt{3}\right)^2 + 0^2} = \sqrt{12}$ and

   $|\mathbf{a}(0)| = \sqrt{2^2} = 2;\ \mathbf{v}(0) \cdot \mathbf{a}(0) = 0 \Rightarrow \cos\theta = 0 \Rightarrow \theta = \frac{\pi}{2}$

17. $\mathbf{v} = \left(\frac{2t}{t^2+1}\right)\mathbf{i} + \left(\frac{1}{t^2+1}\right)\mathbf{j} + t\left(t^2+1\right)^{-1/2}\mathbf{k}$ and $\mathbf{a} = \left[\frac{-2t^2+2}{\left(t^2+1\right)^2}\right]\mathbf{i} - \left[\frac{2t}{\left(t^2+1\right)^2}\right]\mathbf{j} + \left[\frac{1}{\left(t^2+1\right)^{3/2}}\right]\mathbf{k} \Rightarrow \mathbf{v}(0) = \mathbf{j}$ and

   $\mathbf{a}(0) = 2\mathbf{i} + \mathbf{k} \Rightarrow |\mathbf{v}(0)| = 1$ and $|\mathbf{a}(0)| = \sqrt{2^2 + 1^2} = \sqrt{5};\ \mathbf{v}(0) \cdot \mathbf{a}(0) = 0 \Rightarrow \cos\theta = 0 \Rightarrow \theta = \frac{\pi}{2}$

19. $\mathbf{r}(t) = (\sin t)\mathbf{i} + \left(t^2 - \cos t\right)\mathbf{j} + e^t\mathbf{k} \Rightarrow \mathbf{v}(t) = (\cos t)\mathbf{i} + (2t + \sin t)\mathbf{j} + e^t\mathbf{k};\ t_0 = 0 \Rightarrow \mathbf{v}(t_0) = \mathbf{i} + \mathbf{k}$ and

   $\mathbf{r}(t_0) = P_0 = (0, -1, 1) \Rightarrow x = 0 + t = t,\ y = -1,$ and $z = 1 + t$ are parametric equations of the tangent line

21. $\mathbf{r}(t) = (\ln t)\mathbf{i} + \frac{t-1}{t+2}\mathbf{j} + (t \ln t)\mathbf{k} \Rightarrow \mathbf{v}(t) = \frac{1}{t}\mathbf{i} + \frac{3}{(t+2)^2}\mathbf{j} + (\ln t + 1)\mathbf{k};\ t_0 = 1 \Rightarrow \mathbf{v}(1) = \mathbf{i} + \frac{1}{3}\mathbf{j} + \mathbf{k}$ and

   $\mathbf{r}(t_0) = P_0 = (0, 0, 0) \Rightarrow x = 0 + t = t,\ y = 0 + \frac{1}{3}t = \frac{1}{3}t,$ and $z = 0 + t = t$ are parametric equations of the tangent
   line

23. (a) $\mathbf{v}(t) = -(\sin t)\mathbf{i} + (\cos t)\mathbf{j} \Rightarrow \mathbf{a}(t) = -(\cos t)\mathbf{i} - (\sin t)\mathbf{j};$

   (i) $|\mathbf{v}(t)| = \sqrt{(-\sin t)^2 + (\cos t)^2} = 1 \Rightarrow$ constant speed;

   (ii) $\mathbf{v} \cdot \mathbf{a} = (\sin t)(\cos t) - (\cos t)(\sin t) = 0 \Rightarrow$ yes, orthogonal;

   (iii) counterclockwise movement;

   (iv) yes, $\mathbf{r}(0) = \mathbf{i} + 0\mathbf{j}$

   (b) $\mathbf{v}(t) = -(2 \sin 2t)\mathbf{i} + (2 \cos 2t)\mathbf{j} \Rightarrow \mathbf{a}(t) = -(4 \cos 2t)\mathbf{i} - (4 \sin 2t)\mathbf{j};$

   (i) $|\mathbf{v}(t)| = \sqrt{4 \sin^2 2t + 4 \cos^2 2t} = 2 \Rightarrow$ constant speed;

   (ii) $\mathbf{v} \cdot \mathbf{a} = 8 \sin 2t \cos 2t - 8 \cos 2t \sin 2t = 0 \Rightarrow$ yes, orthogonal;

   (iii) counterclockwise movement;

   (iv) yes, $\mathbf{r}(0) = \mathbf{i} + 0\mathbf{j}$

   (c) $\mathbf{v}(t) = -\sin\left(t - \frac{\pi}{2}\right)\mathbf{i} + \cos\left(t - \frac{\pi}{2}\right)\mathbf{j} \Rightarrow \mathbf{a}(t) = -\cos\left(t - \frac{\pi}{2}\right)\mathbf{i} - \sin\left(t - \frac{\pi}{2}\right)\mathbf{j};$

   (i) $|\mathbf{v}(t)| = \sqrt{\sin^2\left(t - \frac{\pi}{2}\right) + \cos^2\left(t - \frac{\pi}{2}\right)} = 1 \Rightarrow$ constant speed;

   (ii) $\mathbf{v} \cdot \mathbf{a} = \sin\left(t - \frac{\pi}{2}\right)\cos\left(t - \frac{\pi}{2}\right) - \cos\left(t - \frac{\pi}{2}\right)\sin\left(t - \frac{\pi}{2}\right) = 0 \Rightarrow$ yes, orthogonal;

   (iii) counterclockwise movement;

   (iv) no, $\mathbf{r}(0) = 0\mathbf{i} - \mathbf{j}$ instead of $\mathbf{i} + 0\mathbf{j}$

   (d) $\mathbf{v}(t) = -(\sin t)\mathbf{i} - (\cos t)\mathbf{j} \Rightarrow \mathbf{a}(t) = -(\cos t)\mathbf{i} + (\sin t)\mathbf{j};$

   (i) $|\mathbf{v}(t)| = \sqrt{(-\sin t)^2 + (-\cos t)^2} = 1 \Rightarrow$ constant speed;

   (ii) $\mathbf{v} \cdot \mathbf{a} = (\sin t)(\cos t) - (\cos t)(\sin t) = 0 \Rightarrow$ yes, orthogonal;

   (iii) clockwise movement;

   (iv) yes, $\mathbf{r}(0) = \mathbf{i} - 0\mathbf{j}$

(e) $\mathbf{v}(t) = -(2t \sin t)\mathbf{i} + (2t \cos t)\mathbf{j} \Rightarrow \mathbf{a}(t) = -(2 \sin t + 2t \cos t)\mathbf{i} + (2 \cos t - 2t \sin t)\mathbf{j}$;

    (i) $|\mathbf{v}(t)| = \sqrt{(-(2t \sin t))^2 + (2t \cos t)^2} = \sqrt{4t^2 \left(\sin^2 t + \cos^2 t\right)} = 2|t| = 2t, t \ge 0 \Rightarrow$ variable speed;

    (ii) $\mathbf{v} \cdot \mathbf{a} = 4\left(t \sin^2 t + t^2 \sin t \cos t\right) + 4\left(t \cos^2 t - t^2 \cos t \sin t\right) = 4t \ne 0$ in general $\Rightarrow$ not orthogonal in general;

    (iii) counterclockwise movement;

    (iv) yes, $\mathbf{r}(0) = \mathbf{i} + 0\mathbf{j}$

25. The velocity vector is tangent to the graph of $y^2 = 2x$ at the point $(2, 2)$, has length 5, and a positive $\mathbf{i}$ component. Now, $y^2 = 2x \Rightarrow 2y\frac{dy}{dx} = 2 \Rightarrow \frac{dy}{dx}\Big|_{(2,2)} = \frac{2}{2 \cdot 2} = \frac{1}{2} \Rightarrow$ the tangent vector lies in the direction of the vector $\mathbf{i} + \frac{1}{2}\mathbf{j} \Rightarrow$ the velocity vector is $\mathbf{v} = \frac{5}{\sqrt{1 + \frac{1}{4}}}\left(\mathbf{i} + \frac{1}{2}\mathbf{j}\right) = \frac{5}{\left(\frac{\sqrt{5}}{2}\right)}\left(\mathbf{i} + \frac{1}{2}\mathbf{j}\right) = 2\sqrt{5}\mathbf{i} + \sqrt{5}\mathbf{j}$

27. $\frac{d}{dt}(\mathbf{r} \cdot \mathbf{r}) = \mathbf{r} \cdot \frac{d\mathbf{r}}{dt} + \frac{d\mathbf{r}}{dt} \cdot \mathbf{r} = 2\mathbf{r} \cdot \frac{d\mathbf{r}}{dt} = 2 \cdot 0 = 0 \Rightarrow \mathbf{r} \cdot \mathbf{r}$ is a constant $\Rightarrow |\mathbf{r}| = \sqrt{\mathbf{r} \cdot \mathbf{r}}$ is constant

29. (a) $\mathbf{u} = f(t)\mathbf{i} + g(t)\mathbf{j} + h(t)\mathbf{k} \Rightarrow c\mathbf{u} = cf(t)\mathbf{i} + cg(t)\mathbf{j} + ch(t)\mathbf{k} \Rightarrow \frac{d}{dt}(c\mathbf{u}) = c\frac{df}{dt}\mathbf{i} + c\frac{dg}{dt}\mathbf{j} + c\frac{dh}{dt}\mathbf{k}$

    $= c\left(\frac{df}{dt}\mathbf{i} + \frac{dg}{dt}\mathbf{j} + \frac{dh}{dt}\mathbf{k}\right) = c\frac{d\mathbf{u}}{dt}$

   (b) $\mathbf{u} = x(t)\mathbf{i} + y(t)\mathbf{j} + z(t)\mathbf{k} \Rightarrow f(t)\mathbf{u} = f(t)x(t)\mathbf{i} + f(t)y(t)\mathbf{j} + f(t)z(t)\mathbf{k}$

    $\Rightarrow \frac{d}{dt}(f(t)\mathbf{u}) = \left[\frac{df}{dt}x(t) + f(t)\frac{dx}{dt}\right]\mathbf{i} + \left[\frac{df}{dt}y(t) + f(t)\frac{dy}{dt}\right]\mathbf{j} + \left[\frac{df}{dt}z(t) + f(t)\frac{dz}{dt}\right]\mathbf{k}$

    $= \frac{df}{dt}[x(t)\mathbf{i} + y(t)\mathbf{j} + z(t)\mathbf{k}] + f(t)\left[\frac{dx}{dt}\mathbf{i} + \frac{dy}{dt}\mathbf{j} + \frac{dz}{dt}\mathbf{k}\right] = \frac{df}{dt}\mathbf{u} + f(t)\frac{d\mathbf{u}}{dt}$

31. Suppose $\mathbf{r}$ is continuous at $t = t_0$. Then $\lim\limits_{t \to t_0} \mathbf{r}(t) = \mathbf{r}(t_0) \Leftrightarrow \lim\limits_{t \to t_0}\left[f(t)\mathbf{i} + g(t)\mathbf{j} + h(t)\mathbf{k}\right]$

    $= f(t_0)\mathbf{i} + g(t_0)\mathbf{j} + h(t_0)\mathbf{k} \Leftrightarrow \lim\limits_{t \to t_0} f(t) = f(t_0), \lim\limits_{t \to t_0} g(t) = g(t_0)$, and $\lim\limits_{t \to t_0} h(t) = h(t_0) \Leftrightarrow f, g$, and h are continuous at $t = t_0$.

33. $r'(t_0)$ exists $\Rightarrow f'(t_0)\mathbf{i} + g'(t_0)\mathbf{j} + h'(t_0)\mathbf{k}$ exists $\Rightarrow f'(t_0), g'(t_0), h'(t_0)$ all exist $\Rightarrow f, g$, and h are continuous at $t = t_0 \Rightarrow \mathbf{r}(t)$ is continuous at $t = t_0$

35-37.   Example CAS commands:

   Maple:

```
> with( plots );

r := t -> [sin(t)-t*cos(t),cos(t)+t*sin(t),t^2];

t0 := 3*Pi/2;

lo := 0;

hi := 6*Pi;

P1 := spacecurve( r(t), t=lo..hi, axes=boxed, thickness=3 ):

display( P1, title="#35(a) (Section 13.1)" );
```

```
Dr := unapply( diff(r(t),t), t );                    # (b)

Dr(t0);                                              # (c)

q1 := expand( r(t0) + Dr(t0)*(t-t0) );

T := unapply( q1, t );

P2 := spacecurve( T(t), t=lo..hi, axes=boxed, thickness=3, color=black ):

display( [P1,P2], title="#35(d) (Section 13.1)" );
```

39. Example CAS commands:

Maple:

```
a := 'a'; b := 'b';

r := (a,b,t) -> [cos(a*t),sin(a*t),b*t];

Dr := unapply( diff(r(a,b,t),t), (a,b,t) );

t0 := 3*Pi/2;

q1 := expand( r(a,b,t0) + Dr(a,b,t0)*(t-t0) );

T := unapply( q1, (a,b,t) );

lo := 0;

hi := 4*Pi;

P := NULL:

for a in [ 1, 2, 4, 6 ] do

   P1 := spacecurve( r(a,1,t), t=lo..hi, thickness=3 ):

   P2 := spacecurve( T(a,1,t), t=lo..hi, thickness=3, color=black ):

   P := P, display( [P1,P2], axes=boxed, title=sprintf("#39 (Section 13.1)\n a=%a",a) );

end do:

display( [P], insequence=true );
```

35-39.   Example CAS commands:

Mathematica: (assigned functions, parameters, and intervals will vary)
The x-y-z components for the curve are entered as a list of functions of t. The unit vectors **i, j, k** are not inserted.
If a graph is too small, highlight it and drag out a corner or side to make it larger.
Only the components of r[t] and values for t0, tmin, and tmax require alteration for each problem.

```
Clear[r, v, t, x, y, z]

r[t_]={ Sin[t] − t Cos[t], Cos[t] + t Sin[t], t^2}

t0 = 3π/2; tmin = 0; tmax = 6π;

ParametricPlot3D[Evaluate[r[t]], {t, tmin, tmax}, AxesLabel → {x, y, z}];

v[t_]= r'[t]

tanline[t_]= v[t0]t + r[t0]

ParametricPlot3D[Evaluate[{r[t], tanline[t]}], {t, tmin, tmax}, AxesLabel → {x, y, z}];
```

For 39 and 40, the curve can be defined as a function of t, a, and b. Leave a space between a and t and b and t.

Clear[r, v, t, x, y, z, a, b]

r[t_,a_,b_]:={Cos[a t], Sin[a t], b t}

t0= 3π/2; tmin= 0; tmax= 4π;

v[t_,a_,b_]= D[r[t, a, b], t]

tanline[t_,a_,b_]= v[t0, a, b] t + r[t0, a, b]

pa1=ParametricPlot3D[Evaluate[{r[t, 1, 1], tanline[t, 1, 1]}], {t,tmin, tmax}, AxesLabel → {x, y, z}];

pa2=ParametricPlot3D[Evaluate[{r[t, 2, 1], tanline[t, 2, 1]}], {t,tmin, tmax}, AxesLabel → {x, y, z}];

pa4=ParametricPlot3D[Evaluate[{r[t, 4, 1], tanline[t, 4, 1]}], {t,tmin, tmax}, AxesLabel → {x, y, z}];

pa6=ParametricPlot3D[Evaluate[{r[t, 6, 1], tanline[t, 6, 1]}], {t,tmin, tmax}, AxesLabel → {x, y, z}];

Show[GraphicsRow[{pa1, pa2, pa4, pa6}]]

12.2 INTEGRALS OF VECTOR FUNCTIONS; PROJECTILE MOTION

1. $\int_0^1 \left[t^3 \mathbf{i} + 7\mathbf{j} + (t+1)\mathbf{k} \right] dt = \left[\frac{t^4}{4} \right]_0^1 \mathbf{i} + [7t]_0^1 \mathbf{j} + \left[\frac{t^2}{2} + t \right]_0^1 \mathbf{k} = \frac{1}{4}\mathbf{i} + 7\mathbf{j} + \frac{3}{2}\mathbf{k}$

3. $\int_{-\pi/4}^{\pi/4} \left[(\sin t)\mathbf{i} + (1+\cos t)\mathbf{j} + \left(\sec^2 t \right)\mathbf{k} \right] dt = [-\cos t]_{-\pi/4}^{\pi/4} \mathbf{i} + [t+\sin t]_{-\pi/4}^{\pi/4} \mathbf{j} + [\tan t]_{-\pi/4}^{\pi/4} \mathbf{k} = \left(\frac{\pi+2\sqrt{2}}{2} \right)\mathbf{j} + 2\mathbf{k}$

5. $\int_1^4 \left(\frac{1}{t}\mathbf{i} + \frac{1}{5-t}\mathbf{j} + \frac{1}{2t}\mathbf{k} \right) dt = [\ln t]_1^4 \mathbf{i} + [-\ln(5-t)]_1^4 \mathbf{j} + \left[\frac{1}{2}\ln t \right]_1^4 \mathbf{k} = (\ln 4)\mathbf{i} + (\ln 4)\mathbf{j} + (\ln 2)\mathbf{k}$

7. $\int_0^1 \left(te^{t^2}\mathbf{i} + e^{-t}\mathbf{j} + \mathbf{k} \right) dt = \left[\frac{1}{2}e^{t^2} \right]_0^1 \mathbf{i} - \left[e^{-t} \right]_0^1 \mathbf{j} + [t]_0^1 \mathbf{k} = \frac{e-1}{2}\mathbf{i} + \frac{e-1}{e}\mathbf{i} + \mathbf{k}$

9. $\int_0^{\pi/2} \left[(\cos t)\mathbf{i} - (\sin 2t)\mathbf{j} + \left(\sin^2 t \right)\mathbf{k} \right] dt = \int_0^{\pi/2} \left[(\cos t)\mathbf{i} - (\sin 2t)\mathbf{j} + \left(\frac{1}{2} - \frac{1}{2}\cos 2t \right)\mathbf{k} \right] dt$

   $= [\sin t]_0^{\pi/2} \mathbf{i} + \left[\frac{1}{2}\cos t \right]_0^{\pi/2} \mathbf{j} + \left[\frac{1}{2}t - \frac{1}{4}\sin 2t \right]_0^{\pi/2} \mathbf{k} = \mathbf{i} - \mathbf{j} + \frac{\pi}{4}\mathbf{k}$

11. $\mathbf{r} = \int (-t\mathbf{i} - t\mathbf{j} - t\mathbf{k}) dt = -\frac{t^2}{2}\mathbf{i} - \frac{t^2}{2}\mathbf{j} - \frac{t^2}{2}\mathbf{k} + \mathbf{C};$ $\mathbf{r}(0) = 0\mathbf{i} - 0\mathbf{j} - 0\mathbf{k} + \mathbf{C} = \mathbf{i} + 2\mathbf{j} + 3\mathbf{k} \Rightarrow \mathbf{C} = \mathbf{i} + 2\mathbf{j} + 3\mathbf{k}$

   $\Rightarrow \mathbf{r} = \left(-\frac{t^2}{2} + 1 \right)\mathbf{i} + \left(-\frac{t^2}{2} + 2 \right)\mathbf{j} + \left(-\frac{t^2}{2} + 3 \right)\mathbf{k}$

13. $\mathbf{r} = \int \left[\left(\frac{3}{2}(t+1)^{1/2} \right)\mathbf{i} + e^{-t}\mathbf{j} + \left(\frac{1}{t+1} \right)\mathbf{k} \right] dt = (t+1)^{3/2}\mathbf{i} - e^{-t}\mathbf{j} + \ln(t+1)\mathbf{k} + \mathbf{C};$

   $\mathbf{r}(0) = (0+1)^{3/2}\mathbf{i} - e^{-0}\mathbf{j} + \ln(0+1)\mathbf{k} + \mathbf{C} = \mathbf{k} \Rightarrow \mathbf{C} = -\mathbf{i} + \mathbf{j} + \mathbf{k} \Rightarrow \mathbf{r} = \left[(t+1)^{3/2} - 1 \right]\mathbf{i} + \left(1 - e^{-t} \right)\mathbf{j} + [1 + \ln(t+1)]\mathbf{k}$

15. $\frac{d\mathbf{r}}{dt} = \int (-32\mathbf{k}) dt = -32t\mathbf{k} + \mathbf{C}_1;$ $\frac{d\mathbf{r}}{dt}(0) = 8\mathbf{i} + 8\mathbf{j} \Rightarrow -32(0)\mathbf{k} + \mathbf{C}_1 = 8\mathbf{i} + 8\mathbf{j} \Rightarrow \mathbf{C}_1 = 8\mathbf{i} + 8\mathbf{j} \Rightarrow \frac{d\mathbf{r}}{dt} = 8\mathbf{i} + 8\mathbf{j} - 32t\mathbf{k};$

   $\mathbf{r} = \int (8\mathbf{i} + 8\mathbf{j} - 32t\mathbf{k}) dt = 8t\mathbf{i} + 8t\mathbf{j} - 16t^2\mathbf{k} + \mathbf{C}_2;$ $\mathbf{r}(0) = 100\mathbf{k} \Rightarrow 8(0)\mathbf{i} + 8(0)\mathbf{j} - 16(0)^2\mathbf{k} + \mathbf{C}_2 = 100\mathbf{k}$

   $\Rightarrow \mathbf{C}_2 = 100\mathbf{k} \Rightarrow \mathbf{r} = 8t\mathbf{i} + 8t\mathbf{j} + \left(100 - 16t^2 \right)\mathbf{k}$

17. $\frac{d\mathbf{v}}{dt} = \mathbf{a} = 3\mathbf{i} - \mathbf{j} + \mathbf{k} \Rightarrow \mathbf{v}(t) = 3t\mathbf{i} - t\mathbf{j} + t\mathbf{k} + \mathbf{C}_1$; the particle travels in the direction of the vector

$(4-1)\mathbf{i} + (1-2)\mathbf{j} + (4-3)\mathbf{k} = 3\mathbf{i} - \mathbf{j} + \mathbf{k}$ (since it travels in a straight line), and at time $t = 0$ it has speed 2

$\Rightarrow \mathbf{v}(0) = \frac{2}{\sqrt{9+1+1}}(3\mathbf{i} - \mathbf{j} + \mathbf{k}) = \mathbf{C}_1 \Rightarrow \frac{d\mathbf{r}}{dt} = \mathbf{v}(t) = \left(3t + \frac{6}{\sqrt{11}}\right)\mathbf{i} - \left(t + \frac{2}{\sqrt{11}}\right)\mathbf{j} + \left(t + \frac{2}{\sqrt{11}}\right)\mathbf{k}$

$\Rightarrow \mathbf{r}(t) = \left(\frac{3}{2}t^2 + \frac{6}{\sqrt{11}}t\right)\mathbf{i} - \left(\frac{1}{2}t^2 + \frac{2}{\sqrt{11}}t\right)\mathbf{j} + \left(\frac{1}{2}t^2 + \frac{2}{\sqrt{11}}t\right)\mathbf{k} + \mathbf{C}_2$; $\mathbf{r}(0) = \mathbf{i} + 2\mathbf{j} + 3\mathbf{k} = \mathbf{C}_2$

$\Rightarrow \mathbf{r}(t) = \left(\frac{3}{2}t^2 + \frac{6}{\sqrt{11}}t + 1\right)\mathbf{i} - \left(\frac{1}{2}t^2 + \frac{2}{\sqrt{11}}t - 2\right)\mathbf{j} + \left(\frac{1}{2}t^2 + \frac{2}{\sqrt{11}}t + 3\right)\mathbf{k} = \left(\frac{1}{2}t^2 + \frac{2}{\sqrt{11}}t\right)(3\mathbf{i} - \mathbf{j} + \mathbf{k}) + (\mathbf{i} - 2\mathbf{j} + 3\mathbf{k})$

19. $x = (v_0 \cos\alpha)t \Rightarrow (21\,\text{km})\left(\frac{1000\,\text{m}}{1\,\text{km}}\right) = (840\,\text{m/s})(\cos 60°)t \Rightarrow t = \frac{21{,}000\,\text{m}}{(840\,\text{m/s})(\cos 60°)} = 50$ seconds

21. (a) $t = \frac{2v_0 \sin\alpha}{g} = \frac{2(500\,\text{m/s})(\sin 45°)}{9.8\,\text{m/s}^2} \approx 72.2$ seconds; $R = \frac{v_0^2}{g}\sin 2\alpha = \frac{(500\,\text{m/s})^2}{9.8\,\text{m/s}^2}(\sin 90°) \approx 25{,}510.2$ m

    (b) $x = (v_0 \cos\alpha)t \Rightarrow 5000\,\text{m} = (500\,\text{m/s})(\cos 45°)t \Rightarrow t = \frac{5000\,\text{m}}{(500\,\text{m/s})(\cos 45°)} \approx 14.14$ s; thus,

    $y = (v_0 \sin\alpha)t - \frac{1}{2}gt^2 \Rightarrow y \approx (500\,\text{m/s})(\sin 45°)(14.14\,\text{s}) - \frac{1}{2}(9.8\,\text{m/s}^2)(14.14\,\text{s})^2 \approx 4020$ m

    (c) $y_{\max} = \frac{(v_0 \sin\alpha)^2}{2g} = \frac{((500\,\text{m/s})(\sin 45°))^2}{2(9.8\,\text{m/s}^2)} \approx 6378$ m

23. (a) $R = \frac{v_0^2}{g}\sin 2\alpha \Rightarrow 10\,\text{m} = \left(\frac{v_0^2}{9.8\,\text{m/s}^2}\right)(\sin 90°) \Rightarrow v_0^2 = 98\,\text{m}^2/\text{s}^2 \Rightarrow v_0 \approx 9.9$ m/s;

    (b) $6\,\text{m} \approx \frac{(9.9\,\text{m/s}^2)}{9.8\,\text{m/s}^2}(\sin 2\alpha) \Rightarrow \sin 2\alpha \approx 0.59999 \Rightarrow 2\alpha \approx 36.87°$ or $143.12° \Rightarrow \alpha \approx 18.4°$ or $71.6°$

25. $R = \frac{v_0^2}{g}\sin 2\alpha \Rightarrow 16{,}000\,\text{m} = \frac{(400\,\text{m/s})^2}{9.8\,\text{m/s}^2}\sin 2\alpha \Rightarrow \sin 2\alpha = 0.98 \Rightarrow 2\alpha \approx 78.5°$ or $2\alpha \approx 101.5° \Rightarrow \alpha \approx 39.3°$

    or $50.7°$

27. The projectile reaches its maximum height when its vertical component of velocity is zero

    $\Rightarrow \frac{dy}{dt} = v_0 \sin\alpha - gt = 0 \Rightarrow t = \frac{v_0 \sin\alpha}{g} \Rightarrow y_{\max} = (v_0 \sin\alpha)\left(\frac{v_0 \sin\alpha}{g}\right) - \frac{1}{2}g\left(\frac{v_0 \sin\alpha}{g}\right)^2 = \frac{(v_0 \sin\alpha)^2}{g} - \frac{(v_0 \sin\alpha)^2}{2g}$

    $= \frac{(v_0 \sin\alpha)^2}{2g}$. To find the flight time we find the time when the projectile lands: $(v_0 \sin\alpha)t - \frac{1}{2}gt^2 = 0$

    $\Rightarrow t\left(v_0 \sin\alpha - \frac{1}{2}gt\right) = 0 \Rightarrow t = 0$ or $t = \frac{2v_0 \sin\alpha}{g}$. Since $t = 0$ is the time when the projectile is fired, then

    $t = \frac{2v_0 \sin\alpha}{g}$ is the time when the projectile strikes the ground. The range is the value of the horizontal

    component when $t = \frac{2v_0 \sin\alpha}{g} \Rightarrow R = x = (v_0 \cos\alpha)\left(\frac{2v_0 \sin\alpha}{g}\right) = \frac{v_0^2}{g}(2\sin\alpha\cos\alpha) = \frac{v_0^2}{g}\sin 2\alpha$. The range is

    largest when $2\alpha = 1 \Rightarrow \alpha = 45°$.

29. $\frac{d\mathbf{r}}{dt} = \int(-g\mathbf{j})dt = -gt\mathbf{j} + \mathbf{C}_1$ and $\frac{d\mathbf{r}}{dt}(0) = (v_0\cos\alpha)\mathbf{i} + (v_0\sin\alpha)\mathbf{j} \Rightarrow -g(0)\mathbf{j} - \mathbf{C}_1 = (v_0\cos\alpha)\mathbf{i} + (v_0\sin\alpha)\mathbf{j}$

$\Rightarrow \mathbf{C}_1 = (v_0\cos\alpha)\mathbf{i} + (v_0\sin\alpha)\mathbf{j} \Rightarrow \frac{d\mathbf{r}}{dt} = (v_0\cos\alpha)\mathbf{i} + (v_0\sin\alpha - gt)\mathbf{j}; \quad \mathbf{r} = \int[(v_0\cos\alpha)\mathbf{i} + (v_0\sin\alpha - gt)\mathbf{j}]dt$

$= (v_0t\cos\alpha)\mathbf{i} + \left(v_0t\sin\alpha - \frac{1}{2}gt^2\right)\mathbf{j} + \mathbf{C}_2$ and $\mathbf{r}(0) = x_0\mathbf{i} + y_0\mathbf{j} \Rightarrow [v_0(0)\cos\alpha]\mathbf{i} + \left[v_0(0)\sin\alpha - \frac{1}{2}g(0)^2\right]\mathbf{j} + \mathbf{C}_2$

$= x_0\mathbf{i} + y_0\mathbf{j} \Rightarrow \mathbf{C}_2 = x_0\mathbf{i} + y_0\mathbf{j} \Rightarrow \mathbf{r} = (x_0 + v_0t\cos\alpha)\mathbf{i} + \left(y_0 + v_0t\sin\alpha - \frac{1}{2}gt^2\right)\mathbf{j} \Rightarrow x = x_0 + v_0t\cos\alpha$ and

$y = y_0 + v_0t\sin\alpha - \frac{1}{2}gt^2$

31. (a) (Assuming that "x" is zero at the point of impact:)

$\mathbf{r}(t) = (x(t))\mathbf{i} + (y(t))\mathbf{j}$; where $x(t) = (35\cos 27°)t$ and $y(t) = 4 + (35\sin 27°)t - 16t^2$.

(b) $y_{max} = \frac{(v_0\sin\alpha)^2}{2g} + 4 = \frac{(35\sin 27°)^2}{64} + 4 \approx 7.945$ feet, which is reached at $t = \frac{v_0\sin\alpha}{g} = \frac{35\sin 27°}{32}$

≈ 0.497 seconds.

(c) For the time, solve $y = 4 + (35\sin 27°)t - 16t^2 = 0$ for t, using the quadratic formula

$t = \frac{35\sin 27° + \sqrt{(-35\sin 27°)^2 + 256}}{32} \approx 1.201$ sec. Then the range is about $x(1.201) = (35\cos 27°)(1.201)$

≈ 37.453 feet.

(d) For the time, solve $y = 4 + (35\sin 27°)t - 16t^2 = 7$ for t, using the quadratic formula

$t = \frac{35\sin 27° + \sqrt{(-35\sin 27°)^2 - 192}}{32} \approx 0.254$ and 0.740 seconds. At those times the ball is about

$x(0.254) = (35\cos 27°)(0.254) \approx 7.921$ feet and $x(0.740) = (35\cos 27°)(0.740) \approx 23.077$ feet from the

impact point, or about $37.453 - 7.921 \approx 29.532$ feet and $37.453 - 23.077 \approx 14.376$ feet from the landing

spot.

(e) Yes. It changes things because the ball won't clear the net ($y_{max} \approx 7.945$).

33. (a) $\int_a^b k\mathbf{r}(t)\,dt = \int_a^b [kf(t)\mathbf{i} + kg(t)\mathbf{j} + kh(t)\mathbf{k}]\,dt = \int_a^b [kf(t)]\,dt\,\mathbf{i} + \int_a^b [kg(t)]\,dt\,\mathbf{j} + \int_a^b [kh(t)]\,dt\,\mathbf{k}$

$= k\left(\int_a^b f(t)\,dt\,\mathbf{i} + \int_a^b g(t)\,dt\,\mathbf{j} + \int_a^b h(t)\,dt\,\mathbf{k}\right) = k\int_a^b \mathbf{r}(t)\,dt$

(b) $\int_a^b [\mathbf{r}_1(t) \pm \mathbf{r}_2(t)]\,dt = \int_a^b ([f_1(t)\mathbf{i} + g_1(t)\mathbf{j} + h_1(t)\mathbf{k}] \pm [f_2(t)\mathbf{i} + g_2(t)\mathbf{j} + h_2(t)\mathbf{k}])\,dt$

$= \int_a^b ([f_1(t) \pm f_2(t)]\mathbf{i} + [g_1(t) \pm g_2(t)]\mathbf{j} + [h_1(t) \pm h_2(t)]\mathbf{k})\,dt$

$= \int_a^b [f_1(t) \pm f_2(t)]\,dt\,\mathbf{i} + \int_a^b [g_1(t) \pm g_2(t)]\,dt\,\mathbf{j} + \int_a^b [h_1(t) \pm h_2(t)]\,dt\,\mathbf{k}$

$= \left[\int_a^b f_1(t)\,dt \pm \int_a^b f_2(t)\,dt\,\mathbf{i}\right] + \left[\int_a^b g_1(t)\,dt\,\mathbf{j} \pm \int_a^b g_2(t)\,dt\,\mathbf{j}\right] + \left[\int_a^b h_1(t)\,dt\,\mathbf{k} \pm \int_a^b h_2(t)\,dt\,\mathbf{k}\right]$

$= \int_a^b \mathbf{r}_1(t)\,dt \pm \int_a^b \mathbf{r}_2(t)\,dt$

(c) Let $\mathbf{C} = c_1\mathbf{i} + c_2\mathbf{j} + c_3\mathbf{k}$. Then $\int_a^b \mathbf{C} \cdot \mathbf{r}(t)\,dt = \int_a^b \left[c_1 f(t) + c_2 g(t) + c_3 h(t) \right] dt$

$$= c_1 \int_a^b f(t)\,dt + c_2 \int_a^b g(t)\,dt + c_3 \int_a^b h(t)\,dt = \mathbf{C} \cdot \int_a^b \mathbf{r}(t)\,dt;$$

$$\int_a^b \mathbf{C} \times \mathbf{r}(t)\,dt = \int_a^b \left[c_2 h(t) - c_3 g(t) \right] \mathbf{i} + \left[c_3 f(t) - c_1 h(t) \right] \mathbf{j} + \left[c_1 g(t) - c_2 f(t) \right] \mathbf{k}\,dt$$

$$= \left[c_2 \int_a^b h(t)\,dt - c_3 \int_a^b g(t)\,dt \right] \mathbf{i} + \left[c_3 \int_a^b f(t)\,dt - c_1 \int_a^b h(t)\,dt \right] \mathbf{j} + \left[c_1 \int_a^b g(t)\,dt - c_2 \int_a^b f(t)\,dt \right] \mathbf{k}$$

$$= \mathbf{C} \times \int_a^b \mathbf{r}(t)\,dt$$

35. (a) If $\mathbf{R}_1(t)$ and $\mathbf{R}_2(t)$ have identical derivatives on I, then $\frac{d\mathbf{R}_1}{dt} = \frac{df_1}{dt}\mathbf{i} + \frac{dg_1}{dt}\mathbf{j} + \frac{dh_1}{dt}\mathbf{k} = \frac{df_2}{dt}\mathbf{i} + \frac{dg_2}{dt}\mathbf{j} + \frac{dh_2}{dt}\mathbf{k}$

$= \frac{d\mathbf{R}_2}{dt} \Rightarrow \frac{df_1}{dt} = \frac{df_2}{dt},\ \frac{dg_1}{dt} = \frac{dg_2}{dt},\ \frac{dh_1}{dt} = \frac{dh_2}{dt} \Rightarrow f_1(t) = f_2(t) + c_1,\ g_1(t) = g_2(t) + c_2,\ h_1(t) = h_2(t) + c_3$

$\Rightarrow f_1(t)\mathbf{i} + g_1(t)\mathbf{j} + h_1(t)\mathbf{k} = \left[f_2(t) + c_1 \right]\mathbf{i} + \left[g_2(t) + c_2 \right]\mathbf{j} + \left[h_2(t) + c_3 \right]\mathbf{k} \Rightarrow \mathbf{R}_1(t) = \mathbf{R}_2(t) + \mathbf{C}$, where

$\mathbf{C} = c_1\mathbf{i} + c_2\mathbf{j} + c_3\mathbf{k}$.

(b) Let $\mathbf{R}(t)$ be an antiderivative of $\mathbf{r}(t)$ on I. Then $\mathbf{R}'(t) = \mathbf{r}(t)$. If $\mathbf{U}(t)$ is an antiderivative of $\mathbf{r}(t)$ on I, then $\mathbf{U}'(t) = \mathbf{r}(t)$. Thus $\mathbf{U}'(t) = \mathbf{R}'(t)$ on $I \Rightarrow \mathbf{U}(t) = \mathbf{R}(t) + \mathbf{C}$.

12.3 ARC LENGTH IN SPACE

1. $\mathbf{r} = (2\cos t)\mathbf{i} + (2\sin t)\mathbf{j} + \sqrt{5}t\mathbf{k} \Rightarrow \mathbf{v} = (-2\sin t)\mathbf{i} + (2\cos t)\mathbf{j} + \sqrt{5}\mathbf{k} \Rightarrow |\mathbf{v}| = \sqrt{(-2\sin t)^2 + (2\cos t)^2 + \left(\sqrt{5}\right)^2}$

$= \sqrt{4\sin^2 t + 4\cos^2 t + 5} = 3;\ \mathbf{T} = \frac{\mathbf{v}}{|\mathbf{v}|} = \left(-\tfrac{2}{3}\sin t\right)\mathbf{i} + \left(\tfrac{2}{3}\cos t\right)\mathbf{j} + \frac{\sqrt{5}}{3}\mathbf{k}$ and Length $= \int_0^\pi |\mathbf{v}|\,dt = \int_0^\pi 3\,dt$

$= \left[3t \right]_0^\pi = 3\pi$

3. $\mathbf{r} = t\mathbf{i} + \tfrac{2}{3}t^{3/2}\mathbf{k} \Rightarrow \mathbf{v} = \mathbf{i} + t^{1/2}\mathbf{k} \Rightarrow |\mathbf{v}| = \sqrt{1^2 + \left(t^{1/2} \right)^2} = \sqrt{1 + t};\ \mathbf{T} = \frac{\mathbf{v}}{|\mathbf{v}|} = \frac{1}{\sqrt{1+t}}\mathbf{i} + \frac{\sqrt{t}}{\sqrt{1+t}}\mathbf{k}$ and

Length $= \int_0^8 \sqrt{1 + t}\,dt = \left[\tfrac{2}{3}(1 + t)^{3/2} \right]_0^8 = \frac{52}{3}$

5. $\mathbf{r} = \left(\cos^3 t \right)\mathbf{j} + \left(\sin^3 t \right)\mathbf{k} \Rightarrow \mathbf{v} = \left(-3\cos^2 t \sin t \right)\mathbf{j} + \left(3\sin^2 t \cos t \right)\mathbf{k}$

$\Rightarrow |\mathbf{v}| = \sqrt{\left(-3\cos^2 t \sin t \right)^2 + \left(3\sin^2 t \cos t \right)^2} = \sqrt{\left(9\cos^2 t \sin^2 t \right)\left(\cos^2 t + \sin^2 t \right)} = 3|\cos t \sin t|;$

$\mathbf{T} = \frac{\mathbf{v}}{|\mathbf{v}|} = \frac{-3\cos^2 t \sin t}{3|\cos t \sin t|}\mathbf{j} + \frac{3\sin^2 t \cos t}{3|\cos t \sin t|}\mathbf{k} = (-\cos t)\mathbf{j} + (\sin t)\mathbf{k}$, if $0 \le t \le \tfrac{\pi}{2}$, and

Length $= \int_0^{\pi/2} 3|\cos t \sin t|\,dt = \int_0^{\pi/2} 3\cos t \sin t\,dt = \int_0^{\pi/2} \tfrac{3}{2}\sin 2t\,dt = \left[-\tfrac{3}{4}\cos 2t \right]_0^{\pi/2} = \frac{3}{2}$

7. $\mathbf{r} = (t\cos t)\mathbf{i} + (t\sin t)\mathbf{j} + \tfrac{2\sqrt{2}}{3}t^{3/2}\mathbf{k} \Rightarrow \mathbf{v} = (\cos t - t\sin t)\mathbf{i} + (\sin t + t\cos t)\mathbf{j} + \left(\sqrt{2}t^{1/2} \right)\mathbf{k}$

$\Rightarrow |\mathbf{v}| = \sqrt{(\cos t - t\sin t)^2 + (\sin t + t\cos t)^2 + \left(\sqrt{2t} \right)^2} = \sqrt{1 + t^2 + 2t} = \sqrt{(t + 1)^2} = |t + 1| = t + 1$, if $t \ge 0$;

$\mathbf{T} = \frac{\mathbf{v}}{|\mathbf{v}|} = \left(\frac{\cos t - t\sin t}{t + 1} \right)\mathbf{i} + \left(\frac{\sin t + t\cos t}{t + 1} \right)\mathbf{j} + \left(\frac{\sqrt{2t}^{1/2}}{t + 1} \right)\mathbf{k}$ and Length $= \int_0^\pi (t + 1)\,dt = \left[\tfrac{t^2}{2} + t \right]_0^\pi = \frac{\pi^2}{2} + \pi$

9. Let $P(t_0)$ denote the point. Then $\mathbf{v} = (5\cos t)\mathbf{i} - (5\sin t)\mathbf{j} + 12\mathbf{k}$ and $26\pi = \int_0^{t_0} \sqrt{25\cos^2 t + 25\sin^2 t + 144}\, dt$

$= \int_0^{t_0} 13\, dt = 13t_0 \Rightarrow t_0 = 2\pi$, and the point is $P(2\pi) = (5\sin 2\pi, 5\cos 2\pi, 24\pi) = (0, 5, 24\pi)$

11. $\mathbf{r} = (4\cos t)\mathbf{i} + (4\sin t)\mathbf{j} + 3t\mathbf{k} \Rightarrow \mathbf{v} = (-4\sin t)\mathbf{i} + (4\cos t)\mathbf{j} + 3\mathbf{k} \Rightarrow |\mathbf{v}| = \sqrt{(-4\sin t)^2 + (4\cos t)^2 + 3^2}$

$= \sqrt{25} = 5 \Rightarrow s(t) = \int_0^t 5\, d\tau = 5t \Rightarrow \text{Length} = s\left(\frac{\pi}{2}\right) = \frac{5\pi}{2}$

13. $\mathbf{r} = \left(e^t \cos t\right)\mathbf{i} + \left(e^t \sin t\right)\mathbf{j} + e^t\mathbf{k} \Rightarrow \mathbf{v} = \left(e^t \cos t - e^t \sin t\right)\mathbf{i} + \left(e^t \sin t + e^t \cos t\right)\mathbf{j} + e^t\mathbf{k}$

$\Rightarrow |\mathbf{v}| = \sqrt{\left(e^t \cos t - e^t \sin t\right)^2 + \left(e^t \sin t + e^t \cos t\right)^2 + \left(e^t\right)^2} = \sqrt{3e^{2t}} = \sqrt{3}e^t \Rightarrow s(t) = \int_0^t \sqrt{3}e^\tau\, d\tau = \sqrt{3}e^t - \sqrt{3}$

$\Rightarrow \text{Length} = s(0) - s(-\ln 4) = 0 - \left(\sqrt{3}e^{-\ln 4} - \sqrt{3}\right) = \frac{3\sqrt{3}}{4}$

15. $\mathbf{r} = \left(\sqrt{2}t\right)\mathbf{i} + \left(\sqrt{2}t\right)\mathbf{j} + \left(1 - t^2\right)\mathbf{k} \Rightarrow \mathbf{v} = \sqrt{2}\mathbf{i} + \sqrt{2}\mathbf{j} - 2t\mathbf{k} \Rightarrow |\mathbf{v}| = \sqrt{\left(\sqrt{2}\right)^2 + \left(\sqrt{2}\right)^2 + (-2t)^2} = \sqrt{4 + 4t^2}$

$= 2\sqrt{1 + t^2} \Rightarrow \text{Length} = \int_0^1 2\sqrt{1 + t^2}\, dt = \left[2\left(\frac{t}{2}\sqrt{1 + t^2} + \frac{1}{2}\ln\left(t + \sqrt{1 + t^2}\right)\right)\right]_0^1 = \sqrt{2} + \ln\left(1 + \sqrt{2}\right)$

17. (a) $\mathbf{r} = (\cos t)\mathbf{i} + (\sin t)\mathbf{j} + (1 - \cos t)\mathbf{k}, 0 \le t \le 2\pi \Rightarrow x = \cos t, y = \sin t, z = 1 - \cos t$

$\Rightarrow x^2 + y^2 = \cos^2 t + \sin^2 t = 1$, a right circular cylinder with the z-axis as the axis and radius $= 1$.

Therefore $P(\cos t, \sin t, 1 - \cos t)$ lies on the cylinder $x^2 + y^2 = 1; t = 0 \Rightarrow P(1, 0, 0)$ is on the curve;

$t = \frac{\pi}{2} \Rightarrow Q(0, 1, 1)$ is on the curve; $t = \pi \Rightarrow R(-1, 0, 2)$ is on the curve. Then $\overrightarrow{PQ} = -\mathbf{i} + \mathbf{j} + \mathbf{k}$ and

$\overrightarrow{PR} = -2\mathbf{i} + 2\mathbf{k} \Rightarrow \overrightarrow{PQ} \times \overrightarrow{PR} = \begin{vmatrix} \mathbf{i} & \mathbf{j} & \mathbf{k} \\ -1 & 1 & 1 \\ -2 & 0 & 2 \end{vmatrix} = 2\mathbf{i} + 2\mathbf{k}$ is a vector normal to the plane of P, Q, and R. Then

the plane containing P, Q, and R has an equation $2x + 2z = 2(1) + 2(0)$ or $x + z = 1$. Any point on the

curve will satisfy this equation since $x + z = \cos t + (1 - \cos t) = 1$. Therefore, any point on the curve lies on

the intersection of the cylinder $x^2 + y^2 = 1$ and the plane $x + z = 1 \Rightarrow$ the curve is an ellipse.

(b) $\mathbf{v} = (-\sin t)\mathbf{i} + (\cos t)\mathbf{j} + (\sin t)\mathbf{k} \Rightarrow |\mathbf{v}| = \sqrt{\sin^2 t + \cos^2 t + \sin^2 t} = \sqrt{1 + \sin^2 t}$

$\Rightarrow \mathbf{T} = \frac{\mathbf{v}}{|\mathbf{v}|} = \frac{(-\sin t)\mathbf{i} + (\cos t)\mathbf{j} + (\sin t)\mathbf{k}}{\sqrt{1 + \sin^2 t}} \Rightarrow \mathbf{T}(0) = \mathbf{j}, \mathbf{T}\left(\frac{\pi}{2}\right) = \frac{-\mathbf{i} + \mathbf{k}}{\sqrt{2}}, \mathbf{T}(\pi) = -\mathbf{j}, \mathbf{T}\left(\frac{3\pi}{2}\right) = \frac{\mathbf{i} - \mathbf{k}}{\sqrt{2}}$

(c) $\mathbf{a} = (-\cos t)\mathbf{i} - (\sin t)\mathbf{j} + (\cos t)\mathbf{k}; \mathbf{n} = \mathbf{i} + \mathbf{k}$ is

normal to the plane $x + z = 1$
$\Rightarrow \mathbf{n} \cdot \mathbf{a} = -\cos t + \cos t = 0$
$\Rightarrow \mathbf{a}$ is orthogonal to $\mathbf{n}$
$\Rightarrow \mathbf{a}$ is parallel to the plane;
$\mathbf{a}(0) = -\mathbf{i} + \mathbf{k}, \mathbf{a}\left(\frac{\pi}{2}\right) = -\mathbf{j}, \mathbf{a}(\pi) = \mathbf{i} - \mathbf{k}, \mathbf{a}\left(\frac{3\pi}{2}\right) = \mathbf{j}$

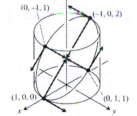

(d) $|\mathbf{v}| = \sqrt{1 + \sin^2 t}$ (See part (b)) $\Rightarrow L = \int_0^{2\pi} \sqrt{1 + \sin^2 t}\, dt$

(e) $L \approx 7.64$ (by *Mathematica*)

19. $\angle PQB = \angle QOB = t$ and $PQ = \text{arc }(AQ) = t$ since

$PQ = $ length of the unwound string $= $ length of arc (AQ);

thus $x = OB + BC = OB + DP = \cos t + t\sin t$, and

$y = PC = QB - QD = \sin t - t\cos t$

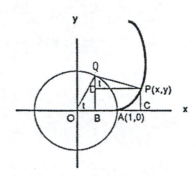

21. $\mathbf{v} = \frac{d}{dt}\left(x_0 + t\,u_1\right)\mathbf{i} + \frac{d}{dt}\left(y_0 + t\,u_2\right)\mathbf{j} + \frac{d}{dt}\left(z_0 + t\,u_3\right)\mathbf{k} = u_1\mathbf{i} + u_2\mathbf{j} + u_3\mathbf{k} = \mathbf{u}, \text{ so}$

$s(t) = \int_0^t |\mathbf{v}|\,dt = \int_0^t |\mathbf{u}|\,d\tau = \int_0^t 1\,d\tau = t$

12.4 CURVATURE AND NORMAL VECTORS OF A CURVE

1. $\mathbf{r} = t\mathbf{i} + \ln(\cos t)\mathbf{j} \Rightarrow \mathbf{v} = \mathbf{i} + \left(\frac{-\sin t}{\cos t}\right)\mathbf{j} = \mathbf{i} - (\tan t)\mathbf{j} \Rightarrow |\mathbf{v}| = \sqrt{1^2 + (-\tan t)^2} = \sqrt{\sec^2 t} = |\sec t| = \sec t,$

since $-\frac{\pi}{2} < t < \frac{\pi}{2} \Rightarrow \mathbf{T} = \frac{\mathbf{v}}{|\mathbf{v}|} = \left(\frac{1}{\sec t}\right)\mathbf{i} - \left(\frac{\tan t}{\sec t}\right)\mathbf{j} = (\cos t)\mathbf{i} - (\sin t)\mathbf{j};\ \frac{d\mathbf{T}}{dt} = (-\sin t)\mathbf{i} - (\cos t)\mathbf{j}$

$\Rightarrow \left|\frac{d\mathbf{T}}{dt}\right| = \sqrt{(-\sin t)^2 + (-\cos t)^2} = 1 \Rightarrow \mathbf{N} = \frac{\left(\frac{d\mathbf{T}}{dt}\right)}{\left|\frac{d\mathbf{T}}{dt}\right|} = (-\sin t)\mathbf{i} - (\cos t)\mathbf{j};\ \kappa = \frac{1}{|\mathbf{v}|}\cdot\left|\frac{d\mathbf{T}}{dt}\right| = \frac{1}{\sec t}\cdot 1 = \cos t.$

3. $\mathbf{r} = (2t+3)\mathbf{i} + \left(5 - t^2\right)\mathbf{j} \Rightarrow \mathbf{v} = 2\mathbf{i} - 2t\mathbf{j} \Rightarrow |\mathbf{v}| = \sqrt{2^2 + (-2t)^2} = 2\sqrt{1+t^2} \Rightarrow \mathbf{T} = \frac{\mathbf{v}}{|\mathbf{v}|} = \frac{2}{2\sqrt{1+t^2}}\mathbf{i} + \frac{-2t}{2\sqrt{1+t^2}}\mathbf{j}$

$= \frac{1}{\sqrt{1+t^2}}\mathbf{i} - \frac{t}{\sqrt{1+t^2}}\mathbf{j};\ \frac{d\mathbf{T}}{dt} = \frac{-t}{\left(\sqrt{1+t^2}\right)^3}\mathbf{i} - \frac{1}{\left(\sqrt{1+t^2}\right)^3}\mathbf{j} \Rightarrow \left|\frac{d\mathbf{T}}{dt}\right| = \sqrt{\left(\frac{-t}{\left(\sqrt{1+t^2}\right)^3}\right)^2 + \left(-\frac{1}{\left(\sqrt{1+t^2}\right)^3}\right)^2} = \sqrt{\frac{1}{\left(1+t^2\right)^2}} = \frac{1}{1+t^2}$

$\Rightarrow \mathbf{N} = \frac{\left(\frac{d\mathbf{T}}{dt}\right)}{\left|\frac{d\mathbf{T}}{dt}\right|} = \frac{-t}{\sqrt{1+t^2}}\mathbf{i} - \frac{1}{\sqrt{1+t^2}}\mathbf{j};\ \kappa = \frac{1}{|\mathbf{v}|}\cdot\left|\frac{d\mathbf{T}}{dt}\right| = \frac{1}{2\sqrt{1+t^2}}\cdot\frac{1}{1+t^2} = \frac{1}{2\left(1+t^2\right)^{3/2}}$

5. (a) $\kappa(x) = \frac{1}{|\mathbf{v}(x)|}\cdot\left|\frac{d\mathbf{T}(x)}{dt}\right|.$ Now, $\mathbf{v} = \mathbf{i} + f'(x)\mathbf{j} \Rightarrow |\mathbf{v}(x)| = \sqrt{1 + \left[f'(x)\right]^2}$

$\Rightarrow \mathbf{T} = \frac{\mathbf{v}}{|\mathbf{v}|} = \left(1 + \left[f'(x)\right]^2\right)^{-1/2}\mathbf{i} + f'(x)\left(1 + \left[f'(x)\right]^2\right)^{-1/2}\mathbf{j}.$ Thus $\frac{d\mathbf{T}}{dt}(x) = \frac{-f'(x)f''(x)}{\left(1 + \left[f'(x)\right]^2\right)^{3/2}}\mathbf{i} + \frac{f''(x)}{\left(1 + \left[f'(x)\right]^2\right)^{3/2}}\mathbf{j}$

$\Rightarrow \left|\frac{d\mathbf{T}(x)}{dt}\right| = \sqrt{\left[\frac{-f'(x)f''(x)}{\left(1 + \left[f'(x)\right]^2\right)^{3/2}}\right]^2 + \left(\frac{f''(x)}{\left(1 + \left[f'(x)\right]^2\right)^{3/2}}\right)^2} = \sqrt{\frac{\left[f''(x)\right]^2\left(1 + \left[f'(x)\right]^2\right)}{\left(1 + \left[f'(x)\right]^2\right)^3}} = \frac{|f''(x)|}{1 + \left[f'(x)\right]^2}$

Thus $\kappa(x) = \frac{1}{\left(1 + \left[f'(x)\right]^2\right)^{1/2}}\cdot\frac{|f''(x)|}{1 + \left[f'(x)\right]^2} = \frac{|f''(x)|}{\left(1 + \left[f'(x)\right]^2\right)^{3/2}}$

(b) $y = \ln(\cos x) \Rightarrow \frac{dy}{dx} = \left(\frac{1}{\cos x}\right)(-\sin x) = -\tan x \Rightarrow \frac{d^2y}{dx^2} = -\sec^2 x \Rightarrow \kappa = \frac{\left|-\sec^2 x\right|}{\left[1+(-\tan x)^2\right]^{3/2}} = \frac{\sec^2 x}{\left|\sec^3 x\right|}$

$= \frac{1}{\sec x} = \cos x$, since $-\frac{\pi}{2} < x < \frac{\pi}{2}$

(c) Note that $f''(x) = 0$ at an inflection point.

7. (a) $\mathbf{r}(t) = f(t)\mathbf{i} + g(t)\mathbf{j} \Rightarrow \mathbf{v} = f'(t)\mathbf{i} + g'(t)\mathbf{j}$ is tangent to the curve at the point $(f(t), g(t))$;

$\mathbf{n} \cdot \mathbf{v} = [-g'(t)\mathbf{i} + f'(t)\mathbf{j}] \cdot [f'(t)\mathbf{i} + g'(t)\mathbf{j}] = -g'(t)f'(t) + f'(t)g'(t) = 0$; $-\mathbf{n} \cdot \mathbf{v} = -(\mathbf{n} \cdot \mathbf{v}) = 0$; thus, $\mathbf{n}$ and $-\mathbf{n}$ are both normal to the curve at the point

(b) $\mathbf{r}(t) = t\mathbf{i} + e^{2t}\mathbf{j} \Rightarrow \mathbf{v} = \mathbf{i} + 2e^{2t}\mathbf{j} \Rightarrow \mathbf{n} = -2e^{2t}\mathbf{i} + \mathbf{j}$ points toward the concave side of the curve; $\mathbf{N} = \frac{\mathbf{n}}{|\mathbf{n}|}$ and

$|\mathbf{n}| = \sqrt{4e^{4t}+1} \Rightarrow \mathbf{N} = \frac{-2e^{2t}}{\sqrt{1+4e^{4t}}}\mathbf{i} + \frac{1}{\sqrt{1+4e^{4t}}}\mathbf{j}$

(c) $\mathbf{r}(t) = \sqrt{4-t^2}\,\mathbf{i} + t\mathbf{j} \Rightarrow \mathbf{v} = \frac{-t}{\sqrt{4-t^2}}\mathbf{i} + \mathbf{j} \Rightarrow \mathbf{n} = -\mathbf{i} - \frac{t}{\sqrt{4-t^2}}\mathbf{j}$ points toward the concave side of the curve;

$\mathbf{N} = \frac{\mathbf{n}}{|\mathbf{n}|}$ and $|\mathbf{n}| = \sqrt{1+\frac{t^2}{4-t^2}} = \frac{2}{\sqrt{4-t^2}} \Rightarrow \mathbf{N} = -\frac{1}{2}\left(\sqrt{4-t^2}\,\mathbf{i} + t\mathbf{j}\right)$

9. $\mathbf{r} = (3\sin t)\mathbf{i} + (3\cos t)\mathbf{j} + 4t\mathbf{k} \Rightarrow \mathbf{v} = (3\cos t)\mathbf{i} + (-3\sin t)\mathbf{j} + 4\mathbf{k} \Rightarrow |\mathbf{v}| = \sqrt{(3\cos t)^2 + (-3\sin t)^2 + 4^2} = \sqrt{25} = 5$

$\Rightarrow \mathbf{T} = \frac{\mathbf{v}}{|\mathbf{v}|} = \left(\frac{3}{5}\cos t\right)\mathbf{i} - \left(\frac{3}{5}\sin t\right)\mathbf{j} + \frac{4}{5}\mathbf{k} \Rightarrow \frac{d\mathbf{T}}{dt} = \left(-\frac{3}{5}\sin t\right)\mathbf{i} - \left(\frac{3}{5}\cos t\right)\mathbf{j} \Rightarrow \left|\frac{d\mathbf{T}}{dt}\right| = \sqrt{\left(-\frac{3}{5}\sin t\right)^2 + \left(-\frac{3}{5}\cos t\right)^2} = \frac{3}{5}$

$\Rightarrow \mathbf{N} = \frac{\left(\frac{d\mathbf{T}}{dt}\right)}{\left|\frac{d\mathbf{T}}{dt}\right|} = (-\sin t)\mathbf{i} - (\cos t)\mathbf{j}; \kappa = \frac{1}{5} \cdot \frac{3}{5} = \frac{3}{25}$

11. $\mathbf{r} = (e^t \cos t)\mathbf{i} + (e^t \sin t)\mathbf{j} + 2\mathbf{k} \Rightarrow \mathbf{v} = (e^t \cos t - e^t \sin t)\mathbf{i} + (e^t \sin t + e^t \cos t)\mathbf{j}$

$\Rightarrow |\mathbf{v}| = \sqrt{(e^t \cos t - e^t \sin t)^2 + (e^t \sin t + e^t \cos t)^2} = \sqrt{2e^{2t}} = e^t\sqrt{2}; \mathbf{T} = \frac{\mathbf{v}}{|\mathbf{v}|} = \left(\frac{\cos t - \sin t}{\sqrt{2}}\right)\mathbf{i} + \left(\frac{\sin t + \cos t}{\sqrt{2}}\right)\mathbf{j}$

$\Rightarrow \frac{d\mathbf{T}}{dt} = \left(\frac{-\sin t - \cos t}{\sqrt{2}}\right)\mathbf{i} + \left(\frac{\cos t - \sin t}{\sqrt{2}}\right)\mathbf{j} \Rightarrow \left|\frac{d\mathbf{T}}{dt}\right| = \sqrt{\left(\frac{-\sin t - \cos t}{\sqrt{2}}\right)^2 + \left(\frac{\cos t - \sin t}{\sqrt{2}}\right)^2} = 1$

$\Rightarrow \mathbf{N} = \frac{\left(\frac{d\mathbf{T}}{dt}\right)}{\left|\frac{d\mathbf{T}}{dt}\right|} = \left(\frac{-\cos t - \sin t}{\sqrt{2}}\right)\mathbf{i} + \left(\frac{-\sin t + \cos t}{\sqrt{2}}\right)\mathbf{j}; \kappa = \frac{1}{|\mathbf{v}|}\left|\frac{d\mathbf{T}}{dt}\right| = \frac{1}{e^t\sqrt{2}} \cdot 1 = \frac{1}{e^t\sqrt{2}}$

13. $\mathbf{r} = \left(\frac{t^3}{3}\right)\mathbf{i} + \left(\frac{t^2}{2}\right)\mathbf{j}, t > 0 \Rightarrow \mathbf{v} = t^2\mathbf{i} + t\mathbf{j} \Rightarrow |\mathbf{v}| = \sqrt{t^4 + t^2} = t\sqrt{t^2 + 1}$, since $t > 0 \Rightarrow \mathbf{T} = \frac{\mathbf{v}}{|\mathbf{v}|} = \frac{t}{\sqrt{t^2+1}}\mathbf{i} + \frac{1}{\sqrt{t^2+1}}\mathbf{j}$

$\Rightarrow \frac{d\mathbf{T}}{dt} = \frac{1}{(t^2+1)^{3/2}}\mathbf{i} - \frac{t}{(t^2+1)^{3/2}}\mathbf{j} \Rightarrow \left|\frac{d\mathbf{T}}{dt}\right| = \sqrt{\left(\frac{1}{(t^2+1)^{3/2}}\right)^2 + \left(\frac{-t}{(t^2+1)^{3/2}}\right)^2} = \sqrt{\frac{1+t^2}{(t^2+1)^3}} = \frac{1}{t^2+1}$

$\Rightarrow \mathbf{N} = \frac{\left(\frac{d\mathbf{T}}{dt}\right)}{\left|\frac{d\mathbf{T}}{dt}\right|} = \frac{1}{\sqrt{t^2+1}}\mathbf{i} - \frac{t}{\sqrt{t^2+1}}\mathbf{j}; \kappa = \frac{1}{|\mathbf{v}|}\left|\frac{d\mathbf{T}}{dt}\right| = \frac{1}{t\sqrt{t^2+1}} \cdot \frac{1}{t^2+1} = \frac{1}{t(t^2+1)^{3/2}}$.

15. $\mathbf{r} = t\mathbf{i} + \left(a\cosh\frac{t}{a}\right)\mathbf{j}, a > 0 \Rightarrow \mathbf{v} = \mathbf{i} + \left(\sinh\frac{t}{a}\right)\mathbf{j} \Rightarrow |\mathbf{v}| = \sqrt{1 + \sinh^2\left(\frac{t}{a}\right)} = \sqrt{\cosh^2\left(\frac{t}{a}\right)} = \cosh\frac{t}{a}$

$\Rightarrow \mathbf{T} = \frac{\mathbf{v}}{|\mathbf{v}|} = \left(\operatorname{sech}\frac{t}{a}\right)\mathbf{i} + \left(\tanh\frac{t}{a}\right)\mathbf{j} \Rightarrow \frac{d\mathbf{T}}{dt} = \left(-\frac{1}{a}\operatorname{sech}\frac{t}{a}\tanh\frac{t}{a}\right)\mathbf{i} + \left(\frac{1}{a}\operatorname{sech}^2\frac{t}{a}\right)\mathbf{j}$

$\Rightarrow \left|\frac{d\mathbf{T}}{dt}\right| = \sqrt{\frac{1}{a^2}\operatorname{sech}^2\left(\frac{t}{a}\right)\tanh^2\left(\frac{t}{a}\right) + \frac{1}{a^2}\operatorname{sech}^4\left(\frac{t}{a}\right)} = \frac{1}{a}\operatorname{sech}\left(\frac{t}{a}\right) \Rightarrow \mathbf{N} = \frac{\left(\frac{d\mathbf{T}}{dt}\right)}{\left|\frac{d\mathbf{T}}{dt}\right|} = \left(-\tanh\frac{t}{a}\right)\mathbf{i} + \left(\operatorname{sech}\frac{t}{a}\right)\mathbf{j};$

$\kappa = \frac{1}{|\mathbf{v}|}\cdot\left|\frac{d\mathbf{T}}{dt}\right| = \frac{1}{\cosh\frac{t}{a}}\cdot\frac{1}{a}\operatorname{sech}\left(\frac{t}{a}\right) = \frac{1}{a}\operatorname{sech}^2\left(\frac{t}{a}\right).$

17. $y = ax^2 \Rightarrow y' = 2ax \Rightarrow y'' = 2a;$ from Exercise 5(a), $\kappa(x) = \frac{|2a|}{\left(1 + 4a^2x^2\right)^{3/2}} = |2a|\left(1 + 4a^2x^2\right)^{-3/2}$

$\Rightarrow \kappa'(x) = -\frac{3}{2}|2a|\left(1 + 4a^2x^2\right)^{-5/2}\left(8a^2x\right);$ thus, $\kappa'(x) = 0 \Rightarrow x = 0.$ Now, $\kappa'(x) > 0$ for $x < 0$ and $\kappa'(x) < 0$ for

$x > 0$ so that $\kappa(x)$ has an absolute maximum at $x = 0$ which is the vertex of the parabola. Since $x = 0$ is the

only critical point for $\kappa(x)$, the curvature has no minimum value.

19. $\kappa = \frac{a}{a^2 + b^2} \Rightarrow \frac{d\kappa}{da} = \frac{-a^2 + b^2}{\left(a^2 + b^2\right)^2}; \frac{d\kappa}{da} = 0 \Rightarrow -a^2 + b^2 = 0 \Rightarrow a = \pm b \Rightarrow a = b$ since $a, b \geq 0.$ Now, $\frac{d\kappa}{da} > 0$ if $a < b$

and $\frac{d\kappa}{da} < 0$ if $a > b \Rightarrow \kappa$ is at a maximum for $a = b$ and $\kappa(b) = \frac{b}{b^2 + b^2} = \frac{1}{2b}$ is the maximum value of κ.

21. $\mathbf{r} = t\mathbf{i} + (\sin t)\mathbf{j} \Rightarrow \mathbf{v} = \mathbf{i} + (\cos t)\mathbf{j} \Rightarrow |\mathbf{v}| = \sqrt{1^2 + (\cos t)^2} = \sqrt{1 + \cos^2 t} \Rightarrow \left|\mathbf{v}\left(\frac{\pi}{2}\right)\right| = \sqrt{1 + \cos^2\left(\frac{\pi}{2}\right)} = 1;$

$\mathbf{T} = \frac{\mathbf{v}}{|\mathbf{v}|} = \frac{\mathbf{i} + (\cos t)\mathbf{j}}{\sqrt{1 + \cos^2 t}} \Rightarrow \frac{d\mathbf{T}}{dt} = \frac{\sin t \cos t}{\left(1 + \cos^2 t\right)^{3/2}}\mathbf{i} + \frac{-\sin t}{\left(1 + \cos^2 t\right)^{3/2}}\mathbf{j} \Rightarrow \left|\frac{d\mathbf{T}}{dt}\right| = \frac{|\sin t|}{1 + \cos^2 t}; \left|\frac{d\mathbf{T}}{dt}\right|_{t = \frac{\pi}{2}} = \frac{\left|\sin\frac{\pi}{2}\right|}{1 + \cos^2\left(\frac{\pi}{2}\right)} = \frac{1}{1} = 1.$ Thus

$\kappa\left(\frac{\pi}{2}\right) = \frac{1}{1}\cdot 1 = 1 \Rightarrow \rho = \frac{1}{1} = 1$ and the center is $\left(\frac{\pi}{2}, 0\right) \Rightarrow \left(x - \frac{\pi}{2}\right)^2 + y^2 = 1$

23. $y = x^2 \Rightarrow f'(x) = 2x$ and $f''(x) = 2$

$\Rightarrow \kappa = \frac{|2|}{\left(1 + (2x)^2\right)^{3/2}} = \frac{2}{\left(1 + 4x^2\right)^{3/2}}$

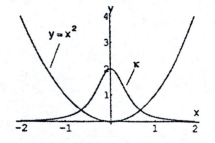

25. $y = \sin x \Rightarrow f'(x) = \cos x$ and $f''(x) = -\sin x$

$\Rightarrow \kappa = \frac{|-\sin x|}{\left(1 + \cos^2 x\right)^{3/2}} = \frac{|\sin x|}{\left(1 + \cos^2 x\right)^{3/2}}$

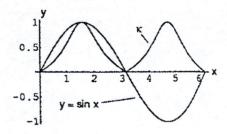

27. We will use the formula $\kappa = \dfrac{1}{|\mathbf{v}(a)|}\left|\dfrac{d\mathbf{T}}{dt}(a)\right|$ to find the curvature at the point $\left(a, a^2\right)$.

By Example 4 in Section 13.4,

$$\mathbf{v}(t) = \sqrt{1+4t^2} \text{ and } \quad \frac{d\mathbf{T}}{dt} = 2\left(1+4t^2\right)^{-3/2}(-2t\mathbf{i}+\mathbf{j}).$$

At $t = a$ this gives $\kappa = \dfrac{1}{|\mathbf{v}(a)|}\left|\dfrac{d\mathbf{T}}{dt}(a)\right| = \dfrac{2}{\sqrt{1+4a^2}}\left(1+4a^2\right)^{-3/2}\sqrt{1+4a^2} = \dfrac{2}{\left(1+4a^2\right)^{3/2}}$. Thus the radius of

the osculating circle is $r = \dfrac{1}{2}\left(1+4a^2\right)^{3/2}$. To show that the given formulas for center and radius are correct we

must first show that the distance between $\left(a, a^2\right)$ and $\;-4a^3, 3a^2 + \dfrac{1}{2}\;$ is r. This distance is

$$\sqrt{\left(-4a^3-a\right)^2 + \left(3a^2+\frac{1}{2}-a^2\right)^2} = \sqrt{16a^6+12a^4+3a^2+\frac{1}{4}} = \frac{1}{2}\sqrt{\left(1+4a^2\right)^3} \text{ as required. Finally we must show}$$

that the line containing the points $\;-4a^3, 3a^2 + \dfrac{1}{2}\;$ and $\left(a, a^2\right)$ is perpendicular to the tangent line at $\left(a, a^2\right)$,

which has slope $2a$. This requires that $\dfrac{3a^2+\dfrac{1}{2}-a^2}{-4a^3-a} = -\dfrac{1}{2a}$, which is correct.

12.5 TANGENTIAL AND NORMAL COMPONENTS OF ACCELERATION

1. $\mathbf{r} = (a\cos t)\mathbf{i} + (a\sin t)\mathbf{j} + bt\mathbf{k} \Rightarrow \mathbf{v} = (-a\sin t)\mathbf{i} + (a\cos t)\mathbf{j} + b\mathbf{k} \Rightarrow |\mathbf{v}| = \sqrt{(-a\sin t)^2 + (a\cos t)^2 + b^2}$

$= \sqrt{a^2+b^2} \Rightarrow a_T = \dfrac{d}{dt}|\mathbf{v}| = 0;\ \mathbf{a} = (-a\cos t)\mathbf{i} + (-a\sin t)\mathbf{j} \Rightarrow |\mathbf{a}| = \sqrt{(-a\cos t)^2 + (-a\sin t)^2} = \sqrt{a^2} = |a|$

$\Rightarrow a_N = \sqrt{|\mathbf{a}|^2 - a_T^2} = \sqrt{|\mathbf{a}|^2 - 0^2} = |\mathbf{a}| = |a| \Rightarrow \mathbf{a} = (0)\mathbf{T} + |a|\mathbf{N} = |a|\mathbf{N}$

3. $\mathbf{r} = (t+1)\mathbf{i} + 2t\mathbf{j} + t^2\mathbf{k} \Rightarrow \mathbf{v} = \mathbf{i} + 2\mathbf{j} + 2t\mathbf{k} \Rightarrow |\mathbf{v}| = \sqrt{1^2 + 2^2 + (2t)^2} = \sqrt{5+4t^2} \Rightarrow a_T = \frac{1}{2}\left(5+4t^2\right)^{-1/2}(8t)$

$= 4t\left(5+4t^2\right)^{-1/2} \Rightarrow a_T(1) = \dfrac{4}{\sqrt{9}} = \dfrac{4}{3};\ \mathbf{a} = 2\mathbf{k} \Rightarrow \mathbf{a}(1) = 2\mathbf{k} \Rightarrow |\mathbf{a}(1)| = 2$

$\Rightarrow a_N = \sqrt{|\mathbf{a}|^2 - a_T^2} = \sqrt{2^2 - \left(\frac{4}{3}\right)^2} = \sqrt{\frac{20}{9}} = \dfrac{2\sqrt{5}}{3} \Rightarrow \mathbf{a}(1) = \dfrac{4}{3}\mathbf{T} + \dfrac{2\sqrt{5}}{3}\mathbf{N}$

5. $\mathbf{r} = t^2\mathbf{i} + \left(t+\frac{1}{3}t^3\right)\mathbf{j} + \left(t-\frac{1}{3}t^3\right)\mathbf{k} \Rightarrow \mathbf{v} = 2t\mathbf{i} + \left(1+t^2\right)\mathbf{j} + \left(1-t^2\right)\mathbf{k} \Rightarrow |\mathbf{v}| = \sqrt{(2t)^2 + \left(1+t^2\right)^2 + \left(1-t^2\right)^2}$

$= \sqrt{2\left(t^4+2t^2+1\right)} = \sqrt{2}\left(1+t^2\right) \Rightarrow a_T = 2t\sqrt{2} \Rightarrow a_T(0) = 0;\ \mathbf{a} = 2\mathbf{i} + 2t\mathbf{j} - 2t\mathbf{k} \Rightarrow \mathbf{a}(0) = 2\mathbf{i} \Rightarrow |\mathbf{a}(0)| = 2$

$\Rightarrow a_N = \sqrt{|\mathbf{a}|^2 - a_T^2} = \sqrt{2^2 - 0^2} = 2 \Rightarrow \mathbf{a}(0) = (0)\mathbf{T} + 2\mathbf{N} = 2\mathbf{N}$

7. $\mathbf{r} = (\cos t)\mathbf{i} + (\sin t)\mathbf{j} - \mathbf{k} \Rightarrow \mathbf{v} = (-\sin t)\mathbf{i} + (\cos t)\mathbf{j} \Rightarrow |\mathbf{v}| = \sqrt{(-\sin t)^2 + (\cos t)^2} = 1$

$\Rightarrow \mathbf{T} = \frac{\mathbf{v}}{|\mathbf{v}|} = (-\sin t)\mathbf{i} + (\cos t)\mathbf{j} \Rightarrow \mathbf{T}\left(\frac{\pi}{4}\right) = -\frac{\sqrt{2}}{2}\mathbf{i} + \frac{\sqrt{2}}{2}\mathbf{j}; \ \frac{d\mathbf{T}}{dt} = (-\cos t)\mathbf{i} - (\sin t)\mathbf{j}$

$\Rightarrow \left|\frac{d\mathbf{T}}{dt}\right| \sqrt{(-\cos t)^2 + (-\sin t)^2} = 1 \Rightarrow \mathbf{N} = \frac{\left(\frac{d\mathbf{T}}{dt}\right)}{\left|\frac{d\mathbf{T}}{dt}\right|} = (-\cos t)\mathbf{i} - (\sin t)\mathbf{j} \Rightarrow \mathbf{N}\left(\frac{\pi}{4}\right) = -\frac{\sqrt{2}}{2}\mathbf{i} - \frac{\sqrt{2}}{2}\mathbf{j};$

$\mathbf{B} = \mathbf{T} \times \mathbf{N} = \begin{vmatrix} \mathbf{i} & \mathbf{j} & \mathbf{k} \\ -\sin t & \cos t & 0 \\ -\cos t & -\sin t & 0 \end{vmatrix} = \mathbf{k} \Rightarrow \mathbf{B}\left(\frac{\pi}{4}\right) = \mathbf{k},$ the normal to the osculating plane; $\mathbf{r}\left(\frac{\pi}{4}\right) = \frac{\sqrt{2}}{2}\mathbf{i} + \frac{\sqrt{2}}{2}\mathbf{j} - \mathbf{k}$

$\Rightarrow P = \left(\frac{\sqrt{2}}{2}, \frac{\sqrt{2}}{2}, -1\right)$ lies on the osculating plane $\Rightarrow 0\left(x - \frac{\sqrt{2}}{2}\right) + 0\left(y - \frac{\sqrt{2}}{2}\right) + (z - (-1)) = 0 \Rightarrow z = -1$ is the

osculating plane; $\mathbf{T}$ is normal to the normal plane $\Rightarrow \left(-\frac{\sqrt{2}}{2}\right)\left(x - \frac{\sqrt{2}}{2}\right) + \left(\frac{\sqrt{2}}{2}\right)\left(y - \frac{\sqrt{2}}{2}\right) + 0(z - (-1)) = 0$

$\Rightarrow -\frac{\sqrt{2}}{2}x + \frac{\sqrt{2}}{2}y = 0 \Rightarrow -x + y = 0$ is the normal plane; $\mathbf{N}$ is normal to the rectifying plane

$\Rightarrow \left(-\frac{\sqrt{2}}{2}\right)\left(x - \frac{\sqrt{2}}{2}\right) + \left(-\frac{\sqrt{2}}{2}\right)\left(y - \frac{\sqrt{2}}{2}\right) + 0(z - (-1)) = 0 \Rightarrow -\frac{\sqrt{2}}{2}x - \frac{\sqrt{2}}{2}y = -1 \Rightarrow x + y = \sqrt{2}$ is the rectifying

plane.

9. Yes. If the car is moving along a curved path, then $\kappa \neq 0$ and $a_N = k|\mathbf{v}|^2 \neq 0 \Rightarrow \mathbf{a} = a_T\mathbf{T} + a_N\mathbf{N} \neq \mathbf{0}$.

11. $\mathbf{a} \perp \mathbf{v} \Rightarrow \mathbf{a} \perp \mathbf{T} \Rightarrow a_T = 0 \Rightarrow \frac{d}{dt}|\mathbf{v}| = 0 \Rightarrow |\mathbf{v}|$ is constant

13. From Example 1, $|\mathbf{v}| = t$ and $a_N = t$ so that $a_N = \kappa|\mathbf{v}|^2 \Rightarrow \kappa = \frac{a_N}{|\mathbf{v}|^2} = \frac{t}{t^2} = \frac{1}{t}, t \neq 0 \Rightarrow \rho = \frac{1}{\kappa} = t$

15. From $\mathbf{a} = \alpha_T\mathbf{T} + \alpha_N\mathbf{N}$ and Using Equations (1) and (2) we have

$$\mathbf{v} \times \mathbf{a} = \left(\frac{ds}{dt}\mathbf{T}\right) \times \left(\frac{d^2s}{dt^2}\mathbf{T} + \kappa\left(\frac{ds}{dt}\right)^2\mathbf{N}\right)$$

$$= \left(\frac{ds}{dt}\left(\frac{ds}{dt}\right)^2\right)(\mathbf{T} \times \mathbf{T}) + \left(\frac{d^2s}{dt^2}\mathbf{T} + \kappa\left(\frac{ds}{dt}\right)^3\right)(\mathbf{T} \times \mathbf{N})$$

$$= \kappa\left(\frac{ds}{dt}\right)^3\mathbf{B}$$

so $\kappa = \frac{|\mathbf{v} \times \mathbf{a}|}{\left(\frac{ds}{dt}\right)^3} = \frac{|\mathbf{v} \times \mathbf{a}|}{|\mathbf{v}|^3}$

17-19.   Example CAS commands:

Maple:

```
with( LinearAlgebra );
r := <t*cos(t) | t*sin(t) | t >;
t0 := sqrt(3);
rr := eval( r, t=t0 );
```

v := map(diff, r, t);

vv := eval(v, t=t0);

a := map(diff, v, t);

aa := eval(a, t=t0);

s := simplify(Norm(v, 2)) assuming t::real;

ss := eval(s, t=t0);

T := v/s;

TT := vv/ss ;

q1 := map(diff, simplify(T), t):

NN := simplify(eval(q1/Norm(q1,2), t=t0));

BB := CrossProduct(TT, NN);

kappa := Norm(CrossProduct(vv,aa),2)/ss^3;

tau := simplify(Determinant(< vv, aa, eval(map(diff,a,t),t=t0) >)/Norm(CrossProduct(vv,aa),2)^3);

a_t := eval(diff(s, t), t=t0);

a_n := evalf[4](kappa*ss^2);

Mathematica: (assigned functions and value for t0 will vary)

Clear[t, v, a, t]

mag[vector_]:=Sqrt[vector.vector]

Print["The position vector is", r[t_]={t Cos[t], t Sin[t], t}]

Print["The velocity vector is", v[t_]=r'[t]]

Print["The acceleration vector is", a[t_]=v'[t]]

Print["The speed is", speed[t_]= mag[v[t]]//Simplify]

Print["The unit tangent vector is", utan[t_]= v[t]/speed[t]//Simplify]

Print["The curvature is", curv[t_]= mag[Cross[v[t],a[t]]] / speed[t]^3//Simplify]

Print["The torsion is", torsion[t_]= Det[{v[t], a[t], a'[t]}] / mag[Cross[v[t],a[t]]]^2//Simplify]

Print["The unit normal vector is", unorm[t_]= utan'[t] / mag[utan'[t]//Simplify]

Print["The unit binormal vector is", ubinorm[t_]= Cross[utan[t],unorm[t]]//Simplify]

Print["The tangential component of the acceleration is", at[t_]=a[t].utan[t]//Simplify]

Print["The normal component of the acceleration is", an[t_]=a[t].unorm[t]//Simplify]

You can evaluate any of these functions at a specified value of t.

t0= Sqrt[3]

{utan[t0], unorm[t0], ubinorm[t0]}

N[{utan[t0], unorm[t0], ubinorm[t0]}]

{curv[t0], torsion[t0]}

N[{curv[t0], torsion[t0]}]

{at[t0], an[t0]}

N[{at[t0], an[t0]}]

To verify that the tangential and normal components of the acceleration agree with the formulas in the book:

at[t]== speed'[t]//Simplify

an[t]==curv [t] speed[t]^2 //Simplify

12.6 VELOCITY AND ACCELERATION IN POLAR COORDINATES

1. $\frac{d\theta}{dt} = 3 = \dot\theta \Rightarrow \ddot\theta = 0,\ r = a(1-\cos\theta) \Rightarrow \dot r = a\sin\theta\,\frac{d\theta}{dt} = 3a\sin\theta \Rightarrow \ddot r = 3a\cos\theta\,\frac{d\theta}{dt} = 9a\cos\theta$

   $\mathbf{v} = (3a\sin\theta)\mathbf{u}_r + (a(1-\cos\theta))(3)\mathbf{u}_\theta = (3a\sin\theta)\mathbf{u}_r + 3a(1-\cos\theta)\mathbf{u}_\theta$

   $\mathbf{a} = \left(9a\cos\theta - a(1-\cos\theta)(3)^2\right)\mathbf{u}_r + \left(a(1-\cos\theta)\cdot 0 + 2(3a\sin\theta)(3)\right)\mathbf{u}_\theta$

   $= (9a\cos\theta - 9a + 9a\cos\theta)\mathbf{u}_r + (18a\sin\theta)\mathbf{u}_\theta = 9a(2\cos\theta - 1)\mathbf{u}_r + (18a\sin\theta)\mathbf{u}_\theta$

3. $\frac{d\theta}{dt} = 2 = \dot\theta \Rightarrow \ddot\theta = 0,\ r = e^{a\theta} \Rightarrow \dot r = e^{a\theta}\cdot a\,\frac{d\theta}{dt} = 2a\,e^{a\theta} \Rightarrow \ddot r = 2a\,e^{a\theta}\cdot a\,\frac{d\theta}{dt} = 4a^2 e^{a\theta}$

   $\mathbf{v} = \left(2a\,e^{a\theta}\right)\mathbf{u}_r + \left(e^{a\theta}\right)(2)\mathbf{u}_\theta = \left(2a\,e^{a\theta}\right)\mathbf{u}_r + \left(2e^{a\theta}\right)\mathbf{u}_\theta$

   $\mathbf{a} = \left[\left(4a^2 e^{a\theta}\right) - \left(e^{a\theta}\right)(2)^2\right]\mathbf{u}_r + \left[\left(e^{a\theta}\right)(0) + 2\left(2a\,e^{a\theta}\right)(2)\right]\mathbf{u}_\theta = \left[4a^2 e^{a\theta} - 4e^{a\theta}\right]\mathbf{u}_r + \left[0 + 8a\,e^{a\theta}\right]\mathbf{u}_\theta$

   $= 4e^{a\theta}\left(a^2 - 1\right)\mathbf{u}_r + \left(8a\,e^{a\theta}\right)\mathbf{u}_\theta$

5. $\theta = 2t \Rightarrow \dot\theta = 2 \Rightarrow \ddot\theta = 0,\ r = 2\cos 4t \Rightarrow \dot r = -8\sin 4t \Rightarrow \ddot r = -32\cos 4t$

   $\mathbf{v} = (-8\sin 4t)\mathbf{u}_r + (2\cos 4t)(2)\mathbf{u}_\theta = -8(\sin 4t)\mathbf{u}_r + 4(\cos 4t)\mathbf{u}_\theta$

   $\mathbf{a} = \left((-32\cos 4t) - (2\cos 4t)(2)^2\right)\mathbf{u}_r + \left((2\cos 4t)\cdot 0 + 2(-8\sin 4t)(2)\right)\mathbf{u}_\theta$

   $= (-32\cos 4t - 8\cos 4t)\mathbf{u}_r + (0 - 32\sin 4t)\mathbf{u}_\theta = -40(\cos 4t)\mathbf{u}_r - 32(\sin 4t)\mathbf{u}_\theta$

7. $r = \frac{GM}{v^2} \Rightarrow v^2 = \frac{GM}{r} \Rightarrow v = \sqrt{\frac{GM}{r}}$ which is constant since G, M, and r (the radius of orbit) are constant

9. $T = \left(\frac{2\pi a^2}{r_0 v_0}\right)\sqrt{1-e^2} \Rightarrow T^2 = \left(\frac{4\pi^2 a^4}{r_0^2 v_0^2}\right)\left(1-e^2\right) = \left(\frac{4\pi^2 a^4}{r_0^2 v_0^2}\right)\left[1 - \left(\frac{r_0 v_0^2}{GM} - 1\right)^2\right]$ (from Equation 5)

   $= \left(\frac{4\pi^2 a^4}{r_0^2 v_0^2}\right)\left[-\frac{r_0^2 v_0^4}{G^2 M^2} + 2\left(\frac{r_0 v_0^2}{GM}\right)\right] = \left(\frac{4\pi^2 a^4}{r_0^2 v_0^2}\right)\left[\frac{2GM r_0 v_0^2 - r_0^2 v_0^4}{G^2 M^2}\right] = \frac{\left(4\pi^2 a^4\right)\left(2GM - r_0 v_0^2\right)}{r_0 G^2 M^2} = \left(4\pi^2 a^4\right)\left(\frac{2GM - r_0 v_0^2}{2r_0 GM}\right)\left(\frac{2}{GM}\right)$

   $= \left(4\pi^2 a^4\right)\left(\frac{1}{2a}\right)\left(\frac{2}{GM}\right)$ (from Equation 10) $\Rightarrow T^2 = \frac{4\pi^2 a^3}{GM} \Rightarrow \frac{T^2}{a^3} = \frac{4\pi^2}{GM}$

11. Solve Kepler's third law for a and double this result: $2\cdot\left[\frac{(365.256\ \text{days})^2}{(4\pi^2)/(GM)}\right]^{1/3} \approx 29.925\times 10^{10}$ m

13. Assuming Earth has a circular orbit with radius 150×10^6 km, the rate of change of area is

    $\dfrac{\pi\left(150\times 10^6\ \text{km}\right)^2}{365.256\ \text{days}} \approx 2.24\times 10^9\ \text{km}^2/\text{sec}.$

15. Solving Kepler's third law for the mass M of the body around which Io is orbiting we find

    $M = 4\pi^2\,\dfrac{a^3}{T^2 G} = 4\pi^2\,\dfrac{\left(0.042\times 10^{10}\ \text{m}\right)^3}{(1.769\ \text{days})^2 G} \approx 1.876\times 10^{27}$ kg.

CHAPTER 12 PRACTICE EXERCISES

1. $\mathbf{r}(t) = (4\cos t)\mathbf{i} + \left(\sqrt{2}\sin t\right)\mathbf{j}$

    $\Rightarrow x = 4\cos t$ and $y = \sqrt{2}\sin t \Rightarrow \frac{x^2}{16} + \frac{y^2}{2} = 1;$

    $\mathbf{v} = (-4\sin t)\mathbf{i} + \left(\sqrt{2}\cos t\right)\mathbf{j}$ and

    $\mathbf{a} = (-4\cos t)\mathbf{i} - \left(\sqrt{2}\sin t\right)\mathbf{j};$

    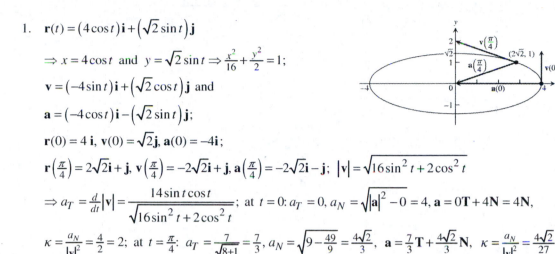

    $\mathbf{r}(0) = 4\mathbf{i}, \mathbf{v}(0) = \sqrt{2}\mathbf{j}, \mathbf{a}(0) = -4\mathbf{i};$

    $\mathbf{r}\left(\frac{\pi}{4}\right) = 2\sqrt{2}\mathbf{i} + \mathbf{j}, \mathbf{v}\left(\frac{\pi}{4}\right) = -2\sqrt{2}\mathbf{i} + \mathbf{j}, \mathbf{a}\left(\frac{\pi}{4}\right) = -2\sqrt{2}\mathbf{i} - \mathbf{j}; |\mathbf{v}| = \sqrt{16\sin^2 t + 2\cos^2 t}$

    $\Rightarrow a_T = \frac{d}{dt}|\mathbf{v}| = \frac{14\sin t\cos t}{\sqrt{16\sin^2 t + 2\cos^2 t}};$ at $t = 0$: $a_T = 0$, $a_N = \sqrt{|\mathbf{a}|^2 - 0} = 4$, $\mathbf{a} = 0\mathbf{T} + 4\mathbf{N} = 4\mathbf{N}$,

    $\kappa = \frac{a_N}{|\mathbf{v}|^2} = \frac{4}{2} = 2;$ at $t = \frac{\pi}{4}$: $a_T = \frac{7}{\sqrt{8+1}} = \frac{7}{3}$, $a_N = \sqrt{9 - \frac{49}{9}} = \frac{4\sqrt{2}}{3}$, $\mathbf{a} = \frac{7}{3}\mathbf{T} + \frac{4\sqrt{2}}{3}\mathbf{N}$, $\kappa = \frac{a_N}{|\mathbf{v}|^2} = \frac{4\sqrt{2}}{27}$

3. $\mathbf{r} = \frac{1}{\sqrt{1+t^2}}\mathbf{i} + \frac{t}{\sqrt{1+t^2}}\mathbf{j} \Rightarrow \mathbf{v} = -t\left(1+t^2\right)^{-3/2}\mathbf{i} + \left(1+t^2\right)^{-3/2}\mathbf{j} \Rightarrow |\mathbf{v}| = \sqrt{\left[-t\left(1+t^2\right)^{-3/2}\right]^2 + \left[\left(1+t^2\right)^{-3/2}\right]^2} = \frac{1}{1+t^2}.$

    We want to maximize $|\mathbf{v}|$: $\frac{d|\mathbf{v}|}{dt} = \frac{-2t}{\left(1+t^2\right)^2}$ and $\frac{d|\mathbf{v}|}{dt} = 0 \Rightarrow \frac{-2t}{\left(1+t^2\right)^2} = 0 \Rightarrow t = 0.$ For $t < 0$, $\frac{-2t}{\left(1+t^2\right)^2} > 0$; for $t > 0$,

    $\frac{-2t}{\left(1+t^2\right)^2} < 0 \Rightarrow |\mathbf{v}|_{max}$ occurs when $t = 0 \Rightarrow |\mathbf{v}|_{max} = 1$

5. $\mathbf{v} = 3\mathbf{i} + 4\mathbf{j}$ and $\mathbf{a} = 5\mathbf{i} + 15\mathbf{j} \Rightarrow \mathbf{v} \times \mathbf{a} = \begin{vmatrix} \mathbf{i} & \mathbf{j} & \mathbf{k} \\ 3 & 4 & 0 \\ 5 & 15 & 0 \end{vmatrix} = 25\mathbf{k} \Rightarrow |\mathbf{v} \times \mathbf{a}| = 25; |\mathbf{v}| = \sqrt{3^2 + 4^2} = 5 \Rightarrow \kappa = \frac{|\mathbf{v} \times \mathbf{a}|}{|\mathbf{v}|^3} = \frac{25}{5^3} = \frac{1}{5}$

7. $\mathbf{r} = x\mathbf{i} + y\mathbf{j} \Rightarrow \mathbf{v} = \frac{dx}{dt}\mathbf{i} + \frac{dy}{dt}\mathbf{j}$ and $\mathbf{v} \cdot \mathbf{i} = y \Rightarrow \frac{dx}{dt} = y.$ Since the particle moves around the unit circle

    $x^2 + y^2 = 1, 2x\frac{dx}{dt} + 2y\frac{dy}{dt} = 0 \Rightarrow \frac{dy}{dt} = -\frac{x}{y}\frac{dx}{dt} \Rightarrow \frac{dy}{dt} = -\frac{x}{y}(y) = -x.$ Since $\frac{dx}{dt} = y$ and $\frac{dy}{dt} = -x,$

    we have $\mathbf{v} = y\mathbf{i} - x\mathbf{j} \Rightarrow$ at $(1, 0)$, $\mathbf{v} = -\mathbf{j}$ and the motion is clockwise.

9. $\frac{d\mathbf{r}}{dt}$ orthogonal to $\mathbf{r} \Rightarrow 0 = \frac{d\mathbf{r}}{dt} \cdot \mathbf{r} = \frac{1}{2}\frac{d\mathbf{r}}{dt} \cdot \mathbf{r} + \frac{1}{2}\mathbf{r} \cdot \frac{d\mathbf{r}}{dt} = \frac{1}{2}\frac{d}{dt}(\mathbf{r} \cdot \mathbf{r}) \Rightarrow \mathbf{r} \cdot \mathbf{r} = K,$ a constant. If $\mathbf{r} = x\mathbf{i} + y\mathbf{j},$ where x

    and y are differentiable functions of t, then $\mathbf{r} \cdot \mathbf{r} = x^2 + y^2 \Rightarrow x^2 + y^2 = K,$ which is the equation of a circle

    centered at the origin.

11. $y = y_0 + \left(v_0 \sin \alpha\right)t - \frac{1}{2}gt^2 \Rightarrow y = 6.5 + (44 \text{ ft/sec})(\sin 45°)(3\sec) - \frac{1}{2}\left(32 \text{ ft/sec}^2\right)(3\sec)^2 = 6.5 + 66\sqrt{2} - 144$

    $\approx -44.16 \text{ ft} \Rightarrow$ the shot put is on the ground. Now, $y = 0 \Rightarrow 6.5 + 22\sqrt{2}t - 16t^2 = 0 \Rightarrow t \approx 2.13 \sec$ (the positive

    root) $\Rightarrow x \approx (44 \text{ ft/sec})(\cos 45°)(2.13\sec) \approx 66.27 \text{ ft}$ or about 66 ft, 3 in. from the stopboard

13. $x = (v_0 \cos\alpha)t$ and $y = (v_0 \sin\alpha)t - \frac{1}{2}gt^2 \Rightarrow \tan\phi = \frac{y}{x} = \frac{(v_0 \sin\alpha)t - \frac{1}{2}gt^2}{(v_0 \cos\alpha)t} = \frac{(v_0 \sin\alpha) - \frac{1}{2}g\,t}{v_0 \cos\alpha}$

$\Rightarrow v_0 \cos\alpha \tan\phi = v_0 \sin\alpha - \frac{1}{2}gt \Rightarrow t = \frac{2v_0 \sin\alpha - 2v_0 \cos\alpha \tan\phi}{g}$, which is the time when the golf ball hits the

upward slope. At this time $x = (v_0 \cos\alpha)\left(\frac{2v_0 \sin\alpha - 2v_0 \cos\alpha \tan\phi}{g}\right) = \left(\frac{2}{g}\right)\left(v_0^2 \sin\alpha \cos\alpha - v_0^2 \cos^2\alpha \tan\phi\right)$. Now

$OR = \frac{x}{\cos\phi} \Rightarrow OR = \left(\frac{2}{g}\right)\left(\frac{v_0^2 \sin\alpha \cos\alpha - v_0^2 \cos^2\alpha \tan\phi}{\cos\phi}\right)$

$= \left(\frac{2v_0^2 \cos\alpha}{g}\right)\left(\frac{\sin\alpha}{\cos\phi} - \frac{\cos\alpha \tan\phi}{\cos\phi}\right) = \left(\frac{2v_0^2 \cos\alpha}{g}\right)\left(\frac{\sin\alpha \cos\phi - \cos\alpha \sin\phi}{\cos^2\phi}\right)$

$= \left(\frac{2v_0^2 \cos\alpha}{g \cos^2\phi}\right)\left[\sin(\alpha - \phi)\right]$. The distance OR is maximized when x

is maximized: $\frac{dx}{d\alpha} = \left(\frac{2v_0^2}{g}\right)(\cos 2\alpha + \sin 2\alpha \tan\phi) = 0$

$\Rightarrow (\cos 2\alpha + \sin 2\alpha \tan\phi) = 0 \Rightarrow \cot 2\alpha + \tan\phi = 0 \Rightarrow \cot 2\alpha = \tan(-\phi) \Rightarrow 2\alpha = \frac{\pi}{2} + \phi \Rightarrow \alpha = \frac{\phi}{2} + \frac{\pi}{4}$

15. $\mathbf{r} = (2\cos t)\mathbf{i} + (2\sin t)\mathbf{j} + t^2 \mathbf{k} \Rightarrow \mathbf{v} = (-2\sin t)\mathbf{i} + (2\cos t)\mathbf{j} + 2t\mathbf{k} \Rightarrow |\mathbf{v}| = \sqrt{(-2\sin t)^2 + (2\cos t)^2 + (2t)^2}$

$= 2\sqrt{1 + t^2} \Rightarrow$ Length $= \int_0^{\pi/4} 2\sqrt{1 + t^2}\, dt = \left[t\sqrt{1 + t^2} + \ln\left|t + \sqrt{1 + t^2}\right|\right]_0^{\pi/4} = \frac{\pi}{4}\sqrt{1 + \frac{\pi^2}{16}} + \ln\left(\frac{\pi}{4} + \sqrt{1 + \frac{\pi^2}{16}}\right)$

17. $\mathbf{r} = \frac{4}{9}(1+t)^{3/2}\,\mathbf{i} + \frac{4}{9}(1-t)^{3/2}\,\mathbf{j} + \frac{1}{3}t\mathbf{k} \Rightarrow \mathbf{v} = \frac{2}{3}(1+t)^{1/2}\,\mathbf{i} - \frac{2}{3}(1-t)^{1/2}\,\mathbf{j} + \frac{1}{3}\mathbf{k}$

$\Rightarrow |\mathbf{v}| = \sqrt{\left[\frac{2}{3}(1+t)^{1/2}\right]^2 + \left[-\frac{2}{3}(1-t)^{1/2}\right]^2 + \left(\frac{1}{3}\right)^2} = 1 \Rightarrow \mathbf{T} = \frac{2}{3}(1+t)^{1/2}\,\mathbf{i} - \frac{2}{3}(1-t)^{1/2}\,\mathbf{j} + \frac{1}{3}\mathbf{k}$

$\Rightarrow \mathbf{T}(0) = \frac{2}{3}\mathbf{i} - \frac{2}{3}\mathbf{j} + \frac{1}{3}\mathbf{k};\ \frac{d\mathbf{T}}{dt} = \frac{1}{3}(1+t)^{-1/2}\,\mathbf{i} + \frac{1}{3}(1-t)^{-1/2}\,\mathbf{j} \Rightarrow \frac{d\mathbf{T}}{dt}(0) = \frac{1}{3}\mathbf{i} + \frac{1}{3}\mathbf{j} \Rightarrow \left|\frac{d\mathbf{T}}{dt}(0)\right| = \frac{\sqrt{2}}{3}$

$\Rightarrow \mathbf{N}(0) = \frac{1}{\sqrt{2}}\mathbf{i} + \frac{1}{\sqrt{2}}\mathbf{j};\ \mathbf{B}(0) = \mathbf{T}(0) \times \mathbf{N}(0) = \begin{vmatrix} \mathbf{i} & \mathbf{j} & \mathbf{k} \\ \frac{2}{3} & -\frac{2}{3} & \frac{1}{3} \\ \frac{1}{\sqrt{2}} & \frac{1}{\sqrt{2}} & 0 \end{vmatrix} = -\frac{1}{3\sqrt{2}}\mathbf{i} + \frac{1}{3\sqrt{2}}\mathbf{j} + \frac{4}{3\sqrt{2}}\mathbf{k};$

$\mathbf{a} = \frac{1}{3}(1+t)^{-1/2}\,\mathbf{i} + \frac{1}{3}(1-t)^{-1/2}\,\mathbf{j} \Rightarrow \mathbf{a}(0) = \frac{1}{3}\mathbf{i} + \frac{1}{3}\mathbf{j}$ and $\mathbf{v}(0) = \frac{2}{3}\mathbf{i} - \frac{2}{3}\mathbf{j} + \frac{1}{3}\mathbf{k} \Rightarrow \mathbf{v}(0) \times \mathbf{a}(0) = \begin{vmatrix} \mathbf{i} & \mathbf{j} & \mathbf{k} \\ \frac{2}{3} & -\frac{2}{3} & \frac{1}{3} \\ \frac{1}{3} & \frac{1}{3} & 0 \end{vmatrix}$

$= -\frac{1}{9}\mathbf{i} + \frac{1}{9}\mathbf{j} + \frac{4}{9}\mathbf{k} \Rightarrow |\mathbf{v} \times \mathbf{a}| = \frac{\sqrt{2}}{3} \Rightarrow \kappa(0) = \frac{|\mathbf{v} \times \mathbf{a}|}{|\mathbf{v}|^3} = \frac{\left(\frac{\sqrt{2}}{3}\right)}{1^3} = \frac{\sqrt{2}}{3};$

$\dot{\mathbf{a}} = -\frac{1}{6}(1+t)^{-3/2}\,\mathbf{i} + \frac{1}{6}(1-t)^{-3/2}\,\mathbf{j} \Rightarrow \dot{\mathbf{a}}(0) = -\frac{1}{6}\mathbf{i} + \frac{1}{6}\mathbf{j} \Rightarrow \tau(0) = \frac{\begin{vmatrix} \frac{2}{3} & -\frac{2}{3} & \frac{1}{3} \\ \frac{1}{3} & \frac{1}{3} & 0 \\ -\frac{1}{6} & \frac{1}{6} & 0 \end{vmatrix}}{|\mathbf{v} \times \mathbf{a}|^2} = \frac{\left(\frac{1}{3}\right)\left(\frac{2}{18}\right)}{\left(\frac{\sqrt{2}}{3}\right)^2} = \frac{1}{6}$

19. $\mathbf{r} = t\mathbf{i} + \frac{1}{2}e^{2t}\mathbf{j} \Rightarrow \mathbf{v} = \mathbf{i} + e^{2t}\mathbf{j} \Rightarrow |\mathbf{v}| = \sqrt{1+e^{4t}} \Rightarrow \mathbf{T} = \frac{1}{\sqrt{1+e^{4t}}}\mathbf{i} + \frac{e^{2t}}{\sqrt{1+e^{4t}}}\mathbf{j} \Rightarrow \mathbf{T}(\ln 2) = \frac{1}{\sqrt{17}}\mathbf{i} + \frac{4}{\sqrt{17}}\mathbf{j};$

$\frac{d\mathbf{T}}{dt} = \frac{-2e^{4t}}{\left(1+e^{4t}\right)^{3/2}}\mathbf{i} + \frac{2e^{2t}}{\left(1+e^{4t}\right)^{3/2}}\mathbf{j} \Rightarrow \frac{d\mathbf{T}}{dt}(\ln 2) = \frac{-32}{17\sqrt{17}}\mathbf{i} + \frac{8}{17\sqrt{17}}\mathbf{j} \Rightarrow \mathbf{N}(\ln 2) = -\frac{4}{\sqrt{17}}\mathbf{i} + \frac{1}{17}\mathbf{j};$

$\mathbf{B}(\ln 2) = \mathbf{T}(\ln 2) \times \mathbf{N}(\ln 2) = \begin{vmatrix} \mathbf{i} & \mathbf{j} & \mathbf{k} \\ \frac{1}{\sqrt{17}} & \frac{4}{\sqrt{17}} & 0 \\ -\frac{4}{\sqrt{17}} & \frac{1}{\sqrt{17}} & 0 \end{vmatrix} = \mathbf{k}; \quad \mathbf{a} = 2e^{2t}\mathbf{j} \Rightarrow \mathbf{a}(\ln 2) = 8\mathbf{j} \text{ and } \mathbf{v}(\ln 2) = \mathbf{i} + 4\mathbf{j}$

$\Rightarrow \mathbf{v}(\ln 2) \times \mathbf{a}(\ln 2) = \begin{vmatrix} \mathbf{i} & \mathbf{j} & \mathbf{k} \\ 1 & 4 & 0 \\ 0 & 8 & 0 \end{vmatrix} = 8\mathbf{k} \Rightarrow |\mathbf{v} \times \mathbf{a}| = 8 \text{ and } |\mathbf{v}(\ln 2)| = \sqrt{17} \Rightarrow \kappa(\ln 2) = \frac{8}{17\sqrt{17}};$

$\dot{\mathbf{a}} = 4e^{2t}\mathbf{j} \Rightarrow \dot{\mathbf{a}}(\ln 2) = 16\mathbf{j} \Rightarrow \tau(\ln 2) = \frac{\begin{vmatrix} 1 & 4 & 0 \\ 0 & 8 & 0 \\ 0 & 16 & 0 \end{vmatrix}}{|\mathbf{v}\times\mathbf{a}|^2} = 0$

21. $\mathbf{r} = \left(2+3t+3t^2\right)\mathbf{i} + \left(4t+4t^2\right)\mathbf{j} - \left(6\cos t\right)\mathbf{k} \Rightarrow \mathbf{v} = (3+6t)\mathbf{i} + (4+8t)\mathbf{j} + (6\sin t)\mathbf{k}$

$\Rightarrow |\mathbf{v}| = \sqrt{(3+6t)^2 + (4+8t)^2 + (6\sin t)^2} = \sqrt{25+100t+100t^2+36\sin^2 t}$

$\Rightarrow \frac{d|\mathbf{v}|}{dt} = \frac{1}{2}\left(25+100t+100t^2+36\sin^2 t\right)^{-1/2}(100+200t+72\sin t\cos t) \Rightarrow a_T(0) = \frac{d|\mathbf{v}|}{dt}(0) = 10;$

$\mathbf{a} = 6\mathbf{i}+8\mathbf{j}+(6\cos t)\mathbf{k} \Rightarrow |\mathbf{a}| = \sqrt{6^2+8^2+(6\cos t)^2} = \sqrt{100+36\cos^2 t} \Rightarrow |\mathbf{a}(0)| = \sqrt{136}$

$\Rightarrow a_N = \sqrt{|\mathbf{a}|^2 - a_T^2} = \sqrt{136-10^2} = \sqrt{36} = 6 \Rightarrow \mathbf{a}(0) = 10\mathbf{T}+6\mathbf{N}$

23. $\mathbf{r} = (\sin t)\mathbf{i} + \left(\sqrt{2}\cos t\right)\mathbf{j} + (\sin t)\mathbf{k} \Rightarrow \mathbf{v} = (\cos t)\mathbf{i} - \left(\sqrt{2}\sin t\right)\mathbf{j} + (\cos t)\mathbf{k}$

$\Rightarrow |\mathbf{v}| = \sqrt{(\cos t)^2 + \left(-\sqrt{2}\sin t\right)^2 + (\cos t)^2} = \sqrt{2} \Rightarrow \mathbf{T} = \frac{\mathbf{v}}{|\mathbf{v}|} = \left(\frac{1}{\sqrt{2}}\cos t\right)\mathbf{i} - (\sin t)\mathbf{j} + \left(\frac{1}{\sqrt{2}}\cos t\right)\mathbf{k};$

$\frac{d\mathbf{T}}{dt} = \left(-\frac{1}{\sqrt{2}}\sin t\right)\mathbf{i} - (\cos t)\mathbf{j} - \left(\frac{1}{\sqrt{2}}\sin t\right)\mathbf{k} \Rightarrow \left|\frac{d\mathbf{T}}{dt}\right| = \sqrt{\left(-\frac{1}{\sqrt{2}}\sin t\right)^2 + (-\cos t)^2 + \left(-\frac{1}{\sqrt{2}}\sin t\right)^2} = 1$

$\Rightarrow \mathbf{N} = \frac{\left(\frac{d\mathbf{T}}{dt}\right)}{\left|\frac{d\mathbf{T}}{dt}\right|} = \left(-\frac{1}{\sqrt{2}}\sin t\right)\mathbf{i} - (\cos t)\mathbf{j} - \left(\frac{1}{\sqrt{2}}\sin t\right)\mathbf{k}; \quad \mathbf{B} = \mathbf{T}\times\mathbf{N} = \begin{vmatrix} \mathbf{i} & \mathbf{j} & \mathbf{k} \\ \frac{1}{\sqrt{2}}\cos t & -\sin t & \frac{1}{\sqrt{2}}\cos t \\ -\frac{1}{\sqrt{2}}\sin t & -\cos t & -\frac{1}{\sqrt{2}}\sin t \end{vmatrix} = \frac{1}{\sqrt{2}}\mathbf{i} - \frac{1}{\sqrt{2}}\mathbf{k};$

$\mathbf{a} = (-\sin t)\mathbf{i} - \left(\sqrt{2}\cos t\right)\mathbf{j} - (\sin t)\mathbf{k} \Rightarrow \mathbf{v}\times\mathbf{a} = \begin{vmatrix} \mathbf{i} & \mathbf{j} & \mathbf{k} \\ \cos t & -\sqrt{2}\sin t & \cos t \\ -\sin t & -\sqrt{2}\cos t & -\sin t \end{vmatrix} = \sqrt{2}\mathbf{i} - \sqrt{2}\mathbf{k}$

$\Rightarrow |\mathbf{v}\times\mathbf{a}| = \sqrt{4} = 2 \Rightarrow \kappa = \frac{|\mathbf{v}\times\mathbf{a}|}{|\mathbf{v}|^3} = \frac{2}{\left(\sqrt{2}\right)^3} = \frac{1}{\sqrt{2}}; \quad \dot{\mathbf{a}} = (-\cos t)\mathbf{i} + \left(\sqrt{2}\sin t\right)\mathbf{j} - (\cos t)\mathbf{k}$

$\Rightarrow \tau = \frac{\begin{vmatrix} \cos t & -\sqrt{2}\sin t & \cos t \\ -\sin t & -\sqrt{2}\cos t & -\sin t \\ -\cos t & \sqrt{2}\sin t & -\cos t \end{vmatrix}}{|\mathbf{v}\times\mathbf{a}|^2} = \frac{(\cos t)\left(\sqrt{2}\right) - \left(\sqrt{2}\sin t\right)(0) + (\cos t)\left(-\sqrt{2}\right)}{4} = 0$

25. $\mathbf{r} = 2\mathbf{i} + \left(4\sin\frac{t}{2}\right)\mathbf{j} + \left(3 - \frac{t}{\pi}\right)\mathbf{k} \Rightarrow 0 = \mathbf{r}\cdot(\mathbf{i}-\mathbf{j}) = 2(1) + \left(4\sin\frac{t}{2}\right)(-1) \Rightarrow 0 = 2 - 4\sin\frac{t}{2} \Rightarrow \sin\frac{t}{2} = \frac{1}{2}$

$\Rightarrow \frac{t}{2} = \frac{\pi}{6} \Rightarrow t = \frac{\pi}{3}$ (for the first time)

27. $\mathbf{r} = e^t\mathbf{i} + (\sin t)\mathbf{j} + \ln(1-t)\mathbf{k} \Rightarrow \mathbf{v} = e^t\mathbf{i} + (\cos t)\mathbf{j} - \left(\frac{1}{1-t}\right)\mathbf{k} \Rightarrow \mathbf{v}(0) = \mathbf{i} + \mathbf{j} - \mathbf{k}; \ \mathbf{r}(0) = \mathbf{i} \Rightarrow (1, 0, 0)$ is on the line

$\Rightarrow x = 1+t, \ y = t, \ \text{and} \ z = -t$ are parametric equations of the line

29. $x^2 = \left(v_0^2\cos^2\alpha\right)t^2$ and $\left(y + \frac{1}{2}gt^2\right)^2 = \left(v_0^2\sin^2\alpha\right)t^2 \Rightarrow x^2 + \left(y + \frac{1}{2}gt^2\right)^2 = v_0^2 t^2$

CHAPTER 12 ADDITIONAL AND ADVANCED EXERCISES

1. (a) $\mathbf{r}(\theta) = (a\cos\theta)\mathbf{i} + (a\sin\theta)\mathbf{j} + b\theta\mathbf{k} \Rightarrow \frac{d\mathbf{r}}{dt} = [(-a\sin\theta)\mathbf{i} + (a\cos\theta)\mathbf{j} + b\mathbf{k}]\frac{d\theta}{dt};$

$|\mathbf{v}| = \sqrt{2gz} = \left|\frac{d\mathbf{r}}{dt}\right| = \sqrt{a^2+b^2}\ \frac{d\theta}{dt} \Rightarrow \frac{d\theta}{dt} = \sqrt{\frac{2gz}{a^2+b^2}} = \sqrt{\frac{2gb\theta}{a^2+b^2}} \Rightarrow \frac{d\theta}{dt}\Big|_{\theta=2\pi} = \sqrt{\frac{4\pi gb}{a^2+b^2}} = 2\sqrt{\frac{\pi gb}{a^2+b^2}}$

(b) $\frac{d\theta}{dt} = \sqrt{\frac{2gb\theta}{a^2+b^2}} \Rightarrow \frac{d\theta}{\sqrt{\theta}} = \sqrt{\frac{2gb}{a^2+b^2}}\,dt \Rightarrow 2\theta^{1/2} = \sqrt{\frac{2gb}{a^2+b^2}}\,t + C; \ t = 0 \Rightarrow \theta = 0 \Rightarrow C = 0 \Rightarrow 2\theta^{1/2} = \sqrt{\frac{2gb}{a^2+b^2}}\,t$

$\Rightarrow \theta = \frac{gbt^2}{2(a^2+b^2)}; \ z = b\theta \Rightarrow z = \frac{gb^2t^2}{2(a^2+b^2)}$

(c) $\mathbf{v}(t) = \frac{d\mathbf{r}}{dt} = [(-a\sin\theta)\mathbf{i} + (a\cos\theta)\mathbf{j} + b\mathbf{k}]\frac{d\theta}{dt} = [(-a\sin\theta)\mathbf{i} + (a\cos\theta)\mathbf{j} + b\mathbf{k}]\left(\frac{gbt}{a^2+b^2}\right),$ from part (b)

$\Rightarrow \mathbf{v}(t) = \left[\frac{(-a\sin\theta)\mathbf{i}+(a\cos\theta)\mathbf{j}+b\mathbf{k}}{\sqrt{a^2+b^2}}\right]\left(\frac{gbt}{\sqrt{a^2+b^2}}\right) = \frac{gbt}{\sqrt{a^2+b^2}}\mathbf{T};$

$\frac{d^2\mathbf{r}}{dt^2} = [(-a\cos\theta)\mathbf{i} - (a\sin\theta)\mathbf{j}]\left(\frac{d\theta}{dt}\right)^2 + [(-a\sin\theta)\mathbf{i} + (a\cos\theta)\mathbf{j} + b\mathbf{k}]\frac{d^2\theta}{dt^2}$

$= \left(\frac{gbt}{a^2+b^2}\right)^2 [(-a\cos\theta)\mathbf{i} - (a\sin\theta)\mathbf{j}] + [(-a\sin\theta)\mathbf{i} + (a\cos\theta)\mathbf{j} + b\mathbf{k}]\left(\frac{gb}{a^2+b^2}\right)$

$= \left[\frac{(-a\sin\theta)\mathbf{i}+(a\cos\theta)\mathbf{j}+b\mathbf{k}}{\sqrt{a^2+b^2}}\right]\left(\frac{gb}{\sqrt{a^2+b^2}}\right) + a\left(\frac{gbt}{a^2+b^2}\right)[(-\cos\theta)\mathbf{i} - (\sin\theta)\mathbf{j}] = \frac{gb}{\sqrt{a^2+b^2}}\mathbf{T} + a\left(\frac{gbt}{a^2+b^2}\right)^2\mathbf{N}$ (there is

no component in the direction of $\mathbf{B}$).

3. $r = \frac{(1+e)r_0}{1+e\cos\theta} \Rightarrow \frac{dr}{d\theta} = \frac{(1+e)r_0(e\sin\theta)}{(1+e\cos\theta)^2}; \ \frac{dr}{d\theta} = 0 \Rightarrow \frac{(1+e)r_0(e\sin\theta)}{(1+e\cos\theta)^2} = 0 \Rightarrow (1+e)r_0(e\sin\theta) = 0 \Rightarrow \sin\theta = 0$

$\Rightarrow \theta = 0$ or π. Note that $\frac{dr}{d\theta} > 0$ when $\sin\theta > 0$ and $\frac{dr}{d\theta} < 0$ when $\sin\theta < 0$. Since $\sin\theta < 0$ on $-\pi < \theta < 0$

and $\sin\theta > 0$ on $0 < \theta < \pi$, r is a minimum when $\theta = 0$ and $r(0) = \frac{(1+e)r_0}{1+e\cos 0} = r_0$

5. (a) $\mathbf{v} = \dot{x}\mathbf{i} + \dot{y}\mathbf{j}$ and $\mathbf{v} = \dot{r}\,\mathbf{u}_r + r\dot{\theta}\,\mathbf{u}_\theta = (\dot{r})[(\cos\theta)\mathbf{i} + (\sin\theta)\mathbf{j}] + (r\dot{\theta})[(-\sin\theta)\mathbf{i} + (\cos\theta)\mathbf{j}] \Rightarrow \mathbf{v}\cdot\mathbf{i} = \dot{x}$ and

$\mathbf{v}\cdot\mathbf{i} = \dot{r}\cos\theta - r\dot{\theta}\sin\theta \Rightarrow \dot{x} = \dot{r}\cos\theta - r\dot{\theta}\sin\theta; \ \mathbf{v}\cdot\mathbf{j} = \dot{y}$ and

$\mathbf{v}\cdot\mathbf{j} = \dot{r}\sin\theta + r\dot{\theta}\cos\theta \Rightarrow \dot{y} = \dot{r}\sin\theta + r\dot{\theta}\cos\theta$

(b) $\mathbf{u}_r = (\cos\theta)\mathbf{i} + (\sin\theta)\mathbf{j} \Rightarrow \mathbf{v}\cdot\mathbf{u}_r = \dot{x}\cos\theta + \dot{y}\sin\theta = (\dot{r}\cos\theta - r\dot{\theta}\sin\theta)(\cos\theta) + (\dot{r}\sin\theta + r\dot{\theta}\cos\theta)(\sin\theta)$

by part (a), $\Rightarrow \mathbf{v}\cdot\mathbf{u}_r = \dot{r}$; therefore, $\dot{r} = \dot{x}\cos\theta + \dot{y}\sin\theta; \ \mathbf{u}_\theta = -(\sin\theta)\mathbf{i} + (\cos\theta)\mathbf{j}$

$\Rightarrow \mathbf{v}\cdot\mathbf{u}_\theta = -\dot{x}\sin\theta + \dot{y}\cos\theta = (\dot{r}\cos\theta - r\dot{\theta}\sin\theta)(-\sin\theta) + (\dot{r}\sin\theta + r\dot{\theta}\cos\theta)(\cos\theta)$ by part (a)

$\Rightarrow \mathbf{v}\cdot\mathbf{u}_\theta = r\dot{\theta}$; therefore, $r\dot{\theta} = -\dot{x}\sin\theta + \dot{y}\cos\theta$

7.   (a) Let $r = 2 - t$ and $\theta = 3t \Rightarrow \frac{dr}{dt} = -1$ and $\frac{d\theta}{dt} = 3 \Rightarrow \frac{d^2 r}{dt^2} = \frac{d^2 \theta}{dt^2} = 0$. The halfway point is $(1, 3) \Rightarrow t = 1$;

$$\mathbf{v} = \frac{dr}{dt}\mathbf{u}_r + r\frac{d\theta}{dt}\mathbf{u}_\theta \Rightarrow \mathbf{v}(1) = -\mathbf{u}_r + 3\mathbf{u}_\theta \;;\quad \mathbf{a} = \left[\frac{d^2 r}{dt^2} - r\left(\frac{d\theta}{dt}\right)^2\right]\mathbf{u}_r + \left[r\frac{d^2 \theta}{dt^2} + 2\frac{dr}{dt}\frac{d\theta}{dt}\right]\mathbf{u}_\theta$$

$$\Rightarrow \mathbf{a}(1) = -9\mathbf{u}_r - 6\mathbf{u}_\theta$$

(b) It takes the beetle 2 min to crawl to the origin $\Rightarrow$ the rod has revolved 6 radians

$$\Rightarrow L = \int_0^6 \sqrt{[f(\theta)]^2 + [f'(\theta)]^2}\, d\theta = \int_0^6 \sqrt{\left(2 - \frac{\theta}{3}\right)^2 + \left(-\frac{1}{3}\right)^2}\, d\theta = \int_0^6 \sqrt{4 - \frac{4\theta}{3} + \frac{\theta^2}{9} + \frac{1}{9}}\, d\theta$$

$$= \int_0^6 \sqrt{\frac{37 - 12\theta + \theta^2}{9}}\, d\theta = \frac{1}{3}\int_0^6 \sqrt{(\theta - 6)^2 + 1}\, d\theta = \frac{1}{3}\left[\frac{(\theta - 6)}{2}\sqrt{(\theta - 6)^2 + 1} + \frac{1}{2}\ln\left|\theta - 6 + \sqrt{(\theta - 6)^2 + 1}\right|\right]_0^6$$

$$= \sqrt{37} - \frac{1}{6}\ln\left(\sqrt{37} - 6\right) \approx 6.5 \text{ in.}$$

CHAPTER 13   PARTIAL DERIVATIVES

13.1 FUNCTIONS OF SEVERAL VARIABLES

1. (a) $f(0,0) = 0$ (b) $f(-1,1) = 0$ (c) $f(2,3) = 58$
   (d) $f(-3,-2) = 33$

3. (a) $f(3,-1,2) = \frac{4}{5}$ (b) $f\left(1, \frac{1}{2}, -\frac{1}{4}\right) = \frac{8}{5}$ (c) $f\left(0, -\frac{1}{3}, 0\right) = 3$
   (d) $f(2,2,100) = 0$

5. Domain: all points (x, y) on or above the line $y = x + 2$

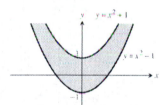

7. Domain: all points (x, y) not lying on the graph of $y = x$ or $y = x^3$

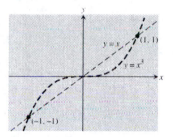

9. Domain: all points (x, y) satisfying $x^2 - 1 \le y \le x^2 + 1$

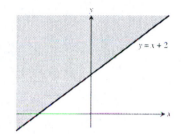

11. Domain: all points (x, y) satisfying $(x-2)(x+2)(y-3)(y+3) \ge 0$

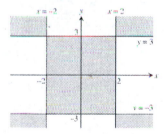

13.

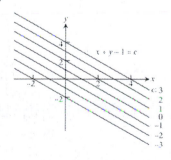

15.

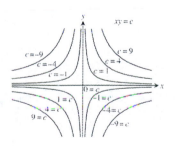

17. (a) Domain: all points in the xy-plane
    (b) Range: all real numbers
    (c) level curves are straight lines $y - x = c$ parallel to the line $y = x$
    (d) no boundary points
    (e) both open and closed
    (f) unbounded

19. (a) Domain: all points in the xy-plane

    (b) Range: $z \geq 0$

    (c) level curves: for $f(x, y) = 0$, the origin; for $f(x, y) = c > 0$, ellipses with center $(0, 0)$ and major and minor axes along the x- and y-axes, respectively

    (d) no boundary points

    (e) both open and closed

    (f) unbounded

21. (a) Domain: all points in the xy-plane

    (b) Range: all real numbers

    (c) level curves are hyperbolas with the x- and y-axes as asymptotes when $f(x, y) \neq 0$, and the x- and y-axes when $f(x, y) = 0$

    (d) no boundary points

    (e) both open and closed

    (f) unbounded

23. (a) Domain: all (x, y) satisfying $x^2 + y^2 < 16$

    (b) Range: $z \geq \frac{1}{4}$

    (c) level curves are circles centered at the origin with radii $r < 4$

    (d) boundary is the circle $x^2 + y^2 = 16$

    (e) open

    (f) bounded

25. (a) Domain: $(x, y) \neq (0, 0)$

    (b) Range: all real numbers

    (c) level curves are circles with center $(0, 0)$ and radii $r > 0$

    (d) boundary is the single point $(0, 0)$

    (e) open

    (f) unbounded

27. (a) Domain: all (x, y) satisfying $-1 \leq y - x \leq 1$

    (b) Range: $-\frac{\pi}{2} \leq z \leq \frac{\pi}{2}$

    (c) level curves are straight lines of the form $y - x = c$ where $-1 \leq c \leq 1$

    (d) boundary is the two straight lines $y = 1 + x$ and $y = -1 + x$

    (e) closed

    (f) unbounded

29. (a) Domain: all points (x, y) outside the circle $x^2 + y^2 = 1$

    (b) Range: all reals

    (c) Circles centered at the origin with radii $r > 1$

    (d) Boundary: the circle $x^2 + y^2 = 1$

    (e) open

    (f) unbounded

31. f   33. a   35. d

37. (a)

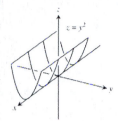

(b)

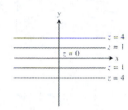

39. (a)

(b)

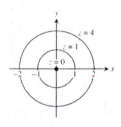

41. (a)

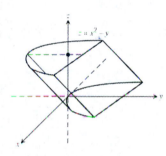

(b)

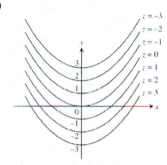

43. (a)

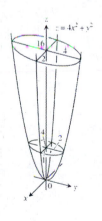

(b)

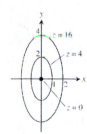

45. (a)

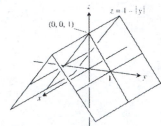

(b)

47. (a)

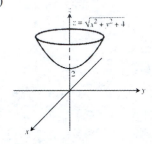

(b)

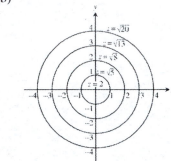

49. $f(x, y) = 16 - x^2 - y^2$ and $\left(2\sqrt{2}, \sqrt{2}\right) \Rightarrow z = 16 - \left(2\sqrt{2}\right)^2 - \left(\sqrt{2}\right)^2 = 6 \Rightarrow 6 = 16 - x^2 - y^2 \Rightarrow x^2 + y^2 = 10$

51. $f(x, y) = \sqrt{x + y^2 - 3}$ and $(3, -1) \Rightarrow z = \sqrt{3 + (-1)^2 - 3} = 1 \Rightarrow x + y^2 - 3 = 1 \Rightarrow x + y^2 = 4$

53.

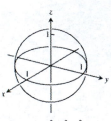

$f(x, y, z) = x^2 + y^2 + z^2 = 1$

55.

$f(x, y, z) = x + z = 1$

57.

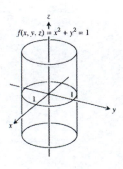

$f(x, y, z) = x^2 + y^2 = 1$

59.

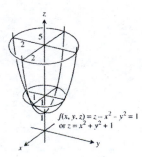

$f(x, y, z) = z - x^2 - y^2 = 1$
or $z = x^2 + y^2 + 1$

61. $f(x, y, z) = \sqrt{x - y} - \ln z$ at $(3, -1, 1) \Rightarrow w = \sqrt{x - y} - \ln z$; at $(3, -1, 1) \Rightarrow w = \sqrt{3 - (-1)} - \ln 1 = 2$
$\Rightarrow \sqrt{x - y} - \ln z = 2$

63. $g(x, y, z) = \sqrt{x^2 + y^2 + z^2}$ at $\left(1, -1, \sqrt{2}\right) \Rightarrow w = \sqrt{x^2 + y^2 + z^2}$; at $\left(1, -1, \sqrt{2}\right) \Rightarrow w = \sqrt{1^2 + (-1)^2 + \left(\sqrt{2}\right)^2} = 2$

$\Rightarrow 2 = \sqrt{x^2 + y^2 + z^2} \Rightarrow x^2 + y^2 + z^2 = 4$

65. $f(x, y) = \sum\limits_{n=0}^{\infty} \left(\frac{x}{y}\right)^n = \frac{1}{1-\left(\frac{x}{y}\right)} = \frac{y}{y-x}$ for $\left|\frac{x}{y}\right| < 1$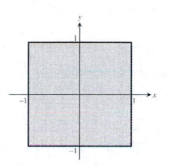

$\Rightarrow$ Domain: all points (x, y) satisfying $|x| < |y|$;

at $(1, 2) \Rightarrow$ since $\left|\frac{1}{2}\right| < 1 \Rightarrow z = \frac{2}{2-1} = 2$

$\Rightarrow \frac{y}{y-x} = 2 \Rightarrow y = 2x$

67. $f(x, y) = \int_x^y \frac{d\theta}{\sqrt{1-\theta^2}} = \sin^{-1} y - \sin^{-1} x$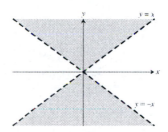

$\Rightarrow$ Domain: all points (x, y) satisfying $-1 \le x \le 1$ and $-1 \le y \le 1$;

at $(0, 1) \Rightarrow \sin^{-1} 1 - \sin^{-1} 0 = \frac{\pi}{2} \Rightarrow \sin^{-1} y - \sin^{-1} x = \frac{\pi}{2}$.

Since $-\frac{\pi}{2} \le \sin^{-1} y \le \frac{\pi}{2}$ and $-\frac{\pi}{2} \le \sin^{-1} x \le \frac{\pi}{2}$, in order

for $\sin^{-1} y - \sin^{-1} x$ to equal $\frac{\pi}{2}$, $0 \le \sin^{-1} y \le \frac{\pi}{2}$ and

$-\frac{\pi}{2} \le \sin^{-1} x \le 0$; that is $0 \le y \le 1$ and $-1 \le x \le 0$. Thus

$y = \sin\left(\frac{\pi}{2} + \sin^{-1} x\right) = \sqrt{1-x^2}$, $x \le 0$

69-71. Example CAS commands:

Maple:

```
with( plots );

f := (x,y) -> x*sin(y/2) + y*sin(2*x);

xdomain := x=0..5*Pi;

ydomain := y=0..5*Pi;

x0,y0 := 3*Pi,3*Pi;

plot3d( f(x,y), xdomain, ydomain, axes=boxed, style=patch, shading=zhue, title="#69(a) (Section 14.1)" );

plot3d( f(x,y), xdomain, ydomain, grid=[50,50], axes=boxed, shading=zhue, style=patchcontour,
    orientation=[-90,0], title="#69(b)(Section 14.1)" )                     # (b)

L := evalf( f(x0,y0) );                                                     # (c)

plot3d( f(x,y), xdomain, ydomain, grid=[50,50], axes=boxed, shading=zhue, style=patchcontour,
    contours=[L], orientation=[-90,0], title="#69(c)(Section 14.1)" );
```

73-75. Example CAS commands:

Maple:

```
eq := 4*ln(x^2+y^2+z^2)=1;

implicitplot3d( eq, x=-2..2, y= -2..2, z=-2..2, grid=[30,30,30], axes=boxed, title="#73 (Section 14.1)");
```

77-79.   Example CAS commands:

   <u>Maple</u>:

   x:= (u,v) -> u*cos(v);

   y:= (u,v) -> u*sin(v);

   z:= (u,v) -> u;

   plot3d([x(u,v),y(u,v),z(u,v)],u=0..2, v=0..2*Pi, axes=boxed, style=patchcontour, contours=[($0..4)/2],
        shading=zhue, title="#77 (Section 14.1)");

69-79.   Example CAS commands:

   <u>Mathematica</u>: (assigned functions and bounds will vary)
   For 69 - 72, the command **ContourPlot** draws 2-dimensional contours that are z-level curves of surfaces
   $z = f(x,y)$.

   Clear[x, y, f]

   f[x_,y_]:= x Sin[y/2]+ y Sin[2x]

   xmin=0; xmax= 5π; ymin=0; ymax= 5π; {x0, y0}={$3\pi, 3\pi$};

   cp= ContourPlot[f[x,y], {x, xmin, xmax},{y, ymin, ymax}, ContourShading $\to$ False];

   cp0= ContourPlot[{f[x,y], {x, xmin, xmax},{y, ymin, ymax}, Contours $\to$ [f[x0,y0]},
        ContourShading $\to$ False, PlotStyle $\to$ {RGBColor[1,0,0]}}];

   Show[cp, cp0]

For 73 - 76, the command **ContourPlot3D** will be used. Write the function f[x, y, z] so that when it is equated
to zero, it represents the level surface given.

For 73, the problem associated with Log[0] can be avoided by rewriting the function as $x^2 + y^2 + z^2 - e^{1/4}$

   Clear[x, y, z, f]

   f[x_, y_, z_]:= $x^2 + y^2 + z^2$ – Exp[1/4]

   ContourPlot3D[f[x, y, z], {x, −5,5},{y, −5,5},{z, −5,5}, PlotPoints $\to$ (7,7)];

For 77 - 80, the command ParametricPlot3D will be used. To get the z-level curves here, we solve x and y in
terms of z and either u or v (v here), create a table of level curves, then plot that table.

   Clear[x, y, z, u, v]

   ParametricPlot3D[{u Cos[v], u Sin[v], u}, {u, 0, 2}, {v, 0, 2Pi}];

   zlevel=Table[{z Cos[v], z sin[v]}, {z, 0, 2, .1}];

   ParametricPlot[Evaluate[zlevel],{v, 0, 2π}];

13.2 LIMITS AND CONTINUITY IN HIGHER DIMENSIONS

1. $\displaystyle\lim_{(x,\,y)\to(0,\,0)} \frac{3x^2-y^2+5}{x^2+y^2+2} = \frac{3(0)^2-0^2+5}{0^2+0^2+2} = \frac{5}{2}$

3. $\displaystyle\lim_{(x,\,y)\to(3,\,4)} \sqrt{x^2+y^2-1} = \sqrt{3^2+4^2-1} = \sqrt{24} = 2\sqrt{6}$

5. $\displaystyle\lim_{(x,\,y)\to\left(0,\,\frac{\pi}{4}\right)} \sec x \, \tan y = (\sec 0)\left(\tan \frac{\pi}{4}\right) = (1)(1) = 1$

7. $\displaystyle\lim_{(x,\,y)\to(0,\,\ln 2)} e^{x-y} = e^{0-\ln 2} = e^{\ln\left(\frac{1}{2}\right)} = \frac{1}{2}$

9. $\displaystyle\lim_{(x,\,y)\to(0,\,0)} \frac{e^y \sin x}{x} = \lim_{(x,\,y)\to(0,\,0)} \left(e^y\right)\left(\frac{\sin x}{x}\right) = e^0 \cdot \lim_{x\to 0}\left(\frac{\sin x}{x}\right) = 1 \cdot 1 = 1$

11. $\displaystyle\lim_{(x,\,y)\to(1,\,\pi/6)} \frac{x\sin y}{x^2+1} = \frac{1\cdot\sin\left(\frac{\pi}{6}\right)}{1^2+1} = \frac{1/2}{2} = \frac{1}{4}$

13. $\displaystyle\lim_{\substack{(x,\,y)\to(1,1)\\ x\neq y}} \frac{x^2-2xy+y^2}{x-y} = \lim_{(x,\,y)\to(1,1)} \frac{(x-y)^2}{x-y} = \lim_{(x,\,y)\to(1,1)} (x-y) = (1-1) = 0$

15. $\displaystyle\lim_{\substack{(x,\,y)\to(1,\,1)\\ x\neq 1}} \frac{xy-y-2x+2}{x-1} = \lim_{\substack{(x,\,y)\to(1,\,1)\\ x\neq 1}} \frac{(x-1)(y-2)}{x-1} = \lim_{(x,\,y)\to(1,1)} (y-2) = (1-2) = -1$

17. $\displaystyle\lim_{\substack{(x,\,y)\to(0,0)\\ x\neq y}} \frac{x-y+2\sqrt{x}-2\sqrt{y}}{\sqrt{x}-\sqrt{y}} = \lim_{\substack{(x,\,y)\to(0,0)\\ x\neq y}} \frac{\left(\sqrt{x}-\sqrt{y}\right)\left(\sqrt{x}+\sqrt{y}+2\right)}{\sqrt{x}-\sqrt{y}} = \lim_{(x,\,y)\to(0,0)} \left(\sqrt{x}+\sqrt{y}+2\right) = \left(\sqrt{0}+\sqrt{0}+2\right) = 2$

Note: (x, y) must approach $(0, 0)$ through the first quadrant only with $x \neq y$.

19. $\displaystyle\lim_{\substack{(x,\,y)\to(2,0)\\ 2x-y\neq 4}} \frac{\sqrt{2x-y}-2}{2x-y-4} = \lim_{\substack{(x,\,y)\to(2,0)\\ 2x-y\neq 4}} \frac{\sqrt{2x-y}-2}{\left(\sqrt{2x-y}+2\right)\left(\sqrt{2x-y}-2\right)} = \lim_{(x,\,y)\to(2,0)} \frac{1}{\sqrt{2x-y}+2} = \frac{1}{\sqrt{(2)(2)-0}+2} = \frac{1}{2+2} = \frac{1}{4}$

21. $\displaystyle\lim_{(x,\,y)\to(0,0)} \frac{\sin\left(x^2+y^2\right)}{x^2+y^2} = \lim_{r\to 0} \frac{\sin\left(r^2\right)}{r^2} = \lim_{r\to 0} \frac{2r\cdot\cos\left(r^2\right)}{2r} = \lim_{r\to 0} \cos\left(r^2\right) = 1$

23. $\displaystyle\lim_{(x,\,y)\to(1,\,-1)} \frac{x^3+y^3}{x+y} = \lim_{(x,\,y)\to(1,\,-1)} \frac{(x+y)\left(x^2-xy+y^2\right)}{x+y} = \lim_{(x,\,y)\to(1,\,-1)} \left(x^2-xy+y^2\right) = \left(1^2-(1)(-1)+(-1)^2\right) = 3$

25. $\displaystyle\lim_{P\to(1,\,3,\,4)} \left(\frac{1}{x}+\frac{1}{y}+\frac{1}{z}\right) = \frac{1}{1}+\frac{1}{3}+\frac{1}{4} = \frac{12+4+3}{12} = \frac{19}{12}$

27. $\lim\limits_{P\to(3,\,3,\,0)}\left(\sin^2 x+\cos^2 y+\sec^2 z\right)=\left(\sin^2 3+\cos^2 3+\sec^2 0\right)=1+1^2=2$

29. $\lim\limits_{P\to(\pi,\,0,\,3)} ze^{-2y}\cos 2x=3e^{-2(0)}\cos 2\pi=(3)(1)(1)=3$

31. (a) All (x,y) (b) All (x,y) except $(0,0)$

33. (a) All (x,y) except where $x=0$ or $y=0$ (b) All (x,y)

35. (a) All (x,y,z) (b) All (x,y,z) except the interior of the cylinder
$$x^2+y^2=1$$

37. (a) All (x,y,z) with $z\neq 0$ (b) All (x,y,z) with $x^2+z^2\neq 1$

39. (a) All (x,y,z) such that $z>x^2+y^2+1$ (b) All (x,y,z) such that $z\neq\sqrt{x^2+y^2}$

41. $\lim\limits_{\substack{(x,y)\to(0,0)\\ \text{along } y=x\\ x>0}}-\dfrac{x}{\sqrt{x^2+y^2}}=\lim\limits_{x\to 0^+}-\dfrac{x}{\sqrt{x^2+x^2}}=\lim\limits_{x\to 0^+}-\dfrac{x}{\sqrt{2}|x|}=\lim\limits_{x\to 0^+}-\dfrac{x}{\sqrt{2}x}=\lim\limits_{x\to 0^+}-\dfrac{1}{\sqrt{2}}=-\dfrac{1}{\sqrt{2}};$

$\lim\limits_{\substack{(x,y)\to(0,0)\\ \text{along } y=x\\ x<0}}-\dfrac{x}{\sqrt{x^2+y^2}}=\lim\limits_{x\to 0^-}-\dfrac{x}{\sqrt{2}|x|}=\lim\limits_{x\to 0^-}-\dfrac{x}{\sqrt{2}(-x)}=\lim\limits_{x\to 0^-}\dfrac{1}{\sqrt{2}}=\dfrac{1}{\sqrt{2}}$

43. $\lim\limits_{\substack{(x,y)\to(0,0)\\ \text{along } y=kx^2}}\dfrac{x^4-y^2}{x^4+y^2}=\lim\limits_{x\to 0}\dfrac{x^4-\left(kx^2\right)^2}{x^4+\left(kx^2\right)^2}=\lim\limits_{x\to 0}\dfrac{x^4-k^2x^4}{x^4+k^2x^4}=\dfrac{1-k^2}{1+k^2}\Rightarrow$ different limits for different values of k

45. $\lim\limits_{\substack{(x,y)\to(0,0)\\ \text{along } y=kx\\ k\neq-1}}\dfrac{x-y}{x+y}=\lim\limits_{x\to 0}\dfrac{x-kx}{x+kx}=\dfrac{1-k}{1+k}\Rightarrow$ different limits for different values of $k,\ k\neq-1$

47. $\lim\limits_{\substack{(x,y)\to(0,0)\\ \text{along } y=kx^2\\ k\neq 0}}\dfrac{x^2+y}{y}=\lim\limits_{x\to 0}\dfrac{x^2+kx^2}{kx^2}=\dfrac{1+k}{k}\Rightarrow$ different limits for different values of $k,\ k\neq 0$

49. $\lim\limits_{\substack{(x,y)\to(1,1)\\ \text{along } x=1}}\dfrac{xy^2-1}{y-1}=\lim\limits_{y\to 1}\dfrac{y^2-1}{y-1}=\lim\limits_{y\to 1}(y+1)=2;\quad \lim\limits_{\substack{(x,y)\to(1,1)\\ \text{along } y=x}}\dfrac{xy^2-1}{y-1}=\lim\limits_{y\to 1}\dfrac{y^3-1}{y-1}=\lim\limits_{y\to 1}\left(y^2+y+1\right)=3$

51. $f(x,y)=\begin{cases}1 & \text{if } y\geq x^4\\ 1 & \text{if } y\leq 0\\ 0 & \text{otherwise}\end{cases}$

     (a) $\lim\limits_{(x,y)\to(0,1)}f(x,y)=1$ since any path through $(0,1)$ that is close to $(0,1)$ satisfies $y\geq x^4$

(b) $\lim\limits_{(x,\,y)\to(2,3)} f(x,y)=0$ since any path through $(2,3)$ that is close to $(2,3)$ does not satisfy either

$y \ge x^4$ or $y \le 0$

(c) $\lim\limits_{\substack{(x,\,y)\to(0,0) \\ \text{along } x=0}} f(x,y)=1$ and $\lim\limits_{\substack{(x,\,y)\to(0,0) \\ \text{along } y=x^2}} f(x,y)=0 \Rightarrow \lim\limits_{(x,\,y)\to(0,0)} f(x,y)$ does not exist

53. First consider the vertical line $x=0 \Rightarrow \lim\limits_{\substack{(x,\,y)\to(0,0) \\ \text{along } x=0}} \dfrac{2x^2 y}{x^4+y^2} = \lim\limits_{y\to 0} \dfrac{2(0)^2 y}{(0)^4+y^2} = \lim\limits_{y\to 0} 0 = 0.$ Now consider any

nonvertical line through $(0,0)$. The equation of any line through $(0,0)$ is of the form $y = mx$

$\Rightarrow \lim\limits_{\substack{(x,\,y)\to(0,0) \\ \text{along } y=mx}} f(x,y) = \lim\limits_{\substack{(x,\,y)\to(0,0) \\ \text{along } y=mx}} \dfrac{2x^2 y}{x^4+y^2} = \lim\limits_{x\to 0} \dfrac{2x^2(mx)}{x^4+(mx)^2} = \lim\limits_{x\to 0} \dfrac{2mx^3}{x^4+m^2 x^2} = \lim\limits_{x\to 0} \dfrac{2mx^3}{x^2\left(x^2+m^2\right)} = \lim\limits_{x\to 0} \dfrac{2mx}{\left(x^2+m^2\right)} = 0.$

Thus $\lim\limits_{\substack{(x,\,y)\to(0,0) \\ \text{any line through }(0,0)}} \dfrac{2x^2 y}{x^4+y^4} = 0.$

55. $\lim\limits_{(x,\,y)\to(0,0)}\left(1-\dfrac{x^2 y^2}{3}\right)=1$ and $\lim\limits_{(x,\,y)\to(0,0)} 1 = 1 \Rightarrow \lim\limits_{(x,\,y)\to(0,0)} \dfrac{\tan^{-1} xy}{xy} = 1$ by the Sandwich Theorem

57. The limit is 0 since $\left|\sin\left(\dfrac{1}{x}\right)\right| \le 1 \Rightarrow -1 \le \sin\left(\dfrac{1}{x}\right) \le 1 \Rightarrow -y \le y\sin\left(\dfrac{1}{x}\right) \le y$ for $y \ge 0$, and $-y \ge y\sin\left(\dfrac{1}{x}\right) \ge y$ for

$y \le 0$. Thus as $(x,y)\to(0,0)$, both $-y$ and y approach $0 \Rightarrow y\sin\left(\dfrac{1}{x}\right)\to 0$, by the Sandwich Theorem.

59. (a) $f(x,y)\big|_{y=mx} = \dfrac{2m}{1+m^2} = \dfrac{2\tan\theta}{1+\tan^2\theta} = \sin 2\theta.$ The value of $f(x,y) = \sin 2\theta$ varies with θ, which is the line's

angle of inclination.

(b)   Since $f(x,y)\big|_{y=mx} = \sin 2\theta$ and since $-1 \le \sin 2\theta \le 1$ for every θ, $\lim\limits_{(x,\,y)\to(0,0)} f(x,y)$ varies from -1 to

1 along $y = mx$.

61. $\lim\limits_{(x,\,y)\to(0,0)} \dfrac{x^3-xy^2}{x^2+y^2} = \lim\limits_{r\to 0} \dfrac{r^3\cos^3\theta - (r\cos\theta)\left(r^2\sin^2\theta\right)}{r^2} = \lim\limits_{r\to 0} \dfrac{r\left(\cos^3\theta - \cos\theta\sin^2\theta\right)}{1} = 0$

63. $\lim\limits_{(x,\,y)\to(0,0)} \dfrac{y^2}{x^2+y^2} = \lim\limits_{r\to 0} \dfrac{r^2\sin^2\theta}{r^2} = \lim\limits_{r\to 0} \left(\sin^2\theta\right) = \sin^2\theta;$ the limit does not exist since $\sin^2\theta$ is between 0 and

1 depending on θ

65. $\lim\limits_{(x,\,y)\to(0,0)} \tan^{-1}\left[\dfrac{|x|+|y|}{x^2+y^2}\right] = \lim\limits_{r\to 0} \tan^{-1}\left[\dfrac{|r\cos\theta|+|r\sin\theta|}{r^2}\right] = \lim\limits_{r\to 0} \tan^{-1}\left[\dfrac{|r|(|\cos\theta|+|\sin\theta|)}{r^2}\right];$ if $r\to 0^+$, then

$\lim\limits_{r\to 0^+} \tan^{-1}\left[\dfrac{|r|(|\cos\theta|+|\sin\theta|)}{r^2}\right] = \lim\limits_{r\to 0^+} \tan^{-1}\left[\dfrac{|\cos\theta|+|\sin\theta|}{r}\right] = \dfrac{\pi}{2};$ if $r\to 0^-$, then $\lim\limits_{r\to 0^-} \tan^{-1}\left[\dfrac{|r|(|\cos\theta|+|\sin\theta|)}{r^2}\right]$

$= \lim\limits_{r\to 0^-} \tan^{-1}\left[\dfrac{|\cos\theta|+|\sin\theta|}{-r}\right] = \dfrac{\pi}{2} \Rightarrow$ the limit is $\dfrac{\pi}{2}$

67. $\lim\limits_{(x,\,y)\to(0,0)} \ln\left(\dfrac{3x^2-x^2 y^2+3y^2}{x^2+y^2}\right) = \lim\limits_{r\to 0} \ln\left(\dfrac{3r^2\cos^2\theta - r^4\cos^2\theta\sin^2\theta+3r^2\sin^2\theta}{r^2}\right) = \lim\limits_{r\to 0} \ln\left(3-r^2\cos^2\theta\sin^2\theta\right) = \ln 3$

$\Rightarrow$ define $f(0,0) = \ln 3$

69. Let $\delta = 0.1$. Then $\sqrt{x^2 + y^2} < \delta \Rightarrow \sqrt{x^2 + y^2} < 0.1 \Rightarrow x^2 + y^2 < 0.01 \Rightarrow |x^2 + y^2 - 0| < 0.01$

   $\Rightarrow |f(x, y) - f(0,0)| < 0.01 = \epsilon.$

71. Let $\delta = 0.005$. Then $|x| < \delta$ and $|y| < \delta \Rightarrow |f(x, y) - f(0,0)| = \left|\frac{x+y}{x^2+1} - 0\right| = \left|\frac{x+y}{x^2+1}\right| \le |x + y| < |x| + |y|$

   $< 0.005 + 0.005 = 0.01 = \epsilon.$

73. Let $\delta = 0.04$. Since $y^2 \le x^2 + y^2 \Rightarrow \frac{y^2}{x^2+y^2} \le 1 \Rightarrow \frac{|x|y^2}{x^2+y^2} \le |x| = \sqrt{x^2} \le \sqrt{x^2 + y^2} < \delta \Rightarrow |f(x, y) - f(0,0)|$

   $= \left|\frac{xy^2}{x^2+y^2} - 0\right| < 0.04 = \epsilon.$

75. Let $\delta = \sqrt{0.015}$. Then $\sqrt{x^2 + y^2 + z^2} < \delta \Rightarrow |f(x, y, z) - f(0, 0, 0)| = |x^2 + y^2 + z^2 - 0| = |x^2 + y^2 + z^2|$

   $= \left(\sqrt{x^2 + y^2 + z^2}\right)^2 < \left(\sqrt{0.015}\right)^2 = 0.015 = \epsilon.$

77. Let $\delta = 0.005$. Then $|x| < \delta, |y| < \delta,$ and $|z| < \delta \Rightarrow |f(x, y, z) - f(0,0,0)| = \left|\frac{x+y+z}{x^2+y^2+z^2+1} - 0\right| = \left|\frac{x+y+z}{x^2+y^2+z^2+1}\right|$

   $\le |x + y + z| \le |x| + |y| + |z| < 0.005 + 0.005 + 0.005 = 0.015 = \epsilon.$

79. $\displaystyle\lim_{(x, y, z) \to (x_0, y_0, z_0)} f(x, y, z) = \lim_{(x, y, z) \to (x_0, y_0, z_0)} (x + y - z) = x_0 + y_0 + z_0 = f(x_0, y_0, z_0) \Rightarrow f$ is continuous at

   every (x_0, y_0, z_0)

13.3 PARTIAL DERIVATIVES

1. $\frac{\partial f}{\partial x} = 4x, \frac{\partial f}{\partial y} = -3$

3. $\frac{\partial f}{\partial x} = 2x(y + 2), \frac{\partial f}{\partial y} = x^2 - 1$

5. $\frac{\partial f}{\partial x} = 2y(xy - 1), \frac{\partial f}{\partial y} = 2x(xy - 1)$

7. $\frac{\partial f}{\partial x} = \frac{x}{\sqrt{x^2+y^2}}, \frac{\partial f}{\partial y} = \frac{y}{\sqrt{x^2+y^2}}$

9. $\frac{\partial f}{\partial x} = -\frac{1}{(x+y)^2} \cdot \frac{\partial}{\partial x}(x + y) = -\frac{1}{(x+y)^2}, \quad \frac{\partial f}{\partial y} = -\frac{1}{(x+y)^2} \cdot \frac{\partial}{\partial y}(x + y) = -\frac{1}{(x+y)^2}$

11. $\frac{\partial f}{\partial x} = \frac{(xy-1)(1)-(x+y)(y)}{(xy-1)^2} = \frac{-y^2-1}{(xy-1)^2}, \quad \frac{\partial f}{\partial y} = \frac{(xy-1)(1)-(x+y)(x)}{(xy-1)^2} = \frac{-x^2-1}{(xy-1)^2}$

13. $\frac{\partial f}{\partial x} = e^{(x+y+1)} \cdot \frac{\partial}{\partial x}(x + y + 1) = e^{(x+y+1)}, \quad \frac{\partial f}{\partial y} = e^{(x+y+1)} \cdot \frac{\partial}{\partial y}(x + y + 1) = e^{(x+y+1)}$

15. $\frac{\partial f}{\partial x} = \frac{1}{x+y} \cdot \frac{\partial}{\partial x}(x + y) = \frac{1}{x+y}, \quad \frac{\partial f}{\partial y} = \frac{1}{x+y} \cdot \frac{\partial}{\partial y}(x + y) = \frac{1}{x+y}$

17. $\frac{\partial f}{\partial x} = 2\sin(x-3y) \cdot \frac{\partial}{\partial x}\sin(x-3y) = 2\sin(x-3y)\cos(x-3y) \cdot \frac{\partial}{\partial x}(x-3y) = 2\sin(x-3y)\cos(x-3y),$

$\frac{\partial f}{\partial y} = 2\sin(x-3y) \cdot \frac{\partial}{\partial y}\sin(x-3y) = 2\sin(x-3y)\cos(x-3y) \cdot \frac{\partial}{\partial y}(x-3y) = -6\sin(x-3y)\cos(x-3y)$

19. $\frac{\partial f}{\partial x} = -yx^{y-1}, \quad \frac{\partial f}{\partial y} = x^y\ln x$

21. $\frac{\partial f}{\partial x} = -g(x), \quad \frac{\partial f}{\partial y} = g(y)$

23. $f_x = y^2, f_y = 2xy, f_z = -4z$

25. $f_x = 1, f_y = -\dfrac{y}{\sqrt{y^2+z^2}}, \quad f_z = -\dfrac{z}{\sqrt{y^2+z^2}}$

27. $f_x = \dfrac{yz}{\sqrt{1-x^2y^2z^2}}, \quad f_y = \dfrac{xz}{\sqrt{1-x^2y^2z^2}}, \quad f_z = \dfrac{xy}{\sqrt{1-x^2y^2z^2}}$

29. $f_x = \dfrac{1}{x+2y+3z}, \quad f_y = \dfrac{2}{x+2y+3z}, \quad f_z = \dfrac{3}{x+2y+3z}$

31. $f_x = -2xe^{-\left(x^2+y^2+z^2\right)}, \quad f_y = -2ye^{-\left(x^2+y^2+z^2\right)}, \quad f_z = -2ze^{-\left(x^2+y^2+z^2\right)}$

33. $f_x = \operatorname{sech}^2(x+2y+3z), \quad f_y = 2\operatorname{sech}^2(x+2y+3z), \quad f_z = 3\operatorname{sech}^2(x+2y+3z)$

35. $\frac{\partial f}{\partial t} = -2\pi\sin(2\pi t-\alpha), \quad \frac{\partial f}{\partial \alpha} = \sin(2\pi t-\alpha)$

37. $\frac{\partial h}{\partial p} = \sin\phi\cos\theta, \quad \frac{\partial h}{\partial \phi} = \rho\cos\phi\cos\theta, \quad \frac{\partial h}{\partial \theta} = -\rho\sin\phi\sin\theta$

39. $W_p = V, \quad W_v = P+\dfrac{\delta v^2}{2g}, \quad W_\delta = \dfrac{Vv^2}{2g}, \quad W_v = \dfrac{2V\delta v}{2g} = \dfrac{V\delta v}{g}, \quad W_g = -\dfrac{V\delta v^2}{2g^2}$

41. $\frac{\partial f}{\partial x} = 1+y, \quad \frac{\partial f}{\partial y} = 1+x, \quad \frac{\partial^2 f}{\partial x^2} = 0, \quad \frac{\partial^2 f}{\partial y^2} = 0, \quad \frac{\partial^2 f}{\partial y\partial x} = \frac{\partial^2 f}{\partial x\partial y} = 1$

43. $\frac{\partial g}{\partial x} = 2xy+y\cos x, \quad \frac{\partial g}{\partial y} = x^2-\sin y+\sin x, \quad \frac{\partial^2 g}{\partial x^2} = 2y-y\sin x, \quad \frac{\partial^2 g}{\partial y^2} = -\cos y, \quad \frac{\partial^2 g}{\partial y\partial x} = \frac{\partial^2 g}{\partial x\partial y} = 2x+\cos x$

45. $\frac{\partial r}{\partial x} = \dfrac{1}{x+y}, \quad \frac{\partial r}{\partial y} = \dfrac{1}{x+y}, \quad \frac{\partial^2 r}{\partial x^2} = \dfrac{-1}{(x+y)^2}, \quad \frac{\partial^2 r}{\partial y^2} = \dfrac{-1}{(x+y)^2}, \quad \frac{\partial^2 r}{\partial y\partial x} = \frac{\partial^2 r}{\partial x\partial y} = \dfrac{-1}{(x+y)^2}$

47. $\frac{\partial w}{\partial x} = 2x\tan(xy) + x^2\sec^2(xy) \cdot y = 2x\tan(xy) + x^2 y\sec^2(xy), \quad \frac{\partial w}{\partial y} = x^2\sec^2(xy) \cdot x = x^3\sec^2(xy),$

$\frac{\partial^2 w}{\partial x^2} = 2\tan(xy) + 2x\sec^2(xy) \cdot y + 2xy\sec^2(xy) + x^2 y\left(2\sec(xy)\sec(xy)\tan(xy) \cdot y\right)$

$= 2\tan(xy) + 4xy\sec^2(xy) + 2x^2 y^2\sec^2(xy)\tan(xy), \quad \frac{\partial^2 w}{\partial y^2} = x^3\left(2\sec(xy)\sec(xy)\tan(xy) \cdot x\right)$

$= 2x^4\sec^2(xy)\tan(xy), \quad \frac{\partial^2 w}{\partial y\partial x} = \frac{\partial^2 w}{\partial x\partial y} = 3x^2\sec^2(xy) + x^3\left(2\sec(xy)\sec(xy)\tan(xy) \cdot y\right)$

$= 3x^2\sec^2(xy) + x^3 y\sec^2(xy)\tan(xy)$

49. $\frac{\partial w}{\partial x} = \sin\left(x^2 y\right) + x\cos\left(x^2 y\right) \cdot 2xy = \sin\left(x^2 y\right) + 2x^2 y\cos\left(x^2 y\right), \quad \frac{\partial w}{\partial y} = x\cos\left(x^2 y\right) \cdot x^2 = x^3\cos\left(x^2 y\right),$

$\frac{\partial^2 w}{\partial x^2} = \cos\left(x^2 y\right) \cdot 2xy + 4xy\cos\left(x^2 y\right) - 2x^2 y\sin\left(x^2 y\right) \cdot 2xy = 6xy\cos\left(x^2 y\right) - 4x^3 y^2\sin\left(x^2 y\right),$

$\frac{\partial^2 w}{\partial y^2} = -x^3\sin\left(x^2 y\right) \cdot x^2 = -x^5\sin\left(x^2 y\right), \quad \frac{\partial^2 w}{\partial y\partial x} = \frac{\partial^2 w}{\partial x\partial y} = 3x^2\cos\left(x^2 y\right) - x^3\sin\left(x^2 y\right) \cdot 2xy$

$= 3x^2\cos\left(x^2 y\right) - 2x^4 y\sin\left(x^2 y\right)$

51. $\frac{\partial w}{\partial x} = \frac{2}{2x+3y}, \quad \frac{\partial w}{\partial y} = \frac{3}{2x+3y}, \quad \frac{\partial^2 w}{\partial y\partial x} = \frac{-6}{(2x+3y)^2}, \quad \text{and} \quad \frac{\partial^2 w}{\partial x\partial y} = \frac{-6}{(2x+3y)^2}$

53. $\frac{\partial w}{\partial x} = y^2 + 2xy^3 + 3x^2 y^4, \quad \frac{\partial w}{\partial y} = 2xy + 3x^2 y^2 + 4x^3 y^3, \quad \frac{\partial^2 w}{\partial y\partial x} = 2y + 6xy^2 + 12x^2 y^3, \quad \text{and}$

$\frac{\partial^2 w}{\partial x\partial y} = 2y + 6xy^2 + 12x^2 y^3$

55. (a) x first   (b) y first   (c) x first   (d) x first   (e) y first   (f) y first

57. $f_x(1,2) = \lim_{h \to 0} \frac{f(1+h,2) - f(1,2)}{h} = \lim_{h \to 0} \frac{\left[1-(1+h)+2-6(1+h)^2\right]-(2-6)}{h} = \lim_{h \to 0} \frac{-h-6\left(1+2h+h^2\right)+6}{h} = \lim_{h \to 0} \frac{-13h-6h^2}{h}$

$= \lim_{h \to 0}(-13-6h) = -13, \quad f_y(1,2) = \lim_{h \to 0} \frac{f(1,2+h) - f(1,2)}{h} = \lim_{h \to 0} \frac{\left[1-1+(2+h)-3(2+h)\right]-(2-6)}{h} = \lim_{h \to 0} \frac{(2-6-2h)-(2-6)}{h}$

$= \lim_{h \to 0}(-2) = -2$

59. $f_x(-2,3) = \lim_{h \to 0} \frac{f(-2+h,3) - f(-2,3)}{h} = \lim_{h \to 0} \frac{\sqrt{2(-2+h)+9-1} - \sqrt{-4+9-1}}{h} = \lim_{h \to 0} \frac{\sqrt{2h+4}-2}{h} = \lim_{h \to 0}\left(\frac{\sqrt{2h+4}-2}{h} \cdot \frac{\sqrt{2h+4}+2}{\sqrt{2h+4}+2}\right)$

$= \lim_{h \to 0} \frac{2}{\sqrt{2h+4}+2} = \frac{1}{2}, \quad f_y(-2,3) = \lim_{h \to 0} \frac{f(-2,3+h) - f(-2,3)}{h} = \lim_{h \to 0} \frac{\sqrt{-4+3(3+h)-1} - \sqrt{-4+9-1}}{h} = \lim_{h \to 0} \frac{\sqrt{3h+4}-2}{h}$

$= \lim_{h \to 0}\left(\frac{\sqrt{3h+4}-2}{h} \cdot \frac{\sqrt{3h+4}+2}{\sqrt{3h+4}+2}\right) = \lim_{h \to 0} \frac{3}{\sqrt{3h+4}+2} = \frac{3}{4}$

61. (a) In the plane $x = 2 \Rightarrow f_y(x,y) = 3 \Rightarrow f_y(2,-1) = 3 \Rightarrow m = 3$

(b) In the plane $y = -1 \Rightarrow f_x(x,y) = 2 \Rightarrow f_y(2,-1) = 2 \Rightarrow m = 2$

63. $f_z(x_0, y_0, z_0) = \lim_{h \to 0} \frac{f(x_0, y_0, z_0 + h) - f(x_0, y_0, z_0)}{h};$

$f_z(1,2,3) = \lim_{h \to 0} \frac{f(1,2,3+h) - f(1,2,3)}{h} = \lim_{h \to 0} \frac{2(3+h)^2 - 2(9)}{h} = \lim_{h \to 0} \frac{12h+2h^2}{h} = \lim_{h \to 0}(12+2h) = 12$

65. $y + \left(3z^2 \frac{\partial z}{\partial x}\right)x + z^3 - 2y\frac{\partial z}{\partial x} = 0 \Rightarrow \left(3xz^2 - 2y\right)\frac{\partial z}{\partial x} = -y - z^3 \Rightarrow$ at $(1, 1, 1)$ we have $(3-2)\frac{\partial z}{\partial x} = -1 - 1$ or $\frac{\partial z}{\partial x} = -2$

67. $a^2 = b^2 + c^2 - 2bc \cos A \Rightarrow 2a = (2bc \sin A)\frac{\partial A}{\partial a} \Rightarrow \frac{\partial A}{\partial a} = \frac{a}{bc \sin A}$; also $0 = 2b - 2c \cos A + (2bc \sin A)\frac{\partial A}{\partial b}$

$\Rightarrow 2c \cos A - 2b = (2bc \sin A)\frac{\partial A}{\partial b} \Rightarrow \frac{\partial A}{\partial b} = \frac{c \cos A - b}{bc \sin A}$

69. Differentiating each equation implicitly gives $1 = v_x \ln u + \left(\frac{v}{u}\right)u_x$ and $0 = u_x \ln v + \left(\frac{u}{v}\right)v_x$ or

$\left. \begin{array}{l} (\ln u)\, v_x + \left(\frac{v}{u}\right)u_x = 1 \\[2mm] \left(\frac{u}{v}\right)v_x + (\ln v)\, u_x = 0 \end{array} \right\} \Rightarrow v_x = \dfrac{\begin{vmatrix} 1 & \frac{v}{u} \\ 0 & \ln v \end{vmatrix}}{\begin{vmatrix} \ln u & \frac{v}{u} \\ \frac{u}{v} & \ln v \end{vmatrix}} = \dfrac{\ln v}{(\ln u)(\ln v) - 1}$

71. $f_x(x, y) = \begin{cases} 0 & \text{if } y \geq 0 \\ 0 & \text{if } y < 0 \end{cases} \Rightarrow f_x(x, y) = 0$ for all points (x, y); at $y = 0$, $f_y(x, 0) = \lim\limits_{h \to 0} \frac{f(x, 0+h) - f(x, 0)}{h}$

$= \lim\limits_{h \to 0} \frac{f(x, h) - 0}{h} = \lim\limits_{h \to 0} \frac{f(x, h)}{h} = 0$ because $\lim\limits_{h \to 0^-} \frac{f(x, h)}{h} = \lim\limits_{h \to 0^-} \frac{h^3}{h} = 0$ and $\lim\limits_{h \to 0^+} \frac{f(x, h)}{h} = \lim\limits_{h \to 0^+} \frac{h^2}{h} = 0$

$\Rightarrow f_y(x, y) = \begin{cases} 3y^2 & \text{if } y \geq 0 \\ -2y & \text{if } y < 0 \end{cases}$; $f_{yx}(x, y) = f_{xy}(x, y) = 0$ for all points (x, y)

73. $\frac{\partial f}{\partial x} = 2x$, $\frac{\partial f}{\partial y} = 2y$, $\frac{\partial f}{\partial z} = -4z \Rightarrow \frac{\partial^2 f}{\partial x^2} = 2$, $\frac{\partial^2 f}{\partial y^2} = 2$, $\frac{\partial^2 f}{\partial z^2} = -4 \Rightarrow \frac{\partial^2 f}{\partial x^2} + \frac{\partial^2 f}{\partial y^2} + \frac{\partial^2 f}{\partial z^2} = 2 + 2 + (-4) = 0$

75. $\frac{\partial f}{\partial x} = -2e^{-2y}\sin 2x$, $\frac{\partial f}{\partial y} = -2e^{-2y}\cos 2x$, $\frac{\partial^2 f}{\partial x^2} = -4e^{-2y}\cos 2x$, $\frac{\partial^2 f}{\partial y^2} = -4e^{-2y}\cos 2x$

$\Rightarrow \frac{\partial^2 f}{\partial x^2} + \frac{\partial^2 f}{\partial y^2} = -4e^{-2y}\cos 2x + 4e^{-2y}\cos 2x = 0$

77. $\frac{\partial f}{\partial x} = 3$, $\frac{\partial f}{\partial y} = 2$, $\frac{\partial^2 f}{\partial x^2} = 0$, $\frac{\partial^2 f}{\partial y^2} = 0 \Rightarrow \frac{\partial^2 f}{\partial x^2} + \frac{\partial^2 f}{\partial y^2} = 0 + 0 = 0$

79. $\frac{\partial f}{\partial x} = -\frac{1}{2}\left(x^2 + y^2 + z^2\right)^{-3/2}(2x) = -x\left(x^2 + y^2 + z^2\right)^{-3/2}$,

$\frac{\partial f}{\partial y} = -\frac{1}{2}\left(x^2 + y^2 + z^2\right)^{-3/2}(2y) = -y\left(x^2 + y^2 + z^2\right)^{-3/2}$,

$\frac{\partial f}{\partial z} = -\frac{1}{2}\left(x^2 + y^2 + z^2\right)^{-3/2}(2z) = -z\left(x^2 + y^2 + z^2\right)^{-3/2}$;

$\frac{\partial^2 f}{\partial x^2} = -\left(x^2 + y^2 + z^2\right)^{-3/2} + 3x^2\left(x^2 + y^2 + z^2\right)^{-5/2}$,

$\frac{\partial^2 f}{\partial y^2} = -\left(x^2 + y^2 + z^2\right)^{-3/2} + 3y^2\left(x^2 + y^2 + z^2\right)^{-5/2}$,

$\frac{\partial^2 f}{\partial z^2} = -\left(x^2 + y^2 + z^2\right)^{-3/2} + 3z^2\left(x^2 + y^2 + z^2\right)^{-5/2}$

$$\Rightarrow \frac{\partial^2 f}{\partial x^2} + \frac{\partial^2 f}{\partial y^2} + \frac{\partial^2 f}{\partial z^2} = \left[-\left(x^2 + y^2 + z^2\right)^{-3/2} + 3x^2 \left(x^2 + y^2 + z^2\right)^{-5/2} \right]$$

$$+ \left[-\left(x^2 + y^2 + z^2\right)^{-3/2} + 3y^2 \left(x^2 + y^2 + z^2\right)^{-5/2} \right] + \left[-\left(x^2 + y^2 + z^2\right)^{-3/2} + 3z^2 \left(x^2 + y^2 + z^2\right)^{-5/2} \right]$$

$$= -3\left(x^2 + y^2 + z^2\right)^{-3/2} + \left(3x^2 + 3y^2 + 3z^2\right)\left(x^2 + y^2 + z^2\right)^{-5/2} = 0$$

81. $\frac{\partial w}{\partial x} = \cos\,(x+ct),\;\; \frac{\partial w}{\partial t} = c\,\cos\,(x+ct);\;\; \frac{\partial^2 w}{\partial x^2} = -\sin\,(x+ct),\;\; \frac{\partial^2 w}{\partial t^2} = -c^2\,\sin\,(x+ct)$

$\Rightarrow \frac{\partial^2 w}{\partial t^2} = c^2 \left[-\sin\,(x+ct) \right] = c^2 \frac{\partial^2 w}{\partial x^2}$

83. $\frac{\partial w}{\partial x} = \cos\,(x+ct) - 2\sin\,(2x+2ct),\;\; \frac{\partial w}{\partial t} = c\,\cos\,(x+ct) - 2c\,\sin\,(2x+2ct);\;\; \frac{\partial^2 w}{\partial x^2} = -\sin\,(x+ct) - 4\cos\,(2x+2ct),$

$\frac{\partial^2 w}{\partial t^2} = -c^2\,\sin\,(x+ct) - 4c^2\,\cos\,(2x+2ct) \Rightarrow \frac{\partial^2 w}{\partial t^2} = c^2 \left[-\sin\,(x+ct) - 4\cos\,(2x+2ct) \right] = c^2 \frac{\partial^2 w}{\partial x^2}$

85. $\frac{\partial w}{\partial x} = 2\sec^2\,(2x-2ct),\;\; \frac{\partial w}{\partial t} = -2c\,\sec^2\,(2x-2ct);\;\; \frac{\partial^2 w}{\partial x^2} = 8\sec^2\,(2x-2ct)\tan\,(2x-2ct),$

$\frac{\partial^2 w}{\partial t^2} = 8c^2\,\sec^2\,(2x-2ct)\tan\,(2x-2ct) \Rightarrow ux\frac{\partial^2 w}{\partial t^2} = c^2 \left[8\sec^2\,(2x-2ct)\tan\,(2x-2ct) \right] = c^2 \frac{\partial^2 w}{\partial x^2}$

87. $\frac{\partial w}{\partial t} = \frac{\partial f}{\partial u}\frac{\partial u}{\partial t} = \frac{\partial f}{\partial u}\,(ac) \Rightarrow \frac{\partial^2 w}{\partial t^2} = (ac)\left(\frac{\partial^2 f}{\partial u^2}\right)(ac) = a^2 c^2 \frac{\partial^2 f}{\partial u^2};\;\; \frac{\partial w}{\partial x} = \frac{\partial f}{\partial u}\frac{\partial u}{\partial x} = \frac{\partial f}{\partial u}\cdot a \Rightarrow \frac{\partial^2 w}{\partial x^2} = \left(a\frac{\partial^2 f}{\partial u^2}\right)\cdot a$

$= a^2 \frac{\partial^2 f}{\partial u^2} \Rightarrow \frac{\partial^2 w}{\partial t^2} = a^2 c^2 \frac{\partial^2 f}{\partial u^2} = c^2 \left(a^2 \frac{\partial^2 f}{\partial u^2}\right) = c^2 \frac{\partial^2 w}{\partial x^2}$

89. Yes, since f_{xx}, f_{yy}, f_{xy}, and f_{yx} are all continuous on R, use the same reasoning as in Exercise 88 with

$f_x(x, y) = f_x(x_0, y_0) + f_{xx}(x_0, y_0)\,\Delta x + f_{xy}(x_0, y_0)\,\Delta y + \epsilon_1 \Delta x + \epsilon_2 \Delta y$ and

$f_y(x, y) = f_y(x_0, y_0) + f_{yx}(x_0, y_0)\,\Delta x + f_{yy}(x_0, y_0)\,\Delta y + \hat{\epsilon}_1 \Delta x + \hat{\epsilon}_2 \Delta y$. Then

$\displaystyle\lim_{(x,\,y)\to(x_0,\,y_0)} f_x(x, y) = f_x(x_0, y_0)$ and $\displaystyle\lim_{(x,\,y)\to(x_0,\,y_0)} f_y(x, y) = f_y(x_0, y_0)$.

91. $f_x(0, 0) = \displaystyle\lim_{h\to 0} \frac{f(0+h, 0) - f(0, 0)}{h} = \lim_{h\to 0} \frac{\frac{h\cdot 0^2}{h^2 + 0^4} - 0}{h} = \lim_{h\to 0} \frac{0}{h} = 0;\;\; f_y(0, 0) = \lim_{h\to 0} \frac{f(0, 0+h) - f(0, 0)}{h} = \lim_{h\to 0} \frac{\frac{0\cdot h^2}{0^2 + h^4} - 0}{h}$

$= \displaystyle\lim_{h\to 0} \frac{0}{h} = 0;\;\; \lim_{\substack{(x,\,y)\to(0,\,0) \\ \text{along } x = ky^2}} f(x, y) = \lim_{y\to 0} \frac{\left(ky^2\right)y^2}{\left(ky^2\right)^2 + y^4} = \lim_{y\to 0} \frac{ky^4}{k^2 y^4 + y^4} = \lim_{y\to 0} \frac{k}{k^2 + 1} = \frac{k}{k^2 + 1} \Rightarrow$ different limits for

different values of $k \Rightarrow \displaystyle\lim_{(x,\,y)\to(0,\,0)} f(x, y)$ does not exist $\Rightarrow f(x, y)$ is not continuous at $(0, 0) \Rightarrow$ by Theorem

4, $f(x, y)$ is not differentiable at $(0, 0)$.

13.4 THE CHAIN RULE

1. (a) $\frac{\partial w}{\partial x} = 2x$, $\frac{\partial w}{\partial y} = 2y$, $\frac{dx}{dt} = -\sin t$, $\frac{dy}{dt} = \cos t \Rightarrow \frac{dw}{dt} = -2x \sin t + 2y \cos t = -2 \cos t \sin t + 2 \sin t \cos t = 0$;

$w = x^2 + y^2 = \cos^2 t + \sin^2 t = 1 \Rightarrow \frac{dw}{dt} = 0$

   (b) $\frac{dw}{dt}(\pi) = 0$

3. (a) $\frac{\partial w}{\partial x} = \frac{1}{z}$, $\frac{\partial w}{\partial y} = \frac{1}{z}$, $\frac{\partial w}{\partial z} = \frac{-(x+y)}{z^2}$, $\frac{dx}{dt} = -2 \cos t \sin t$, $\frac{dy}{dt} = 2 \sin t \cos t$, $\frac{dz}{dt} = -\frac{1}{t^2}$

$\Rightarrow \frac{dw}{dt} = -\frac{2}{z} \cos t \sin t + \frac{2}{z} \sin t \cos t + \frac{x+y}{z^2 t^2} = \frac{\cos^2 t + \sin^2 t}{\left(\frac{1}{t^2}\right)(t^2)} = 1$; $w = \frac{x}{z} + \frac{y}{z} = \frac{\cos^2 t}{\left(\frac{1}{t}\right)} + \frac{\sin^2 t}{\left(\frac{1}{t}\right)} = t \Rightarrow \frac{dw}{dt} = 1$

   (b) $\frac{dw}{dt}(3) = 1$

5. (a) $\frac{\partial w}{\partial x} = 2ye^x$, $\frac{\partial w}{\partial y} = 2e^x$, $\frac{\partial w}{\partial z} = -\frac{1}{z}$, $\frac{dx}{dt} = \frac{2t}{t^2+1}$, $\frac{dy}{dt} = \frac{1}{t^2+1}$, $\frac{dz}{dt} = e^t \Rightarrow \frac{dw}{dt} = \frac{4yte^x}{t^2+1} + \frac{2e^x}{t^2+1} - \frac{e^t}{z}$

$= \frac{(4t)\left(\tan^{-1} t\right)\left(t^2+1\right)}{t^2+1} + \frac{2\left(t^2+1\right)}{t^2+1} - \frac{e^t}{e^t} = 4t \tan^{-1} t + 1$; $w = 2ye^x - \ln z = \left(2 \tan^{-1} t\right)\left(t^2+1\right) - t$

$\Rightarrow \frac{dw}{dt} = \left(\frac{2}{t^2+1}\right)\left(t^2+1\right) + \left(2 \tan^{-1} t\right)(2t) - 1 = 4t \tan^{-1} t + 1$

   (b) $\frac{dw}{dt}(1) = (4)(1)\left(\frac{\pi}{4}\right) + 1 = \pi + 1$

7. (a) $\frac{\partial z}{\partial u} = \frac{\partial z}{\partial x}\frac{\partial x}{\partial u} + \frac{\partial z}{\partial y}\frac{\partial y}{\partial u} = \left(4e^x \ln y\right)\left(\frac{\cos v}{u \cos v}\right) + \left(\frac{4e^x}{y}\right)(\sin v) = \frac{4e^x \ln y}{u} + \frac{4e^x \sin v}{y}$

$= \frac{4(u \cos v) \ln (u \sin v)}{u} + \frac{4(u \cos v)(\sin v)}{u \sin v} = (4 \cos v) \ln (u \sin v) + 4 \cos v$;

$\frac{\partial z}{\partial v} = \frac{\partial z}{\partial x}\frac{\partial x}{\partial v} + \frac{\partial z}{\partial y}\frac{\partial y}{\partial v} = \left(4e^x \ln y\right)\left(\frac{-u \sin v}{u \cos v}\right) + \left(\frac{4e^x}{y}\right)(u \cos v) = -\left(4e^x \ln y\right)(\tan v) + \frac{4e^x u \cos v}{y}$

$= \left[-4(u \cos v) \ln (u \sin v)\right](\tan v) + \frac{4(u \cos v)(u \cos v)}{u \sin v} = (-4u \sin v) \ln (u \sin v) + \frac{4u \cos^2 v}{\sin v}$;

$z = 4e^x \ln y = 4(u \cos v) \ln (u \sin v) \Rightarrow \frac{\partial z}{\partial u} = (4 \cos v) \ln (u \sin v) + 4(u \cos v)\left(\frac{\sin v}{u \sin v}\right)$

$= (4 \cos v) \ln (u \sin v) + 4 \cos v$; also $\frac{\partial z}{\partial v} = (-4u \sin v) \ln (u \sin v) + 4(u \cos v)\left(\frac{u \cos v}{u \sin v}\right)$

$= (-4u \sin v) \ln (u \sin v) + \frac{4u \cos^2 v}{\sin v}$

   (b) At $\left(2, \frac{\pi}{4}\right)$: $\frac{\partial z}{\partial u} = 4 \cos \frac{\pi}{4} \ln \left(2 \sin \frac{\pi}{4}\right) + 4 \cos \frac{\pi}{4} = 2\sqrt{2} \ln \sqrt{2} + 2\sqrt{2} = \sqrt{2}(\ln 2 + 2)$;

$\frac{\partial z}{\partial v} = (-4)(2) \sin \frac{\pi}{4} \ln \left(2 \sin \frac{\pi}{4}\right) + \frac{(4)(2)\left(\cos^2 \frac{\pi}{4}\right)}{\left(\sin \frac{\pi}{4}\right)} = -4\sqrt{2} \ln \sqrt{2} + 4\sqrt{2} = -2\sqrt{2} \ln 2 + 4\sqrt{2}$

9. (a) $\frac{\partial w}{\partial u} = \frac{\partial w}{\partial x}\frac{\partial x}{\partial u} + \frac{\partial w}{\partial y}\frac{\partial y}{\partial u} + \frac{\partial w}{\partial z}\frac{\partial z}{\partial u} = (y+z)(1) + (x+z)(1) + (y+x)(v) = x + y + 2z + v(y+x)$

$= (u+v) + (u-v) + 2uv + v(2u) = 2u + 4uv$; $\frac{\partial w}{\partial v} = \frac{\partial w}{\partial x}\frac{\partial x}{\partial v} + \frac{\partial w}{\partial y}\frac{\partial y}{\partial v} + \frac{\partial w}{\partial z}\frac{\partial z}{\partial v}$

$= (y+z)(1) + (x+z)(-1) + (y+x)(u) = y - x + (y+x)u = -2v + (2u)u = -2v + 2u^2$; $w = xy + yz + xz$

$= \left(u^2 - v^2\right) + \left(u^2 v - vu^2\right) + \left(u^2 v + uv^2\right) = u^2 - v^2 + 2u^2 v \Rightarrow \frac{\partial w}{\partial u} = 2u + 4uv$ and $\frac{\partial w}{\partial v} = -2v + 2u^2$

   (b) At $\left(\frac{1}{2}, 1\right)$: $\frac{\partial w}{\partial u} = 2\left(\frac{1}{2}\right) + 4\left(\frac{1}{2}\right)(1) = 3$ and $\frac{\partial w}{\partial v} = -2(1) + 2\left(\frac{1}{2}\right)^2 = -\frac{3}{2}$

11. (a) $\dfrac{\partial u}{\partial x} = \dfrac{\partial u}{\partial p}\dfrac{\partial p}{\partial x} + \dfrac{\partial u}{\partial q}\dfrac{\partial q}{\partial x} + \dfrac{\partial u}{\partial r}\dfrac{\partial r}{\partial x} = \dfrac{1}{q-r} + \dfrac{r-p}{(q-r)^2} + \dfrac{p-q}{(q-r)^2} = \dfrac{q-r+r-p+p-q}{(q-p)^2} = 0;$

$\dfrac{\partial u}{\partial y} = \dfrac{\partial u}{\partial p}\dfrac{\partial p}{\partial y} + \dfrac{\partial u}{\partial q}\dfrac{\partial q}{\partial y} + \dfrac{\partial u}{\partial r}\dfrac{\partial r}{\partial y} = \dfrac{1}{q-r} - \dfrac{r-p}{(q-r)^2} + \dfrac{p-q}{(q-r)^2} = \dfrac{q-r-r+p+p-q}{(q-r)^2} = \dfrac{2p-2r}{(q-r)^2}$

$= \dfrac{(2x+2y+2z)-(2x+2y-2z)}{(2z-2y)^2} = \dfrac{z}{(z-y)^2};\quad \dfrac{\partial u}{\partial z} = \dfrac{\partial u}{\partial p}\dfrac{\partial p}{\partial z} + \dfrac{\partial u}{\partial q}\dfrac{\partial q}{\partial z} + \dfrac{\partial u}{\partial r}\dfrac{\partial r}{\partial z} = \dfrac{1}{q-r} + \dfrac{r-p}{(q-r)^2} - \dfrac{p-q}{(q-r)^2}$

$= \dfrac{q-r+r-p-p+q}{(q-r)^2} = \dfrac{2q-2p}{(q-r)^2} = \dfrac{-4y}{(2z-2y)^2} = -\dfrac{y}{(z-y)^2};\quad u = \dfrac{p-q}{q-r} = \dfrac{2y}{2z-2y} = \dfrac{y}{z-y} \Rightarrow \dfrac{\partial u}{\partial x} = 0;$

$\dfrac{\partial u}{\partial y} = \dfrac{(z-y)-y(-1)}{(z-y)^2} = \dfrac{z}{(z-y)^2},$ and $\dfrac{\partial u}{\partial z} = \dfrac{(z-y)(0)-y(1)}{(z-y)^2} = -\dfrac{y}{(z-y)^2}$

   (b) At $\left(\sqrt{3}, 2, 1\right):$ $\dfrac{\partial u}{\partial x} = 0, \dfrac{\partial u}{\partial y} = \dfrac{1}{(1-2)^2} = 1,$ and $\dfrac{\partial u}{\partial z} = \dfrac{-2}{(1-2)^2} = -2$

13. $\dfrac{dz}{dt} = \dfrac{\partial z}{\partial x}\dfrac{dx}{dt} + \dfrac{\partial z}{\partial y}\dfrac{dy}{dt}$

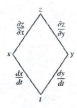

15. $\dfrac{\partial w}{\partial u} = \dfrac{\partial w}{\partial x}\dfrac{\partial x}{\partial u} + \dfrac{\partial w}{\partial y}\dfrac{\partial y}{\partial u} + \dfrac{\partial w}{\partial z}\dfrac{\partial z}{\partial u}$ $\dfrac{\partial w}{\partial v} = \dfrac{\partial w}{\partial x}\dfrac{\partial x}{\partial v} + \dfrac{\partial w}{\partial y}\dfrac{\partial y}{\partial v} + \dfrac{\partial w}{\partial z}\dfrac{\partial z}{\partial v}$

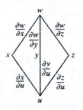 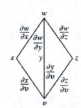

17. $\dfrac{\partial w}{\partial u} = \dfrac{\partial w}{\partial x}\dfrac{\partial x}{\partial u} + \dfrac{\partial w}{\partial y}\dfrac{\partial y}{\partial u}$ $\dfrac{\partial w}{\partial v} = \dfrac{\partial w}{\partial x}\dfrac{\partial x}{\partial v} + \dfrac{\partial w}{\partial y}\dfrac{\partial y}{\partial v}$

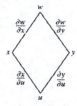 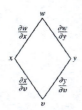

19. $\dfrac{\partial z}{\partial t} = \dfrac{\partial z}{\partial x}\dfrac{\partial x}{\partial t} + \dfrac{\partial z}{\partial y}\dfrac{\partial y}{\partial t}$ $\dfrac{\partial z}{\partial s} = \dfrac{\partial z}{\partial x}\dfrac{\partial x}{\partial s} + \dfrac{\partial z}{\partial y}\dfrac{\partial y}{\partial s}$

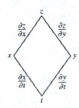 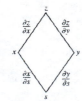

21. $\dfrac{\partial w}{\partial s} = \dfrac{dw}{du}\dfrac{\partial u}{\partial s}$ $\dfrac{\partial w}{\partial t} = \dfrac{dw}{du}\dfrac{\partial u}{\partial t}$

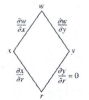

23. $\dfrac{\partial w}{\partial r} = \dfrac{\partial w}{\partial x}\dfrac{dx}{dr} + \dfrac{\partial w}{\partial y}\dfrac{dy}{dr} = \dfrac{\partial w}{\partial x}\dfrac{dx}{dr}$ since $\dfrac{dy}{dr} = 0$ $\dfrac{\partial w}{\partial s} = \dfrac{\partial w}{\partial x}\dfrac{dx}{ds} + \dfrac{\partial w}{\partial y}\dfrac{dy}{ds} = \dfrac{\partial w}{\partial y}\dfrac{dy}{ds}$ since $\dfrac{dx}{ds} = 0$

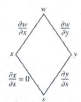

25. Let

$$F(x, y) = x^3 - 2y^2 + xy = 0 \Rightarrow F_x(x, y) = 3x^2 + y$$

and $F_y(x, y) = -4y + x \Rightarrow \dfrac{dy}{dx} = -\dfrac{F_x}{F_y} = -\dfrac{3x^2 + y}{(-4y + x)}$

$$\Rightarrow \dfrac{dy}{dx}(1, 1) = \dfrac{4}{3}$$

27. Let $F(x, y) = x^2 + xy + y^2 - 7 = 0 \Rightarrow F_x(x, y) = 2x + y$ and $F_y(x, y) = x + 2y \Rightarrow \dfrac{dy}{dx} = -\dfrac{F_x}{F_y} = -\dfrac{2x + y}{x + 2y}$

$$\Rightarrow \dfrac{dy}{dx}(1, 2) = -\dfrac{4}{5}$$

29. Let $F(x, y, z) = z^3 - xy + yz + y^3 - 2 = 0 \Rightarrow F_x(x, y, z) = -y, \; F_y(x, y, z) = -x + z + 3y^2,$

$F_z(x, y, z) = 3z^2 + y \Rightarrow \dfrac{\partial z}{\partial x} = -\dfrac{F_x}{F_z} = -\dfrac{-y}{3z^2 + y} = \dfrac{y}{3z^2 + y} \Rightarrow \dfrac{\partial z}{\partial x}(1, 1, 1) = \dfrac{1}{4}; \; \dfrac{\partial z}{\partial y} = -\dfrac{F_y}{F_z} = -\dfrac{-x + z + 3y^2}{3z^2 + y} = \dfrac{x - z - 3y^2}{3z^2 + y}$

$$\Rightarrow \dfrac{\partial z}{\partial y}(1, 1, 1) = -\dfrac{3}{4}$$

31. Let $F(x, y, z) = \sin(x + y) + \sin(y + z) + \sin(x + z) = 0 \Rightarrow F_x(x, y, z) = \cos(x + y) + \cos(x + z),$

$F_y(x, y, z) = \cos(x + y) + \cos(y + z), \; F_z(x, y, z) = \cos(y + z) + \cos(x + z)$

$\Rightarrow \dfrac{\partial z}{\partial x} = -\dfrac{F_x}{F_z} = -\dfrac{\cos(x+y) + \cos(x+z)}{\cos(y+z) + \cos(x+z)} \Rightarrow \dfrac{\partial z}{\partial x}(\pi, \pi, \pi) = -1; \; \dfrac{\partial z}{\partial y} = -\dfrac{F_y}{F_z} = -\dfrac{\cos(x+y) + \cos(y+z)}{\cos(y+z) + \cos(x+z)} \Rightarrow \dfrac{\partial z}{\partial y}(\pi, \pi, \pi) = -1$

33. $\dfrac{\partial w}{\partial r} = \dfrac{\partial w}{\partial x}\dfrac{\partial x}{\partial r} + \dfrac{\partial w}{\partial y}\dfrac{\partial y}{\partial r} + \dfrac{\partial w}{\partial z}\dfrac{\partial z}{\partial r} = 2(x + y + z)(1) + 2(x + y + z)[-\sin(r + s)] + 2(x + y + z)[\cos(r + s)]$

$= 2(x + y + z)[1 - \sin(r + s) + \cos(r + s)] = 2[r - s + \cos(r + s) + \sin(r + s)][1 - \sin(r + s) + \cos(r + s)]$

$$\Rightarrow \left.\dfrac{\partial w}{\partial r}\right|_{r=1, \, s=-1} = 2(3)(2) = 12$$

35. $\dfrac{\partial w}{\partial v} = \dfrac{\partial w}{\partial x}\dfrac{\partial x}{\partial v} + \dfrac{\partial w}{\partial y}\dfrac{\partial y}{\partial v} = \left(2x - \dfrac{y}{x^2}\right)(-2) + \left(\dfrac{1}{x}\right)(1) = \left[2(u - 2v + 1) - \dfrac{2u + v - 2}{(u - 2v + 1)^2}\right](-2) + \dfrac{1}{u - 2v + 1} \Rightarrow \left.\dfrac{\partial w}{\partial v}\right|_{u=0, \, v=0} = -7$

37. $\frac{\partial z}{\partial u} = \frac{dz}{dx}\frac{\partial x}{\partial u} = \left(\frac{5}{1+x^2}\right)e^u = \left[\frac{5}{1+(e^u+\ln v)^2}\right]e^u \Rightarrow \frac{\partial z}{\partial u}\Big|_{u=\ln 2,\, v=1} = \left[\frac{5}{1+(2)^2}\right](2) = 2;$

$\frac{\partial z}{\partial v} = \frac{dz}{dx}\frac{\partial x}{\partial v} = \left(\frac{5}{1+x^2}\right)\left(\frac{1}{v}\right) = \left[\frac{5}{1+(e^u+\ln v)^2}\right]\left(\frac{1}{v}\right) \Rightarrow \frac{\partial z}{\partial v}\Big|_{u=\ln 2,\, v=1} = \left[\frac{5}{1+(2)^2}\right](1) = 1$

39. Let $x = s^3 + t^2 \Rightarrow w = f\left(s^3 + t^2\right) = f(x) \Rightarrow \frac{\partial w}{\partial s} = \frac{dw}{dx}\frac{\partial x}{\partial s} = f'(x)\cdot 3s^2 = 3s^2 e^{s^3+t^2},$

$\frac{\partial w}{\partial t} = \frac{dw}{dx}\frac{\partial x}{\partial t} = f'(x)\cdot 2t = 2t\, e^{s^3+t^2}$

41. $V = IR \Rightarrow \frac{\partial V}{\partial I} = R$ and $\frac{\partial V}{\partial R} = I;\ \frac{dV}{dt} = \frac{\partial V}{\partial I}\frac{dI}{dt} + \frac{\partial V}{\partial R}\frac{dR}{dt} = R\frac{dI}{dt} + I\frac{dR}{dt} \Rightarrow -0.01$ volts/sec

$= (600\text{ ohms})\frac{dI}{dt} + (0.04\text{ amps})(0.5\text{ ohms/sec}) \Rightarrow \frac{dI}{dt} = -0.00005$ amps/sec

43. $\frac{\partial f}{\partial x} = \frac{\partial f}{\partial u}\frac{\partial u}{\partial x} + \frac{\partial f}{\partial v}\frac{\partial v}{\partial x} + \frac{\partial f}{\partial w}\frac{\partial w}{\partial x} = \frac{\partial f}{\partial u}(1) + \frac{\partial f}{\partial v}(0) + \frac{\partial f}{\partial w}(-1) = \frac{\partial f}{\partial u} - \frac{\partial f}{\partial w},$

$\frac{\partial f}{\partial y} = \frac{\partial f}{\partial u}\frac{\partial u}{\partial y} + \frac{\partial f}{\partial v}\frac{\partial v}{\partial y} + \frac{\partial f}{\partial w}\frac{\partial w}{\partial y} = \frac{\partial f}{\partial u}(-1) + \frac{\partial f}{\partial v}(1) + \frac{\partial f}{\partial w}(0) = -\frac{\partial f}{\partial u} + \frac{\partial f}{\partial v},$ and

$\frac{\partial f}{\partial z} = \frac{\partial f}{\partial u}\frac{\partial u}{\partial z} + \frac{\partial f}{\partial v}\frac{\partial v}{\partial z} + \frac{\partial f}{\partial w}\frac{\partial w}{\partial z} = \frac{\partial f}{\partial u}(0) + \frac{\partial f}{\partial v}(-1) + \frac{\partial f}{\partial w}(1) = -\frac{\partial f}{\partial v} + \frac{\partial f}{\partial w} \Rightarrow \frac{\partial f}{\partial x} + \frac{\partial f}{\partial y} + \frac{\partial f}{\partial z} = 0$

45. $w_x = \frac{\partial w}{\partial x} = \frac{\partial w}{\partial u}\frac{\partial u}{\partial x} + \frac{\partial w}{\partial v}\frac{\partial v}{\partial x} = x\frac{\partial w}{\partial u} + y\frac{\partial w}{\partial v} \Rightarrow w_{xx} = \frac{\partial w}{\partial u} + x\frac{\partial}{\partial x}\left(\frac{\partial w}{\partial u}\right) + y\frac{\partial}{\partial x}\left(\frac{\partial w}{\partial v}\right)$

$= \frac{\partial w}{\partial u} + x\left(\frac{\partial^2 w}{\partial u^2}\frac{\partial u}{\partial x} + \frac{\partial^2 w}{\partial v\partial u}\frac{\partial v}{\partial x}\right) + y\left(\frac{\partial^2 w}{\partial u\partial v}\frac{\partial u}{\partial x} + \frac{\partial^2 w}{\partial v^2}\frac{\partial v}{\partial x}\right) = \frac{\partial w}{\partial u} + x\left(x\frac{\partial^2 w}{\partial u^2} + y\frac{\partial^2 w}{\partial v\partial u}\right) + y\left(x\frac{\partial^2 w}{\partial u\partial v} + y\frac{\partial^2 w}{\partial v^2}\right)$

$= \frac{\partial w}{\partial u} + x^2\frac{\partial^2 w}{\partial u^2} + 2xy\frac{\partial^2 w}{\partial v\partial u} + y^2\frac{\partial^2 w}{\partial v^2};$

$w_y = \frac{\partial w}{\partial y} = \frac{\partial w}{\partial u}\frac{\partial u}{\partial y} + \frac{\partial w}{\partial v}\frac{\partial v}{\partial y} = -y\frac{\partial w}{\partial u} + x\frac{\partial w}{\partial v} \Rightarrow w_{yy} = -\frac{\partial w}{\partial u} - y\left(\frac{\partial^2 w}{\partial u^2}\frac{\partial u}{\partial y} + \frac{\partial^2 w}{\partial v\partial u}\frac{\partial v}{\partial y}\right) + x\left(\frac{\partial^2 w}{\partial u\partial v}\frac{\partial u}{\partial y} + \frac{\partial^2 w}{\partial v^2}\frac{\partial v}{\partial y}\right)$

$= -\frac{\partial w}{\partial u} - y\left(-y\frac{\partial^2 w}{\partial u^2} + x\frac{\partial^2 w}{\partial v\partial u}\right) + x\left(-y\frac{\partial^2 w}{\partial u\partial v} + x\frac{\partial^2 w}{\partial v^2}\right) = -\frac{\partial w}{\partial u} + y^2\frac{\partial^2 w}{\partial u^2} - 2xy\frac{\partial^2 w}{\partial v\partial u} + x^2\frac{\partial^2 w}{\partial v^2};$ thus

$w_{xx} + w_{yy} = \left(x^2 + y^2\right)\frac{\partial^2 w}{\partial u^2} + \left(x^2 + y^2\right)\frac{\partial^2 w}{\partial v^2} = \left(x^2 + y^2\right)(w_{uu} + w_{vv}) = 0,$ since $w_{uu} + w_{vv} = 0$

47. $f_x(x, y, z) = \cos t,\ f_y(x, y, z) = \sin t,$ and $f_z(x, y, z) = t^2 + t - 2 \Rightarrow \frac{df}{dt} = \frac{\partial f}{\partial x}\frac{dx}{dt} + \frac{\partial f}{\partial y}\frac{dy}{dt} + \frac{\partial f}{\partial z}\frac{dz}{dt}$

$= (\cos t)(-\sin t) + (\sin t)(\cos t) + \left(t^2 + t - 2\right)(1) = t^2 + t - 2;\ \frac{df}{dt} = 0 \Rightarrow t^2 + t - 2 = 0 \Rightarrow t = -2$ or $t = 1;$

$t = -2 \Rightarrow x = \cos(-2),\ y = \sin(-2),\ z = -2$ for the point $(\cos(-2), \sin(-2), -2);$

$t = 1 \Rightarrow x = \cos 1,\ y = \sin 1,\ z = 1$ for the point $(\cos 1, \sin 1, 1)$

49. (a) $\frac{\partial T}{\partial x} = 8x - 4y$ and $\frac{\partial T}{\partial y} = 8y - 4x \Rightarrow \frac{dT}{dt} = \frac{\partial T}{\partial x}\frac{dx}{dt} + \frac{\partial T}{\partial y}\frac{dy}{dt} = (8x - 4y)(-\sin t) + (8y - 4x)(\cos t)$

$= (8\cos t - 4\sin t)(-\sin t) + (8\sin t - 4\cos t)(\cos t) = 4\sin^2 t - 4\cos^2 t \Rightarrow \frac{d^2 T}{dt^2} = 16\sin t\cos t;$

$\frac{dT}{dt} = 0 \Rightarrow 4\sin^2 t - 4\cos^2 t = 0 \Rightarrow \sin^2 t = \cos^2 t \Rightarrow \sin t = \cos t$ or $\sin t = -\cos t \Rightarrow t = \frac{\pi}{4}, \frac{5\pi}{4}, \frac{3\pi}{4}, \frac{7\pi}{4}$ on

the interval $0 \le t \le 2\pi;$

$\left.\frac{d^2 T}{dt^2}\right|_{t=\frac{\pi}{4}} = 16\sin\frac{\pi}{4}\cos\frac{\pi}{4} > 0 \Rightarrow T$ has a minimum at $(x, y) = \left(\frac{\sqrt{2}}{2}, \frac{\sqrt{2}}{2}\right);$

$\left.\frac{d^2 T}{dt^2}\right|_{t=\frac{3\pi}{4}} = 16\sin\frac{3\pi}{4}\cos\frac{3\pi}{4} < 0 \Rightarrow T$ has a maximum at $(x, y) = \left(-\frac{\sqrt{2}}{2}, \frac{\sqrt{2}}{2}\right);$

$\left.\frac{d^2 T}{dt^2}\right|_{t=\frac{5\pi}{4}} = 16\sin\frac{5\pi}{4}\cos\frac{5\pi}{4} > 0 \Rightarrow T$ has a minimum at $(x, y) = \left(-\frac{\sqrt{2}}{2}, -\frac{\sqrt{2}}{2}\right);$

$\left.\frac{d^2 T}{dt^2}\right|_{t=\frac{7\pi}{4}} = 16\sin\frac{7\pi}{4}\cos\frac{7\pi}{4} < 0 \Rightarrow T$ has a maximum at $(x, y) = \left(\frac{\sqrt{2}}{2}, -\frac{\sqrt{2}}{2}\right)$

(b) $T = 4x^2 - 4xy + 4y^2 \Rightarrow \frac{\partial T}{\partial x} = 8x - 4y,$ and $\frac{\partial T}{\partial y} = 8y - 4x$ so the extreme values occur at the four points

found in part (a): $T\left(-\frac{\sqrt{2}}{2}, \frac{\sqrt{2}}{2}\right) = T\left(\frac{\sqrt{2}}{2}, -\frac{\sqrt{2}}{2}\right) = 4\left(\frac{1}{2}\right) - 4\left(-\frac{1}{2}\right) + 4\left(\frac{1}{2}\right) = 6,$ the maximum and

$T\left(\frac{\sqrt{2}}{2}, \frac{\sqrt{2}}{2}\right) = T\left(-\frac{\sqrt{2}}{2}, -\frac{\sqrt{2}}{2}\right) = 4\left(\frac{1}{2}\right) - 4\left(\frac{1}{2}\right) + 4\left(\frac{1}{2}\right) = 2,$ the minimum

51. $G(u, x) = \int_a^u g(t, x)\, dt$ where $u = f(x) \Rightarrow \frac{dG}{dx} = \frac{\partial G}{\partial u}\frac{du}{dx} + \frac{\partial G}{\partial x}\frac{dx}{dx} = g(u, x)f'(x) + \int_a^u g_x(t, x)\, dt;$ thus

$F(x) = \int_0^{x^2} \sqrt{t^4 + x^3}\, dt \Rightarrow F'(x) = \sqrt{\left(x^2\right)^4 + x^3}\,(2x) + \int_0^{x^2} \frac{\partial}{\partial x}\sqrt{t^4 + x^3}\, dt = 2x\sqrt{x^8 + x^3} + \int_0^{x^2} \frac{3x^2}{2\sqrt{t^4 + x^3}}\, dt$

52. Using the result in Exercise 51, $F(x) = \int_{x^2}^1 \sqrt{t^3 + x^2}\, dt = -\int_1^{x^2} \sqrt{t^3 + x^2}\, dt$

$\Rightarrow F'(x) = \left[-\sqrt{\left(x^2\right)^3 + x^2}\, x^2 - \int_1^{x^2} \frac{\partial}{\partial x}\sqrt{t^3 + x^2}\, dt\right] = -x^2\sqrt{x^6 + x^2} + \int_{x^2}^1 \frac{x}{\sqrt{t^3 + x^2}}\, dt$

13.5 DIRECTIONAL DERIVATIVES AND GRADIENT VECTORS

1. $\frac{\partial f}{\partial x} = -1,\ \frac{\partial f}{\partial y} = 1 \Rightarrow \nabla f = -\mathbf{i} + \mathbf{j};\ f(2, 1) = -1$

$\Rightarrow -1 = y - x$ is the level curve

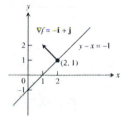

3. $\frac{\partial g}{\partial x} = y^2 \Rightarrow \frac{\partial g}{\partial x}(2, -1) = 1;$

$\frac{\partial g}{\partial y} = 2xy \Rightarrow \frac{\partial g}{\partial x}(2, -1) = -4 \Rightarrow \nabla g = \mathbf{i} - 4\mathbf{j};$

$g(2, -1) = 2 \Rightarrow x = \frac{2}{y^2}$ is the level curve

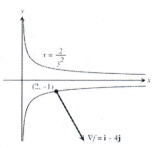

5. $\frac{\partial f}{\partial x} = \frac{1}{\sqrt{2x+3y}} \Rightarrow \frac{\partial f}{\partial x}(-1, 2) = \frac{1}{2}$;

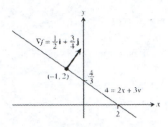

$\frac{\partial f}{\partial y} = \frac{3}{2\sqrt{2x+3y}} \Rightarrow \frac{\partial f}{\partial x}(-1, 2) = \frac{3}{4} \Rightarrow \nabla f = \frac{1}{2}\mathbf{i} + \frac{3}{4}\mathbf{j}$;

$f(-1, 2) = 2 \Rightarrow 4 = 2x+3y$ is the level curve

7. $\frac{\partial f}{\partial x} = 2x + \frac{z}{x} \Rightarrow \frac{\partial f}{\partial x}(1, 1, 1) = 3$; $\frac{\partial f}{\partial y} = 2y \Rightarrow \frac{\partial f}{\partial y}(1, 1, 1) = 2$; $\frac{\partial f}{\partial z} = -4z + \ln x \Rightarrow \frac{\partial f}{\partial z}(1, 1, 1) = -4$; thus

$\nabla f = 3\mathbf{i} + 2\mathbf{j} - 4\mathbf{k}$

9. $\frac{\partial f}{\partial x} = -\frac{x}{\left(x^2+y^2+z^2\right)^{3/2}} + \frac{1}{x} \Rightarrow \frac{\partial f}{\partial x}(-1, 2, -2) = -\frac{26}{27}$; $\frac{\partial f}{\partial y} = -\frac{y}{\left(x^2+y^2+z^2\right)^{3/2}} + \frac{1}{y} \Rightarrow \frac{\partial f}{\partial y}(-1, 2, -2) = \frac{23}{54}$;

$\frac{\partial f}{\partial z} = -\frac{z}{\left(x^2+y^2+z^2\right)^{3/2}} + \frac{1}{z} \Rightarrow \frac{\partial f}{\partial z}(-1, 2, -2) = -\frac{23}{54}$; thus $\nabla f = -\frac{26}{27}\mathbf{i} + \frac{23}{54}\mathbf{j} - \frac{23}{54}\mathbf{k}$

11. $\mathbf{u} = \frac{\mathbf{A}}{|\mathbf{A}|} = \frac{4\mathbf{i}+3\mathbf{j}}{\sqrt{4^2+3^2}} = \frac{4}{5}\mathbf{i} + \frac{3}{5}\mathbf{j}$; $f_x(x, y) = 2y \Rightarrow f_x(5, 5) = 10$; $f_y(x, y) = 2x-6y \Rightarrow f_y(5, 5) = -20$

$\Rightarrow \nabla f = 10\mathbf{i} - 20\mathbf{j} \Rightarrow (D_{\mathbf{u}}f)_{P_0} = \nabla f \cdot \mathbf{u} = 10\left(\frac{4}{5}\right) - 20\left(\frac{3}{5}\right) = -4$

13. $\mathbf{u} = \frac{\mathbf{A}}{|\mathbf{A}|} = \frac{12\mathbf{i}+5\mathbf{j}}{\sqrt{12^2+5^2}} = \frac{12}{13}\mathbf{i} + \frac{5}{13}\mathbf{j}$; $g_x(x, y) = \frac{y^2+2}{(xy+2)^2} \Rightarrow g_x(1, -1) = 3$; $g_y(x, y) = -\frac{x^2+2}{(xy+2)^2} \Rightarrow g_y(1, -1) = -3$

$\Rightarrow \nabla g = 3\mathbf{i} - 3\mathbf{j} \Rightarrow (D_{\mathbf{u}}g)_{P_0} = \nabla g \cdot \mathbf{u} = \frac{36}{13} - \frac{15}{13} = \frac{21}{13}$

15. $\mathbf{u} = \frac{\mathbf{A}}{|\mathbf{A}|} = \frac{3\mathbf{i}+6\mathbf{j}-2\mathbf{k}}{\sqrt{3^2+6^2+(-2)^2}} = \frac{3}{7}\mathbf{i} + \frac{6}{7}\mathbf{j} - \frac{2}{7}\mathbf{k}$; $f_x(x, y, z) = y+z \Rightarrow f_x(1, -1, 2) = 1$;

$f_y(x, y, z) = x+z \Rightarrow f_y(1, -1, 2) = 3$; $f_z(x, y, z) = y+x \Rightarrow f_z(1, -1, 2) = 0 \Rightarrow \nabla f = \mathbf{i} + 3\mathbf{j}$

$\Rightarrow (D_{\mathbf{u}}f)_{P_0} = \nabla f \cdot \mathbf{u} = \frac{3}{7} + \frac{18}{7} = 3$

17. $\mathbf{u} = \frac{\mathbf{A}}{|\mathbf{A}|} = \frac{2\mathbf{i}+\mathbf{j}-2\mathbf{k}}{\sqrt{2^2+1^2+(-2)^2}} = \frac{2}{3}\mathbf{i} + \frac{1}{3}\mathbf{j} - \frac{2}{3}\mathbf{k}$; $g_x(x, y, z) = 3e^x \cos yz \Rightarrow g_x(0, 0, 0) = 3$;

$g_y(x, y, z) = -3ze^x \sin yz \Rightarrow g_y(0, 0, 0) = 0$; $g_z(x, y, z) = -3ye^x \sin yz \Rightarrow g_z(0, 0, 0) = 0 \Rightarrow \nabla g = 3\mathbf{i}$

$\Rightarrow (D_{\mathbf{u}}g)_{P_0} = \nabla g \cdot \mathbf{u} = 2$

19. $\nabla f = (2x+y)\mathbf{i} + (x+2y)\mathbf{j} \Rightarrow \nabla f(-1, 1) = -\mathbf{i} + \mathbf{j} \Rightarrow \mathbf{u} = \frac{\nabla f}{|\nabla f|} = \frac{-\mathbf{i}+\mathbf{j}}{\sqrt{(-1)^2+1^2}} = -\frac{1}{\sqrt{2}}\mathbf{i} + \frac{1}{\sqrt{2}}\mathbf{j}$; f increases most

rapidly in the direction $\mathbf{u} = -\frac{1}{\sqrt{2}}\mathbf{i} + \frac{1}{\sqrt{2}}\mathbf{j}$ and decreases most rapidly in the direction $-\mathbf{u} = \frac{1}{\sqrt{2}}\mathbf{i} - \frac{1}{\sqrt{2}}\mathbf{j}$;

$(D_{\mathbf{u}}f)_{P_0} = \nabla f \cdot \mathbf{u} = |\nabla f| = \sqrt{2}$ and $(D_{-\mathbf{u}}f)_{P_0} = -\sqrt{2}$

21. $\nabla f = \frac{1}{y}\mathbf{i} - \left(\frac{x}{y^2} + z\right)\mathbf{j} - y\mathbf{k} \Rightarrow \nabla f(4, 1, 1) = \mathbf{i} - 5\mathbf{j} - \mathbf{k} \Rightarrow \mathbf{u} = \frac{\nabla f}{|\nabla f|} = \frac{\mathbf{i}-5\mathbf{j}-\mathbf{k}}{\sqrt{1^2+(-5)^2+(-1)^2}} = \frac{1}{3\sqrt{3}}\mathbf{i} - \frac{5}{3\sqrt{3}}\mathbf{j} - \frac{1}{3\sqrt{3}}\mathbf{k}$; f

increases most rapidly in the direction of $\mathbf{u} = \frac{1}{3\sqrt{3}}\mathbf{i} - \frac{5}{3\sqrt{3}}\mathbf{j} - \frac{1}{3\sqrt{3}}\mathbf{k}$ and decreases most rapidly in the direction

$-\mathbf{u} = -\frac{1}{3\sqrt{3}}\mathbf{i} + \frac{5}{3\sqrt{3}}\mathbf{j} + \frac{1}{3\sqrt{3}}\mathbf{k}$; $(D_{\mathbf{u}}f)_{P_0} = \nabla f \cdot \mathbf{u} = |\nabla f| = 3\sqrt{3}$ and $(D_{-\mathbf{u}}f)_{P_0} = -3\sqrt{3}$

23. $\nabla f = \left(\frac{1}{x} + \frac{1}{x}\right)\mathbf{i} + \left(\frac{1}{y} + \frac{1}{y}\right)\mathbf{j} + \left(\frac{1}{z} + \frac{1}{z}\right)\mathbf{k} \Rightarrow \nabla f(1,1,1) = 2\mathbf{i} + 2\mathbf{j} + 2\mathbf{k} \Rightarrow \mathbf{u} = \frac{\nabla f}{|\nabla f|} = \frac{1}{\sqrt{3}}\mathbf{i} + \frac{1}{\sqrt{3}}\mathbf{j} + \frac{1}{\sqrt{3}}\mathbf{k}$; f increases

most rapidly in the direction $\mathbf{u} = \frac{1}{\sqrt{3}}\mathbf{i} + \frac{1}{\sqrt{3}}\mathbf{j} + \frac{1}{\sqrt{3}}\mathbf{k}$ and decreases most rapidly in the direction

$-\mathbf{u} = -\frac{1}{\sqrt{3}}\mathbf{i} - \frac{1}{\sqrt{3}}\mathbf{j} - \frac{1}{\sqrt{3}}\mathbf{k}$; $(D_{\mathbf{u}}f)_{P_0} = \nabla f \cdot \mathbf{u} = |\nabla f| = 2\sqrt{3}$ and $(D_{-\mathbf{u}}f)_{P_0} = -2\sqrt{3}$

25. $\nabla f = 2x\mathbf{i} + 2y\mathbf{j} \Rightarrow \nabla f\left(\sqrt{2}, \sqrt{2}\right) = 2\sqrt{2}\mathbf{i} + 2\sqrt{2}\mathbf{j}$

$\Rightarrow$ Tangent line: $2\sqrt{2}\left(x - \sqrt{2}\right) + 2\sqrt{2}\left(y - \sqrt{2}\right) = 0$

$\Rightarrow \sqrt{2}x + \sqrt{2}y = 4$

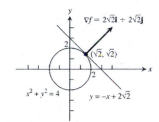

27. $\nabla f = y\mathbf{i} + x\mathbf{j} \Rightarrow \nabla f(2, -2) = -2\mathbf{i} + 2\mathbf{j}$

$\Rightarrow$ Tangent line: $-2(x - 2) + 2(y + 2) = 0$

$\Rightarrow y = x - 4$

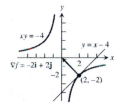

29. $\nabla f = (2x - y)\mathbf{i} + (-x + 2y - 1)\mathbf{j}$

(a) $\nabla f(1, -1) = 3\mathbf{i} - 4\mathbf{j} \Rightarrow |\nabla f(1, -1)| = 5 \Rightarrow D_{\mathbf{u}}f(1, -1) = 5$ in the direction of $\mathbf{u} = \frac{3}{5}\mathbf{i} - \frac{4}{5}\mathbf{j}$

(b) $-\nabla f(1, -1) = -3\mathbf{i} + 4\mathbf{j} \Rightarrow |\nabla f(1, -1)| = 5 \Rightarrow D_{\mathbf{u}}f(1, -1) = -5$ in the direction of $\mathbf{u} = -\frac{3}{5}\mathbf{i} + \frac{4}{5}\mathbf{j}$

(c) $D_{\mathbf{u}}f(1, -1) = 0$ in the direction of $\mathbf{u} = \frac{4}{5}\mathbf{i} + \frac{3}{5}\mathbf{j}$ or $\mathbf{u} = -\frac{4}{5}\mathbf{i} - \frac{3}{5}\mathbf{j}$

(d) Let $\mathbf{u} = u_1\mathbf{i} + u_2\mathbf{j} \Rightarrow |\mathbf{u}| = \sqrt{u_1^2 + u_2^2} = 1 \Rightarrow u_1^2 + u_2^2 = 1$; $D_{\mathbf{u}}f(1, -1) = \nabla f(1, -1) \cdot \mathbf{u} = (3\mathbf{i} - 4\mathbf{j}) \cdot (u_1\mathbf{i} + u_2\mathbf{j})$

$= 3u_1 - 4u_2 = 4 \Rightarrow u_2 = \frac{3}{4}u_1 - 1 \Rightarrow u_1^2 + \left(\frac{3}{4}u_1 - 1\right)^2 = 1 \Rightarrow \frac{25}{16}u_1^2 - \frac{3}{2}u_1 = 0 \Rightarrow u_1 = 0$ or

$u_1 = \frac{24}{25}$; $u_1 = 0 \Rightarrow u_2 = -1 \Rightarrow \mathbf{u} = -\mathbf{j}$, or $u_1 = \frac{24}{25} \Rightarrow u_2 = -\frac{7}{25} \Rightarrow \mathbf{u} = \frac{24}{25}\mathbf{i} - \frac{7}{25}\mathbf{j}$

(e) Let $\mathbf{u} = u_1\mathbf{i} + u_2\mathbf{j} \Rightarrow |\mathbf{u}| = \sqrt{u_1^2 + u_2^2} = 1 \Rightarrow u_1^2 + u_2^2 = 1$; $D_{\mathbf{u}}f(1, -1) = \nabla f(1, -1) \cdot \mathbf{u} = (3\mathbf{i} - 4\mathbf{j}) \cdot (u_1\mathbf{i} + u_2\mathbf{j})$

$= 3u_1 - 4u_2 = -3 \Rightarrow u_1 = \frac{4}{3}u_2 - 1 \Rightarrow \left(\frac{4}{3}u_2 - 1\right)^2 + u_2^2 = 1 \Rightarrow \frac{25}{9}u_2^2 - \frac{8}{3}u_2 = 0 \Rightarrow u_2 = 0$ or $u_2 = \frac{24}{25}$;

$u_2 = 0 \Rightarrow u_1 = -1 \Rightarrow \mathbf{u} = -\mathbf{i}$, or $u_2 = \frac{24}{25} \Rightarrow u_2 = \frac{7}{25} \Rightarrow \mathbf{u} = \frac{7}{25}\mathbf{i} + \frac{24}{25}\mathbf{j}$

31. $\nabla f = y\mathbf{i} + (x + 2y)\mathbf{j} \Rightarrow \nabla f(3, 2) = 2\mathbf{i} + 7\mathbf{j}$; a vector orthogonal to ∇f is $\mathbf{v} = 7\mathbf{i} - 2\mathbf{j} \Rightarrow \mathbf{u} = \frac{\mathbf{v}}{|\mathbf{v}|} = \frac{7\mathbf{i} - 2\mathbf{j}}{\sqrt{7^2 + (-2)^2}}$

$= \frac{7}{\sqrt{53}}\mathbf{i} - \frac{2}{\sqrt{53}}\mathbf{j}$ and $-\mathbf{u} = -\frac{7}{\sqrt{53}}\mathbf{i} + \frac{2}{\sqrt{53}}\mathbf{j}$ are the directions where the derivative is zero

33. $\nabla f = (2x - 3y)\mathbf{i} + (-3x + 8y)\mathbf{j} \Rightarrow \nabla f(1, 2) = -4\mathbf{i} + 13\mathbf{j} \Rightarrow |\nabla f(1, 2)| = \sqrt{(-4)^2 + (13)^2} = \sqrt{185}$; no, the maximum

rate of change is $\sqrt{185} < 14$

35. $\nabla f = f_x(1, 2)\mathbf{i} + f_y(1, 2)\mathbf{j}$ and $\mathbf{u}_1 = \frac{\mathbf{i} + \mathbf{j}}{\sqrt{1^2 + 1^2}} = \frac{1}{\sqrt{2}}\mathbf{i} + \frac{1}{\sqrt{2}}\mathbf{j} \Rightarrow (D_{\mathbf{u}_1}f)(1, 2) = f_x(1, 2)\left(\frac{1}{\sqrt{2}}\right) + f_y(1, 2)\left(\frac{1}{\sqrt{2}}\right) = 2\sqrt{2}$

$\Rightarrow f_x(1, 2) + f_y(1, 2) = 4$; $\mathbf{u}_2 = -\mathbf{j} \Rightarrow (D_{\mathbf{u}_2}f)(1, 2) = f_x(1, 2)(0) + f_y(1, 2)(-1) = -3 \Rightarrow -f_y(1, 2) = -3$

$\Rightarrow f_y(1, 2) = 3$; then $f_x(1, 2) + 3 = 4 \Rightarrow f_x(1, 2) = 1$; thus $\nabla f(1, 2) = \mathbf{i} + 3\mathbf{j}$ and $\mathbf{u} = \dfrac{\mathbf{v}}{|\mathbf{v}|} = \dfrac{-\mathbf{i} - 2\mathbf{j}}{\sqrt{(-1)^2 + (-2)^2}}$

$= -\dfrac{1}{\sqrt{5}}\mathbf{i} - \dfrac{2}{\sqrt{5}}\mathbf{j} \Rightarrow (D_{\mathbf{u}}f)_{P_0} = \nabla f \cdot \mathbf{u} = -\dfrac{1}{\sqrt{5}} - \dfrac{6}{\sqrt{5}} = -\dfrac{7}{\sqrt{5}}$

37. The directional derivative is the scalar component. With ∇f evaluated at P_0, the scalar component of ∇f in the direction of $\mathbf{u}$ is $\nabla f \cdot \mathbf{u} = (D_{\mathbf{u}}f)_{P_0}$.

39. If (x, y) is a point on the line, then $\mathbf{T}(x, y) = (x - x_0)\mathbf{i} + (y - y_0)\mathbf{j}$ is a vector parallel to the line
$\Rightarrow \mathbf{T} \cdot \mathbf{N} = 0 \Rightarrow A(x - x_0) + B(y - y_0) = 0$, as claimed.

13.6 TANGENT PLANES AND DIFFERENTIALS

1. (a) $\nabla f = 2x\mathbf{i} + 2y\mathbf{j} + 2z\mathbf{k} \Rightarrow \nabla f(1, 1, 1) = 2\mathbf{i} + 2\mathbf{j} + 2\mathbf{k}$
    $\Rightarrow$ Tangent plane: $2(x - 1) + 2(y - 1) + 2(z - 1) = 0 \Rightarrow x + y + z = 3$;
   (b) Normal line: $x = 1 + 2t$, $y = 1 + 2t$, $z = 1 + 2t$

3. (a) $\nabla f = -2x\mathbf{i} + 2\mathbf{k} \Rightarrow \nabla f(2, 0, 2) = -4\mathbf{i} + 2\mathbf{k}$
    $\Rightarrow$ Tangent plane: $-4(x - 2) + 2(z - 2) = 0 \Rightarrow -4x + 2z + 4 = 0 \Rightarrow -2x + z + 2 = 0$;
   (b) Normal line: $x = 2 - 4t$, $y = 0$, $z = 2 + 2t$

5. (a) $\nabla f = \left(-\pi \sin \pi x - 2xy + ze^{xz}\right)\mathbf{i} + \left(-x^2 + z\right)\mathbf{j} + \left(xe^{xz} + y\right)\mathbf{k} \Rightarrow \nabla f(0, 1, 2) = 2\mathbf{i} + 2\mathbf{j} + \mathbf{k}$
    $\Rightarrow$ Tangent plane: $2(x - 0) + 2(y - 1) + 1(z - 2) = 0 \Rightarrow 2x + 2y + z - 4 = 0$;
   (b) Normal line: $x = 2t$, $y = 1 + 2t$, $z = 2 + t$

7. (a) $\nabla f = \mathbf{i} + \mathbf{j} + \mathbf{k}$ for all points $\Rightarrow \nabla f(0, 1, 0) = \mathbf{i} + \mathbf{j} + \mathbf{k}$
    $\Rightarrow$ Tangent plane: $1(x - 0) + 1(y - 1) + 1(z - 0) = 0 \Rightarrow x + y + z - 1 = 0$;
   (b) Normal line: $x = t$, $y = 1 + t$, $z = t$

9. $z = f(x, y) = \ln\left(x^2 + y^2\right) \Rightarrow f_x(x, y) = \dfrac{2x}{x^2 + y^2}$ and $f_y(x, y) = \dfrac{2y}{x^2 + y^2} \Rightarrow f_x(1, 0) = 2$ and $f_y(1, 0) = 0$
    $\Rightarrow$ from Eq. (4) the tangent plane at $(1, 0, 0)$ is $2(x - 1) - z = 0$ or $2x - z - 2 = 0$

11. $z = f(x, y) = \sqrt{y - x} \Rightarrow f_x(x, y) = -\dfrac{1}{2}(y - x)^{-1/2}$ and $f_y(x, y) = -\dfrac{1}{2}(y - x)^{-1/2} \Rightarrow f_x(1, 2) = -\dfrac{1}{2}$ and
    $f_y(1, 2) = \dfrac{1}{2} \Rightarrow$ from Eq. (4) the tangent plane at $(1, 2, 1)$ is $-\dfrac{1}{2}(x - 1) + \dfrac{1}{2}(y - 2) - (z - 1) = 0$
    $\Rightarrow x - y + 2z - 1 = 0$

13. $\nabla f = \mathbf{i} + 2y\mathbf{j} + 2\mathbf{k} \Rightarrow \nabla f(1, 1, 1) = \mathbf{i} + 2\mathbf{j} + 2\mathbf{k}$ and $\nabla g = \mathbf{i}$ for all points; $\mathbf{v} = \nabla f \times \nabla g \Rightarrow \mathbf{v} = \begin{vmatrix} \mathbf{i} & \mathbf{j} & \mathbf{k} \\ 1 & 2 & 2 \\ 1 & 0 & 0 \end{vmatrix} = 2\mathbf{j} - 2\mathbf{k}$

    $\Rightarrow$ Tangent line: $x = 1$, $y = 1 + 2t$, $z = 1 - 2t$

15. $\nabla f = 2x\mathbf{i} + 2\mathbf{j} + 2\mathbf{k} \Rightarrow \nabla f\left(1, 1, \frac{1}{2}\right) = 2\mathbf{i} + 2\mathbf{j} + 2\mathbf{k}$ and $\nabla g = \mathbf{j}$ for all points ;

$$\mathbf{v} = \nabla f \times \nabla g \Rightarrow \mathbf{v} = \begin{vmatrix} \mathbf{i} & \mathbf{j} & \mathbf{k} \\ 2 & 2 & 2 \\ 0 & 1 & 0 \end{vmatrix} = -2\mathbf{i} + 2\mathbf{k} \Rightarrow \text{Tangent line: } x = 1 - 2t, \ y = 1, \ z = \frac{1}{2} + 2t$$

17. $\nabla f = \left(3x^2 + 6xy^2 + 4y\right)\mathbf{i} + \left(6x^2 y + 3y^2 + 4x\right)\mathbf{j} - 2z\mathbf{k} \Rightarrow \nabla f(1, 1, 3) = 13\mathbf{i} + 13\mathbf{j} - 6\mathbf{k};$

$$\nabla g = 2x\mathbf{i} + 2y\mathbf{j} + 2z\mathbf{k} \Rightarrow \nabla g(1, 1, 3) = 2\mathbf{i} + 2\mathbf{j} + 6\mathbf{k}; \quad \mathbf{v} = \nabla f \times \nabla g \Rightarrow \mathbf{v} = \begin{vmatrix} \mathbf{i} & \mathbf{j} & \mathbf{k} \\ 13 & 13 & -6 \\ 2 & 2 & 6 \end{vmatrix} = 90\mathbf{i} - 90\mathbf{j}$$

$\Rightarrow$ Tangent line: $x = 1 + 90t, \ y = 1 - 90t, \ z = 3$

19. $\nabla f = \left(\frac{x}{x^2 + y^2 + z^2}\right)\mathbf{i} + \left(\frac{y}{x^2 + y^2 + z^2}\right)\mathbf{j} + \left(\frac{z}{x^2 + y^2 + z^2}\right)\mathbf{k} \Rightarrow \nabla f(3, 4, 12) = \frac{3}{169}\mathbf{i} + \frac{4}{169}\mathbf{j} + \frac{12}{169}\mathbf{k};$

$\mathbf{u} = \frac{\mathbf{v}}{|\mathbf{v}|} = \frac{3\mathbf{i} + 6\mathbf{j} - 2\mathbf{k}}{\sqrt{3^2 + 6^2 + (-2)^2}} = \frac{3}{7}\mathbf{i} + \frac{6}{7}\mathbf{j} - \frac{2}{7}\mathbf{k} \Rightarrow \nabla f \cdot \mathbf{u} = \frac{9}{1183}$ and $df = (\nabla f \cdot \mathbf{u}) \, ds = \left(\frac{9}{1183}\right)(0.1) \approx 0.0008$

21. $\nabla g = (1 + \cos z)\mathbf{i} + (1 - \sin z)\mathbf{j} + (-x\sin z - y\cos z)\mathbf{k} \Rightarrow \nabla g(2, -1, 0) = 2\mathbf{i} + \mathbf{j} + \mathbf{k}; \quad \mathbf{A} = \overline{P_0 P_1} = -2\mathbf{i} + 2\mathbf{j} + 2\mathbf{k}$

$\Rightarrow \mathbf{u} = \frac{\mathbf{v}}{|\mathbf{v}|} = \frac{-2\mathbf{i} + 2\mathbf{j} + 2\mathbf{k}}{\sqrt{(-2)^2 + 2^2 + 2^2}} = -\frac{1}{\sqrt{3}}\mathbf{i} + \frac{1}{\sqrt{3}}\mathbf{j} + \frac{1}{\sqrt{3}}\mathbf{k} \Rightarrow \nabla g \cdot \mathbf{u} = 0$ and $dg = (\nabla g \cdot \mathbf{u}) \, ds = (0)(0.2) = 0$

23. (a) The unit tangent vector at $\left(\frac{1}{2}, \frac{\sqrt{3}}{2}\right)$ in the direction of motion is $\mathbf{u} = \frac{\sqrt{3}}{2}\mathbf{i} - \frac{1}{2}\mathbf{j};$

$$\nabla T = (\sin 2y)\mathbf{i} + (2x\cos 2y)\mathbf{j} \Rightarrow \nabla T\left(\frac{1}{2}, \frac{\sqrt{3}}{2}\right) = \left(\sin\sqrt{3}\right)\mathbf{i} + \left(\cos\sqrt{3}\right)\mathbf{j} \Rightarrow D_\mathbf{u} T\left(\frac{1}{2}, \frac{\sqrt{3}}{2}\right) = \nabla T \cdot \mathbf{u}$$

$$= \frac{\sqrt{3}}{2}\sin\sqrt{3} - \frac{1}{2}\cos\sqrt{3} \approx 0.935^\circ \, \text{C/ft}$$

(b) $\mathbf{r}(t) = (\sin 2t)\mathbf{i} + (\cos 2t)\mathbf{j} \Rightarrow \mathbf{v}(t) = (2\cos 2t)\mathbf{i} - (2\sin 2t)\mathbf{j}$ and $|\mathbf{v}| = 2;$

$\frac{dT}{dt} = \frac{\partial T}{\partial x}\frac{dx}{dt} + \frac{\partial T}{\partial y}\frac{dy}{dt} = \nabla T \cdot \mathbf{v} = \left(\nabla T \cdot \frac{\mathbf{v}}{|\mathbf{v}|}\right)|\mathbf{v}| = (D_\mathbf{u}T)|\mathbf{v}|$, where $\mathbf{u} = \frac{\mathbf{v}}{|\mathbf{v}|}$; at $\left(\frac{1}{2}, \frac{\sqrt{3}}{2}\right)$ we have $\mathbf{u} = \frac{\sqrt{3}}{2}\mathbf{i} - \frac{1}{2}\mathbf{j}$

from part (a) $\Rightarrow \frac{dT}{dt} = \left(\frac{\sqrt{3}}{2}\sin\sqrt{3} - \frac{1}{2}\cos\sqrt{3}\right) \cdot 2 = \sqrt{3}\sin\sqrt{3} - \cos\sqrt{3} \approx 1.87^\circ$ C/sec

25. (a) $f(0, 0) = 1, \ f_x(x, y) = 2x \Rightarrow f_x(0,0) = 0, \ f_y(x, y) = 2y \Rightarrow f_y(0,0) = 0$

$\Rightarrow L(x, y) = 1 + 0(x - 0) + 0(y - 0) = 1$

(b) $f(1, 1) = 3, \ f_x(1, 1) = 2, \ f_y(1, 1) = 2 \Rightarrow L(x, y) = 3 + 2(x - 1) + 2(y - 1) = 2x + 2y - 1$

27. (a) $f(0, 0) = 5, \ f_x(x, y) = 3$ for all $(x, y), \ f_y(x, y) = -4$ for all $(x, y) \Rightarrow L(x, y) = 5 + 3(x - 0) - 4(y - 0)$

$= 3x - 4y + 5$

(b) $f(1, 1) = 4, \ f_x(1, 1) = 3, \ f_y(1, 1) = -4 \Rightarrow L(x, y) = 4 + 3(x - 1) - 4(y - 1) = 3x - 4y + 5$

29. (a) $f(0, 0) = 1, \ f_x(x, y) = e^x \cos y = f_x(0, 0) = 1, \ f_y(x, y) = -e^x \sin y \Rightarrow f_y(0, 0) = 0$

$\Rightarrow L(x, y) = 1 + 1(x - 0) + 0(y - 0) = x + 1$

(b) $f\left(0, \frac{\pi}{2}\right) = 0, f_x\left(0, \frac{\pi}{2}\right) = 0, \ f_y\left(0, \frac{\pi}{2}\right) = -1 \Rightarrow L(x, y) = 0 + 0(x - 0) - 1\left(y - \frac{\pi}{2}\right) = -y + \frac{\pi}{2}$

31. (a) $W(20, 25) = 11°F; W(30, -10) = -39°F; W(15,15) = 0°F$

   (b) $W(10, -40) = -65.5°F; W(50, -40) = -88°F; W(60, 30) = 10.2°F;$

   (c) $W(25, 5) = -17.4088°F; \frac{\partial W}{\partial V} = -\frac{5.72}{v^{0.84}} + \frac{0.0684t}{v^{0.84}} \Rightarrow \frac{\partial W}{\partial V}(25, 5) = -0.36; \frac{\partial W}{\partial T} = 0.6215 + 0.4275v^{0.16}$

   $\Rightarrow \frac{\partial W}{\partial T}(25, 5) = 1.3370 \Rightarrow L(V, T) = -17.4088 - 0.36(V - 25) + 1.337(T - 5) = 1.337T - 0.36V - 15.0938$

   (d) i) $W(24, 6) \approx L(24, 6) = -15.7118 \approx -15.7°F$

   ii) $W(27, 2) \approx L(27, 2) = -22.1398 \approx -22.1°F$

   iii) $W(5, -10) \approx L(5, -10) = -30.2638 \approx -30.2°F$ This value is very different because the point $(5, -10)$ is not close to the point $(25, 5)$.

33. $f(2, 1) = 3, f_x(x, y) = 2x - 3y \Rightarrow f_x(2, 1) = 1, f_y(x, y) = -3x \Rightarrow f_y(2, 1) = -6 \Rightarrow L(x, y) = 3 + 1(x - 2) - 6(y - 1)$

   $= 7 + x - 6y; f_{xx}(x, y) = 2, f_{yy}(x, y) = 0, f_{xy}(x, y) = -3 \Rightarrow M = 3; \text{ thus } |E(x, y)| \leq \left(\frac{1}{2}\right)(3)\left(|x - 2| + |y - 1|\right)^2$

   $\leq \left(\frac{3}{2}\right)(0.1 + 0.1)^2 = 0.06$

35. $f(0, 0) = 1, f_x(x, y) = \cos y \Rightarrow f_x(0, 0) = 1, f_y(x, y) = 1 - x\sin y \Rightarrow f_y(0, 0) = 1$

   $\Rightarrow L(x, y) = 1 + 1(x - 0) + 1(y - 0) = x + y + 1;\ f_{xx}(x, y) = 0, f_{yy}(x, y) = -x\cos y, f_{xy}(x, y) = -\sin y$

   $\Rightarrow M = 1; \text{ thus } |E(x, y)| \leq \left(\frac{1}{2}\right)(1)\left(|x| + |y|\right)^2 \leq \left(\frac{1}{2}\right)(0.2 + 0.2)^2 = 0.08$

37. $f(0, 0) = 1, f_x(x, y) = e^x \cos y \Rightarrow f_x(0, 0) = 1, f_y(x, y) = -e^x \sin y \Rightarrow f_y(0, 0) = 0$

   $\Rightarrow L(x, y) = 1 + 1(x - 0) + 0(y - 0) = 1 + x;\ f_{xx}(x, y) = e^x \cos y, f_{yy}(x, y) = -e^x \cos y, f_{xy}(x, y) = -e^x \sin y;$

   $|x| \leq 0.1 \Rightarrow -0.1 \leq x \leq 0.1 \text{ and } |y| \leq 0.1 \Rightarrow -0.1 \leq y \leq 0.1; \text{ thus the max of } |f_{xx}(x, y)| \text{ on } R \text{ is}$

   $e^{0.1} \cos(0.1) \leq 1.11, \text{ the max of } |f_{yy}(x, y)| \text{ on } R \text{ is } e^{0.1} \cos(0.1) \leq 1.11, \text{ and the max of } |f_{xy}(x, y)| \text{ on } R \text{ is}$

   $e^{0.1} \sin(0.1) \leq 0.12 \Rightarrow M = 1.11; \text{ thus } |E(x, y)| \leq \left(\frac{1}{2}\right)(1.11)\left(|x| + |y|\right)^2 \leq (0.555)(0.1 + 0.1)^2 = 0.0222$

39. (a) $f(1, 1, 1) = 3, f_x(1, 1, 1) = y + z|_{(1,1,1)} = 2, f_y(1, 1, 1) = x + z|_{(1,1,1)} = 2, f_z(1, 1, 1) = y + x|_{(1,1,1)} = 2$

   $\Rightarrow L(x, y, z) = 3 + 2(x - 1) + 2(y - 1) + 2(z - 1) = 2x + 2y + 2z - 3$

   (b) $f(1, 0, 0) = 0, f_x(1, 0, 0) = 0, f_y(1, 0, 0) = 1, f_z(1, 0, 0) = 1$

   $\Rightarrow L(x, y, z) = 0 + 0(x - 1) + (y - 0) + (z - 0) = y + z$

   (c) $f(0, 0, 0) = 0, f_x(0, 0, 0) = 0, f_y(0, 0, 0) = 0, f_z(0, 0, 0) = 0 \Rightarrow L(x, y, z) = 0$

41. (a) $f(1, 0, 0) = 1, f_x(1, 0, 0) = \left.\frac{x}{\sqrt{x^2 + y^2 + z^2}}\right|_{(1, 0, 0)} = 1,\ f_y(1, 0, 0) = \left.\frac{y}{\sqrt{x^2 + y^2 + z^2}}\right|_{(1, 0, 0)} = 0,$

   $f_z(1, 0, 0) = \left.\frac{z}{\sqrt{x^2 + y^2 + z^2}}\right|_{(1, 0, 0)} = 0 \Rightarrow L(x, y, z) = 1 + 1(x - 1) + 0(y - 0) + 0(z - 0) = x$

   (b) $f(1, 1, 0) = \sqrt{2}, f_x(1, 1, 0) = \frac{1}{\sqrt{2}}, f_y(1, 1, 0) = \frac{1}{\sqrt{2}},\ f_z(1, 1, 0) = 0$

   $\Rightarrow L(x, y, z) = \sqrt{2} + \frac{1}{\sqrt{2}}(x - 1) + \frac{1}{\sqrt{2}}(y - 1) + 0(z - 0) = \frac{1}{\sqrt{2}}x + \frac{1}{\sqrt{2}}y$

   (c) $f(1, 2, 2) = 3, f_x(1, 2, 2) = \frac{1}{3}, f_y(1, 2, 2) = \frac{2}{3},\ f_z(1, 2, 2) = \frac{2}{3}$

   $\Rightarrow L(x, y, z) = 3 + \frac{1}{3}(x - 1) + \frac{2}{3}(y - 2) + \frac{2}{3}(z - 2) = \frac{1}{3}x + \frac{2}{3}y + \frac{2}{3}z$

43. (a) $f(0,0,0) = 2$, $f_x(0,0,0) = e^x\big|_{(0,0,0)} = 1$, $f_y(0,0,0) = -\sin(y+z)\big|_{(0,0,0)} = 0$,

$f_z(0,0,0) = -\sin(y+z)\big|_{(0,0,0)} = 0 \Rightarrow L(x,y,z) = 2 + 1(x-0) + 0(y-0) + 0(z-0) = 2 + x$

(b) $f(0,\frac{\pi}{2},0) = 1$, $f_x(0,\frac{\pi}{2},0) = 1$, $f_y(0,\frac{\pi}{2},0) = -1$, $f_z(0,\frac{\pi}{2},0) = -1$,

$\Rightarrow L(x,y,z) = 1 + 1(x-0) - 1\left(y - \frac{\pi}{2}\right) - 1(z-0) = x - y - z + \frac{\pi}{2} + 1$

(c) $f(0,\frac{\pi}{4},\frac{\pi}{4}) = 1$, $f_x(0,\frac{\pi}{4},\frac{\pi}{4}) = 1$, $f_y(0,\frac{\pi}{4},\frac{\pi}{4}) = -1$, $f_z(0,\frac{\pi}{4},\frac{\pi}{4}) = -1$,

$\Rightarrow L(x,y,z) = 1 + 1(x-0) - 1\left(y - \frac{\pi}{4}\right) - 1\left(z - \frac{\pi}{4}\right) = x - y - z + \frac{\pi}{2} + 1$

45. $f(x,y,z) = xz - 3yz + 2$ at $P_0(1,1,2) \Rightarrow f(1,1,2) = -2$; $f_x = z$, $f_y = -3z$, $f_z = x - 3y$

$\Rightarrow L(x,y,z) = -2 + 2(x-1) - 6(y-1) - 2(z-2) = 2x - 6y - 2z + 6$; $f_{xx} = 0$, $f_{yy} = 0$, $f_{zz} = 0$, $f_{xy} = 0$, $f_{yz} = -3$

$\Rightarrow M = 3$; thus, $|E(x,y,z)| \leq \left(\frac{1}{2}\right)(3)(0.01 + 0.01 + 0.02)^2 = 0.0024$

47. $f(x,y,z) = xy + 2yz - 3xz$ at $P_0(1,1,0) \Rightarrow f(1,1,0) = 1$; $f_x = y - 3z$, $f_y = x + 2z$, $f_z = 2y - 3x$

$\Rightarrow L(x,y,z) = 1 + (x-1) + (y-1) - (z-0) = x + y - z - 1$; $f_{xx} = 0$, $f_{yy} = 0$, $f_{zz} = 0$, $f_{xy} = 1$, $f_{xz} = -3$, $f_{yz} = 2$

$\Rightarrow M = 3$; thus, $|E(x,y,z)| \leq \left(\frac{1}{2}\right)(3)(0.01 + 0.01 + 0.01)^2 = 0.00135$

49. $T_x(x,y) = e^y + e^{-y}$ and $T_y(x,y) = x\left(e^y - e^{-y}\right) \Rightarrow dT = T_x(x,y)\,dx + T_y(x,y)\,dy$

$= \left(e^y + e^{-y}\right)dx + x\left(e^y - e^{-y}\right)dy \Rightarrow dT\big|_{(2,\ln 2)} = 2.5\,dx + 3.0\,dy$. If $|dx| \leq 0.1$ and $|dy| \leq 0.02$, then the

maximum possible error in the computed value of T is $(2.5)(0.1) + (3.0)(0.02) = 0.31$ in magnitude.

51. $A = xy \Rightarrow dA = x\,dy + y\,dx$; if $x > y$ then a 1-unit change in y gives a greater change in dA than a 1-unit change in x. Thus, pay more attention to y which is the smaller of the two dimensions.

53. $f(a,b,c,d) = \begin{vmatrix} a & b \\ c & d \end{vmatrix} = ad - bc \Rightarrow f_a = d$, $f_b = -c$, $f_c = -b$, $f_d = a \Rightarrow df = d\,da - c\,db - b\,dc + a\,dd$;

since $|a|$ is much greater than $|b|, |c|$, and $|d|$, the function f is most sensitive to a change in d.

55. $z = f(x,y) \Rightarrow g(x,y,z) = f(x,y) - z = 0 \Rightarrow g_x(x,y,z) = f_x(x,y)$, $g_y(x,y,z) = f_y(x,y)$ and

$g_z(x,y,z) = -1 \Rightarrow g_x(x_0, y_0, f(x_0, y_0)) = f_x(x_0, y_0)$, $g_y(x_0, y_0, f(x_0, y_0)) = f_y(x_0, y_0)$ and

$g_z(x_0, y_0, f(x_0, y_0)) = -1 \Rightarrow$ the tangent plane at the point P_0 is

$f_x(x_0, y_0)(x - x_0) + f_y(x_0, y_0)(y - y_0) - [z - f(x_0, y_0)] = 0$ or

$z = f_x(x_0, y_0)(x - x_0) + f_y(x_0, y_0)(y - y_0) + f(x_0, y_0)$

57. $\mathbf{r} = \sqrt{t}\,\mathbf{i} + \sqrt{t}\,\mathbf{j} + (2t-1)\mathbf{k} \Rightarrow \mathbf{v} = \frac{1}{2}t^{-1/2}\mathbf{i} + \frac{1}{2}t^{-1/2}\mathbf{j} + 2\mathbf{k}$; $t = 1 \Rightarrow x = 1$, $y = 1$, $z = 1 \Rightarrow P_0 = (1,1,1)$ and

$\mathbf{v}(1) = \frac{1}{2}\mathbf{i} + \frac{1}{2}\mathbf{j} + 2\mathbf{k}$; $f(x,y,z) = x^2 + y^2 - z - 1 = 0 \Rightarrow \nabla f = 2x\mathbf{i} + 2y\mathbf{j} - \mathbf{k} \Rightarrow \nabla f(1,1,1) = 2\mathbf{i} + 2\mathbf{j} - \mathbf{k}$;

now $\mathbf{v}(1) \cdot \nabla f(1,1,1) = 0$, thus the curve is tangent to the surface when $t = 1$

13.7 EXTREME VALUES AND SADDLE POINTS

1. $f_x(x, y) = 2x + y + 3 = 0$ and $f_y(x, y) = x + 2y - 3 = 0 \Rightarrow x = -3$ and $y = 3 \Rightarrow$ critical point is $(-3, 3)$;

   $f_{xx}(-3, 3) = 2, f_{yy}(-3, 3) = 2, f_{xy}(-3, 3) = 1 \Rightarrow f_{xx}f_{yy} - f_{xy}^2 = 3 > 0$ and $f_{xx} > 0 \Rightarrow$ local minimum of

   $f(-3, 3) = -5$

3. $f_x(x, y) = 2x + y + 3 = 0$ and $f_y(x, y) = x + 2 = 0 \Rightarrow x = -2$ and $y = 1 \Rightarrow$ critical point is $(-2, 1)$;

   $f_{xx}(-2, 1) = 2, f_{yy}(-2, 1) = 0, f_{xy}(-2, 1) = 1 \Rightarrow f_{xx}f_{yy} - f_{xy}^2 = -1 < 0 \Rightarrow$ saddle point

5. $f_x(x, y) = 2y - 2x + 3 = 0$ and $f_y(x, y) = 2x - 4y = 0 \Rightarrow x = 3$ and $y = \frac{3}{2} \Rightarrow$ critical point is $\left(3, \frac{3}{2}\right)$;

   $f_{xx}\left(3, \frac{3}{2}\right) = -2, f_{yy}\left(3, \frac{3}{2}\right) = -4, f_{xy}\left(3, \frac{3}{2}\right) = 2 \Rightarrow f_{xx}f_{yy} - f_{xy}^2 = 4 > 0$ and $f_{xx} < 0 \Rightarrow$ local maximum of

   $f\left(3, \frac{3}{2}\right) = \frac{17}{2}$

7. $f_x(x, y) = 4x + 3y - 5 = 0$ and $f_y(x, y) = 3x + 8y + 2 = 0 \Rightarrow x = 2$ and $y = -1 \Rightarrow$ critical point is $(2, -1)$;

   $f_{xx}(2, -1) = 4, f_{yy}(2, -1) = 8, f_{xy}(2, -1) = 3 \Rightarrow f_{xx}f_{yy} - f_{xy}^2 = 23 > 0$ and $f_{xx} > 0 \Rightarrow$ local minimum of

   $f(2, -1) = -6$

9. $f_x(x, y) = 2x - 2 = 0$ and $f_y(x, y) = -2y + 4 = 0 \Rightarrow x = 1$ and $y = 2 \Rightarrow$ critical point is $(1, 2)$;

   $f_{xx}(1, 2) = 2, f_{yy}(1, 2) = -2, f_{xy}(1, 2) = 0 \Rightarrow f_{xx}f_{yy} - f_{xy}^2 = -4 < 0 \Rightarrow$ saddle point

11. $f_x(x, y) = \frac{112x - 8x}{\sqrt{56x^2 - 8y^2 - 16x - 31}} - 8 = 0$ and $f_y(x, y) = \frac{-8y}{\sqrt{56x^2 - 8y^2 - 16x - 31}} = 0 \Rightarrow$ critical point is $\left(\frac{16}{7}, 0\right)$;

   $f_{xx}\left(\frac{16}{7}, 0\right) = -\frac{8}{15}, f_{yy}\left(\frac{16}{7}, 0\right) = -\frac{8}{15}, f_{xy}\left(\frac{16}{7}, 0\right) = 0 \Rightarrow f_{xx}f_{yy} - f_{xy}^2 = \frac{64}{225} > 0$ and $f_{xx} < 0 \Rightarrow$ local maximum

   of $f\left(\frac{16}{7}, 0\right) = -\frac{16}{7}$

13. $f_x(x, y) = 3x^2 - 2y = 0$ and $f_y(x, y) = -3y^2 - 2x = 0 \Rightarrow x = 0$ and $y = 0$, or $x = -\frac{2}{3}$ and $y = \frac{2}{3} \Rightarrow$ critical

   points are $(0, 0)$ and $\left(-\frac{2}{3}, \frac{2}{3}\right)$; for $(0, 0)$: $f_{xx}(0, 0) = 6x|_{(0, 0)} = 0, f_{yy}(0, 0) = -6y|_{(0, 0)} = 0, f_{xy}(0, 0) = -2$

   $\Rightarrow f_{xx}f_{yy} - f_{xy}^2 = -4 < 0 \Rightarrow$ saddle point; for $\left(-\frac{2}{3}, \frac{2}{3}\right)$: $f_{xx}\left(-\frac{2}{3}, \frac{2}{3}\right) = -4, f_{yy}\left(-\frac{2}{3}, \frac{2}{3}\right) = -4, f_{xy}\left(-\frac{2}{3}, \frac{2}{3}\right) = -2$

   $\Rightarrow f_{xx}f_{yy} - f_{xy}^2 = 12 > 0$ and $f_{xx} < 0 \Rightarrow$ local maximum of $f\left(-\frac{2}{3}, \frac{2}{3}\right) = \frac{170}{27}$

15. $f_x(x, y) = 12x - 6x^2 + 6y = 0$ and $f_y(x, y) = 6y + 6x = 0 \Rightarrow x = 0$ and $y = 0$, or $x = 1$ and $y = -1 \Rightarrow$ critical

   points are $(0, 0)$ and $(1, -1)$; for $(0, 0)$: $f_{xx}(0, 0) = 12 - 12x|_{(0, 0)} = 12, f_{yy}(0, 0) = 6, f_{xy}(0, 0) = 6$

   $\Rightarrow f_{xx}f_{yy} - f_{xy}^2 = 36 > 0$ and $f_{xx} > 0 \Rightarrow$ local minimum of $f(0, 0) = 0$; for $(1, -1)$: $f_{xx}(1, -1) = 0$,

   $f_{yy}(1, -1) = 6, f_{xy}(1, -1) = 6 \Rightarrow f_{xx}f_{yy} - f_{xy}^2 = -36 < 0 \Rightarrow$ saddle point

17. $f_x(x, y) = 3x^2 + 3y^2 - 15 = 0$ and $f_y(x, y) = 6xy + 3y^2 - 15 = 0 \Rightarrow$ critical points are $(2, 1), (-2, -1), \left(0, \sqrt{5}\right)$,

and $\left(0, -\sqrt{5}\right)$; for $(2, 1)$: $f_{xx}(2, 1) = 6x|_{(2,1)} = 12$, $f_{yy}(2, 1) = (6x + 6y)|_{(2,1)} = 18$, $f_{xy}(2, 1) = 6y|_{(2,1)} = 6$

$\Rightarrow f_{xx}f_{yy} - f_{xy}^2 = 180 > 0$ and $f_{xx} > 0 \Rightarrow$ local minimum of $f(2, 1) = -30$; for $(-2, -1)$:

$f_{xx}(-2, -1) = 6x|_{(-2,-1)} = -12$, $f_{yy}(-2, -1) = (6x + 6y)|_{(-2,-1)} = -18$, $f_{xy}(-2, -1) = 6y|_{(-2,-1)} = -6$

$\Rightarrow f_{xx}f_{yy} - f_{xy}^2 = 180 > 0$ and $f_{xx} < 0 \Rightarrow$ local maximum of $f(-2, -1) = 30$; for $\left(0, -\sqrt{5}\right)$:

$f_{xx}\left(0, -\sqrt{5}\right) = 6x|_{\left(0, \sqrt{5}\right)} = 0$, $f_{yy}\left(0, \sqrt{5}\right) = (6x + 6y)|_{\left(0, \sqrt{5}\right)} = 6\sqrt{5}$, $f_{xy}\left(0, \sqrt{5}\right) = 6y|_{\left(0, \sqrt{5}\right)} = 6\sqrt{5}$

$\Rightarrow f_{xx}f_{yy} - f_{xy}^2 = -180 < 0 \Rightarrow$ saddle point; for $\left(0, -\sqrt{5}\right)$: $f_{xx}\left(0, -\sqrt{5}\right) = 6x|_{\left(0, -\sqrt{5}\right)} = 0$

$f_{yy}\left(0, -\sqrt{5}\right) = (6x + 6y)|_{\left(0, -\sqrt{5}\right)} = -6\sqrt{5}$, $f_{xy}\left(0, -\sqrt{5}\right) = 6y|_{\left(0, -\sqrt{5}\right)} = -6\sqrt{5} \Rightarrow f_{xx}f_{yy} - f_{xy}^2 = -180 < 0$

$\Rightarrow$ saddle point.

19. $f_x(x, y) = 4y - 4x^3 = 0$ and $f_y(x, y) = 4x - 4y^3 = 0 \Rightarrow x = y \Rightarrow x\left(1 - x^2\right) = 0 \Rightarrow x = 0, 1, -1 \Rightarrow$ the critical

points are $(0, 0), (1, 1),$ and $(-1, -1)$; for $(0, 0)$: $f_{xx}(0, 0) = -12x^2|_{(0,0)} = 0$, $f_{yy}(0, 0) = -12y^2|_{(0,0)} = 0$,

$f_{xy}(0, 0) = 4 \Rightarrow f_{xx}f_{yy} - f_{xy}^2 = -16 < 0 \Rightarrow$ saddle point; for $(1, 1)$: $f_{xx}(1, 1) = -12$, $f_{yy}(1, 1) = -12$, $f_{xy}(1, 1) = 4$

$\Rightarrow f_{xx}f_{yy} - f_{xy}^2 = 128 > 0$ and $f_{xx} < 0 \Rightarrow$ local maximum of $f(1, 1) = 2$; for $(-1, -1)$: $f_{xx}(-1, -1) = -12$,

$f_{yy}(-1, -1) = -12$, $f_{xy}(-1, -1) = 4 \Rightarrow f_{xx}f_{yy} - f_{xy}^2 = 128 > 0$ and $f_{xx} < 0 \Rightarrow$ local maximum of $f(-1, -1) = 2$

21. $f_x(x, y) = \dfrac{-2x}{\left(x^2 + y^2 - 1\right)^2} = 0$ and $f_y(x, y) = \dfrac{-2y}{\left(x^2 + y^2 - 1\right)^2} = 0 \Rightarrow x = 0$ and $y = 0 \Rightarrow$ the critical points is $(0, 0)$;

$f_{xx} = \dfrac{4x^2 - 2y^2 + 2}{\left(x^2 + y^2 - 1\right)^3}$, $f_{yy} = \dfrac{-2x^2 + 4y^2 + 2}{\left(x^2 + y^2 - 1\right)^3}$, $f_{xy} = \dfrac{8xy}{\left(x^2 + y^2 - 1\right)^3}$; $f_{xx}(0, 0) = -2$, $f_{yy}(0, 0) = -2$, $f_{xy}(0, 0) = 0$

$\Rightarrow f_{xx}f_{yy} - f_{xy}^2 = 4 > 0$ and $f_{xx} < 0 \Rightarrow$ local maximum of $f(0, 0) = -1$

23. $f_x(x, y) = y\cos x = 0$ and $f_y(x, y) = \sin x = 0 \Rightarrow x = n\pi, n$ an integer, and $y = 0 \Rightarrow$ the critical points are

$(n\pi, 0), n$ an integer (Note: $\cos x$ and $\sin x$ cannot both be 0 for the same x, so $\sin x$ must be 0 and $y = 0$);

$f_{xx} = -y\sin x$, $f_{yy} = 0$, $f_{xy} = \cos x$; $f_{xx}(n\pi, 0) = 0$, $f_{yy}(n\pi, 0) = 0$, $f_{xy}(n\pi, 0) = 1$ if n is even and

$f_{xy}(n\pi, 0) = -1$ if n is odd $\Rightarrow f_{xx}f_{yy} - f_{xy}^2 = -1 < 0 \Rightarrow$ saddle point.

25. $f_x(x, y) = (2x - 4)e^{x^2 + y^2 - 4x} = 0$ and $f_y(x, y) = 2ye^{x^2 + y^2 - 4x} = 0 \Rightarrow$ critical point is $(2, 0)$; $f_{xx}(2, 0) = \dfrac{2}{e^4}$,

$f_{xy}(2, 0) = 0$, $f_{yy}(2, 0) = \dfrac{2}{e^4} \Rightarrow f_{xx}f_{yy} - f_{xy}^2 = \dfrac{4}{e^8} > 0$ and $f_{xx} > 0 \Rightarrow$ local minimum of $f(2, 0) = \dfrac{1}{e^4}$

27. $f_x(x, y) = 2xe^{-y} = 0$ and $f_y(x, y) = 2ye^{-y} - e^{-y}\left(x^2 + y^2\right) = 0 \Rightarrow$ critical points are $(0, 0)$ and $(0, 2)$;

for $(0, 0)$: $f_{xx}(0, 0) = 2e^{-y}\big|_{(0,0)} = 2$, $f_{yy}(0, 0) = \left(2e^{-y} - 4ye^{-y} + e^{-y}\left(x^2 + y^2\right)\right)\Big|_{(0,0)} = 2$,

$f_{xy}(0, 0) = -2xe^{-y}|_{(0,0)} = 0, \Rightarrow f_{xx}f_{yy} - f_{xy}^2 = 4 > 0 \Rightarrow$ and $f_{xx} > 0 \Rightarrow$ local minimum of $f(0, 0) = 0$;

for $(0, 2)$: $f_{xx}(0, 2) = 2e^{-y}\big|_{(0,2)} = \frac{2}{e^2}$, $f_{yy}(0, 2) = \left(2e^{-y} - 4ye^{-y} + e^{-y}\left(x^2 + y^2\right)\right)\big|_{(0,2)} = -\frac{2}{e^2}$,

$f_{xy}(0, 2) = -2xe^{-y}\big|_{(0,2)} = 0 \Rightarrow f_{xx}f_{yy} - f_{xy}^2 = -\frac{4}{e^4} < 0 \Rightarrow$ saddle point

29. $f_x(x, y) = -4 + \frac{2}{x} = 0$ and $f_y(x, y) = -1 + \frac{1}{y} = 0 \Rightarrow$ critical point is $\left(\frac{1}{2}, 1\right)$; $f_{xx}\left(\frac{1}{2}, 1\right) = -8$, $f_{yy}\left(\frac{1}{2}, 1\right) = -1$,

$f_{xy}\left(\frac{1}{2}, 1\right) = 0 \Rightarrow f_{xx}f_{yy} - f_{xy}^2 = 8 > 0$ and $f_{xx} < 0 \Rightarrow$ local maximum of $f\left(\frac{1}{2}, 1\right) = -3 - 2\ln 2$

31. (i)   On OA, $f(x, y) = f(0, y) = y^2 - 4y + 1$ on

$0 \le y \le 2$; $f'(0, y) = 2y - 4 = 0 \Rightarrow y = 2$;

$f(0, 0) = 1$ and $f(0, 2) = -3$

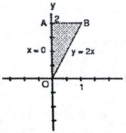

   (ii) On AB, $f(x, y) = f(x, 2) = 2x^2 - 4x - 3$ on

$0 \le x \le 1$; $f'(x, 2) = 4x - 4 = 0 \Rightarrow x = 1$;

$f'(0, 2) = -3$ and $f(1, 2) = -5$

   (iii) On OB, $f(x, y) = f(x, 2x) = 6x^2 - 12x + 1$ on $0 \le x \le 1$; endpoint values have been found above;

$f'(x, 2x) = 12x - 12 = 0 \Rightarrow x = 1$ and $y = 2$, but $(1, 2)$ is not an interior point of OB

   (iv) For interior points of the triangular region, $f_x(x, y) = 4x - 4 = 0$ and $f_y(x, y) = 2y - 4 = 0 \Rightarrow x = 1$ and

$y = 2$, but $(1, 2)$ is not an interior point of the region. Therefore, the absolute maximum is 1 at $(0, 0)$

and the absolute minimum is -5 at $(1, 2)$.

33. (i)   On OA, $f(x, y) = f(0, y) = y^2$ on $0 \le y \le 2$;

$f'(0, y) = 2y = 0 \Rightarrow y = 0$ and $x = 0$; $f(0, 0) = 0$

and $f(0, 2) = 4$

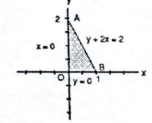

   (ii) On OB, $f(x, y) = f(x, 0) = x^2$ on $0 \le x \le 1$;

$f'(x, 0) = 2x = 0 \Rightarrow x = 0$ and $y = 0$; $f(0, 0) = 0$

and $f(1, 0) = 1$

   (iii) On AB, $f(x, y) = f(x, -2x + 2) = 5x^2 - 8x + 4$ on $0 \le x \le 1$; $f'(x, -2x + 2) = 10x - 8 = 0 \Rightarrow x = \frac{4}{5}$ and

$y = \frac{2}{5}$; $f\left(\frac{4}{5}, \frac{2}{5}\right) = \frac{4}{5}$; endpoint values have been found above.

   (iv) For interior points of the triangular region, $f_x(x, y) = 2x = 0$ and $f_y(x, y) = 2y = 0 \Rightarrow x = 0$ and $y = 0$,

but $(0, 0)$ is not an interior point of the region. Therefore the absolute maximum is 4 at $(0, 2)$ and the

absolute minimum is 0 at $(0, 0)$.

35. (i)   On OC, $T(x, y) = T(x, 0) = x^2 - 6x + 2$ on

$0 \le x \le 5$; $T'(x, 0) = 2x - 6 = 0 \Rightarrow x = 3$ and $y = 0$;

$T(3, 0) = -7$, $T(0, 0) = 2$, and $T(5, 0) = -3$

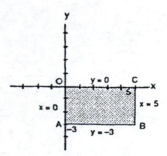

   (ii) On CB, $T(x, y) = T(5, y) = y^2 + 5y - 3$ on

$-3 \le y \le 0$; $T'(5, y) = 2y + 5 = 0 \Rightarrow y = -\frac{5}{2}$ and

$x = 5$; $T\left(5, -\frac{5}{2}\right) = -\frac{37}{4}$ and $T(5, -3) = -9$

(iii) On $AB, T(x, y) = T(x, -3) = x^2 - 9x + 11$ on $0 \le x \le 5; T'(x, -3) = 2x - 9 = 0 \Rightarrow x = \frac{9}{2}$ and $y = -3$;

$T\left(\frac{9}{2}, -3\right) = -\frac{37}{4}$ and $T(0, -3) = 11$

(iv) On $AO, T(x, y) = T(0, y) = y^2 + 2$ on $-3 \le y \le 0; T'(0, y) = 2y = 0 \Rightarrow y = 0$ and $x = 0,$ but $(0, 0)$ is not an interior point of AO

(v) For interior points of the rectangular region, $T_x(x, y) = 2x + y - 6 = 0$ and $T_y(x, y) = x + 2y = 0 \Rightarrow x = 4$ and $y = -2,$ an interior critical point with $T(4, -2) = -10.$ Therefore the absolute maximum is 11 at $(0, -3)$ and the absolute minimum is -10 at $(4, -2).$

37. (i) On $AB, f(x, y) = f(1, y) = 3\cos y$ on

$-\frac{\pi}{4} \le y \le \frac{\pi}{4}; f'(1, y) = -3\sin y = 0 \Rightarrow y = 0$ and

$x = 1; f(1, 0) = 3, f\left(1, -\frac{\pi}{4}\right) = \frac{3\sqrt{2}}{2},$ and

$f\left(1, \frac{\pi}{4}\right) = \frac{3\sqrt{2}}{2}$

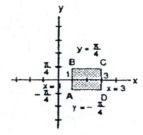

(ii) On $CD, f(x, y) = f(3, y) = 3\cos y$ on $-\frac{\pi}{4} \le y \le \frac{\pi}{4}; f'(3, y) = -3\sin y = 0 \Rightarrow y = 0$ and $x = 3;$

$f(3, 0) = 3, f\left(3, -\frac{\pi}{4}\right) = \frac{3\sqrt{2}}{2}$ and $f\left(3, \frac{\pi}{4}\right) = \frac{3\sqrt{2}}{2}$

(iii) On $BC, f(x, y) = f\left(x, \frac{\pi}{4}\right) = \frac{\sqrt{2}}{2}\left(4x - x^2\right)$ on $1 \le x \le 3; f'\left(x, \frac{\pi}{4}\right) = \sqrt{2}(2 - x) = 0 \Rightarrow x = 2$ and $y = \frac{\pi}{4};$

$f\left(2, \frac{\pi}{4}\right) = 2\sqrt{2}, f\left(1, \frac{\pi}{4}\right) = \frac{3\sqrt{2}}{2},$ and $f\left(3, \frac{\pi}{4}\right) = \frac{3\sqrt{2}}{2}$

(iv) On $AD, f(x, y) = f\left(x, -\frac{\pi}{4}\right) = \frac{\sqrt{2}}{2}\left(4x - x^2\right)$ on $1 \le x \le 3; f'\left(x, -\frac{\pi}{4}\right) = \sqrt{2}(2 - x) = 0 \Rightarrow x = 2$ and

$y = -\frac{\pi}{4}; f\left(2, -\frac{\pi}{4}\right) = 2\sqrt{2}, f\left(1, -\frac{\pi}{4}\right) = \frac{3\sqrt{2}}{2},$ and $f\left(3, -\frac{\pi}{4}\right) = \frac{3\sqrt{2}}{2}$

(v) For interior points of the region, $f_x(x, y) = (4 - 2x)\cos y = 0$ and $f_y(x, y) = -\left(4x - x^2\right)\sin y = 0$

$\Rightarrow x = 2$ and $y = 0,$ which is an interior critical point with $f(2, 0) = 4.$ Therefore the absolute maximum

is 4 at $(2, 0)$ and the absolute minimum is $\frac{3\sqrt{2}}{2}$ at $\left(3, -\frac{\pi}{4}\right), \left(3, \frac{\pi}{4}\right), \left(1, -\frac{\pi}{4}\right),$ and $\left(1, \frac{\pi}{4}\right).$

39. Let $F(a, b) = \int_a^b \left(6 - x - x^2\right) dx$ where $a \le b.$ The boundary of the domain of F is the line $a = b$ in the ab-plane, and $F(a, a) = 0,$ so F is identically 0 on the boundary of its domain. For interior critical points we have:

$\frac{\partial F}{\partial a} = -\left(6 - a - a^2\right) = 0 \Rightarrow a = -3, 2$ and $\frac{\partial F}{\partial b} = \left(6 - b - b^2\right) = 0 \Rightarrow b = -3, 2.$ Since $a \le b,$ there is only one

interior critical point $(-3, 2)$ and $F(-3, 2) = \int_{-3}^2 \left(6 - x - x^2\right) dx$ gives the area under the parabola

$y = 6 - x - x^2$ that is above the x axis. Therefore, $a = -3$ and $b = 2.$

41. $T_x(x, y) = 2x - 1 = 0$ and $T_y(x, y) = 4y = 0 \Rightarrow x = \frac{1}{2}$ and $y = 0$ with $T\left(\frac{1}{2}, 0\right) = -\frac{1}{4};$ on the boundary

$x^2 + y^2 = 1: T(x, y) = -x^2 - x + 2$ for $-1 \le x \le 1 \Rightarrow T'(x, y) = -2x - 1 = 0 \Rightarrow x = -\frac{1}{2}$ and $y = \pm\frac{\sqrt{3}}{2};$

$T\left(-\frac{1}{2}, \frac{\sqrt{3}}{2}\right) = \frac{9}{4}, T\left(-\frac{1}{2}, -\frac{\sqrt{3}}{2}\right) = \frac{9}{4}, T(-1, 0) = 2,$ and $T(1, 0) = 0 \Rightarrow$ the hottest is $\left(2\frac{1}{4}\right)^\circ$ at $\left(-\frac{1}{2}, -\frac{\sqrt{3}}{2}\right)$ and

$\left(-\frac{1}{2}, \frac{\sqrt{3}}{2}\right);$ the coldest is $\left(-\frac{1}{4}\right)^\circ$ at $\left(\frac{1}{2}, 0\right).$

43. (a) $f_x(x, y) = 2x - 4y = 0$ and $f_y(x, y) = 2y - 4x = 0 \Rightarrow x = 0$ and $y = 0; f_{xx}(0, 0) = 2, f_{yy}(0, 0) = 2,$

    $f_{xy}(0, 0) = -4 \Rightarrow f_{xx}f_{yy} - f_{xy}^2 = -12 < 0 \Rightarrow$ saddle point at $(0, 0)$

    (b) $f_x(x, y) = 2x - 2 = 0$ and $f_y(x, y) = 2y - 4 = 0 \Rightarrow x = 1$ and $y = 2; f_{xx}(1, 2) = 2; f_{yy}(1, 2) = 2,$

    $f_{xy}(1, 2) = 0 \Rightarrow f_{xx}f_{yy} - f_{xy}^2 = 4 > 0$ and $f_{xx} > 0 \Rightarrow$ local minimum at $(1, 2)$

    (c) $f_x(x, y) = 9x^2 - 9 = 0$ and $f_y(x, y) = 2y + 4 = 0 \Rightarrow x = \pm 1$ and $y = -2; f_{xx}(1, -2) = 18x|_{(1, -2)} = 18,$

    $f_{yy}(1, -2) = 2, f_{xy}(1, -2) = 0 \Rightarrow f_{xx}f_{yy} - f_{xy}^2 = 36 > 0$ and $f_{xx} > 0 \Rightarrow$ local minimum at $(1, -2)$;

    $f_{xx}(-1, -2) = -18, f_{yy}(-1, -2) = 2, f_{xy}(-1, -2) = 0 \Rightarrow f_{xx}f_{yy} - f_{xy}^2 = -36 < 0 \Rightarrow$ saddle point at $(-1, -2)$

45. If $k = 0$, then $f(x, y) = x^2 + y^2 \Rightarrow f_x(x, y) = 2x = 0$ and $f_y(x, y) = 2y = 0 \Rightarrow x = 0$ and $y = 0 \Rightarrow (0, 0)$ is

    the only critical point. If $k \neq 0, f_x(x, y) = 2x + ky = 0 \Rightarrow y = -\frac{2}{k}x; \; f_y(x, y) = kx + 2y = 0$

    $\Rightarrow kx + 2\left(-\frac{2}{k}x\right) = 0 \Rightarrow kx - \frac{4x}{k} = 0 \Rightarrow \left(k - \frac{4}{k}\right)x = 0 \Rightarrow x = 0$ or $k = \pm 2 \Rightarrow y = \left(-\frac{2}{k}\right)(0) = 0$ or $y = \pm x$; in any

    case $(0, 0)$ is a critical point.

47. No; for example $f(x, y) = xy$ has a saddle point at $(a, b) = (0, 0)$ where $f_x = f_y = 0$.

49. We want the point on $z = 10 - x^2 - y^2$ where the tangent plane is parallel to the plane $x + 2y + 3z = 0$. To find

    a normal vector to $z = 10 - x^2 - y^2$ let $w = z + x^2 + y^2 - 10$. Then $\nabla w = 2x\mathbf{i} + 2y\mathbf{j} + \mathbf{k}$ is normal to

    $z = 10 - x^2 - y^2$ at (x, y). The vector ∇w is parallel to $\mathbf{i} + 2\mathbf{j} + 3\mathbf{k}$ which is normal to the plane

    $x + 2y + 3z = 0$ if $6x\mathbf{i} + 6y\mathbf{j} + 3\mathbf{k} = \mathbf{i} + 2\mathbf{j} + 3\mathbf{k}$ or $x = \frac{1}{6}$ and $y = \frac{1}{3}$. Thus the point is $\left(\frac{1}{6}, \frac{1}{3}, 10 - \frac{1}{36} - \frac{1}{9}\right)$ or

    $\left(\frac{1}{6}, \frac{1}{3}, \frac{355}{36}\right)$.

51. $d(x, y, z) = \sqrt{(x-0)^2 + (y-0)^2 + (z-0)^2} \Rightarrow$ we can minimize $d(x, y, z)$ by minimizing

    $D(x, y, z) = x^2 + y^2 + z^2; \; 3x + 2y + z = 6 \Rightarrow z = 6 - 3x - 2y \Rightarrow D(x, y) = x^2 + y^2 + \left(6 - 3x - 2y^2\right)$

    $\Rightarrow D_x(x, y) = 2x - 6(6 - 3x - 2y) = 0$ and $D_y(x, y) = 2y - 4(6 - 3x - 2y) = 0 \Rightarrow$ critical point is

    $\left(\frac{9}{7}, \frac{6}{7}\right) \Rightarrow z = \frac{3}{7}; \; D_{xx}\left(\frac{9}{7}, \frac{6}{7}\right) = 20, D_{yy}\left(\frac{9}{7}, \frac{6}{7}\right) = 10, D_{xy}\left(\frac{9}{7}, \frac{6}{7}\right) = 12 \Rightarrow D_{xx}D_{yy} - D_{xy}^2 = 56 > 0$ and $D_{xx} > 0$

    $\Rightarrow$ local minimum of $d\left(\frac{9}{7}, \frac{6}{7}, \frac{3}{7}\right) = \frac{3\sqrt{14}}{7}$

53. $s(x, y, z) = x^2 + y^2 + z^2; \; x + y + z = 9 \Rightarrow z = 9 - x - y \Rightarrow s(x, y) = x^2 + y^2 + (9 - x - y)^2$

    $\Rightarrow s_x(x, y) = 2x - 2(9 - x - y) = 0$ and $s_y(x, y) = 2y - 2(9 - x - y) = 0 \Rightarrow$ critical point is $(3, 3) \Rightarrow z = 3;$

    $s_{xx}(3, 3) = 4, s_{yy}(3, 3) = 4, s_{xy}(3, 3) = 2 \Rightarrow s_{xx}s_{yy} - s_{xy}^2 = 12 > 0$ and $s_{xx} > 0 \Rightarrow$ local minimum of

    $s(3, 3, 3) = 27$

55. $s(x, y, z) = xy + yz + xz; \; x + y + z = 6 \Rightarrow z = 6 - x - y \Rightarrow s(x, y) = xy + y(6 - x - y) + x(6 - x - y)$

    $= 6x + 6y - xy - x^2 - y^2 \Rightarrow s_x(x, y) = 6 - 2x - y = 0$ and $s_y(x, y) = 6 - x - 2y = 0 \Rightarrow$ critical point is $(2, 2)$

    $\Rightarrow z = 2; \; s_{xx}(2, 2) = -2, s_{yy}(2, 2) = -2, s_{xy}(2, 2) = -1 \Rightarrow s_{xx}s_{yy} - s_{xy}^2 = 3 > 0$ and $s_{xx} < 0 \Rightarrow$ local maximum

    of $s(2, 2, 2) = 12$

57. $V(x, y, z) = (2x)(2y)(2z) = 8xyz;\ x^2 + y^2 + z^2 = 4 \Rightarrow z = \sqrt{4 - x^2 - y^2} \Rightarrow V(x, y) = 8xy\sqrt{4 - x^2 - y^2},\ x \geq 0$

and $y \geq 0 \Rightarrow V_x(x, y) = \dfrac{32y - 16x^2 y - 8y^3}{\sqrt{4 - x^2 - y^2}} = 0$ and $V_y(x, y) = \dfrac{32x - 16xy^2 - 8x^3}{\sqrt{4 - x^2 - y^2}} = 0 \Rightarrow$ critical points are $(0, 0)$,

$\left(\frac{2}{\sqrt{3}}, \frac{2}{\sqrt{3}}\right), \left(\frac{2}{\sqrt{3}}, -\frac{2}{\sqrt{3}}\right), \left(-\frac{2}{\sqrt{3}}, \frac{2}{\sqrt{3}}\right)$, and $\left(-\frac{2}{\sqrt{3}}, -\frac{2}{\sqrt{3}}\right)$. Only $(0, 0)$ and $\left(\frac{2}{\sqrt{3}}, \frac{2}{\sqrt{3}}\right)$ satisfy $x \geq 0$ and $y \geq 0$

$V(0, 0) = 0$ and $V\left(\frac{2}{\sqrt{3}}, \frac{2}{\sqrt{3}}\right) = \frac{64}{3\sqrt{3}}$; On $x = 0, 0 \leq y \leq 2 \Rightarrow V(0, y) = 8(0)y\sqrt{4 - 0^2 - y^2} = 0$, no critical points,

$V(0, 0) = 0, V(0, 2) = 0$; On $y = 0, 0 \leq x \leq 2 \Rightarrow V(x, 0) = 8x(0)\sqrt{4 - x^2 - 0^2} = 0$, no critical points,

$V(0, 0) = 0, V(0, 2) = 0$; On $y = \sqrt{4 - x^2}, 0 \leq x \leq 2 \Rightarrow V\left(x, \sqrt{4 - x^2}\right) = 8x\sqrt{4 - x^2}\sqrt{4 - x^2 - \left(\sqrt{4 - x^2}\right)^2} = 0$

no critical points, $V(0, 2) = 0, V(2, 0) = 0$. Thus, there is a maximum volume of $\frac{64}{3\sqrt{3}}$ if the box is

$\frac{2}{\sqrt{3}} \times \frac{2}{\sqrt{3}} \times \frac{2}{\sqrt{3}}$.

59. Let $x =$ height of the box, $y =$ width, and $z =$ length, cut out squares of length x from corner of the material See diagram at right. Fold along the dashed lines to form the box. From the diagram we see that the length of the material is $2x + y$ and the width is $2x + z$. Thus

$(2x + y)(2x + z) = 12 \Rightarrow z = \dfrac{2(6 - 2x^2 + xy)}{2x + y}$. Since

$V(x, y, z) = x\,y\,z \Rightarrow V(x, y) = \dfrac{2xy(6 - 2x^2 + xy)}{2x + y}$, where

$x > 0, y > 0$. $V_x(x, y) = \dfrac{4(3y^2 - 4x^3 y - 4x^2 y^2 - xy^3)}{(2x + y)^2} = 0$ and $V_y(x, y) = \dfrac{2(12x^2 - 4x^4 - 4x^3 y - x^2 y^2)}{(2x + y)^2} = 0 \Rightarrow$ critical points

are $\left(\sqrt{3}, 0\right), \left(-\sqrt{3}, 0\right), \left(\frac{1}{\sqrt{3}}, \frac{4}{\sqrt{3}}\right)$ and $\left(-\frac{1}{\sqrt{3}}, -\frac{4}{\sqrt{3}}\right)$. Only $\left(\sqrt{3}, 0\right)$ and $\left(\frac{1}{\sqrt{3}}, \frac{4}{\sqrt{3}}\right)$ satisfy $x > 0$ and $y > 0$.

For $\left(\sqrt{3}, 0\right): z = 0$; $V_{xx}\left(\sqrt{3}, 0\right) = 0, V_{yy}\left(\sqrt{3}, 0\right) = -2\sqrt{3}, V_{xy}\left(\sqrt{3}, 0\right) = -4\sqrt{3} \Rightarrow V_{xx}V_{yy} - V_{xy}^2 = -48 < 0$

$\Rightarrow$ saddle point. For $\left(\frac{1}{\sqrt{3}}, \frac{4}{\sqrt{3}}\right): z = \frac{4}{\sqrt{3}}$; $V_{xx}\left(\frac{1}{\sqrt{3}}, \frac{4}{\sqrt{3}}\right) = -\frac{80}{3\sqrt{3}}, V_{yy}\left(\frac{1}{\sqrt{3}}, \frac{4}{\sqrt{3}}\right) = -\frac{2}{3\sqrt{3}}, V_{xy}\left(\frac{1}{\sqrt{3}}, \frac{4}{\sqrt{3}}\right) = -\frac{4}{3\sqrt{3}}$

$\Rightarrow V_{xx}V_{yy} - V_{xy}^2 = \frac{16}{3} > 0$ and $V_{xx} < 0 \Rightarrow$ local maximum of $V\left(\frac{1}{\sqrt{3}}, \frac{4}{\sqrt{3}}, \frac{4}{\sqrt{3}}\right) = \frac{16}{3\sqrt{3}}$

61. (a) $\dfrac{df}{dt} = \dfrac{\partial f}{\partial x}\dfrac{dx}{dt} + \dfrac{\partial f}{\partial y}\dfrac{dy}{dt} = \dfrac{dx}{dt} + \dfrac{dy}{dt} = -2\sin t + 2\cos t = 0 \Rightarrow \cos t = \sin t \Rightarrow x = y$

   (i) On the semicircle $x^2 + y^2 = 4,\ y \geq 0$, we have $t = \frac{\pi}{4}$ and $x = y = \sqrt{2} \Rightarrow f\left(\sqrt{2}, \sqrt{2}\right) = 2\sqrt{2}$. At the endpoints, $f(-2, 0) = -2$ and $f(2, 0) = 2$. Therefore the absolute minimum is $f(-2, 0) = -2$ when $t = \pi$; the absolute maximum is $f\left(\sqrt{2}, \sqrt{2}\right) = 2\sqrt{2}$ when $t = \frac{\pi}{4}$.

   (ii) On the quartercircle $x^2 + y^2 = 4,\ x \geq 0$ and $y \geq 0$, the endpoints give $f(0, 2) = 2$ and $f(2, 0) = 2$. Therefore the absolute minimum is $f(2, 0) = 2$ and $f(0, 2) = 2$ when $t = 0, \frac{\pi}{2}$ respectively; the absolute maximum is $f\left(\sqrt{2}, \sqrt{2}\right) = 2\sqrt{2}$ when $t = \frac{\pi}{4}$.

   (b) $\dfrac{dg}{dt} = \dfrac{\partial g}{\partial x}\dfrac{dx}{dt} + \dfrac{\partial g}{\partial y}\dfrac{dy}{dt} = y\dfrac{dx}{dt} + x\dfrac{dy}{dt} = -4\sin^2 t + 4\cos^2 t = 0 \Rightarrow \cos t = \pm\sin t \Rightarrow x = \pm y$.

(i)   On the semicircle $x^2 + y^2 = 4$, $y \geq 0$, we obtain $x = y = \sqrt{2}$ at $t = \frac{\pi}{4}$ and $x = -\sqrt{2}$, $y = \sqrt{2}$ at

$t = \frac{3\pi}{4}$. Then $g\left(\sqrt{2}, \sqrt{2}\right) = 2$ and $g\left(-\sqrt{2}, \sqrt{2}\right) = -2$. At the endpoints, $g(-2, 0) = g(2, 0) = 0$.

Therefore the absolute minimum is $g\left(-\sqrt{2}, \sqrt{2}\right) = -2$ when $t = \frac{3\pi}{4}$; the absolute maximum is

$g\left(\sqrt{2}, \sqrt{2}\right) = 2$ when $t = \frac{\pi}{4}$.

(ii) On the quartercircle $x^2 + y^2 = 4$, $x \geq 0$ and $y \geq 0$, the endpoints give $g(0, 2) = 0$ and $g(2, 0) = 0$.

Therefore the absolute minimum is $g(2, 0) = 0$ and $g(0, 2) = 0$ when $t = 0, \frac{\pi}{2}$ respectively; the

absolute maximum is $g\left(\sqrt{2}, \sqrt{2}\right) = 2$ when $t = \frac{\pi}{4}$.

(c) $\frac{dh}{dt} = \frac{\partial h}{\partial x}\frac{dx}{dt} + \frac{\partial h}{\partial y}\frac{dy}{dt} = 4x\frac{dx}{dt} + 2y\frac{dy}{dt} = (8\cos t)(-2\sin t) + (4\sin t)(2\cos t) = -8\cos t \sin t = 0$

$\Rightarrow t = 0, \frac{\pi}{2}, \pi$ yielding the points $(2, 0), (0, 2)$ for $0 \leq t \leq \pi$.

(i)   On the semicircle $x^2 + y^2 = 4$, $y \geq 0$ we have $h(2, 0) = 8, h(0, 2) = 4$, and $h(-2, 0) = 8$. Therefore,

the absolute minimum is $h(0, 2) = 4$ when $t = \frac{\pi}{2}$; the absolute maximum is $h(2, 0) = 8$ and

$h(-2, 0) = 8$ when $t = 0, \pi$ respectively.

(ii) On the quartercircle $x^2 + y^2 = 4$, $x \geq 0$ and $y \geq 0$ the absolute minimum is $h(0, 2) = 4$ when $t = \frac{\pi}{2}$;

the absolute maximum is $h(2, 0) = 8$ when $t = 0$.

63. $\frac{df}{dt} = \frac{\partial f}{\partial x}\frac{dx}{dt} + \frac{\partial f}{\partial y}\frac{dy}{dt} = y\frac{dx}{dt} + x\frac{dy}{dt}$

(i) $x = 2t$ and $y = t + 1 \Rightarrow \frac{df}{dt} = (t+1)(2) + (2t)(1) = 4t + 2 = 0 \Rightarrow t = -\frac{1}{2} \Rightarrow x = -1$ and $y = \frac{1}{2}$ with

$f\left(-1, \frac{1}{2}\right) = -\frac{1}{2}$. The absolute minimum is $f\left(-1, \frac{1}{2}\right) = -\frac{1}{2}$ when $t = -\frac{1}{2}$; there is no absolute maximum.

(ii) For the endpoints: $t = -1 \Rightarrow x = -2$ and $y = 0$ with $f(-2, 0) = 0; t = 0 \Rightarrow x = 0$ and $y = 1$ with

$f(0, 1) = 0$. The absolute minimum is $f\left(-1, \frac{1}{2}\right) = -\frac{1}{2}$ when $t = -\frac{1}{2}$; the absolute maximum is

$f(0, 1) = 0$ and $f(-2, 0) = 0$ when $t = -1, 0$ respectively.

(iii) There are no interior critical points. For the endpoints: $t = 0 \Rightarrow x = 0$ and $y = 1$ with $f(0, 1) = 0$;

$t = 1 \Rightarrow x = 2$ and $y = 2$ with $f(2, 2) = 4$. The absolute minimum is $f(0, 1) = 0$ when $t = 0$; the absolute

maximum is $f(2, 2) = 4$ when $t = 1$.

65. $w = (mx_1 + b - y_1)^2 + (mx_2 + b - y_2)^2 + \cdots + (mx_n + b - y_n)^2$

$\Rightarrow \frac{\partial w}{\partial m} = 2(mx_1 + b - y_1)(x_1) + 2(mx_2 + b - y_2)(x_2) + \cdots + 2(mx_n + b - y_n)(x_n) = 0$

$\Rightarrow 2\left[(mx_1 + b - y_1)(x_1) + (mx_2 + b - y_2)(x_2) + \cdots + (mx_n + b - y_n)(x_n)\right] = 0$

$\Rightarrow mx_1^2 + bx_1 - x_1 y_1 + mx_2^2 + bx_2 - x_2 y_2 + \cdots + mx_n^2 + bx_n - x_n y_n = 0$

$\Rightarrow m\left(x_1^2 + x_2^2 + \cdots + x_n^2\right) + b(x_1 + x_2 + \cdots + x_n) - (x_1 y_1 + x_2 y_2 + \cdots + x_n y_n) = m\sum_{k=1}^{n}\left(x_k^2\right) + b\sum_{k=1}^{n}x_k - \sum_{k=1}^{n}(x_k y_k) = 0$

$\Rightarrow \frac{\partial w}{\partial b} = 2(mx_1 + b - y_1)(1) + 2(mx_2 + b - y_2)(1) + \cdots + 2(mx_n + b - y_n)(1) = 0$

$\Rightarrow 2\left[(mx_1 + b - y_1) + (mx_2 + b - y_2) + \cdots + (mx_n + b - y_n)\right] = 0$

$\Rightarrow mx_1 + b - y_1 + mx_2 + b - y_2 + \cdots + mx_n + b - y_n = 0$

$$\Rightarrow m\left(x_1 + x_2 + \cdots + x_n\right) + \left(b + b + \cdots + b\right) - \left(y_1 + y_2 + \cdots + y_n\right) = m\sum_{k=1}^{n} x_k + bn - \sum_{k=1}^{n} y_k = 0$$

$$\Rightarrow b = \frac{1}{n}\left(\sum_{k=1}^{n} y_k - m\sum_{k=1}^{n} x_k\right). \text{ Substituting for } b \text{ in the equation obtained for } \frac{\partial w}{\partial m} \text{ we get}$$

$$m\sum_{k=1}^{n}\left(x_k^2\right) + \frac{1}{n}\left(\sum_{k=1}^{n} y_k - m\sum_{k=1}^{n} x_k\right)\sum_{k=1}^{n} x_k - \sum_{k=1}^{n}\left(x_k y_k\right) = 0. \text{ Multiply both sides by } n \text{ to obtain}$$

$$mn\sum_{k=1}^{n}\left(x_k^2\right) + \left(\sum_{k=1}^{n} y_k - m\sum_{k=1}^{n} x_k\right)\sum_{k=1}^{n} x_k - n\sum_{k=1}^{n}\left(x_k y_k\right) = mn\sum_{k=1}^{n}\left(x_k^2\right) + \left(\sum_{k=1}^{n} x_k\right)\left(\sum_{k=1}^{n} y_k\right) - m\left(\sum_{k=1}^{n} x_k\right) - n\sum_{k=1}^{n}\left(x_k y_k\right) = 0$$

$$\Rightarrow m\left[n\sum_{k=1}^{n}\left(x_k^2\right) - \left(\sum_{k=1}^{n} x_k\right)^2\right] = n\sum_{k=1}^{n}\left(x_k y_k\right) - \left(\sum_{k=1}^{n} x_k\right)\left(\sum_{k=1}^{n} y_k\right)$$

$$\Rightarrow m = \frac{n\sum_{k=1}^{n}\left(x_k y_k\right) - \left(\sum_{k=1}^{n} x_k\right)\left(\sum_{k=1}^{n} y_k\right)}{n\sum_{k=1}^{n}\left(x_k^2\right) - \left(\sum_{k=1}^{n} x_k\right)^2} = \frac{\left(\sum_{k=1}^{n} x_k\right)\left(\sum_{k=1}^{n} y_k\right) - n\sum_{k=1}^{n}\left(x_k y_k\right)}{\left(\sum_{k=1}^{n} x_k\right)^2 - n\sum_{k=1}^{n}\left(x_k^2\right)}$$

To show that these values for m and b minimize the sum of the squares of the distances, use second derivative

test. $\dfrac{\partial^2 w}{\partial m^2} = 2x_1^2 + 2x_2^2 + \cdots + 2x_n^2 = 2\sum_{k=1}^{n}\left(x_k^2\right);$ $\dfrac{\partial^2 w}{\partial m\, \partial b} = 2x_1 + 2x_2 + \cdots + 2x_n = 2\sum_{k=1}^{n} x_k;$ $\dfrac{\partial^2 w}{\partial b^2} = 2 + 2 + \cdots + 2 = 2n$

The discriminant is: $\left(\dfrac{\partial^2 w}{\partial m^2}\right)\left(\dfrac{\partial^2 w}{\partial b^2}\right) - \left(\dfrac{\partial^2 w}{\partial m\, \partial b}\right)^2 = \left[2\sum_{k=1}^{n}\left(x_k^2\right)\right](2n) - \left[2\sum_{k=1}^{n} x_k\right]^2 = 4\left[n\sum_{k=1}^{n}\left(x_k^2\right) - \left(\sum_{k=1}^{n} x_k\right)^2\right].$

Now, $n\sum_{k=1}^{n}\left(x_k^2\right) - \left(\sum_{k=1}^{n} x_k\right)^2 = n\left(x_1^2 + x_2^2 + \cdots + x_n^2\right) - \left(x_1 + x_2 + \cdots + x_n\right)\left(x_1 + x_2 + \cdots + x_n\right)$

$= nx_1^2 + nx_2^2 + \cdots + nx_n^2 - x_1^2 - x_1 x_2 - \cdots - x_1 x_n - x_2 x_1 - x_2^2 - \cdots - x_2 x_n - \cdots - x_n x_1 - x_n x_2 - \cdots - x_n^2$

$= (n-1)x_1^2 + (n-1)x_2^2 + \cdots + (n-1)x_n^2 - 2x_1 x_2 - 2x_1 x_3 - \cdots - 2x_1 x_n - 2x_2 x_3 - \cdots - 2x_2 x_n - \cdots - 2x_{n-1}x_n$

$= \left(x_1^2 - 2x_1 x_2 + x_2^2\right) + \left(x_1^2 - 2x_1 x_3 + x_3^2\right) + \cdots + \left(x_1^2 - 2x_1 x_n + x_n^2\right) + \left(x_2^2 - 2x_2 x_3 + x_3^2\right) + \cdots$

$\qquad + \left(x_2^2 - 2x_2 x_n + x_n^2\right) + \cdots + \left(x_{n-1}^2 - 2x_{n-1}x_n + x_n^2\right)$

$= \left(x_1 - x_2\right)^2 + \left(x_1 - x_3\right)^2 + \cdots + \left(x_1 - x_n\right)^2 + \left(x_2 - x_3\right)^2 + \cdots + \left(x_2 - x_n\right)^2 + \cdots + \left(x_{n-1} - x_n\right)^2 \geq 0.$

Thus we have: $\left(\dfrac{\partial^2 w}{\partial m^2}\right)\left(\dfrac{\partial^2 w}{\partial b^2}\right) - \left(\dfrac{\partial^2 w}{\partial m\, \partial b}\right)^2 = 4\left[n\sum_{k=1}^{n}\left(x_k^2\right) - \left(\sum_{k=1}^{n} x_k\right)^2\right] \geq 4(0) = 0.$ If $x_1 = x_2 = \cdots = x_n$ then

$\left(\dfrac{\partial^2 w}{\partial m^2}\right)\left(\dfrac{\partial^2 w}{\partial b^2}\right) - \left(\dfrac{\partial^2 w}{\partial m\, \partial b}\right)^2 = 0.$ Also, $\dfrac{\partial^2 w}{\partial m^2} = 2\sum_{k=1}^{n}\left(x_k^2\right) \geq 0.$ If $x_1 = x_2 = \cdots = x_n = 0$, then $\dfrac{\partial^2 w}{\partial m^2} = 0.$ Provided that at

least one x_i is nonzero and different from the rest of x_j, $j \neq i$, then $\left(\dfrac{\partial^2 w}{\partial m^2}\right)\left(\dfrac{\partial^2 w}{\partial b^2}\right) - \left(\dfrac{\partial^2 w}{\partial m\, \partial b}\right)^2 > 0$ and

$\dfrac{\partial^2 w}{\partial m^2} > 0 \Rightarrow$ the values given above for m and b minimize w.

69-73. Example CAS commands:

Maple:

```
f := (x,y) -> x^2+y^3-3*x*y;

x0,x1 := -5,5;

y0,y1 := -5,5;

plot3d( f(x,y), x=x0..x1, y=y0..y1, axes=boxed, shading=zhue, title="#69(a) (Section 14.7)" );

plot3d( f(x,y), x=x0..x1, y=y0..y1, grid=[40,40], axes=boxed, shading=zhue, style=patchcontour,
    title="#69(b) (Section 14.7)" );

fx := D[1](f);                              # (c)

fy := D[2](f);

crit_pts := solve( {fx(x,y) = 0, fy(x,y)=0}, {x,y} );

fxx := D[1](fx);                            # (d)

fxy := D[2](fx);

fyy := D[2](fy);

discr := unapply( fxx(x,y)*fyy(x,y)-fxy(x,y)^2, (x,y) );

for CP in { crit_pts} do                    # (e)

eval( [x,y,fxx(x,y),discr(x,y)], CP );

end do;

# (0,0) is a saddle point

# ( 9/4,  3/2) is a local minimum
```

Mathematica: (assigned functions and bounds will vary)

```
Clear[x,y,f]

f[x_,y_]:= x^2 + y^3 - 3x y

xmin=-5; xmax=5; ymin=-5; ymax=5;

Plot3D[f[x,y], {x, xmin, xmax}, {y, ymin, ymax}, AxesLabel → {x, y, z}]

ContourPlot[f[x,y], {x, xmin, xmax}, {y, ymin, ymax}, ContourShading → False, Contours → 40]

fx= D[f[x,y], x];

fy= D[f[x,y], y];

critical=Solve[{fx==0, fy==0},{x, y}]

fxx= D[fx, x];

fxy= D[fx, y];

fyy= D[fy, y];

discriminant= fxx fyy - fxy^2

{{x, y}, f[x, y], discriminant, fxx} /.critical
```

13.8 LAGRANGE MULTIPLIERS

1. $\nabla f = y\mathbf{i} + x\mathbf{j}$ and $\nabla g = 2x\mathbf{i} + 4y\mathbf{j}$ so that $\nabla f = \lambda \nabla g \Rightarrow y\mathbf{i} + x\mathbf{j} = \lambda(2x\mathbf{i} + 4y\mathbf{j}) \Rightarrow y = 2x\lambda$ and $x = 4y\lambda$
   $\Rightarrow x = 8x\lambda^2 \Rightarrow \lambda = \pm\frac{\sqrt{2}}{4}$ or $x = 0$.

   CASE 1: If $x = 0$, then $y = 0$. But $(0,0)$ is not on the ellipse so $x \neq 0$.

   CASE 2: $x \neq 0 \Rightarrow \lambda = \pm\frac{\sqrt{2}}{4} \Rightarrow x = \pm\sqrt{2}\,y \Rightarrow \left(\pm\sqrt{2}\,y\right)^2 + 2y^2 = 1 \Rightarrow y = \pm\frac{1}{2}$.

   Therefore f takes on its extreme values at $\left(\pm\frac{\sqrt{2}}{2}, \frac{1}{2}\right)$ and $\left(\pm\frac{\sqrt{2}}{2}, -\frac{1}{2}\right)$. The extreme values of f on the ellipse
   are $\pm\frac{\sqrt{2}}{2}$.

3. $\nabla f = -2x\mathbf{i} - 2y\mathbf{j}$ and $\nabla g = \mathbf{i} + 3\mathbf{j}$ so that $\nabla f = \lambda \nabla g \Rightarrow -2x\mathbf{i} - 2y\mathbf{j} = \lambda(\mathbf{i} + 3\mathbf{j}) \Rightarrow x = -\frac{\lambda}{2}$ and $y = -\frac{3\lambda}{2}$
   $\Rightarrow \left(-\frac{\lambda}{2}\right) + 3\left(-\frac{3\lambda}{2}\right) = 10 \Rightarrow \lambda = -2 \Rightarrow x = 1$ and $y = 3 \Rightarrow f$ takes on its extreme value at $(1, 3)$ on the line. The
   extreme value is $f(1, 3) = 49 - 1 - 9 = 39$.

5. We optimize $f(x, y) = x^2 + y^2$, the square of the distance to the origin, subject to the constraint
   $g(x, y) = xy^2 - 54 = 0$. Thus $\nabla f = 2x\mathbf{i} + 2y\mathbf{j}$ and $\nabla g = y^2\mathbf{i} + 2xy\mathbf{j}$ so that $\nabla f = \lambda \nabla g \Rightarrow 2x\mathbf{i} + 2y\mathbf{j}$
   $= \lambda\left(y^2\mathbf{i} + 2xy\mathbf{j}\right) \Rightarrow 2x = \lambda y^2$ and $2y = 2\lambda xy$.

   CASE 1: If $y = 0$, then $x = 0$. But $(0, 0)$ does not satisfy the constraint $xy^2 = 54$ so $y \neq 0$.

   CASE 2: If $y \neq 0$, then $2 = 2\lambda x \Rightarrow x = \frac{1}{\lambda} \Rightarrow 2\left(\frac{1}{\lambda}\right) = \lambda y^2 \Rightarrow y^2 = \frac{2}{\lambda^2}$. Then $xy^2 = 54 \Rightarrow \left(\frac{1}{\lambda}\right)\left(\frac{2}{\lambda^2}\right) = 54$
   $\Rightarrow \lambda^3 = \frac{1}{27} \Rightarrow \lambda = \frac{1}{3} \Rightarrow x = 3$ and $y^2 = 18 \Rightarrow x = 3$ and $y = \pm 3\sqrt{2}$.

   Therefore $\left(3, \pm 3\sqrt{2}\right)$ are the points on the curve $xy^2 = 54$ nearest the origin (since $xy^2 = 54$ has points
   increasingly far away as y gets close to 0, no points are farthest away).

7. (a) $\nabla f = \mathbf{i} + \mathbf{j}$ and $\nabla g = y\mathbf{i} + x\mathbf{j}$ so that $\nabla f = \lambda \nabla g \Rightarrow \mathbf{i} + \mathbf{j} = \lambda(y\mathbf{i} + x\mathbf{j}) \Rightarrow 1 = \lambda y$ and $1 = \lambda x \Rightarrow y = \frac{1}{\lambda}$ and
   $x = \frac{1}{\lambda} \Rightarrow \frac{1}{\lambda^2} = 16 \Rightarrow \lambda = \pm\frac{1}{4}$. Use $\lambda = \frac{1}{4}$ since $x > 0$ and $y > 0$. Then $x = 4$ and $y = 4 \Rightarrow$ the minimum
   value is 8 at the point $(4, 4)$. Now, $xy = 16$, $x > 0$, $y > 0$ is a branch of a hyperbola in the first quadrant
   with the x- and y-axes as asymptotes. The equations $x + y = c$ give a family of parallel lines with
   $m = -1$. As these lines move away from the origin, the number c increases. Thus the minimum value of c
   occurs where $x + y = c$ is tangent to the hyperbola's branch.

   (b) $\nabla f = y\mathbf{i} + x\mathbf{j}$ and $\nabla g = \mathbf{i} + \mathbf{j}$ so that s $\nabla f = \lambda \nabla g \Rightarrow y\mathbf{i} + x\mathbf{j} = \lambda(\mathbf{i} + \mathbf{j}) \Rightarrow y = \lambda = x \Rightarrow y + y = 16 \Rightarrow y = 8$
   $\Rightarrow x = 8 \Rightarrow f(8, 8) = 64$ is the maximum value. The equations $xy = c$ ($x > 0$ and $y > 0$ or $x < 0$ and
   $y < 0$ to get a maximum value) give a family of hyperbolas in the first and third quadrants with the x-
   and y-axes as asymptotes. The maximum value of c occurs where the hyperbola $xy = c$ is tangent to the
   line $x + y = 16$.

9. $V = \pi r^2 h \Rightarrow 16\pi = \pi r^2 h \Rightarrow 16 = r^2 h \Rightarrow g(r, h) = r^2 h - 16;$ $S = 2\pi rh + 2\pi r^2 \Rightarrow \nabla S = (2\pi h + 4\pi r)\mathbf{i} + 2\pi r\mathbf{j}$ and

$\nabla g = 2rh\mathbf{i} + r^2\mathbf{j}$ so that $\nabla S = \lambda \nabla g \Rightarrow (2\pi rh + 4\pi r)\mathbf{i} + 2\pi r\mathbf{j} = \lambda\left(2rh\mathbf{i} + r^2\mathbf{j}\right) \Rightarrow 2\pi rh + 4\pi r = 2rh\lambda$ and

$2\pi r = \lambda r^2 \Rightarrow r = 0$ or $\lambda = \frac{2\pi}{r}$. But $r = 0$ gives no physical can, so $r \neq 0 \Rightarrow \lambda = \frac{2\pi}{r} \Rightarrow 2\pi h + 4\pi r = 2rh\left(\frac{2\pi}{r}\right)$

$\Rightarrow 2r = h \Rightarrow 16 = r^2(2r) \Rightarrow r = 2 \Rightarrow h = 4;$ thus $r = 2$ cm and $h = 4$ cm give the only extreme surface area of

24π cm^2. Since $r = 4$ cm and $h = 1$ cm $\Rightarrow V = 16\pi$ cm^3 and $S = 40\pi$ cm^2, which is a larger surface area,

then 24π cm^2 must be the minimum surface area.

11. $A = (2x)(2y) = 4xy$ subject to $g(x, y) = \frac{x^2}{16} + \frac{y^2}{9} - 1 = 0;$ $\nabla A = 4y\mathbf{i} + 4x\mathbf{j}$ and $\nabla g = \frac{x}{8}\mathbf{i} + \frac{2y}{9}\mathbf{j}$ so that

$\nabla A = \lambda \nabla g \Rightarrow 4y\mathbf{i} + 4x\mathbf{j} = \lambda\left(\frac{x}{8}\mathbf{i} + \frac{2y}{9}\mathbf{j}\right) \Rightarrow 4y = \left(\frac{x}{8}\right)\lambda$ and $4x = \left(\frac{2y}{9}\right)\lambda \Rightarrow \lambda = \frac{32y}{x}$ and $4x = \left(\frac{2y}{9}\right)\left(\frac{32y}{x}\right)$

$\Rightarrow y = \pm\frac{3}{4}x \Rightarrow \frac{x^2}{16} + \frac{\left(\pm\frac{3}{4}x\right)^2}{9} = 1 \Rightarrow x^2 = 8 \Rightarrow x = \pm 2\sqrt{2}.$ We use $x = 2\sqrt{2}$ since x represents distance. Then

$y = \frac{3}{4}(2\sqrt{2}) = \frac{3\sqrt{2}}{2}$, so the length is $2x = 4\sqrt{2}$ and the width is $2y = 3\sqrt{2}$.

13. $\nabla f = 2x\mathbf{i} + 2y\mathbf{j}$ and $\nabla g = (2x - 2)\mathbf{i} + (2y - 4)\mathbf{j}$ so that $\nabla f = \lambda \nabla g = 2x\mathbf{i} + 2y\mathbf{j} = \lambda[(2x - 2)\mathbf{i} + (2y - 4)\mathbf{j}]$

$\Rightarrow 2x = \lambda(2x - 2)$ and $2y = \lambda(2y - 4) \Rightarrow x = \frac{\lambda}{\lambda - 1}$, and $y = \frac{2\lambda}{\lambda - 1}, \lambda \neq 1 \Rightarrow y = 2x$

$\Rightarrow x^2 - 2x + (2x)^2 - 4(2x) = 0 \Rightarrow x = 0$ and $y = 0$, or $x = 2$ and $y = 4$. $f(0, 0) = 0$ is the minimum value and

$f(2, 4) = 20$ is the maximum value. (Note that $\lambda = 1$ gives $2x = 2x - 2$ or $0 = -2$, which is impossible.)

15. $\nabla T = (8x - 4y)\mathbf{i} + (-4x + 2y)\mathbf{j}$ and $g(x, y) = x^2 + y^2 - 25 = 0 \Rightarrow \nabla g = 2x\mathbf{i} + 2y\mathbf{j}$ so that $\nabla T = \lambda \nabla g$

$\Rightarrow (8x - 4y)\mathbf{i} + (-4x + 2y)\mathbf{j} = \lambda(2x\mathbf{i} + 2y\mathbf{j}) \Rightarrow 8x - 4y = 2\lambda x$ and $-4x + 2y = 2\lambda y \Rightarrow y = \frac{-2x}{\lambda - 1}, \lambda \neq 1$

$\Rightarrow 8x - 4\left(\frac{-2x}{\lambda - 1}\right) = 2\lambda x \Rightarrow x = 0$, or $\lambda = 0$, or $\lambda = 5$.

CASE 1: $x = 0 \Rightarrow y = 0$; but $(0, 0)$ is not on $x^2 + y^2 = 25$ so $x \neq 0$.

CASE 2: $\lambda = 0 \Rightarrow y = 2x \Rightarrow x^2 + (2x)^2 = 25 \Rightarrow x = \pm\sqrt{5}$ and $y = 2x$.

CASE 3: $\lambda = 5 \Rightarrow y = -\frac{2x}{4} = -\frac{x}{2} \Rightarrow x^2 + \left(-\frac{x}{2}\right)^2 = 25 \Rightarrow x = \pm 2\sqrt{5} \Rightarrow x = 2\sqrt{5}$ and $y = -\sqrt{5}$, or $x = -2\sqrt{5}$

and $y = \sqrt{5}$.

Therefore $T\left(\sqrt{5}, 2\sqrt{5}\right) = 0° = T\left(-\sqrt{5}, -2\sqrt{5}\right)$ is the minimum value and $T\left(2\sqrt{5}, -\sqrt{5}\right) = 125° = T\left(-2\sqrt{5}, \sqrt{5}\right)$

is the maximum value. (Note: $\lambda = 1 \Rightarrow x = 0$ from the equation $-4x + 2y = 2\lambda y$; but we found $x \neq 0$ in

CASE 1.)

17. Let $f(x, y, z) = (x - 1)^2 + (y - 1)^2 + (z - 1)^2$ be the square of the distance from $(1, 1, 1)$. Then

$\nabla f = 2(x - 1)\mathbf{i} + 2(y - 1)\mathbf{j} + 2(z - 1)\mathbf{k}$ and $\nabla g = \mathbf{i} + 2\mathbf{j} + 3\mathbf{k}$ so that $\nabla f = \lambda \nabla g$

$\Rightarrow 2(x - 1)\mathbf{i} + 2(y - 1)\mathbf{j} + 2(z - 1)\mathbf{k} = \lambda(\mathbf{i} + 2\mathbf{j} + 3\mathbf{k}) \Rightarrow 2(x - 1) = \lambda, 2(y - 1) = 2\lambda, 2(z - 1) = 3\lambda$

$\Rightarrow 2(y - 1) = 2[2(x - 1)]$ and $2(z - 1) = 3[2(x - 1)] \Rightarrow x = \frac{y + 1}{2} \Rightarrow z + 2 = 3\left(\frac{y + 1}{2}\right)$ or $z = \frac{3y - 1}{2}$; thus

$\frac{y + 1}{2} + 2y + 3\left(\frac{3y - 1}{2}\right) - 13 = 0 \Rightarrow y = 2 \Rightarrow x = \frac{3}{2}$ and $z = \frac{5}{2}$. Therefore the point $\left(\frac{3}{2}, 2, \frac{5}{2}\right)$ is closest (since no

point on the plane is farthest from the point $(1, 1, 1)$).

19. Let $f(x, y, z) = x^2 + y^2 + z^2$ be the square of the distance from the origin. Then $\nabla f = 2x\mathbf{i} + 2y\mathbf{j} + 2z\mathbf{k}$ and $\nabla g = 2x\mathbf{i} - 2y\mathbf{j} - 2z\mathbf{k}$ so that $\nabla f = \lambda \nabla g \Rightarrow 2x\mathbf{i} + 2y\mathbf{j} + 2z\mathbf{k} = \lambda(2x\mathbf{i} - 2y\mathbf{j} - 2z\mathbf{k}) \Rightarrow 2x = 2x\lambda, 2y = -2y\lambda$, and $2z = -2z\lambda \Rightarrow x = 0$ or $\lambda = 1$.
   CASE 1: $\lambda = 1 \Rightarrow 2y = -2y \Rightarrow y = 0$; $2z = -2z \Rightarrow z = 0 \Rightarrow x^2 - 1 = 0 \Rightarrow x^2 - 1 = 0 \Rightarrow x = \pm 1$ and $y = z = 0$.
   CASE 2: $x = 0 \Rightarrow y^2 - z^2 = 1$, which has no solution.
   Therefore the points on the unit circle $x^2 + y^2 = 1$, are the points on the surface $x^2 + y^2 - z^2 = 1$ closest to the origin. The minimum distance is 1.

21. Let $f(x, y, z) = x^2 + y^2 + z^2$ be the square of the distance to the origin. Then $\nabla f = 2x\mathbf{i} + 2y\mathbf{j} + 2z\mathbf{k}$ and $\nabla g = -y\mathbf{i} - x\mathbf{j} + 2z\mathbf{k}$ so that $\nabla f = \lambda \nabla g \Rightarrow 2x\mathbf{i} + 2y\mathbf{j} + 2z\mathbf{k} = \lambda(-y\mathbf{i} - x\mathbf{j} + 2z\mathbf{k}) \Rightarrow 2x = -y\lambda, 2y = -x\lambda$, and $2z = 2z\lambda \Rightarrow \lambda = 1$ or $z = 0$.
   CASE 1: $\lambda = 1 \Rightarrow 2x = -y$ and $2y = -x \Rightarrow y = 0$ and $x = 0 \Rightarrow z^2 - 4 = 0 \Rightarrow z = \pm 2$ and $x = y = 0$.
   CASE 2: $z = 0 \Rightarrow -xy - 4 = 0 \Rightarrow y = -\frac{4}{x}$. Then $2x = \frac{4}{x}\lambda \Rightarrow \lambda = \frac{x^2}{2}$, and $-\frac{8}{x} = -x\lambda \Rightarrow -\frac{8}{x} = -x\left(\frac{x^2}{2}\right)$
   $\Rightarrow x^4 = 16 \Rightarrow x = \pm 2$. Thus, $x = 2$ and $y = -2$, or $x = -2$ and $y = 2$.
   Therefore we get four points: $(2, -2, 0), (-2, 2, 0), (0, 0, -2)$. and $(0, 0, -2)$ But the points $(0, 0, 2)$ and $(0, 0, -2)$ are closest to the origin since they are 2 units away and the others are $2\sqrt{2}$ units away.

23. $\nabla f = \mathbf{i} - 2\mathbf{j} + 5\mathbf{k}$ and $\nabla g = 2x\mathbf{i} + 2y\mathbf{j} + 2z\mathbf{k}$ so that $\nabla f = \lambda \nabla g \Rightarrow \mathbf{i} - 2\mathbf{j} + 5\mathbf{k} = \lambda(2x\mathbf{i} + 2y\mathbf{j} + 2z\mathbf{k}) \Rightarrow 1 = 2x\lambda$,
   $-2 = 2y\lambda$, and $5 = 2z\lambda \Rightarrow x = \frac{1}{2\lambda}, y = -\frac{1}{\lambda} = -2x$, and $z = \frac{5}{2\lambda} = 5x \Rightarrow x^2 + (-2x)^2 + (5x)^2 = 30 \Rightarrow x = \pm 1$.
   Thus, $x = 1, y = -2, z = 5$ or $x = -1, y = 2, z = -5$. Therefore $f(1, -2, 5) = 30$ is the maximum value and $f(-1, 2, -5) = -30$ is the minimum value.

25. $f(x, y, z) = x^2 + y^2 + z^2$ and $g(x, y, z) = x + y + z - 9 = 0 \Rightarrow \nabla f = 2x\mathbf{i} + 2y\mathbf{j} + 2z\mathbf{k}$ and $\nabla g = \mathbf{i} + \mathbf{j} + \mathbf{k}$ so that $\nabla f = \lambda \nabla g \Rightarrow 2x\mathbf{i} + 2y\mathbf{j} + 2z\mathbf{k} = \lambda(\mathbf{i} + \mathbf{j} + \mathbf{k}) \Rightarrow 2x = \lambda, 2y = \lambda$, and $2z = \lambda \Rightarrow x = y = z \Rightarrow x + x + x - 9 = 0$ $\Rightarrow x = 3, y = 3$, and $z = 3$.

27. $V = xyz$ and $g(x, y, z) = x^2 + y^2 + z^2 - 1 = 0 \Rightarrow \nabla V = yz\mathbf{i} + xz\mathbf{j} + xy\mathbf{k}$ and $\nabla g = 2x\mathbf{i} + 2y\mathbf{j} + 2z\mathbf{k}$ so that $\nabla V = \lambda \nabla g \Rightarrow yz = \lambda x, xz = \lambda y$, and $xy = \lambda z \Rightarrow xyz = \lambda x^2$ and $xyz = \lambda y^2 \Rightarrow y = \pm x \Rightarrow z = \pm x$ $\Rightarrow x^2 + x^2 + x^2 = 1 \Rightarrow x = \frac{1}{\sqrt{3}}$ since $x > 0 \Rightarrow$ the dimensions of the box are $\frac{1}{\sqrt{3}}$ by $\frac{1}{\sqrt{3}}$ by $\frac{1}{\sqrt{3}}$ for maximum volume. (Note that there is no minimum volume since the box could be made arbitrarily thin.)

29. $\nabla T = 16x\mathbf{i} + 4z\mathbf{j} + (4y - 16)\mathbf{k}$ and $\nabla g = 8x\mathbf{i} + 2y\mathbf{j} + 8z\mathbf{k}$ so that $\nabla T = \lambda \nabla g \Rightarrow 16x\mathbf{i} + 4z\mathbf{j} + (4y - 16)\mathbf{k}$ $= \lambda(8x\mathbf{i} + 2y\mathbf{j} + 8z\mathbf{k}) \Rightarrow 16x = 8x\lambda, 4z = 2y\lambda$, and $4y - 16 = 8z\lambda \Rightarrow \lambda = 2$ or $x = 0$.
   CASE 1: $\lambda = 2 \Rightarrow 4z = 2y(2) \Rightarrow z = y$. Then $4z - 16 = 16z \Rightarrow z = -\frac{4}{3} \Rightarrow y = -\frac{4}{3}$. Then
   $$4x^2 + \left(-\frac{4}{3}\right)^2 + 4\left(-\frac{4}{3}\right)^2 = 16 \Rightarrow x = \pm\frac{4}{3}.$$
   CASE 2: $x = 0 \Rightarrow \lambda = \frac{2z}{y} \Rightarrow 4y - 16 = 8z\left(\frac{2z}{y}\right) \Rightarrow y^2 - 4y = 4z^2 \Rightarrow 4(0)^2 + y^2 + \left(y^2 - 4y\right) - 16 = 0$
   $\Rightarrow y^2 - 2y - 8 = 0 \Rightarrow (y - 4)(y + 2) = 0 \Rightarrow y = 4$ or $y = -2$. Now $y = 4 \Rightarrow 4z^2 = 4^2 - 4(4) \Rightarrow z = 0$
   and $y = -2 \Rightarrow 4z^2 = (-2)^2 - 4(-2) \Rightarrow z = \pm\sqrt{3}$.

The temperatures are $T\left(\pm\frac{4}{3}, -\frac{4}{3}, -\frac{4}{3}\right) = \left(642\frac{2}{3}\right)^{\circ}$, $T(0, 4, 0) = 600°$, $T\left(0, -2, \sqrt{3}\right) = \left(600 - 24\sqrt{3}\right)^{\circ}$, and

$T\left(0, -2, -\sqrt{3}\right) = \left(600 + 24\sqrt{3}\right)^{\circ} \approx 641.6°$. Therefore $\left(\pm\frac{4}{3}, -\frac{4}{3}, -\frac{4}{3}\right)$ are the hottest points on the space probe.

31. (a) If we replace x by $2x$ and y by $2y$ in the formula for P, P changes by a factor of $2^{\alpha} \cdot 2^{1-\alpha} = 2$.

   (b) For the given information, the production function is $P(x, y) = 120x^{3/4}y^{1/4}$ and the constraint is
   $G(x, y) = 250x + 400y - 100,000 = 0$. $\nabla P = \lambda \nabla G$ implies

   $$90 \left(\frac{x}{y}\right)^{-1/4} = 250\lambda$$

   $$30 \left(\frac{x}{y}\right)^{3/4} = 400\lambda$$

   Eliminating λ yields $\dfrac{x}{y} = \dfrac{24}{5}$ or $5x = 24y$. Now we solve the system

   $$5x - 24y = 0$$
   $$250x + 400y = 100,000$$

   and find $x = 300$, $y = 62.5$. $P(300, 62.5) \approx 24,322$ units.

33. $\nabla U = (y+2)\mathbf{i} + x\mathbf{j}$ and $\nabla g = 2\mathbf{i} + \mathbf{j}$ so that $\nabla U = \lambda \nabla g \Rightarrow (y+2)\mathbf{i} + x\mathbf{j} = \lambda(2\mathbf{i} + \mathbf{j}) \Rightarrow y + 2 = 2\lambda$ and $x = \lambda$
   $\Rightarrow y + 2 = 2x \Rightarrow y = 2x - 2 \Rightarrow 2x + (2x - 2) = 30 \Rightarrow x = 8$ and $y = 14$. Therefore $U(8, 14) = \$128$ is the
   maximum value of U under the constraint.

35. The following diagram shows $L, h,$ and k. We also name the lengths u and v as shown; we assume all named
   lengths are positive. We will minimize the square of L, which is equal to $(u+k)^2 + (v+h)^2$.

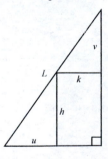

   Our function to minimize is $S(u, v) = (u+k)^2 + (v+h)^2$. Using similar triangles, $\dfrac{v}{k} = \dfrac{h}{u}$, so our constraint
   equation is $G(u, v) = uv - hk = 0$. $\nabla S = \lambda \nabla G$ implies

   $$2(u+k) = \lambda v$$
   $$2(v+h) = \lambda u$$

   Eliminating λ by dividing the first equation by the second yields $\dfrac{u+k}{v+h} = \dfrac{v}{u}$. We combine this with the
   constraint equation and solve the system

   $$(u+k) \cdot u = (v+h) \cdot v$$
   $$hk = uv$$

Substituting hk/v for u in the first equation, expanding, and moving all variables to the left side yields the equation $u^4 + ku^3 - h^2ku - h^2k^2 = 0$. Factoring by grouping we get $(u^3 - h^2k)(u+k) = 0$. Since no lengths are negative this gives us the solution $u = (h^2k)^{1/3}$. Substituting this value for u in $hk = uv$ we find $v = (hk^2)^{1/3}$. (Note that throughout this calculation interchanging h and k has the effect of interchanging u and v. This must be true, since if we rotate the diagram 90 degrees clockwise we don't change the solution to the minimization problem, but now h and v are the horizontal distances and k and u are the vertical distances. All of our equations reflect this symmetry, and we could exploit it to save some computation.)

Now $L^2 = (u+k)^2 + (v+h)^2 = \left((h^2k)^{1/3} + k \right)^2 + \left((hk^2)^{1/3} + h \right)^2$

$$= h^{4/3}k^{2/3} + 2h^{2/3}k^{4/3} + k^2 + h^{2/3}k^{4/3} + 2h^{4/3}k^{2/3} + h^2$$

$$= \left(h^{2/3} + k^{2/3} \right)^3$$

Thus the minimum L is given by $\left(h^{2/3} + k^{2/3} \right)^{3/2}$.

37. Let $g_1(x, y, z) = 2x - y = 0$ and $g_2(x, y, z) = y + z = 0 \Rightarrow \nabla g_1 = 2\mathbf{i} - \mathbf{j}, \nabla g_2 = \mathbf{j} + \mathbf{k}$, and $\nabla f = 2x\mathbf{i} + 2\mathbf{j} - 2z\mathbf{k}$ so that $\nabla f = \lambda \nabla g_1 + \mu \nabla g_2 \Rightarrow 2x\mathbf{i} + 2\mathbf{j} - 2z\mathbf{k} = \lambda(2\mathbf{i} - \mathbf{j}) + \mu(\mathbf{j} + \mathbf{k}) \Rightarrow 2x\mathbf{i} + 2\mathbf{j} - 2z\mathbf{k} = 2\lambda\mathbf{i} + (\mu - \lambda)\mathbf{j} + \mu\mathbf{k}$ $\Rightarrow 2x = 2\lambda, 2 = \mu - \lambda$, and $-2z = \mu \Rightarrow x = \lambda$. Then $2 = -2z - x \Rightarrow x = -2z - 2$ so that $2x - y = 0$ $\Rightarrow 2(-2z - 2) - y = 0 \Rightarrow -4z - 4 - y = 0$. This equation coupled with $y + z = 0$ implies $z = -\frac{4}{3}$ and $y = \frac{4}{3}$. Then $x = \frac{2}{3}$ so that $\left(\frac{2}{3}, \frac{4}{3}, -\frac{4}{3} \right)$ is the point that gives the maximum value $f\left(\frac{2}{3}, \frac{4}{3}, -\frac{4}{3} \right) = \left(\frac{2}{3} \right)^2 + 2\left(\frac{4}{3} \right) - \left(-\frac{4}{3} \right)^2 = \frac{4}{3}$.

39. Let $f(x, y, z) = x^2 + y^2 + z^2$ be the square of the distance from the origin. We want to minimize $f(x, y, z)$ subject to the constraints $g_1(x, y, z) = y + 2z - 12 = 0$ and $g_2(x, y, z) = x + y - 6 = 0$. Thus $\nabla f = 2x\mathbf{i} + 2y\mathbf{j} + 2z\mathbf{k}, \nabla g_1 = \mathbf{j} + 2\mathbf{k}$, and $\nabla g_2 = \mathbf{i} + \mathbf{j}$ so that $\nabla f = \lambda \nabla g_1 + \mu \nabla g_2 \Rightarrow 2x = \mu, 2y = \lambda + \mu$, and $2z = 2\lambda$. Then $0 = y + 2z - 12 = \left(\frac{\lambda}{2} + \frac{\mu}{2} \right) + 2\lambda - 12 \Rightarrow \frac{5}{2}\lambda + \frac{1}{2}\mu = 12 \Rightarrow 5\lambda + \mu = 24$; $0 = x + y - 6 = \frac{\mu}{2} + \left(\frac{\lambda}{2} + \frac{\mu}{2} \right) - 6 \Rightarrow \frac{1}{2}\lambda + \mu = 6 \Rightarrow \lambda + 2\mu = 12$. Solving these two equations for λ and μ gives $\lambda = 4$ and $\mu = 4 \Rightarrow x = \frac{\mu}{2} = 2$, $y = \frac{\lambda + \mu}{2} = 4$, and $z = \lambda = 4$. The point $(2, 4, 4)$ on the line of intersection is closest to the origin. (There is no maximum distance from the origin since points on the line can be arbitrarily far away.)

41. Let $g_1(x, y, z) = z - 1 = 0$ and $g_2(x, y, z) = x^2 + y^2 + z^2 - 10 = 0 \Rightarrow \nabla g_1 = \mathbf{k}, \nabla g_2 = 2x\mathbf{i} + 2y\mathbf{j} + 2z\mathbf{k}$, and $\nabla f = 2xyz\mathbf{i} + x^2z\mathbf{j} + x^2y\mathbf{k}$ so that $\nabla f = \lambda \nabla g_1 + \mu \nabla g_2 \Rightarrow 2xyz\mathbf{i} + x^2z\mathbf{j} + x^2y\mathbf{k} = \lambda(\mathbf{k}) + \mu(2x\mathbf{i} + 2y\mathbf{j} + 2z\mathbf{k})$ $\Rightarrow 2xyz = 2x\mu, x^2z = 2y\mu$, and $x^2y = 2z\mu + \lambda \Rightarrow xyz = x\mu \Rightarrow x = 0$ or $yz = \mu \Rightarrow \mu = y$ since $z = 1$.
CASE 1: $x = 0$ and $z = 1 \Rightarrow y^2 - 9 = 0$ (from g_2) $\Rightarrow y = \pm 3$ yielding the points $(0, \pm 3, 1)$.
CASE 2: $\mu = y \Rightarrow x^2z = 2y^2 \Rightarrow x^2 = 2y^2$ (since $z = 1$) $\Rightarrow 2y^2 + y^2 + 1 - 10 = 0$ (from g_2) $\Rightarrow 3y^2 - 9 = 0$
$\Rightarrow y = \pm\sqrt{3} \Rightarrow x^2 = 2\left(\pm\sqrt{3} \right)^2 \Rightarrow x = \pm\sqrt{6}$ yielding the points $\left(\pm\sqrt{6}, \pm\sqrt{3}, 1 \right)$.
Now $f(0, \pm 3, 1) = 1$ and $f\left(\pm\sqrt{6}, \pm\sqrt{3}, 1 \right) = 6\left(\pm\sqrt{3} \right) + 1 = 1 \pm 6\sqrt{3}$. Therefore the maximum of f is $1 + 6\sqrt{3}$ at $\left(\pm\sqrt{6}, \sqrt{3}, 1 \right)$, and the minimum of f is $1 - 6\sqrt{3}$ at $\left(\pm\sqrt{6}, -\sqrt{3}, 1 \right)$.

43. Let $g_1(x, y, z) = y - x = 0$ and $g_2(x, y, z) = x^2 + y^2 + z^2 - 4 = 0$. Then $\nabla f = y\mathbf{i} + x\mathbf{j} + 2z\mathbf{k}$, $\nabla g_1 = -\mathbf{i} + \mathbf{j}$, and
$\nabla g_2 = 2x\mathbf{i} + 2y\mathbf{j} + 2z\mathbf{k}$ so that $\nabla f = \lambda\nabla g_1 + \mu\nabla g_2 \Rightarrow y\mathbf{i} + x\mathbf{j} + 2z\mathbf{k} = \lambda(-\mathbf{i} + \mathbf{j}) + \mu(2x\mathbf{i} + 2y\mathbf{j} + 2z\mathbf{k})$
$\Rightarrow y = -\lambda + 2x\mu$, $x = \lambda + 2y\mu$, and $2z = 2z\mu \Rightarrow z = 0$ or $\mu = 1$.
CASE 1: $z = 0 \Rightarrow x^2 + y^2 - 4 = 0 \Rightarrow 2x^2 - 4 = 0$ (since $x = y$) $\Rightarrow x = \pm\sqrt{2}$ and $y = \pm\sqrt{2}$ yielding the points
$\left(\pm\sqrt{2}, \pm\sqrt{2}, 0\right)$.
CASE 2: $\mu = 1 \Rightarrow y = -\lambda + 2x$ and $x = \lambda + 2y \Rightarrow x + y = 2(x + y) \Rightarrow 2x = 2(2x)$ since $x = y \Rightarrow x = 0 \Rightarrow y = 0$
$\Rightarrow z^2 - 4 = 0 \Rightarrow z = \pm2$ yielding the points $(0, 0, \pm2)$.
Now, $f(0, 0, \pm2) = 4$ and $f\left(\pm\sqrt{2}, \pm\sqrt{2}, 0\right) = 2$. Therefore the maximum value of f is 4 at $(0, 0, \pm2)$ and the
minimum value of f is 2 at $f\left(\pm\sqrt{2}, \pm\sqrt{2}, 0\right)$.

45. $\nabla f = \mathbf{i} + \mathbf{j}$ and $\nabla g = y\mathbf{i} + x\mathbf{j}$ so that $\nabla f = \lambda\nabla g \Rightarrow \mathbf{i} + \mathbf{j} = \lambda(y\mathbf{i} + x\mathbf{j}) \Rightarrow 1 = y\lambda$ and $1 = x\lambda \Rightarrow y = x \Rightarrow y^2 = 16$
$\Rightarrow y = \pm4 \Rightarrow (4, 4)$ and $(-4, -4)$ are candidates for the location of extreme values. But as $x \to \infty$, $y \to \infty$
and $f(x, y) \to \infty$; as $x \to -\infty$, $y \to 0$ and $f(x, y) \to -\infty$. Therefore no maximum or minimum value exists
subject to the constraint.

47. (a) Maximize $f(a, b, c) = a^2b^2c^2$ subject to $a^2 + b^2 + c^2 = r^2$. Thus $\nabla f = 2ab^2c^2\mathbf{i} + 2a^2bc^2\mathbf{j} + 2a^2b^2c\mathbf{k}$
and $\nabla g = 2a\mathbf{i} + 2b\mathbf{j} + 2c\mathbf{k}$ so that $\nabla f = \lambda\nabla g \Rightarrow 2ab^2c^2 = 2a\lambda$, $2a^2bc^2 = 2b\lambda$, and $2a^2b^2c = 2c\lambda$
$\Rightarrow 2a^2b^2c^2 = 2a^2\lambda = 2b^2\lambda = 2c^2\lambda \Rightarrow \lambda = 0$ or $a^2 = b^2 = c^2$.
CASE 1: $\lambda = 0 \Rightarrow a^2b^2c^2 = 0$.
CASE 2: $a^2 = b^2 = c^2 \Rightarrow f(a, b, c) = a^2a^2a^2$ and $3a^2 = r^2 \Rightarrow f(a, b, c) = \left(\frac{r^2}{3}\right)^3$ is the maximum value.

(b) The point $\left(\sqrt{a}, \sqrt{b}, \sqrt{c}\right)$ is on the sphere if $a + b + c = r^2$. Moreover, by part (a),
$abc = f\left(\sqrt{a}, \sqrt{b}, \sqrt{c}\right) \le \left(\frac{r^2}{3}\right)^3 \Rightarrow (abc)^{1/3} \le \frac{r^2}{3} = \frac{a+b+c}{3}$, as claimed.

49-53. Example CAS commands:

Maple:

```
f := (x,y,z) -> x*y+y*z;

g1 := (x,y,z) -> x^2+y^2-2;

g2 := (x,y,z) -> x^2+z^2-2;

h := unapply( f(x,y,z)-lambda[1]*g1(x,y,z)-lambda[2]*g2(x,y,z), (x,y,z,lambda[1],lambda[2]) );    # (a)

hx := diff( h(x,y,z,lambda[1],lambda[2]), x );                                                     # (b)

hy := diff( h(x,y,z,lambda[1],lambda[2]), y );

hz := diff( h(x,y,z,lambda[1],lambda[2]), z );

h11 := diff( h(x,y,z,lambda[1],lambda[2]), lambda[1] );

h12 := diff( h(x,y,z,lambda[1],lambda[2]), lambda[2] );

sys := { hx=0, hy=0, hz=0, h11=0, h12=0 };

q1 := solve( sys, {x,y,z,lambda[1],lambda[2]} );                                                   # (c)

q2 := map(allvalues,{q1});                                                                         # (d)

for p in q2 do

 eval( [x,y,z,f(x,y,z)], p );

  ``=evalf(eval( [x,y,z,f(x,y,z)], p ));

end do;
```

Mathematica: (assigned functions will vary)

```
Clear[x, y, z, lambda1, lambda2]

f[x_, y_, z_]:= x y + y z

g1[x_, y_, z_]:= x^2 + y^2 - 2

g2[x_, y_, z_]:= x^2 + z^2 - 2

h = f[x, y, z] - lambda1 g1[x, y, z] - lambda2 g2[x, y, z];

hx= D[h, x]; hy= D[h, y]; hz= D[h,z]; hL1=D[h, lambda1]; hL2= D[h, lambda2];

critical=Solve[{hx==0, hy==0, hz==0, hL1==0, hL2==0, g1[x,y,z]==0, g2[x,y,z]==0},

{x, y, z, lambda1, lambda2}]//N

{{x, y, z}, f[x, y, z]}/.critical
```

CHAPTER 13 PRACTICE EXERCISES

1. Domain: All points in the xy-plane
    Range: $z \geq 0$

    Level curves are ellipses with major axis along the
    y-axis. and minor axis along the x-axis.

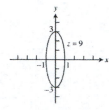

3. Domain: All (x, y) so that $x \neq 0$ and $y \neq 0$
    Range: $z \neq 0$

    Level curves are hyperbolas with the x- and y-axes
    as asymptotes.

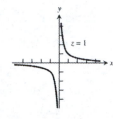

5. Domain: All points (x, y, z) in space
    Range: All real numbers

    Level surfaces are paraboloids of revolution with
    the z-axis as axis.

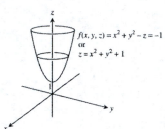

7. Domain: All (x, y, z) such that $(x, y, z) \neq (0, 0, 0)$
    Range: Positive real numbers

    Level surfaces are spheres with center $(0, 0, 0)$ and
    radius $r > 0$.

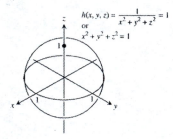

9. $\displaystyle\lim_{(x,\,y)\to(\pi,\,\ln 2)} e^y \cos x = e^{\ln 2} \cos \pi = (2)(-1) = -2$

11. $\displaystyle\lim_{\substack{(x,\,y)\to(1,\,1)\\ x \neq \pm y}} \frac{x-y}{x^2 - y^2} = \lim_{\substack{(x,\,y)\to(1,\,1)\\ x \neq \pm y}} \frac{x-y}{(x-y)(x+y)} = \lim_{(x,\,y)\to(1,\,1)} \frac{1}{x+y} = \frac{1}{1+1} = \frac{1}{2}$

13. $\displaystyle\lim_{P\to(1,\,-1,\,e)} \ln|x+y+z| = \ln|1+(-1)+e| = \ln e = 1$

15. Let $y = kx^2, k \neq 1$. Then $\displaystyle\lim_{\substack{(x,\,y)\to(0,\,0)\\ y\neq x^2}} \frac{y}{x^2-y} = \lim_{(x,\,kx^2)\to(0,\,0)} \frac{kx^2}{x^2-kx^2} = \frac{k}{1-k}$ which gives different limits for

    different values of $k \Rightarrow$ the limit does not exist.

17. Let $y = kx$. Then $\displaystyle\lim_{(x,\,y)\to(0,\,0)} \frac{x^2-y^2}{x^2+y^2} = \frac{x^2-k^2x^2}{x^2+k^2x^2} = \frac{1-k^2}{1+k^2}$ which gives different limits for different values of

    $k \Rightarrow$ the limit does not exist so $f(0, 0)$ cannot be defined in a way that makes f continuous at the origin.

19. $\frac{\partial g}{\partial r} = \cos\theta + \sin\theta, \frac{\partial g}{\partial \theta} = -r\sin\theta + r\cos\theta$

21. $\frac{\partial f}{\partial R_1} = -\frac{1}{R_1^2}, \frac{\partial f}{\partial R_2} = -\frac{1}{R_2^2}, \frac{\partial f}{\partial R_3} = -\frac{1}{R_3^2}$

23. $\frac{\partial P}{\partial n} = \frac{RT}{V}, \frac{\partial P}{\partial R} = \frac{nT}{V}, \frac{\partial P}{\partial T} = \frac{nR}{V}, \frac{\partial P}{\partial V} = -\frac{nRT}{V^2}$

25. $\frac{\partial g}{\partial x} = \frac{1}{y}, \frac{\partial g}{\partial y} = 1 - \frac{x}{y^2} \Rightarrow \frac{\partial^2 g}{\partial x^2} = 0, \frac{\partial^2 g}{\partial y^2} = \frac{2x}{y^3}, \frac{\partial^2 g}{\partial y \partial x} = \frac{\partial^2 g}{\partial x \partial y} = -\frac{1}{y^2}$

27. $\frac{\partial f}{\partial x} = 1 + y - 15x^2 + \frac{2x}{x^2+1}, \frac{\partial f}{\partial y} = x \Rightarrow \frac{\partial^2 f}{\partial x^2} = -30x + \frac{2-2x^2}{(x^2+1)^2}, \frac{\partial^2 f}{\partial y^2} = 0, \frac{\partial^2 f}{\partial y \partial x} = \frac{\partial^2 f}{\partial x \partial y} = 1$

29. $\frac{\partial w}{\partial x} = y\cos(xy + \pi), \frac{\partial w}{\partial y} = x\cos(xy + \pi), \frac{dx}{dt} = e^t, \frac{dy}{dt} = \frac{1}{t+1} \Rightarrow \frac{dw}{dt} = [y\cos(xy+\pi)]e^t + [x\cos(xy+\pi)]\left(\frac{1}{t+1}\right);$

$t = 0 \Rightarrow x = 1$ and $y = 0 \Rightarrow \left.\frac{dw}{dt}\right|_{t=0} = 0 \cdot 1 + [1 \cdot (-1)]\left(\frac{1}{0+1}\right) = -1$

31. $\frac{\partial w}{\partial x} = 2\cos(2x - y), \frac{\partial w}{\partial y} = -\cos(2x - y), \frac{\partial x}{\partial r} = 1, \frac{\partial x}{\partial s} = \cos s, \frac{\partial y}{\partial r} = s, \frac{\partial y}{\partial s} = r$

$\Rightarrow \frac{\partial w}{\partial r} = [2\cos(2x - y)](1) + [-\cos(2x - y)](s);$

$r = \pi$ and $s = 0 \Rightarrow x = \pi$ and $y = 0 \Rightarrow \left.\frac{\partial w}{\partial r}\right|_{(\pi,0)} = (2\cos 2\pi) - (\cos 2\pi)(0) = 2;$

$\frac{\partial w}{\partial s} = [2\cos(2x - y)](\cos s) + [-\cos(2x - y)](r) \Rightarrow \left.\frac{\partial w}{\partial s}\right|_{(\pi,0)} = (2\cos 2\pi)(\cos 0) - (\cos 2\pi)(\pi) = 2 - \pi$

33. $\frac{\partial f}{\partial x} = y + z, \frac{\partial f}{\partial y} = x + z, \frac{\partial f}{\partial z} = y + x, \frac{dx}{dt} = -\sin t, \frac{dy}{dt} = \cos t, \frac{dz}{dt} = -2\sin 2t$

$\Rightarrow \frac{df}{dt} = -(y + z)(\sin t) + (x + z)(\cos t) - 2(y + x)(\sin 2t); \ t = 1 \Rightarrow x = \cos 1, \ y = \sin 1, \text{ and } z = \cos 2$

$\Rightarrow \left.\frac{df}{dt}\right|_{t=1} = -(\sin 1 + \cos 2)(\sin 1) + (\cos 1 + \cos 2)(\cos 1) - 2(\sin 1 + \cos 1)(\sin 2)$

35. $F(x, y) = 1 - x - y^2 - \sin xy \Rightarrow F_x = -1 - y\cos xy$ and $F_y = -2y - x\cos xy \Rightarrow \frac{dy}{dx} = -\frac{F_x}{F_y} = -\frac{-1 - y\cos xy}{-2y - x\cos xy}$

$= \frac{1 + y\cos xy}{-2y - x\cos xy} \Rightarrow$ at $(x, y) = (0, 1)$ we have $\left.\frac{dy}{dx}\right|_{(0,1)} = \frac{1+1}{-2} = -1$

37. $\nabla f = (-\sin x\cos y)\mathbf{i} - (\cos x\sin y)\mathbf{j} \Rightarrow \left.\nabla f\right|_{\left(\frac{\pi}{4},\frac{\pi}{4}\right)} = -\frac{1}{2}\mathbf{i} - \frac{1}{2}\mathbf{j} \Rightarrow |\nabla f| = \sqrt{\left(-\frac{1}{2}\right)^2 + \left(-\frac{1}{2}\right)^2} = \frac{1}{\sqrt{2}} = \frac{\sqrt{2}}{2};$

$\mathbf{u} = \frac{\nabla f}{|\nabla f|} = -\frac{\sqrt{2}}{2}\mathbf{i} - \frac{\sqrt{2}}{2}\mathbf{j} \Rightarrow f$ increases most rapidly in the direction $\mathbf{u} = -\frac{\sqrt{2}}{2}\mathbf{i} - \frac{\sqrt{2}}{2}\mathbf{j}$ and decreases most rapidly

in the direction $-\mathbf{u} = \frac{\sqrt{2}}{2}\mathbf{i} + \frac{\sqrt{2}}{2}\mathbf{j}; \ (D_\mathbf{u} f)_{P_0} = |\nabla f| = \frac{\sqrt{2}}{2}$ and $(D_{-\mathbf{u}} f)_{P_0} = -\frac{\sqrt{2}}{2}; \ \mathbf{u}_1 = \frac{\mathbf{v}}{|\mathbf{v}|} = \frac{3\mathbf{i} + 4\mathbf{j}}{\sqrt{3^2 + 4^2}} = \frac{3}{5}\mathbf{i} + \frac{4}{5}\mathbf{j}$

$\Rightarrow (D_{\mathbf{u}_1} f)_{P_0} = f \cdot \mathbf{u}_1 = \left(-\frac{1}{2}\right)\left(\frac{3}{5}\right) + \left(-\frac{1}{2}\right)\left(\frac{4}{5}\right) = -\frac{7}{10}$

39. $\nabla f = \left(\frac{2}{2x+3y+6z}\right)\mathbf{i} + \left(\frac{3}{2x+3y+6z}\right)\mathbf{j} + \left(\frac{6}{2x+3y+6z}\right)\mathbf{k} \Rightarrow \left.\nabla f\right|_{(-1,-1,1)} = 2\mathbf{i} + 3\mathbf{j} + 6\mathbf{k}; \ \mathbf{u} = \frac{\nabla f}{|\nabla f|} = \frac{2\mathbf{i} + 3\mathbf{j} + 6\mathbf{k}}{\sqrt{2^2 + 3^2 + 6^2}}$

$= \frac{2}{7}\mathbf{i} + \frac{3}{7}\mathbf{j} + \frac{6}{7}\mathbf{k} \Rightarrow f$ increases most rapidly in the direction $\mathbf{u} = \frac{2}{7}\mathbf{i} + \frac{3}{7}\mathbf{j} + \frac{6}{7}\mathbf{k}$ and decreases most rapidly in the direction $-\mathbf{u} = -\frac{2}{7}\mathbf{i} - \frac{3}{7}\mathbf{j} - \frac{6}{7}\mathbf{k}$; $(D_{\mathbf{u}}f)_{P_0} = \left|\nabla f\right| = 7$, $(D_{-\mathbf{u}}f)_{P_0} = -7$; $\mathbf{u}_1 = \frac{\mathbf{v}}{|\mathbf{v}|} = \frac{2}{7}\mathbf{i} + \frac{3}{7}\mathbf{j} + \frac{6}{7}\mathbf{k}$

$\Rightarrow (D_{\mathbf{u}_1}f)_{P_0} = (D_{\mathbf{u}}f)_{P_0} = 7$

41. $\mathbf{r} = (\cos 3t)\mathbf{i} + (\sin 3t)\mathbf{j} + 3t\mathbf{k} \Rightarrow \mathbf{v}(t) = (-3\sin 3t)\mathbf{i} + (3\cos 3t)\mathbf{j} + 3\mathbf{k} \Rightarrow \mathbf{v}\left(\frac{\pi}{3}\right) = -3\mathbf{j} + 3\mathbf{k} \Rightarrow \mathbf{u} = -\frac{1}{\sqrt{2}}\mathbf{j} + \frac{1}{\sqrt{2}}\mathbf{k}$;

$f(x, y, z) = xyz \Rightarrow \nabla f = yz\mathbf{i} + xz\mathbf{j} + xy\mathbf{k}$; $t = \frac{\pi}{3}$ yields the point on the helix $(-1, 0, \pi)$

$\Rightarrow \nabla f\big|_{(-1,0,\pi)} = -\pi\mathbf{j} \Rightarrow \nabla f \cdot \mathbf{u} = (-\pi\mathbf{j}) \cdot \left(-\frac{1}{\sqrt{2}}\mathbf{j} + \frac{1}{\sqrt{2}}\mathbf{k}\right) = \frac{\pi}{\sqrt{2}}$

43. (a) Let $\nabla f = a\mathbf{i} + b\mathbf{j}$ at $(1, 2)$. The direction toward $(2, 2)$ is determined by $\mathbf{v}_1 = (2-1)\mathbf{i} + (2-2)\mathbf{j} = \mathbf{i} = \mathbf{u}$ so

that $\nabla f \cdot \mathbf{u} = 2 \Rightarrow a = 2$. The direction toward $(1, 1)$ is determined by $\mathbf{v}_2 = (1-1)\mathbf{i} + (1-2)\mathbf{j} = -\mathbf{j} = \mathbf{u}$ so

that $\nabla f \cdot \mathbf{u} = -2 \Rightarrow -b = -2 \Rightarrow b = 2$. Therefore $\nabla f = 2\mathbf{i} + 2\mathbf{j}$; $f_x(1, 2) = f_y(1, 2) = 2$.

(b) The direction toward $(4, 6)$ is determined by $\mathbf{v}_3 = (4-1)\mathbf{i} + (6-2)\mathbf{j} = 3\mathbf{i} + 4\mathbf{j} \Rightarrow \mathbf{u} = \frac{3}{5}\mathbf{i} + \frac{4}{5}\mathbf{j}$

$\Rightarrow \nabla f \cdot \mathbf{u} = \frac{14}{5}$.

45. $\nabla f = 2x\mathbf{i} + \mathbf{j} + 2z\mathbf{k} \Rightarrow$

$\nabla f\big|_{(0,-1,-1)} = -\mathbf{j} - 2\mathbf{k}$,

$\nabla f\big|_{(0,0,0)} = \mathbf{j}$,

$\nabla f\big|_{(0,-1,1)} = \mathbf{j} + 2\mathbf{k}$

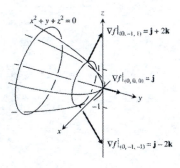

47. $\nabla f = 2x\mathbf{i} - \mathbf{j} - 5\mathbf{k} \Rightarrow \nabla f\big|_{(2,-1,1)} = 4\mathbf{i} - \mathbf{j} - 5\mathbf{k} \Rightarrow$ Tangent Plane: $4(x-2) - (y+1) - 5(z-1) = 0$

$\Rightarrow 4x - y - 5z = 4$; Normal Line: $x = 2 + 4t$, $y = -1 - t$, $z = 1 - 5t$

49. $\frac{\partial z}{\partial x} = \frac{2x}{x^2 + y^2} \Rightarrow \frac{\partial z}{\partial x}\big|_{(0,1,0)} = 0$ and $\frac{\partial z}{\partial y} = \frac{2y}{x^2 + y^2} \Rightarrow \frac{\partial z}{\partial y}\big|_{(0,1,0)} = 2$; thus the tangent plane is $2(y-1) - (z-0) = 0$ or

$2y - z - 2 = 0$

51. $\nabla F = (-\cos x)\mathbf{i} + \mathbf{j} \Rightarrow \nabla f\big|_{(\pi,1)} = \mathbf{i} + \mathbf{j} \Rightarrow$ the tangent

line is $(x - \pi) + (y - 1) = 0 \Rightarrow x + y = \pi + 1$; the normal

line is $y - 1 = 1(x - \pi) \Rightarrow y = x - \pi + 1$

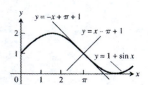

53. Let $f(x, y, z) = x^2 + 2y + 2z - 4$ and $g(x, y, z) = y - 1$. Then $\nabla f = 2x\mathbf{i} + 2\mathbf{j} + 2\mathbf{k}\big|_{\left(1,1,\frac{1}{2}\right)} = 2\mathbf{i} + 2\mathbf{j} + 2\mathbf{k}$ and

$\nabla g = \mathbf{j} \Rightarrow \nabla f \times \nabla g = \begin{vmatrix} \mathbf{i} & \mathbf{j} & \mathbf{k} \\ 2 & 2 & 2 \\ 0 & 1 & 0 \end{vmatrix} = -2\mathbf{i} + 2\mathbf{k} \Rightarrow$ the line is $x = 1 - 2t$, $y = 1$, $z = \frac{1}{2} + 2t$

55. $f\left(\frac{\pi}{4}, \frac{\pi}{4}\right) = \frac{1}{2}$, $f_x\left(\frac{\pi}{4}, \frac{\pi}{4}\right) = \cos x \cos y\big|_{(\pi/4, \pi/4)} = \frac{1}{2}$, $f_y\left(\frac{\pi}{4}, \frac{\pi}{4}\right) = -\sin x \sin y\big|_{(\pi/4, \pi/4)} = -\frac{1}{2}$

$\Rightarrow L(x, y) = \frac{1}{2} + \frac{1}{2}\left(x - \frac{\pi}{4}\right) - \frac{1}{2}\left(y - \frac{\pi}{4}\right) = \frac{1}{2} + \frac{1}{2}x - \frac{1}{2}y$; $f_{xx}(x, y) = -\sin x \cos y$, $f_{yy}(x, y) = -\sin x \cos y$, and

$f_{xy}(x, y) = -\cos x \sin y$. Thus an upper bound for E depends on the bound M used for $|f_{xx}|, |f_{xy}|,$ and $|f_{yy}|$.

With $M = \frac{\sqrt{2}}{2}$ we have $|E(x, y)| \leq \frac{1}{2}\left(\frac{\sqrt{2}}{2}\right)\left(\left|x - \frac{\pi}{4}\right| + \left|y - \frac{\pi}{4}\right|\right)^2 \leq \frac{\sqrt{2}}{4}(0.2)^2 \leq 0.0142;$

with $M = 1, |E(x, y)| \leq \frac{1}{2}(1)\left(\left|x - \frac{\pi}{4}\right| + \left|y - \frac{\pi}{4}\right|\right)^2 = \frac{1}{2}(0.2)^2 = 0.02.$

57. $f(1, 0, 0) = 0, f_x(1, 0, 0) = y - 3z\big|_{(1,0,0)} = 0, f_y(1, 0, 0) = x + 2z\big|_{(1,0,0)} = 1, f_z(1, 0, 0) = 2y - 3x\big|_{(1,0,0)} = -3$

$\Rightarrow L(x, y, z) = 0(x-1) + (y - 0) - 3(z - 0) = y - 3z; \ f(1, 1, 0) = 1, f_x(1, 1, 0) = 1, f_y(1, 1, 0) = 1, f_z(1, 1, 0) = -1$

$\Rightarrow L(x, y, z) = 1 + (x - 1) + (y - 1) - 1(z - 0) = x + y - z - 1$

59. $V = \pi r^2 h \Rightarrow dV = 2\pi r h \, dr + \pi r^2 \, dh \Rightarrow dV\big|_{(1.5,5280)} = 2\pi(1.5)(5280) \, dr + \pi(1.5)^2 \, dh = 15,840\pi \, dr + 2.25\pi \, dh.$
You should be more careful with the diameter since it has a greater effect on dV.

61. $dI = \frac{1}{R}dV - \frac{V}{R^2}dR \Rightarrow dI\big|_{(24,100)} = \frac{1}{100}dV - \frac{24}{100^2}dR \Rightarrow dI\big|_{dV=-1, dR=-20} = -0.01 + (480)(.0001) = 0.038,$ or

increases by 0.038 amps; % change in $V = (100)\left(-\frac{1}{24}\right) \approx -4.17\%;$ % change in $R = \left(-\frac{20}{100}\right)(100) = -20\%;$

$I = \frac{24}{100} = 0.24 \Rightarrow$ estimated % change in $I = \frac{dI}{I} \times 100 = \frac{0.038}{0.24} \times 100 \approx 15.83\% \Rightarrow$ more sensitive to voltage
change.

63. (a) $y = uv \Rightarrow dy = v \, du + u \, dv;$ percentage change in $u \leq 2\% \Rightarrow |du| \leq 0.02,$ and percentage change in $v \leq 3\%$

$\Rightarrow |dv| \leq 0.03; \frac{dy}{y} = \frac{v \, du + u \, dv}{uv} = \frac{du}{u} + \frac{dv}{v} \Rightarrow \left|\frac{dy}{y} \times 100\right| = \left|\frac{du}{u} \times 100 + \frac{dv}{v} \times 100\right| \leq \left|\frac{du}{u} \times 100\right| + \left|\frac{dv}{v} \times 100\right|$

$\leq 2\% + 3\% = 5\%$

(b) $z = u + v \Rightarrow \frac{dz}{z} = \frac{du + dv}{u + v} = \frac{du}{u + v} + \frac{dv}{u + v} \leq \frac{du}{u} + \frac{dv}{v}$ (since $u > 0, v > 0$)

$\Rightarrow \left|\frac{dz}{z} \times 100\right| \leq \left|\frac{du}{u} \times 100 + \frac{dv}{v} \times 100\right| = \left|\frac{dy}{y} \times 100\right|$

65. $f_x(x, y) = 2x - y + 2 = 0$ and $f_y(x, y) = -x + 2y + 2 = 0 \Rightarrow x = -2$ and $y = -2 \Rightarrow (-2, -2)$ is the critical

point; $f_{xx}(-2, -2) = 2, f_{yy}(-2, -2) = 2, f_{xy}(-2, -2) = -1 \Rightarrow f_{xx}f_{yy}f_{xy}^2 = 3 > 0$ and $f_{xx} > 0 \Rightarrow$ local minimum

value of $f(-2, -2) = -8$

67. $f_x(x, y) = 6x^2 + 3y = 0$ and $f_y(x, y) = 3x + 6y^2 = 0 \Rightarrow y = -2x^2$ and $3x + 6(4x^4) = 0 \Rightarrow x(1 + 8x^3) = 0$

$\Rightarrow x = 0$ and $y = 0,$ or $x = -\frac{1}{2}$ and $y = -\frac{1}{2} \Rightarrow$ the critical points are $(0, 0)$ and $\left(-\frac{1}{2}, -\frac{1}{2}\right).$ For $(0, 0):$

$f_{xx}(0, 0) = 12x\big|_{(0,0)} = 0, f_{yy}(0, 0) = 12y\big|_{(0,0)} = 0, f_{xy}(0, 0) = 3 \Rightarrow f_{xx}f_{yy} - f_{xy}^2 = -9 < 0 \Rightarrow$ saddle point width

$f(0, 0) = 0.$ For $\left(-\frac{1}{2}, -\frac{1}{2}\right): f_{xx} = -6, f_{yy} = -6, f_{xy} = 3 \Rightarrow f_{xx}f_{yy} - f_{xy}^2 = 27 > 0$ and $f_{xx} < 0 \Rightarrow$ local

maximum value of $f\left(-\frac{1}{2}, -\frac{1}{2}\right) = \frac{1}{4}$

69. $f_x(x, y) = 3x^2 + 6x = 0$ and $f_y(x, y) = 3y^2 - 6y = 0 \Rightarrow x(x + 2) = 0$ and $y(y - 2) = 0 \Rightarrow x = 0$ or $x = -2$ and

$y = 0$ or $y = 2 \Rightarrow$ the critical points are $(0, 0), (0, 2), (-2, 0),$ and $(-2, 2).$ For $(0, 0):$

$f_{xx}(0, 0) = 6x + 6\big|_{(0,0)} = 6, f_{yy}(0, 0) = 6y - 6\big|_{(0,0)} = -6, f_{xy}(0, 0) = 0 \Rightarrow f_{xx}f_{yy} - f_{xy}^2 = -36 < 0 \Rightarrow$ saddle point with $f(0, 0) = 0$. For $(0, 2)$: $f_{xx}(0, 2) = 6$, $f_{yy}(0, 2) = 6$, $f_{xy}(0, 2) = 0 \Rightarrow f_{xx}f_{yy} - f_{xy}^2 = 36 > 0$ and

$f_{xx} > 0 \Rightarrow$ local minimum value of $f(0, 2) = -4$. For $(-2, 0)$: $f_{xx}(-2, 0) = -6$, $f_{yy}(-2, 0) = -6$

$f_{xy}(-2, 0) = 0 \Rightarrow f_{xx}f_{yy} - f_{xy}^2 = 36 > 0$ and $f_{xx} < 0 \Rightarrow$ local maximum value of $f(-2, 0) = 4$. For $(-2, 2)$:

$f_{xx}(-2, 2) = -6$, $f_{yy}(-2, 2) = 6$, $f_{xy}(-2, 2) = 0 \Rightarrow f_{xx}f_{yy} - f_{xy}^2 = -36 < 0 \Rightarrow$ saddle point with $f(-2, 2) = 0$.

71. (i) On OA $f(x, y) = f(0, y) = y^2 + 3y$ for $0 \le y \le 4$

   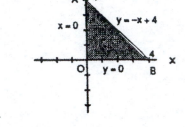

   $\Rightarrow f'(0, y) = 2y + 3 = 0 \Rightarrow y = -\frac{3}{2}$. But $\left(0, -\frac{3}{2}\right)$
   is not in the region.
   Endpoints: $f(0, 0) = 0$ and $f(0, 4) = 28$.

   (ii) On AB, $f(x, y) = f(x, -x + 4) = x^2 - 10x + 28$
   for $0 \le x \le 4 \Rightarrow f'(x, -x + 4) = 2x - 10 = 0$
   $\Rightarrow x = 5$, $y = -1$. But $(5, -1)$ is not in the region.
   Endpoints: $f(4, 0) = 4$ and $f(0, 4) = 28$.

   (iii) On OB, $f(x, y) = f(x, 0) = x^2 - 3x$ for $0 \le x \le 4 \Rightarrow f'(x, 0) = 2x - 3 \Rightarrow x = \frac{3}{2}$ and $y = 0 \Rightarrow \left(\frac{3}{2}, 0\right)$ is a

   critical point with $f\left(\frac{3}{2}, 0\right) = -\frac{9}{4}$. Endpoints: $f(0, 0) = 0$ and $f(4, 0) = 4$.

   (iv) For the interior of the triangular region, $f_x(x, y) = 2x + y - 3 = 0$ and $f_y(x, y) = x + 2y + 3 = 0 \Rightarrow x = 3$

   and $y = -3$. But $(3, -3)$ is not in the region. Therefore the absolute maximum is 28 at $(0, 4)$ and the

   absolute minimum is $-\frac{9}{4}$ at $\left(\frac{3}{2}, 0\right)$.

73. (i) On AB, $f(x, y) = f(-2, y) = y^2 - y - 4$ for

   $-2 \le y \le 2 \Rightarrow f'(-2, y) = 2y - 1 \Rightarrow y = \frac{1}{2}$ and

   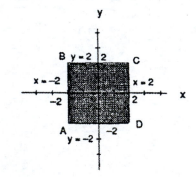

   $x = -2 \Rightarrow \left(-2, \frac{1}{2}\right)$ is an interior critical point in

   AB with $f\left(-2, \frac{1}{2}\right) = -\frac{17}{4}$.
   Endpoints: $f(-2, -2) = 2$ and $f(2, 2) = -2$.

   (ii) On BC, $f(x, y) = f(x, 2) = -2$ for $-2 \le x \le 2$
   $\Rightarrow f'(x, 2) = 0 \Rightarrow$ no critical points in the interior
   of BC.
   Endpoints: $f(-2, 2) = -2$ and $f(2, 2) = -2$.

   (iii) On CD, $f(x, y) = f(2, y) = y^2 - 5y + 4$ for $-2 \le y \le 2 \Rightarrow f'(2, y) = 2y - 5 = 0 \Rightarrow y = \frac{5}{2}$ and $x = 2$. But

   $\left(2, \frac{5}{2}\right)$ is not in the region. Endpoints: $f(2, -2) = 18$ and $f(2, 2) = -2$.

   (iv) On AD, $f(x, y) = f(x, -2) = 4x + 10$ for $-2 \le x \le 2 \Rightarrow f'(x, -2) = 4 \Rightarrow$ no critical points in the interior
   of AD. Endpoints: $f(-2, -2) = 2$ and $f(2, -2) = 18$.

   (v) For the interior of the square, $f_x(x, y) = -y + 2 = 0$ and $f_y(x, y) = 2y - x - 3 = 0 \Rightarrow y = 2$ and $x = 1$

   $\Rightarrow (1, 2)$ is an interior critical point of the square with $f(1, 2) = -2$. Therefore the absolute maximum is

   18 at $(2, -2)$ and the absolute minimum is $-\frac{17}{4}$ at $\left(-2, \frac{1}{2}\right)$.

75. (i) On AB, $f(x, y) = f(x, x+2) = -2x+4$ for
    $-2 \le x \le 2 \Rightarrow f'(x, x+2) = -2 = 0 \Rightarrow$ no critical
    points in the interior of AB.
    Endpoints: $f(-2, 0) = 8$ and $f(2, 4) = 0$.

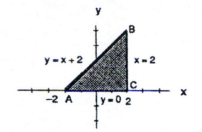

(ii) On BC, $f(x, y) = f(2, y) = -y^2 + 4y$ for
    $0 \le y \le 4 \Rightarrow f'(2, y) = -2y+4 = 0 \Rightarrow y = 2$ and
    $x = 2 \Rightarrow (2, 2)$ is and interior critical point of BC
    with $f(2, 2) = 4$.
    Endpoints: $f(2, 0) = 0$ and $f(2, 4) = 0$.

(iii) On AC, $f(x, y) = f(x, 0) = x^2 - 2x$ for $-2 \le x \le 2 \Rightarrow f'(x, 0) = 2x-2 \Rightarrow x = 1$ and $y = 0 \Rightarrow (1, 0)$ is an
    interior critical point of AC with $f(1, 0) = -1$. Endpoints: $f(-2, 0) = 8$ and $f(2, 0) = 0$.

(iv) For the interior of the triangular region, $f_x(x, y) = 2x-2 = 0$ and $f_y(x, y) = -2y+4 = 0 \Rightarrow x = 1$ and
    $y = 2 \Rightarrow (1, 2)$ is an interior critical point of the region with $f(1, 2) = 3$. Therefore the absolute maximum
    is 8 at $(-2, 0)$ and the absolute minimum is -1 at $(1, 0)$.

77. (i) On AB, $f(x, y) = f(-1, y) = y^3 - 3y^2 + 2$ for
    $-1 \le y \le 1 \Rightarrow f'(-1, y) = 3y^2 - 6y = 0 \Rightarrow y = 0$
    and $x = -1$, or $y = 2$ and $x = -1 \Rightarrow (-1, 0)$ is an
    interior critical point of AB with $f(-1, 0) = $
    $2; (-1, 2)$ is outside the boundary.
    Endpoints: $f(-1, -1) = -2$ and $f(-1, 1) = 0$.

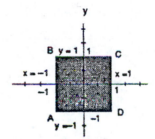

(ii) On BC, $f(x, y) = f(x, 1) = x^3 + 3x^2 - 2$ for $-1 \le x \le 1 \Rightarrow f'(x, 1) = 3x^2 + 6x = 0 \Rightarrow x = 0$ and $y = 1$, or
    $x = -2$ and $y = 1 \Rightarrow (0, 1)$ is an interior critical point of BC with $f(0, 1) = -2; (-2, 1)$ is outside the
    boundary. Endpoints: $f(-1, 1) = 0$ and $f(1, 1) = 2$.

(iii) On CD, $f(x, y) = f(1, y) = y^3 - 3y^2 + 4$ for $-1 \le y \le 1 \Rightarrow f'(1, y) = 3y^2 - 6y = 0 \Rightarrow y = 0$ and $x = 1$, or
    $y = 2$ and $x = 1 \Rightarrow (1, 0)$ is an interior critical point of CD with $f(1, 0) = 4; (1, 2)$ is outside the boundary.
    Endpoints: $f(1, 1) = 2$ and $f(1, -1) = 0$.

(iv) On AD, $f(x, y) = f(x, -1) = x^3 + 3x^2 - 4$ for $-1 \le x \le 1 \Rightarrow f'(x, -1) = 3x^2 + 6x = 0 \Rightarrow x = 0$ and
    $y = -1$, or $x = -2$ and $y = -1 \Rightarrow (0, -1)$ is an interior critical point of AD with $f(0, -1) = -4; (-2, -1)$ is
    outside the boundary. Endpoints: $f(-1, -1) = -2$ and $f(1, -1) = 0$.

(v) For the interior of the square, $f_x(x, y) = 3x^2 + 6x = 0$ and $f_y(x, y) = 3y^2 - 6y = 0 \Rightarrow x = 0$ or $x = -2$,
    and $y = 0$ or $y = 2 \Rightarrow (0, 0)$ is an interior critical point of the square region with $f(0, 0) = 0$; the points
    $(0, 2), (-2, 0)$, and $(-2, 2)$ are outside the region. Therefore the absolute maximum is 4 at $(1, 0)$ and the
    absolute minimum is -4 at $(0, -1)$.

79. $\nabla f = 3x^2 \mathbf{i} + 2y\mathbf{j}$ and $\nabla g = 2x\mathbf{i} + 2y\mathbf{j}$ so that $\nabla f = \lambda \nabla g \Rightarrow 3x^2 \mathbf{i} + 2y\mathbf{j} = \lambda(2x\mathbf{i} + 2y\mathbf{j}) \Rightarrow 3x^2 = 2x\lambda$ and
    $2y = 2y\lambda \Rightarrow \lambda = 1$ or $y = 0$.
    CASE 1: $\lambda = 1 \Rightarrow 3x^2 = 2x \Rightarrow x = 0$ or $x = \frac{2}{3}$; $x = 0 \Rightarrow y = \pm 1$ yielding the points $(0, 1)$ and $(0, -1)$;

$x = \frac{2}{3} \Rightarrow y = \pm\frac{\sqrt{5}}{3}$ yielding the points $\left(\frac{2}{3}, \frac{\sqrt{5}}{3}\right)$ and $\left(\frac{2}{3}, -\frac{\sqrt{5}}{3}\right)$.

CASE 2: $y = 0 \Rightarrow x^2 - 1 = 0 \Rightarrow x = \pm 1$ yielding the points $(1, 0)$ and $(-1, 0)$.

Evaluations give $f(0, \pm 1) = 1$, $f\left(\frac{2}{3}, \pm\frac{\sqrt{5}}{3}\right) = \frac{23}{27}$, $f(1, 0) = 1$, and $f(-1, 0) = -1$. Therefore the absolute maximum is 1 at $(0, \pm 1)$ and $(1, 0)$, and the absolute minimum is -1 at $(-1, 0)$.

81. (i) $f(x, y) = x^2 + 3y^2 + 2y$ on $x^2 + y^2 = 1 \Rightarrow \nabla f = 2x\mathbf{i} + (6y + 2)\mathbf{j}$ and $\nabla g = 2x\mathbf{i} + 2y\mathbf{j}$ so that
$\nabla f = \lambda \nabla g \Rightarrow 2x\mathbf{i} + (6y + 2)\mathbf{j} = \lambda(2x\mathbf{i} + 2y\mathbf{j}) \Rightarrow 2x = 2x\lambda$ and $6y + 2 = 2y\lambda \Rightarrow \lambda = 1$ or $x = 0$.

CASE 1: $\lambda = 1 \Rightarrow 6y + 2 = 2y \Rightarrow y = -\frac{1}{2}$ and $x = \pm\frac{\sqrt{3}}{2}$ yielding the points $\left(\pm\frac{\sqrt{3}}{2}, -\frac{1}{2}\right)$.

CASE 2: $x = 0 \Rightarrow y^2 = 1 \Rightarrow y = \pm 1$ yielding the points $(0, \pm 1)$.

Evaluations give $f\left(\pm\frac{\sqrt{3}}{2}, -\frac{1}{2}\right) = \frac{1}{2}$, $f(0, 1) = 5$, and $f(0, -1) = 1$. Therefore $\frac{1}{2}$ and 5 are the extreme values on the boundary of the disk.

(ii) For the interior of the disk, $f_x(x, y) = 2x = 0$ and $f_y(x, y) = 6y + 2 = 0 \Rightarrow x = 0$ and $y = -\frac{1}{3} \Rightarrow \left(0, -\frac{1}{3}\right)$ is an interior critical point with $f\left(0, -\frac{1}{3}\right) = -\frac{1}{3}$. Therefore the absolute maximum of f on the disk is 5 at $(0, 1)$ and the absolute minimum of f on the disk is $-\frac{1}{3}$ at $\left(0, -\frac{1}{3}\right)$.

83. $\nabla f = \mathbf{i} - \mathbf{j} + \mathbf{k}$ and $\nabla g = 2x\mathbf{i} + 2y\mathbf{j} + 2z\mathbf{k}$ so that $\nabla f = \lambda \nabla g \Rightarrow \mathbf{i} - \mathbf{j} + \mathbf{k} = \lambda(2x\mathbf{i} + 2y\mathbf{j} + 2z\mathbf{k}) \Rightarrow 1 = 2x\lambda$,
$-1 = 2y\lambda, 1 = 2z\lambda \Rightarrow x = -y = z = \frac{1}{\lambda}$. Thus $x^2 + y^2 + z^2 = 1 \Rightarrow 3x^2 = 1 \Rightarrow x = \pm\frac{1}{\sqrt{3}}$ yielding the points $\left(\frac{1}{\sqrt{3}}, -\frac{1}{\sqrt{3}}, \frac{1}{\sqrt{3}}\right)$ and $\left(-\frac{1}{\sqrt{3}}, \frac{1}{\sqrt{3}}, -\frac{1}{\sqrt{3}}\right)$. Evaluations give the absolute maximum value of $f\left(\frac{1}{\sqrt{3}}, -\frac{1}{\sqrt{3}}, \frac{1}{\sqrt{3}}\right) = \frac{3}{\sqrt{3}} = \sqrt{3}$ and the absolute minimum value of $f\left(-\frac{1}{\sqrt{3}}, \frac{1}{\sqrt{3}}, -\frac{1}{\sqrt{3}}\right) = -\sqrt{3}$.

85. The cost is $f(x, y, z) = 2axy + 2bxz + 2cyz$ subject to the constraint $xyz = V$. Then $\nabla f = \lambda \nabla g$
$\Rightarrow 2ay + 2bz = \lambda yz, 2ax + 2cz = \lambda xz$, and $2bx + 2cy = \lambda xy \Rightarrow 2axy + 2bxz = \lambda xyz, 2axy + 2cyz = \lambda xyz$, and $2bxz + 2cyz = \lambda xyz \Rightarrow 2axy + 2bxz = 2axy + 2cyz \Rightarrow y = \left(\frac{b}{c}\right)x$. Also $2axy + 2bxz = 2bxz + 2cyz \Rightarrow z = \left(\frac{a}{c}\right)x$.

Then $x\left(\frac{b}{c}x\right)\left(\frac{a}{c}x\right) = V \Rightarrow x^3 = \frac{c^2 V}{ab} \Rightarrow$ width $= x = \left(\frac{c^2 V}{ab}\right)^{1/3}$, Depth $= y = \left(\frac{b}{c}\right)\left(\frac{c^2 V}{ab}\right)^{1/3} = \left(\frac{b^2 V}{ac}\right)^{1/3}$, and

Height $= z = \left(\frac{a}{c}\right)\left(\frac{c^2 V}{ab}\right)^{1/3} = \left(\frac{a^2 V}{bc}\right)^{1/3}$.

87. $\nabla f = (y + z)\mathbf{i} + x\mathbf{j} + x\mathbf{k}$, $\nabla g = 2x\mathbf{i} + 2y\mathbf{j}$, and $\nabla h = z\mathbf{i} + x\mathbf{k}$ so that $\nabla f = \lambda \nabla g + \mu \nabla h \Rightarrow (y + z)\mathbf{i} + x\mathbf{j} + x\mathbf{k}$
$= \lambda(2x\mathbf{i} + 2y\mathbf{j}) + \mu(z\mathbf{i} + x\mathbf{k}) \Rightarrow y + z = 2\lambda x + \mu z, x = 2\lambda y, x = \mu x \Rightarrow x = 0$ or $\mu = 1$.
CASE 1: $x = 0$ which is impossible since $xz = 1$.
CASE 2: $\mu = 1 \Rightarrow y + z = 2\lambda x + z \Rightarrow y = 2\lambda x$ and $x = 2\lambda y \Rightarrow y = (2\lambda)(2\lambda y) \Rightarrow y = 0$ or $4\lambda^2 = 1$. If $y = 0$,
then $x^2 = 1 \Rightarrow x = \pm 1$ so with $xz = 1$ we obtain the points $(1, 0, 1)$ and $(-1, 0, -1)$. If $4\lambda^2 = 1$,
then $\lambda = \pm\frac{1}{2}$. For $\lambda = -\frac{1}{2}$, $y = -x$ so $x^2 + y^2 = 1 \Rightarrow x^2 = \frac{1}{2} \Rightarrow x = \pm\frac{1}{\sqrt{2}}$ with $xz = 1 \Rightarrow z = \pm\sqrt{2}$,
and we obtain the points $\left(\frac{1}{\sqrt{2}}, -\frac{1}{\sqrt{2}}, \sqrt{2}\right)$ and $\left(-\frac{1}{\sqrt{2}}, \frac{1}{\sqrt{2}}, -\sqrt{2}\right)$. For $\lambda = \frac{1}{2}$, $y = x$

$\Rightarrow x^2 = \frac{1}{2} \Rightarrow x = \pm\frac{1}{\sqrt{2}}$ with $xz = 1 \Rightarrow z = \pm\sqrt{2}$, and we obtain the points $\left(\frac{1}{\sqrt{2}}, \frac{1}{\sqrt{2}}, \sqrt{2}\right)$ and

$\left(-\frac{1}{\sqrt{2}}, -\frac{1}{\sqrt{2}}, -\sqrt{2}\right)$.

Evaluations give $f(1, 0, 1) = 1$, $f(-1, 0, -1) = 1$, $f\left(\frac{1}{\sqrt{2}}, -\frac{1}{\sqrt{2}}, \sqrt{2}\right) = \frac{1}{2}$, $f\left(-\frac{1}{\sqrt{2}}, \frac{1}{\sqrt{2}}, -\sqrt{2}\right) = \frac{1}{2}$,

$f\left(\frac{1}{\sqrt{2}}, \frac{1}{\sqrt{2}}, \sqrt{2}\right) = \frac{3}{2}$, and $f\left(-\frac{1}{\sqrt{2}}, -\frac{1}{\sqrt{2}}, -\sqrt{2}\right) = \frac{3}{2}$. Therefore the absolute maximum is $\frac{3}{2}$ at $\left(\frac{1}{\sqrt{2}}, \frac{1}{\sqrt{2}}, \sqrt{2}\right)$

and $\left(-\frac{1}{\sqrt{2}}, -\frac{1}{\sqrt{2}}, -\sqrt{2}\right)$, and the absolute minimum is $\frac{1}{2}$ at $\left(-\frac{1}{\sqrt{2}}, \frac{1}{\sqrt{2}}, -\sqrt{2}\right)$ and $\left(\frac{1}{\sqrt{2}}, -\frac{1}{\sqrt{2}}, \sqrt{2}\right)$.

89. Note that $x = r\cos\theta$ and $y = r\sin\theta \Rightarrow r = \sqrt{x^2 + y^2}$ and $\theta = \tan^{-1}\left(\frac{y}{x}\right)$. Thus,

$\frac{\partial w}{\partial x} = \frac{\partial w}{\partial r}\frac{\partial r}{\partial x} + \frac{\partial w}{\partial \theta}\frac{\partial \theta}{\partial x} = \left(\frac{\partial w}{\partial r}\right)\left(\frac{x}{\sqrt{x^2+y^2}}\right) + \left(\frac{\partial w}{\partial \theta}\right)\left(\frac{-y}{x^2+y^2}\right) = (\cos\theta)\frac{\partial w}{\partial r} - \left(\frac{\sin\theta}{r}\right)\frac{\partial w}{\partial \theta}$;

$\frac{\partial w}{\partial y} = \frac{\partial w}{\partial r}\frac{\partial r}{\partial y} + \frac{\partial w}{\partial \theta}\frac{\partial \theta}{\partial y} = \left(\frac{\partial w}{\partial r}\right)\left(\frac{y}{\sqrt{x^2+y^2}}\right) + \left(\frac{\partial w}{\partial \theta}\right)\left(\frac{x}{x^2+y^2}\right) = (\sin\theta)\frac{\partial w}{\partial r} + \left(\frac{\cos\theta}{r}\right)\frac{\partial w}{\partial \theta}$

91. $\frac{\partial u}{\partial y} = b$ and $\frac{\partial u}{\partial x} = a \Rightarrow \frac{\partial w}{\partial x} = \frac{dw}{du}\frac{\partial u}{\partial x} = a\frac{dw}{du}$ and $\frac{\partial w}{\partial y} = \frac{dw}{du}\frac{\partial u}{\partial y} = b\frac{dw}{du} \Rightarrow \frac{1}{a}\frac{\partial w}{\partial x} = \frac{dw}{du}$ and $\frac{1}{b}\frac{\partial w}{\partial y} = \frac{dw}{du}$

$\Rightarrow \frac{1}{a}\frac{\partial w}{\partial x} = \frac{1}{b}\frac{\partial w}{\partial y} \Rightarrow b\frac{\partial w}{\partial x} = a\frac{\partial w}{\partial y}$

93. $e^u\cos v - x = 0 \Rightarrow \left(e^u\cos v\right)\frac{\partial u}{\partial x} - \left(e^u\sin v\right)\frac{\partial v}{\partial x} = 1$; $e^u\sin v - y = 0 \Rightarrow \left(e^u\sin v\right)\frac{\partial u}{\partial x} + \left(e^u\cos v\right)\frac{\partial v}{\partial x} = 0$. Solving

this system yields $\frac{\partial u}{\partial x} = e^{-u}\cos v$ and $\frac{\partial v}{\partial x} = -e^{-u}\sin v$. Similarly, $e^u\cos v - x = 0$

$\Rightarrow \left(e^u\cos v\right)\frac{\partial u}{\partial y} - \left(e^u\sin v\right)\frac{\partial v}{\partial y} = 0$ and $e^u\sin v - y = 0 \Rightarrow \left(e^u\sin v\right)\frac{\partial u}{\partial y} + \left(e^u\cos v\right)\frac{\partial v}{\partial y} = 1$. Solving this second

system yields $\frac{\partial u}{\partial y} = e^{-u}\sin v$ and $\frac{\partial v}{\partial y} = e^{-u}\cos v$. Therefore $\left(\frac{\partial u}{\partial x}\mathbf{i} + \frac{\partial u}{\partial y}\mathbf{j}\right) \cdot \left(\frac{\partial v}{\partial x}\mathbf{i} + \frac{\partial v}{\partial y}\mathbf{j}\right)$

$= \left[\left(e^{-u}\cos v\right)\mathbf{i} + \left(e^{-u}\sin v\right)\mathbf{j}\right] \cdot \left[\left(-e^{-u}\sin v\right)\mathbf{i} + \left(e^{-u}\cos v\right)\mathbf{j}\right] = 0 \Rightarrow$ the vectors are orthogonal $\Rightarrow$ the angle

between the vectors is the constant $\frac{\pi}{2}$.

95. $(y+z)^2 + (z-x)^2 = 16 \Rightarrow \nabla f = -2(z-x)\mathbf{i} + 2(y+z)\mathbf{j} + 2(y+2z-x)\mathbf{k}$; if the normal line is parallel to the

yz-plane, then x is constant $\Rightarrow \frac{\partial f}{\partial x} = 0 \Rightarrow -2(z-x) = 0 \Rightarrow z = x \Rightarrow (y+z)^2 + (z-z)^2 = 16 \Rightarrow y+z = \pm 4$.

Let $x = t \Rightarrow z = t \Rightarrow y = -t \pm 4$. Therefore the points are $(t, -t \pm 4, t)$, t a real number.

97. $\nabla f = \lambda(x\mathbf{i} + y\mathbf{j} + z\mathbf{k}) \Rightarrow \frac{\partial f}{\partial x} = \lambda x \Rightarrow f(x, y, z) = \frac{1}{2}\lambda x^2 + g(y, z)$ for some function $g \Rightarrow \lambda y = \frac{\partial f}{\partial y} = \frac{\partial g}{\partial y}$

$\Rightarrow g(y, z) = \frac{1}{2}\lambda y^2 + h(z)$ for some function $h \Rightarrow \lambda z = \frac{\partial f}{\partial z} = \frac{\partial g}{\partial z} = h'(z) \Rightarrow h(z) = \frac{1}{2}\lambda z^2 + C$ for some arbitrary

constant $C \Rightarrow g(y, z) = \frac{1}{2}\lambda y^2 + \left(\frac{1}{2}\lambda z^2 + C\right) \Rightarrow f(x, y, z) = \frac{1}{2}\lambda x^2 + \frac{1}{2}\lambda y^2 + \frac{1}{2}\lambda z^2 + C$

$\Rightarrow f(0, 0, a) = \frac{1}{2}\lambda a^2 + C$ and $f(0, 0, -a) = \frac{1}{2}\lambda(-a)^2 + C \Rightarrow f(0, 0, a) = f(0, 0, -a)$ for any constant a, as

claimed.

99. Let $f(x, y, z) = xy + z - 2 \Rightarrow \nabla f = y\mathbf{i} + x\mathbf{j} + \mathbf{k}$. At $(1, 1, 1)$, we have $\nabla f = \mathbf{i} + \mathbf{j} + \mathbf{k} \Rightarrow$ the normal line is

$x = 1 + t$, $y = 1 + t$, $z = 1 + t$, so at $t = -1 \Rightarrow x = 0, y = 0, z = 0$ and the normal line passes through the origin.

CHAPTER 13 ADDITIONAL AND ADVANCED EXERCISES

1. By definition, $f_{xy}(0,0) = \lim\limits_{h\to 0} \frac{f_x(0,h)-f_x(0,0)}{h}$ so we need to calculate the first partial derivatives in the

   numerator. For $(x,y) \neq (0,0)$ we calculate $f_x(x,y)$ by applying the differentiation rules to the formula for

   $f(x,y)$: $f_x(x,y) = \frac{x^2 y - y^3}{x^2+y^2} + (xy)\frac{\left(x^2+y^2\right)(2x)-\left(x^2-y^2\right)(2x)}{\left(x^2+y^2\right)^2} = \frac{x^2 y - y^3}{x^2+y^2} + \frac{4x^2 y^3}{\left(x^2+y^2\right)^2} \Rightarrow f_x(0,h) = \frac{h^3}{h^2} = -h.$ For

   $(x,y) = (0,0)$ we apply the definition: $f_x(0,0) = \lim\limits_{h\to 0} \frac{f(h,0)-f(0,0)}{h} = \lim\limits_{h\to 0} \frac{0-0}{h} = 0.$ Then by definition

   $f_{xy}(0,0) = \lim\limits_{h\to 0} \frac{-h-0}{h} = -1.$ Similarly, $f_{yx}(0,0) = \lim\limits_{h\to 0} \frac{f_y(h,0)-f_y(0,0)}{h}$, so for $(x,y) \neq (0,0)$ we have

   $f_y(x,y) = \frac{x^3 - xy^2}{x^2+y^2} - \frac{4x^3 y^2}{\left(x^2+y^2\right)^2} \Rightarrow f_y(h,0) = \frac{h^3}{h^2} = h;$ for $(x,y) = (0,0)$ we obtain $f_y(0,0) = \lim\limits_{h\to 0} \frac{f(0,h)-f(0,0)}{h}$

   $= \lim\limits_{h\to 0} \frac{0-0}{h} = 0.$ Then by definition $f_{yx}(0,0) = \lim\limits_{h\to 0} \frac{h-0}{h} = 1.$ Note that $f_{xy}(0,0) \neq f_{yx}(0,0)$ in this case.

3. Substitution of $u + u(x)$ and $v = v(x)$ in $g(u,v)$ gives $g(u(x), v(x))$ which is a function of the independent

   variable x. Then, $g(u,v) = \int_u^v f(t)dt \Rightarrow \frac{dg}{dx} = \frac{\partial g}{\partial u}\frac{du}{dx} + \frac{\partial g}{\partial v}\frac{dv}{dx} = \left(\frac{\partial}{\partial u}\int_u^v f(t)\,dt\right)\frac{du}{dx} + \left(\frac{\partial}{\partial v}\int_u^v f(t)\,dt\right)\frac{dv}{dx}$

   $= \left(-\frac{\partial}{\partial u}\int_v^u f(t)\,dt\right)\frac{du}{dx} + \left(\frac{\partial}{\partial v}\int_u^v f(t)\,dt\right)\frac{dv}{dx} = -f(u(x))\frac{du}{dx} + f(v(x))\frac{dv}{dx} = f(v(x))\frac{dv}{dx} - f(u(x))\frac{du}{dx}$

5. (a) Let $u = tx, v = ty$, and $w = f(u,v) = f(u(t,x), v(t,y)) = f(tx, ty) = t^n f(x,y)$, where t, x, and y are

      independent variables. Then $nt^{n-1}f(x,y) = \frac{\partial w}{\partial t} = \frac{\partial w}{\partial u}\frac{\partial u}{\partial t} + \frac{\partial w}{\partial v}\frac{\partial v}{\partial t} = x\frac{\partial w}{\partial u} + y\frac{\partial w}{\partial v}$. Now, $\frac{\partial w}{\partial x} = \frac{\partial w}{\partial u}\frac{\partial u}{\partial x} + \frac{\partial w}{\partial v}\frac{\partial v}{\partial x}$

      $= \left(\frac{\partial w}{\partial u}\right)(t) + \left(\frac{\partial w}{\partial v}\right)(0) = t\frac{\partial w}{\partial u} \Rightarrow \frac{\partial w}{\partial u} = \left(\frac{1}{t}\right)\left(\frac{\partial w}{\partial x}\right)$. Likewise, $\frac{\partial w}{\partial y} = \frac{\partial w}{\partial u}\frac{\partial u}{\partial y} + \frac{\partial w}{\partial v}\frac{\partial v}{\partial y} = \left(\frac{\partial w}{\partial u}\right)(0) + \left(\frac{\partial w}{\partial v}\right)(t)$

      $\Rightarrow \frac{\partial w}{\partial v} = \left(\frac{1}{t}\right)\left(\frac{\partial w}{\partial y}\right)$. Therefore, $nt^{n-1}f(x,y) = x\frac{\partial w}{\partial u} + y\frac{\partial w}{\partial v} = \left(\frac{x}{t}\right)\left(\frac{\partial w}{\partial x}\right) + \left(\frac{y}{t}\right)\left(\frac{\partial w}{\partial y}\right)$. When $t = 1, u = x, v = y$,

      and $w = f(x,y) \Rightarrow \frac{\partial w}{\partial x} = \frac{\partial f}{\partial x}$ and $\frac{\partial w}{\partial y} = \frac{\partial f}{\partial x} \Rightarrow nf(x,y) = x\frac{\partial f}{\partial x} + y\frac{\partial f}{\partial y}$, as claimed.

   (b) From part (a), $nt^{n-1}f(x,y) = x\frac{\partial w}{\partial u} + y\frac{\partial w}{\partial v}$. Differentiating with respect to t again we obtain

      $n(n-1)t^{n-2}f(x,y) = x\frac{\partial^2 w}{\partial u^2}\frac{\partial u}{\partial t} + x\frac{\partial^2 w}{\partial v\partial u}\frac{\partial v}{\partial t} + y\frac{\partial^2 w}{\partial u\partial v}\frac{\partial v}{\partial t} + y\frac{\partial^2 w}{\partial v^2}\frac{\partial v}{\partial t} = x^2\frac{\partial^2 w}{\partial u^2} + 2xy\frac{\partial^2 w}{\partial u\partial v} + y^2\frac{\partial^2 w}{\partial v^2}$. Also from

      part (a), $\frac{\partial^2 w}{\partial x^2} = \frac{\partial}{\partial x}\left(\frac{\partial w}{\partial x}\right) = \frac{\partial}{\partial x}\left(t\frac{\partial w}{\partial u}\right) = t\frac{\partial^2 w}{\partial u^2}\frac{\partial u}{\partial x} + t\frac{\partial^2 w}{\partial v\partial u}\frac{\partial v}{\partial x} = t^2\frac{\partial^2 w}{\partial u^2}, \frac{\partial^2 w}{\partial y^2} = \frac{\partial}{\partial y}\left(\frac{\partial w}{\partial y}\right) = \frac{\partial}{\partial y}\left(t\frac{\partial w}{\partial v}\right)$

      $= t\frac{\partial^2 w}{\partial u\partial v}\frac{\partial u}{\partial y} + t\frac{\partial^2 w}{\partial v^2}\frac{\partial v}{\partial y} = t^2\frac{\partial^2 w}{\partial v^2}$, and $\frac{\partial^2 w}{\partial y\partial x} = \frac{\partial}{\partial y}\left(\frac{\partial w}{\partial x}\right) = \frac{\partial}{\partial y}\left(t\frac{\partial w}{\partial u}\right) = t\frac{\partial^2 w}{\partial u^2}\frac{\partial u}{\partial y} + t\frac{\partial^2 w}{\partial v\partial u}\frac{\partial v}{\partial y} = t^2\frac{\partial^2 w}{\partial v\partial u}$

      $\Rightarrow \left(\frac{1}{t^2}\right)\frac{\partial^2 w}{\partial x^2} = \frac{\partial^2 w}{\partial u^2}, \left(\frac{1}{t^2}\right)\frac{\partial^2 w}{\partial y^2} = \frac{\partial^2 w}{\partial v^2}$, and $\left(\frac{1}{t^2}\right)\frac{\partial^2 w}{\partial y\partial x} = \frac{\partial^2 w}{\partial v\partial u}$

      $\Rightarrow n(n-1)t^{n-2}f(x,y) = \left(\frac{x^2}{t^2}\right)\left(\frac{\partial^2 w}{\partial x^2}\right) + \left(\frac{2xy}{t^2}\right)\left(\frac{\partial^2 w}{\partial y\partial x}\right) + \left(\frac{y^2}{t^2}\right)\left(\frac{\partial^2 w}{\partial y^2}\right)$ for $t \neq 0$.

      When $t = 1, w = f(x,y)$ and we have $n(n-1)f(x,y) = x^2\left(\frac{\partial^2 f}{\partial x^2}\right) + 2xy\left(\frac{\partial^2 f}{\partial x\partial y}\right) + y^2\left(\frac{\partial^2 f}{\partial y^2}\right)$ as claimed.

7. (a) $\mathbf{r} = x\mathbf{i} + y\mathbf{j} + z\mathbf{k} \Rightarrow r = |\mathbf{r}| = \sqrt{x^2 + y^2 + z^2}$ and $\nabla r = \frac{x}{\sqrt{x^2+y^2+z^2}}\mathbf{i} + \frac{y}{\sqrt{x^2+y^2+z^2}}\mathbf{j} + \frac{z}{\sqrt{x^2+y^2+z^2}}\mathbf{k} = \frac{\mathbf{r}}{r}$

   (b) $r^n = \left(\sqrt{x^2 + y^2 + z^2}\right)^n$

   $\Rightarrow \nabla(r^n) = nx\left(x^2 + y^2 + z^2\right)^{(n/2)-1}\mathbf{i} + ny\left(x^2 + y^2 + z^2\right)^{(n/2)-1}\mathbf{j} + nz\left(x^2 + y^2 + z^2\right)^{(n/2)-1}\mathbf{k} = nr^{n-2}\mathbf{r}$

   (c) Let $n = 2$ in part (b). Then $\frac{1}{2}\nabla(r^2) = \mathbf{r} \Rightarrow \nabla\left(\frac{1}{2}r^2\right) = \mathbf{r} \Rightarrow \frac{r^2}{2} = \frac{1}{2}\left(x^2 + y^2 + z^2\right)$ is the function.

   (d) $d\mathbf{r} = dx\mathbf{i} + dy\mathbf{j} + dz\mathbf{k} \Rightarrow \mathbf{r} \cdot d\mathbf{r} = x\,dx + y\,dy + z\,dz$, and $dr = r_x\,dx + r_y\,dy + r_z\,dz = \frac{x}{r}dx + \frac{y}{r}dy + \frac{z}{r}dz$

   $\Rightarrow r\,dr = x\,dx + y\,dy + z\,dz = \mathbf{r} \cdot d\mathbf{r}$

   (e) $\mathbf{A} = a\mathbf{i} + b\mathbf{j} + c\mathbf{k} \Rightarrow \mathbf{A} \cdot \mathbf{r} = ax + by + cz \Rightarrow \nabla(\mathbf{A} \cdot \mathbf{r}) = a\mathbf{i} + b\mathbf{j} + c\mathbf{k} = \mathbf{A}$

9. $f(x, y, z) = xy^2 - yz + \cos xy - 1 \Rightarrow \nabla f = \left(z^2 - y\sin xy\right)\mathbf{i} + (-z - x\sin xy)\mathbf{j} + (2xz - y)\mathbf{k} \Rightarrow \nabla f(0, 0, 1) = \mathbf{i} - \mathbf{j}$

   $\Rightarrow$ the tangent plane is $x - y = 0$; $\mathbf{r} = (\ln t)\mathbf{i} + (t\ln t)\mathbf{j} + t\mathbf{k} \Rightarrow \mathbf{r}' = \left(\frac{1}{t}\right)\mathbf{i} + (\ln t + 1)\mathbf{j} + \mathbf{k}$; $x = y = 0, z = 1$

   $\Rightarrow t = 1 \Rightarrow \mathbf{r}'(1) = \mathbf{i} + \mathbf{j} + \mathbf{k}$. Since $(\mathbf{i} + \mathbf{j} + \mathbf{k}) \cdot (\mathbf{i} - \mathbf{j}) = \mathbf{r}'(1) \cdot \nabla f = 0$, $\mathbf{r}$ is parallel to the plane, and

   $\mathbf{r}(1) = 0\mathbf{i} + 0\mathbf{j} + \mathbf{k} \Rightarrow \mathbf{r}$ is contained in the plane.

11. $\frac{\partial z}{\partial x} = 3x^2 - 9y = 0$ and $\frac{\partial z}{\partial x} = 3y^2 - 9x = 0 \Rightarrow y = \frac{1}{3}x^2$ and $3\left(\frac{1}{3}x^2\right)^2 - 9x = 0 \Rightarrow \frac{1}{3}x^4 - 9x = 0$

    $\Rightarrow x\left(x^3 - 27\right) = 0 \Rightarrow x = 0$ or $x = 3$. Now $x = 0 \Rightarrow y = 0$ or $(0, 0)$ and $x = 3 \Rightarrow y = 3$ or $(3, 3)$. Next

    $\frac{\partial^2 z}{\partial x^2} = 6x$, $\frac{\partial^2 z}{\partial y^2} = 6y$, and $\frac{\partial^2 z}{\partial x\partial y} = -9$. For $(0, 0)$, $\frac{\partial^2 z}{\partial x^2}\frac{\partial^2 z}{\partial y^2} - \left(\frac{\partial^2 z}{\partial x\partial y}\right)^2 = -81 \Rightarrow$ no extremum (a saddle point), and

    for $(3, 3)$, $\frac{\partial^2 z}{\partial x^2}\frac{\partial^2 z}{\partial y^2} - \left(\frac{\partial^2 z}{\partial x\partial y}\right)^2 = 243 > 0$ and $\frac{\partial^2 z}{\partial x^2} = 18 > 0 \Rightarrow$ a local minimum.

13. Let $f(x, y, z) = \frac{x^2}{a^2} + \frac{y^2}{b^2} + \frac{z^2}{c^2} - 1 \Rightarrow \nabla f = \frac{2x}{a^2}\mathbf{i} + \frac{2y}{b^2}\mathbf{j} + \frac{2z}{c^2}\mathbf{k} \Rightarrow$ an equation of the plane tangent at the point

    $P_0(x_0, y_0, y_0)$ is $\left(\frac{2x_0}{a^2}\right)x + \left(\frac{2y_0}{b^2}\right)y + \left(\frac{2z_0}{c^2}\right)z = \frac{2x_0^2}{a^2} + \frac{2y_0^2}{b^2} + \frac{2z_0^2}{c^2} = 2$ or $\left(\frac{x_0}{a^2}\right)x + \left(\frac{y_0}{b^2}\right)y + \left(\frac{z_0}{c^2}\right)z = 1$. The

    intercepts of the plane are $\left(\frac{a^2}{x_0}, 0, 0\right), \left(0, \frac{b^2}{y_0}, 0\right)$ and $\left(0, 0, \frac{c^2}{z_0}\right)$. The volume of the tetrahedron formed by the

    plane and the coordinate planes is $V = \left(\frac{1}{3}\right)\left(\frac{1}{2}\right)\left(\frac{a^2}{x_0}\right)\left(\frac{b^2}{y_0}\right)\left(\frac{c^2}{z_0}\right) \Rightarrow$ we need to maximize

    $V(x, y, z) = \frac{(abc)^2}{6}(xyz)^{-1}$ subject to the constraint $f(x, y, z) = \frac{x^2}{a^2} + \frac{y^2}{b^2} + \frac{z^2}{c^2} = 1$. Thus,

    $\left[-\frac{(abc)^2}{6}\right]\left(\frac{1}{x^2yz}\right) = \frac{2x}{a^2}\lambda$, $\left[-\frac{(abc)^2}{6}\right]\left(\frac{1}{xy^2z}\right) = \frac{2y}{b^2}\lambda$, and $\left[-\frac{(abc)^2}{6}\right]\left(\frac{1}{xyz^2}\right) = \frac{2z}{c^2}\lambda$. Multiply the first equation by

    a^2yz, the second by b^2xz, and the third by c^2xy. Then equate the first and second $\Rightarrow a^2y^2 = b^2x^2$

    $\Rightarrow y = \frac{b}{a}x, x > 0$; equate the first and third $\Rightarrow a^2z^2 = c^2x^2 \Rightarrow z = \frac{c}{a}x, x > 0$; substitute into

    $f(x, y, z) = 0 \Rightarrow x = \frac{a}{\sqrt{3}} \Rightarrow y = \frac{b}{\sqrt{3}} \Rightarrow z = \frac{c}{\sqrt{3}} \Rightarrow V = \frac{\sqrt{3}}{2}abc$.

15. Let (x_0, y_0) be any point in R. We must show $\lim\limits_{(x,\,y)\to(x_0,\,y_0)} f(x, y) = f(x_0, y_0)$ or, equivalently that

$\lim\limits_{(h,\,k)\to(0,\,0)} |f(x_0 + h, y_0 + k) - f(x_0, y_0)| = 0$. Consider $f(x_0 + h, y_0 + k) - f(x_0, y_0)$

$= [f(x_0 + h, y_0 + k) - f(x_0, y_0 + k)] + [f(x_0, y_0 + k) - f(x_0, y_0)]$. Let $F(x) = f(x, y_0 + k)$ and apply the

Mean Value Theorem: there exists ξ with $x_0 < \xi < x_0 + h$ such that $F'(\xi)h = F(x_0 + h) - F(x_0)$

$\Rightarrow hf_x(\xi, y_0 + k) = f(x_0 + h, y_0 + k) - f(x_0, y_0 + k)$. Similarly, $k\, f_y(x_0, \eta) = f(x_0, y_0 + k) - f(x_0, y_0)$ for

some η with $y_0 < \eta < y_0 + k$. Then $|f(x_0 + h, y_0 + k) - f(x_0, y_0)| \le |hf_x(\xi, y_0 + k)| + |k\, f_y(x_0, \eta)|$. If M, N are

positive real numbers such that $|f_x| \le M$ and $|f_y| \le N$ for all (x, y) in the xy-plane, then

$|f(x_0 + h, y_0 + k) - f(x_0, y_0)| \le M|h| + N|k|$. As $|(h, k)| \to 0, |f(x_0 + h, y_0 + k) - f(x_0, y_0)| \to 0$

$\Rightarrow \lim\limits_{(h,\,k)\to(0,\,0)} |f(x_0 + h, y_0 + k) - f(x_0, y_0)| = 0 \Rightarrow f$ is continuous at (x_0, y_0).

17. $\frac{\partial f}{\partial x} = 0 \Rightarrow f(x, y) = h(y)$ is a function of y only. Also, $\frac{\partial g}{\partial y} = \frac{\partial f}{\partial x} = 0 \Rightarrow g(x, y) = k(x)$ is a function of x only.

Moreover, $\frac{\partial f}{\partial y} = \frac{\partial g}{\partial x} \Rightarrow h'(y) = k'(x)$ for all x and y. This can happen only if $h'(y) = k'(x) = c$ is a constant.

Integration gives $h(y) = cy + c_1$ and $k(x) = cx + c_2$, where c_1 and c_2 are constants. Therefore $f(x, y) = cy + c_1$

and $g(x, y) = cx + c_2$. Then $f(1, 2) = g(1, 2) = 5 \Rightarrow 5 = 2c + c_1 = c + c_2$, and $f(0, 0) = 4 \Rightarrow c_1 = 4 \Rightarrow c = \frac{1}{2}$

$\Rightarrow c_2 = \frac{9}{2}$. Thus, $f(x, y) = \frac{1}{2}y + 4$ and $g(x, y) = \frac{1}{2}x + \frac{9}{2}$.

19. Since the particle is heat-seeking, at each point (x, y) it moves in the direction of maximal temperature

increase, that is in the direction of $\nabla T(x, y) = \left(e^{-2y} \sin x\right)\mathbf{i} + \left(2e^{-2y} \cos x\right)\mathbf{j}$. Since $\nabla T(x, y)$ is parallel to the

particle's velocity vector, it is tangent to the path $y = f(x)$ of the particle $\Rightarrow f'(x) = \frac{2e^{-2y}\cos x}{e^{-2y}\sin x} = 2\cot x$.

Integration gives $f(x) = 2\ln|\sin x| + C$ and $f\left(\frac{\pi}{4}\right) = 0 \Rightarrow 0 = 2\ln\left|\sin \frac{\pi}{4}\right| + C \Rightarrow C = -2\ln\frac{\sqrt{2}}{2} = \ln\left(\frac{2}{\sqrt{2}}\right)^2 = \ln 2$.

Therefore, the path of the particle is the graph of $y = 2\ln|\sin x| + \ln 2$.

21. (a) $\mathbf{k}$ is a vector normal to $z = 10 - x^2 - y^2$ at the point $(0, 0, 10)$. So directions tangential to S at $(0, 0, 10)$

will be unit vectors $\mathbf{u} = a\mathbf{i} + b\mathbf{j}$. Also, $\nabla T(x, y, z) = (2xy + 4)\mathbf{i} + \left(x^2 + 2yz + 14\right)\mathbf{j} + \left(y^2 + 1\right)\mathbf{k}$

$\Rightarrow \nabla T(0, 0, 10) = 4\mathbf{i} + 14\mathbf{j} + \mathbf{k}$. We seek the unit vector $\mathbf{u} = a\mathbf{i} + b\mathbf{j}$ such that $D_{\mathbf{u}}T(0, 0, 10)$

$= (4\mathbf{i} + 14\mathbf{j} + \mathbf{k})\cdot(a\mathbf{i} + b\mathbf{j}) = (4\mathbf{i} + 14\mathbf{j})\cdot(a\mathbf{i} + b\mathbf{j})$ is a maximum. The maximum will occur when $a\mathbf{i} + b\mathbf{j}$ has

the same direction as $4\mathbf{i} + 14\mathbf{j}$, or $\mathbf{u} = \frac{1}{\sqrt{53}}(2\mathbf{i} + 7\mathbf{j})$.

(b) A vector normal to S at $(1, 1, 8)$ is $\mathbf{n} = 2\mathbf{i} + 2\mathbf{j} + \mathbf{k}$. Now, $\nabla T(1, 1, 8) = 6\mathbf{i} + 31\mathbf{j} + 2\mathbf{k}$ and we seek the unit

vector $\mathbf{u}$ such that $D_{\mathbf{u}}T(1, 1, 8) = \nabla T\cdot\mathbf{u}$ has its largest value. Now write $\nabla T = \mathbf{v} + \mathbf{w}$, where $\mathbf{v}$ is parallel

to ∇T and $\mathbf{w}$ is orthogonal to ∇T. Then $D_{\mathbf{u}}T = \nabla T\cdot\mathbf{u} = (\mathbf{v} + \mathbf{w})\cdot\mathbf{u} = \mathbf{v}\cdot\mathbf{u} + \mathbf{w}\cdot\mathbf{u} = \mathbf{w}\cdot\mathbf{u}$. Thus

$D_{\mathbf{u}}T(1, 1, 8)$ is a maximum when $\mathbf{u}$ has the same direction as $\mathbf{w}$. Now, $\mathbf{w} = \nabla T - \left(\frac{\nabla T\cdot\mathbf{n}}{|\mathbf{n}|^2}\right)\mathbf{n}$

$= (6\mathbf{i} + 31\mathbf{j} + 2\mathbf{k}) - \left(\frac{12 + 62 + 2}{4 + 4 + 1}\right)(2\mathbf{i} + 2\mathbf{j} + \mathbf{k}) = \left(6 - \frac{152}{9}\right)\mathbf{i} + \left(31 - \frac{152}{9}\right)\mathbf{j} + \left(2 - \frac{76}{9}\right)\mathbf{k} = -\frac{98}{9}\mathbf{i} + \frac{127}{9}\mathbf{j} - \frac{58}{9}\mathbf{k}$

$\Rightarrow \mathbf{u} = \frac{\mathbf{w}}{|\mathbf{w}|} = -\frac{1}{\sqrt{29{,}097}}(98\mathbf{i} - 127\mathbf{j} + 58\mathbf{k})$.

23. $w = e^{rt} \sin \pi x \Rightarrow w_t = re^{rt} \sin \pi x$ and $w_x = \pi e^{rt} \cos \pi x \Rightarrow w_{xx} = -\pi^2 e^{rt} \sin \pi x$; $w_{xx} = \frac{1}{c^2} w_t$, where c^2 is the

positive constant determined by the material of the rod $\Rightarrow -\pi^2 e^{rt} \sin \pi x = \frac{1}{c^2}\left(re^{rt} \sin \pi x \right)$

$\Rightarrow \left(r + c^2 \pi^2 \right) e^{rt} \sin \pi x = 0 \Rightarrow r = -c^2 \pi^2 \Rightarrow w = e^{-c^2 \pi^2 t} \sin \pi x$

CHAPTER 14   MULTIPLE INTEGRALS

14.1 DOUBLE AND ITERATED INTEGRALS OVER RECTANGLES

1. $\int_1^2 \int_0^4 2xy \, dy \, dx = \int_1^2 \left[xy^2 \right]_0^4 dx = \int_1^2 16x \, dx = \left[8x^2 \right]_1^2 = 24$

3. $\int_{-1}^0 \int_{-1}^1 (x+y+1) \, dx \, dy = \int_{-1}^0 \left[\frac{x^2}{2} + yx + x \right]_{-1}^1 dy = \int_{-1}^0 (2y+2) \, dy = \left[y^2 + 2y \right]_{-1}^0 = 1$

5. $\int_0^3 \int_0^2 \left(4 - y^2 \right) dy \, dx = \int_0^3 \left[4y - \frac{y^3}{3} \right]_0^2 dx = \int_0^3 \frac{16}{3} \, dx = \left[\frac{16}{3} x \right]_0^3 = 16$

7. $\int_0^1 \int_0^1 \frac{y}{1+xy} \, dx \, dy = \int_0^1 \left[\ln|1+xy| \right]_0^1 dy = \int_0^1 \ln|1+y| dy = \left[y \ln|1+y| - y + \ln|1+y| \right]_0^1 = 2\ln 2 - 1$

9. $\int_0^{\ln 2} \int_1^{\ln 5} e^{2x+y} dy \, dx = \int_0^{\ln 2} \left[e^{2x+y} \right]_1^{\ln 5} dx = \int_0^{\ln 2} \left(5e^{2x} - e^{2x+1} \right) dx = \left[\frac{5}{2} e^{2x} - \frac{1}{2} e^{2x+1} \right]_0^{\ln 2} = \frac{3}{2}(5-e)$

11. $\int_{-1}^2 \int_0^{\pi/2} y \sin x \, dx \, dy = \int_{-1}^2 \left[-y\cos x \right]_0^{\pi/2} dy = \int_{-1}^2 y \, dy = \left[\frac{1}{2} y^2 \right]_{-1}^2 = \frac{3}{2}$

13. $\int_1^4 \int_1^e \frac{\ln x}{xy} dx \, dy = \int_1^4 \left. \frac{(\ln x)^2}{2y} \right|_{x=1}^{x=e} dy = \int_1^4 \frac{1}{2y} \, dy = \left. \frac{\ln y}{2} \right|_1^4 = \ln 2$

15. $\iint\limits_R \left(6y^2 - 2x \right) dA = \int_0^1 \int_0^2 \left(6y^2 - 2x \right) dy \, dx = \int_0^1 [2y^3 - 2xy]_0^2 \, dx = \int_0^1 (16 - 4x) \, dx = [16x - 2x^2]_0^1 = 14$

17. $\iint\limits_R xy \cos y \, dA = \int_{-1}^1 \int_0^\pi xy \cos y \, dy \, dx = \int_{-1}^1 [xy \sin y + x\cos y]_0^\pi \, dx = \int_{-1}^1 (-2x) \, dx = \left[-x^2 \right]_{-1}^1 = 0$

19. $\iint\limits_R e^{x-y} dA = \int_0^{\ln 2} \int_0^{\ln 2} e^{x-y} dy \, dx = \int_0^{\ln 2} \left[-e^{x-y} \right]_0^{\ln 2} dx = \int_0^{\ln 2} \left(-e^{x-\ln 2} + e^x \right) dx = \left[-e^{x-\ln 2} + e^x \right]_0^{\ln 2} = \frac{1}{2}$

21. $\iint\limits_R \frac{xy^3}{x^2+1} dA = \int_0^1 \int_0^2 \frac{xy^3}{x^2+1} dy \, dx = \int_0^1 \left[\frac{xy^4}{4(x^2+1)} \right]_0^2 dx = \int_0^1 \frac{4x}{x^2+1} \, dx = \left[2\ln|x^2 + 1| \right]_0^1 = 2\ln 2$

23. $\int_1^2 \int_1^2 \frac{1}{xy} \, dy \, dx = \int_1^2 \frac{1}{x}(\ln 2 - \ln 1) \, dx = (\ln 2) \int_1^2 \frac{1}{x} \, dx = (\ln 2)^2$

25. $V = \iint\limits_R f(x, y) \, dA = \int_{-1}^1 \int_{-1}^1 \left(x^2 + y^2 \right) dy \, dx = \int_{-1}^1 \left[x^2 y + \frac{1}{3} y^3 \right]_{-1}^1 dx = \int_{-1}^1 \left(2x^2 + \frac{2}{3} \right) dx = \left[\frac{2}{3} x^3 + \frac{2}{3} x \right]_{-1}^1 = \frac{8}{3}$

27. $V = \iint\limits_{R} f(x,y)\, dA = \int_0^1\int_0^1 (2-x-y)\, dy\, dx = \int_0^1 \left[2y - xy - \tfrac{1}{2}y^2\right]_0^1 dx = \int_0^1\left(\tfrac{3}{2}-x\right)dx = \left[\tfrac{3}{2}x - \tfrac{1}{2}x^2\right]_0^1 = 1$

29. $V = \iint\limits_{R} f(x,y)\, dA = \int_0^{\pi/2}\int_0^{\pi/4} 2\sin x\cos y\, dy\, dx = \int_0^{\pi/2}\left[2\sin x\sin y\right]_0^{\pi/4} dx = \int_0^{\pi/2}\left(\sqrt{2}\sin x\right)dx$

$= \left[-\sqrt{2}\cos x\right]_0^{\pi/2} = \sqrt{2}$

31. $\int_1^2\int_0^3 kx^2 y\, dx\, dy = \int_1^2 \left.\tfrac{k}{3}x^3 y\right|_{x=0}^{x=3} dy = \int_1^2 9ky\, dy = \tfrac{9}{2}ky^2\Big|_1^2 = \tfrac{27}{2}k$

Thus we choose $k = 2/27$.

33. By Fubini's Theorem,

$\int_0^2\int_0^1 \frac{x}{1+xy}\, dx\, dy = \int_0^1\int_0^2 \frac{x}{1+xy}\, dy\, dx$

$= \int_0^1 \left(\ln(1+xy)\right]_{y=0}^{y=2}\right) dx = \int_0^1 \ln(1+2x)\, dx = \frac{(1+2x)}{2}\left[\ln(1+2x)-1\right]_0^1 = \frac{3}{2}\ln 3 - 1$

35. (a) MAPLE gives $\int_0^1\int_0^2 \frac{y-x}{(x+y)^3}\, dx\, dy = \frac{1}{3}$ and $\int_0^2\int_0^1 \frac{y-x}{(x+y)^3}\, dy\, dx = -\frac{2}{3}$. This does not contradict

Fubini's Theorem since the integrand is not continuous on the region $R: 0 \le x \le 2, 0 \le y \le 1$.

14.2 DOUBLE INTEGRALS OVER GENERAL REGIONS

1.
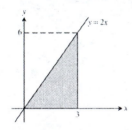

3.
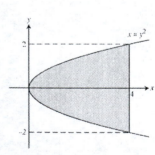

5.
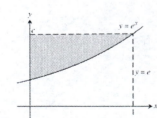

7.
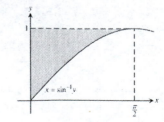

9. (a) $\int_0^2 \int_{x^3}^8 dy\, dx$ (b) $\int_0^8 \int_0^{y^{1/3}} dx\, dy$

11. (a) $\int_0^3 \int_{x^2}^{3x} dy\, dx$ (b) $\int_0^9 \int_{y/3}^{\sqrt{y}} dx\, dy$

13. (a) $\int_0^9 \int_0^{\sqrt{x}} dy\, dx$

    (b) $\int_0^3 \int_{y^2}^9 dx\, dy$

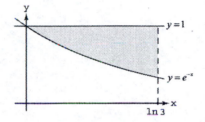

15. (a) $\int_0^{\ln 3} \int_{e^{-x}}^1 dy\, dx$

    (b) $\int_{1/3}^1 \int_{-\ln y}^{\ln 3} dx\, dy$

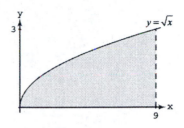

17. (a) $\int_0^1 \int_x^{3-2x} dy\, dx$

    (b) $\int_0^1 \int_0^y dx\, dy + \int_1^3 \int_0^{(3-y)/2} dx\, dy$

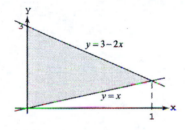

19. $\int_0^\pi \int_0^\pi (x\sin y)\, dy\, dx = \int_0^\pi \left[-x\cos y\right]_0^x dx$

    $= \int_0^\pi (x - x\cos x)\, dx = \left[\frac{x^2}{2} - (\cos x + x\sin x)\right]_0^\pi$

    $= \frac{\pi^2}{2} + 2$

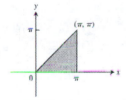

21. $\int_1^{\ln 8} \int_0^{\ln y} e^{x+y}\, dx\, dy = \int_1^{\ln 8} \left[e^{x+y}\right]_0^{\ln y} dy$

    $= \int_1^{\ln 8} \left(ye^y - e^y\right) dy = \left[(y-1)e^y - e^y\right]_1^{\ln 8}$

    $= 8(\ln 8 - 1) - 8 + e = 8\ln 8 - 16 + e$

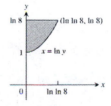

23. $\int_0^1 \int_0^{y^2} 3y^3 e^{xy}\, dx\, dy = \int_0^1 \left[3y^2 e^{xy}\right]_0^{y^2} dy$

    $= \int_0^1 \left(3y^2 e^{y^3} - 3y^2\right) dy = \left[e^{y^3} - y^3\right]_0^1 = e - 2$

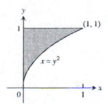

25. $\int_1^2 \int_x^{2x} \frac{x}{y}\, dy\, dx = \int_1^2 \left[x \ln y\right]_x^{2x} dx = (\ln 2)\int_1^2 x\, dx = \frac{3}{2}\ln 2$

27. $\int_0^1 \int_0^{1-u} \left(v - \sqrt{u}\right) dv\, du = \int_0^1 \left[\frac{v^2}{2} - v\sqrt{u}\right]_0^{1-u} du = \int_0^1 \left[\frac{1-2u+u^2}{2} - \sqrt{u}(1-u)\right] du$

$= \int_0^1 \left(\frac{1}{2} - u + \frac{u^2}{2} - u^{1/2} + u^{3/2}\right) du = \left[\frac{u}{2} - \frac{u^2}{2} + \frac{u^3}{6} - \frac{2}{3}u^{3/2} + \frac{2}{5}u^{5/2}\right]_0^1 = \frac{1}{2} - \frac{1}{2} + \frac{1}{6} - \frac{2}{3} + \frac{2}{5} = -\frac{1}{2} + \frac{2}{5} = -\frac{1}{10}$

29. $\int_{-2}^0 \int_v^{-v} 2\, dp\, dv = 2\int_{-2}^0 \left[p\right]_v^{-v} dv = 2\int_{-2}^0 -2v\, dv$

$= -2\left[v^2\right]_{-2}^0 = 8$

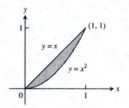

31. $\int_{-\pi/3}^{\pi/3} \int_0^{\sec t} 3\cos t\, du\, dt = \int_{-\pi/3}^{\pi/3} \left[(3\cos t)u\right]_0^{\sec t}$

$= \int_{-\pi/3}^{\pi/3} 3\, dt = 2\pi$

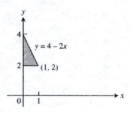

33. $\int_2^4 \int_0^{(4-y)/2} dx\, dy$

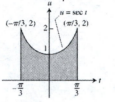

35. $\int_0^1 \int_{x^2}^x dy\, dx$

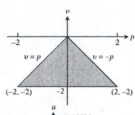

37. $\int_1^e \int_{\ln y}^1 dx\, dy$

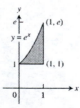

39. $\int_0^9 \int_0^{\frac{1}{2}\sqrt{9-y}} 16x\, dx\, dy$

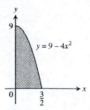

41. $\int_{-1}^{1}\int_{0}^{\sqrt{1-x^2}} 3y \, dy \, dx$

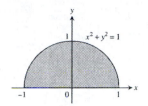

43. $\int_{0}^{1}\int_{e^y}^{e} x \, y \, dx \, dy$

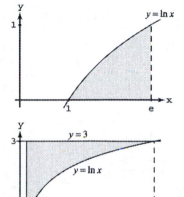

45. $\int_{1}^{e^3}\int_{\ln x}^{3} (x+y) \, dy \, dx$

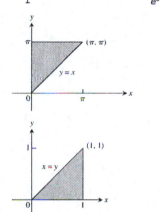

47. $\int_{0}^{\pi}\int_{x}^{\pi} \frac{\sin y}{y} \, dy \, dx = \int_{0}^{\pi}\int_{0}^{y} \frac{\sin y}{y} \, dx \, dy = \int_{0}^{\pi} \sin y \, dy = 2$

49. $\int_{0}^{1}\int_{y}^{1} x^2 e^{xy} \, dx \, dy = \int_{0}^{1}\int_{0}^{x} x^2 e^{xy} \, dy \, dx = \int_{0}^{1} \left[x e^{xy} \right]_{0}^{x} dx$

$= \int_{0}^{1} \left(x e^{x^2} - x \right) dx = \left[\frac{1}{2} e^{x^2} - \frac{x^2}{2} \right]_{0}^{1} = \frac{e-2}{2}$

51. $\int_{0}^{2\sqrt{\ln 3}}\int_{y/2}^{\sqrt{\ln 3}} e^{x^2} \, dx \, dy = \int_{0}^{\sqrt{\ln 3}}\int_{0}^{2x} e^{x^2} \, dy \, dx$

$= \int_{0}^{\sqrt{\ln 3}} 2x e^{x^2} \, dx = \left[e^{x^2} \right]_{0}^{\sqrt{\ln 3}} = e^{\ln 3} - 1 = 2$

53. $\int_{0}^{1/16}\int_{y^{1/4}}^{1/2} \cos\left(16\pi x^5\right) dx \, dy$

$= \int_{0}^{1/2}\int_{0}^{x^4} \cos\left(16\pi x^5\right) dy \, dx = \int_{0}^{1/2} x^4 \cos\left(16\pi x^5\right) dx$

$= \left[\frac{\sin\left(16\pi x^5\right)}{80\pi} \right]_{0}^{1/2} = \frac{1}{80\pi}$

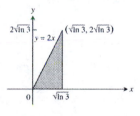

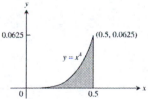

55. $\displaystyle\iint\limits_{R}\left(y-2x^2\right)dA$

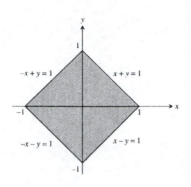

$$= \int_{-1}^{0}\int_{-x-1}^{x+1}\left(y-2x^2\right)dy\,dx + \int_{0}^{1}\int_{x-1}^{1-x}\left(y-2x^2\right)dy\,dx$$

$$= \int_{-1}^{0}\left[\tfrac{1}{2}y^2-2x^2y\right]_{-x-1}^{x+1}dx + \int_{0}^{1}\left[\tfrac{1}{2}y^2-2x^2y\right]_{x-1}^{1-x}dx$$

$$= \int_{-1}^{0}\left(\tfrac{1}{2}(x+1)^2-2x^2(x+1)-\tfrac{1}{2}(-x-1)^2+2x^2(-x-1)\right)dx$$

$$\qquad + \int_{0}^{1}\left[\tfrac{1}{2}(1-x)^2-2x^2(1-x)-\tfrac{1}{2}(x-1)^2+2x^2(x-1)\right]dx$$

$$= -4\int_{-1}^{0}\left(x^3+x^2\right)dx + 4\int_{0}^{1}\left(x^3+x^2\right)dx$$

$$= -4\left[\tfrac{x^4}{4}+\tfrac{x^3}{3}\right]_{-1}^{0} + 4\left[\tfrac{x^4}{4}-\tfrac{x^3}{3}\right]_{0}^{1}$$

$$= 4\left[\tfrac{(-1)^4}{4}+\tfrac{(-1)^3}{3}\right]+4\left(\tfrac{1}{4}-\tfrac{1}{3}\right)=8\left(\tfrac{3}{12}-\tfrac{4}{12}\right)=-\tfrac{8}{12}=-\tfrac{2}{3}$$

57. $\displaystyle V = \int_{0}^{1}\int_{x}^{2-x}\left(x^2+y^2\right)dy\,dx = \int_{0}^{1}\left[x^2y+\tfrac{y^3}{3}\right]_{x}^{2-x}dx = \int_{0}^{1}\left[2x^2-\tfrac{7x^3}{3}+\tfrac{(2-x)^3}{3}\right]dx = \left[\tfrac{2x^3}{3}-\tfrac{7x^4}{12}-\tfrac{(2-x)^4}{12}\right]_{0}^{1}$

$= \left(\tfrac{2}{3}-\tfrac{7}{12}-\tfrac{1}{12}\right)-\left(0-0-\tfrac{16}{12}\right)=\tfrac{4}{3}$

59. $\displaystyle V = \int_{-4}^{1}\int_{3x}^{4-x^2}(x+4)\,dy\,dx = \int_{-4}^{1}\left[xy+4y\right]_{3x}^{4-x^2}dx = \int_{-4}^{1}\left[x\left(4-x^2\right)+4\left(4-x^2\right)-3x^2-12x\right]dx$

$= \int_{-4}^{1}\left(-x^3-7x^2-8x+16\right)dx = \left[-\tfrac{1}{4}x^4-\tfrac{7}{3}x^3-4x^2+16x\right]_{-4}^{1} = \left(-\tfrac{1}{4}-\tfrac{7}{3}+12\right)-\left(\tfrac{64}{3}-64\right)=\tfrac{157}{3}-\tfrac{1}{4}=\tfrac{625}{12}$

61. $\displaystyle V = \int_{0}^{2}\int_{0}^{3}\left(4-y^2\right)dx\,dy = \int_{0}^{2}\left[4x-y^2x\right]_{0}^{3}dy = \int_{0}^{2}\left(12-3y^2\right)dy = \left[12y-y^3\right]_{0}^{2} = 24-8 = 16$

63. $\displaystyle V = \int_{0}^{2}\int_{0}^{2-x}\left(12-3y^2\right)dy\,dx = \int_{0}^{2}\left[12y-y^3\right]_{0}^{2-x}dx = \int_{0}^{2}\left[24-12x-(2-x)^3\right]dx = \left[24x-6x^2+\tfrac{(2-x)^4}{4}\right]_{0}^{2} = 20$

65. $\displaystyle V = \int_{1}^{2}\int_{-1/x}^{1/x}(x+1)\,dy\,dx = \int_{1}^{2}\left[xy+y\right]_{-1/x}^{1/x}dx = \int_{1}^{2}\left[1+\tfrac{1}{x}-\left(-1-\tfrac{1}{x}\right)\right]dx = 2\int_{1}^{2}\left(1+\tfrac{1}{x}\right)dx = 2\left[x+\ln x\right]_{1}^{2}$

$= 2(1+\ln 2)$

67.

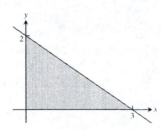

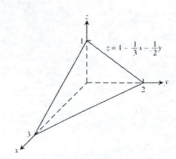

69. $\int_1^\infty \int_{e^{-x}}^1 \frac{1}{x^3 y} \, dy \, dx = \int_1^\infty \left[\frac{\ln y}{x^3} \right]_{e^{-x}}^1 dx = \int_1^\infty - \left(\frac{-x}{x^3} \right) dx = -\lim_{b \to \infty} \left[\frac{1}{x} \right]_1^b = -\lim_{b \to \infty} \left(\frac{1}{b} - 1 \right) = 1$

71. $\int_{-\infty}^\infty \int_{-\infty}^\infty \frac{1}{(x^2+1)(y^2+1)} \, dx \, dy = 2 \int_0^\infty \left(\frac{2}{y^2+1} \right) \left(\lim_{b \to \infty} \tan^{-1} b - \tan^{-1} 0 \right) dy = 2\pi \lim_{b \to \infty} \int_0^b \frac{1}{y^2+1} \, dy$

    $= 2\pi \left(\lim_{b \to \infty} \tan^{-1} b - \tan^{-1} 0 \right) = (2\pi) \left(\frac{\pi}{2} \right) = \pi^2$

73. $\iint\limits_R f(x,y) \, dA \approx \frac{1}{4} f\left(-\frac{1}{2}, 0\right) + \frac{1}{8} f(0,0) + \frac{1}{8} f\left(\frac{1}{4}, 0\right) = \frac{1}{4}\left(-\frac{1}{2}\right) + \frac{1}{8}\left(0 + \frac{1}{4}\right) = -\frac{3}{32}$

75. The ray $\theta = \frac{\pi}{6}$ meets the circle $x^2 + y^2 = 4$ at the point $(\sqrt{3}, 1) \Rightarrow$ the ray is represented by the line $y = \frac{x}{\sqrt{3}}$.

    Thus, $\iint\limits_R f(x,y) \, dA = \int_0^{\sqrt{3}} \int_{x/\sqrt{3}}^{\sqrt{4-x^2}} \sqrt{4-x^2} \, dy \, dx = \int_0^{\sqrt{3}} \left[\left(4 - x^2\right) - \frac{x}{\sqrt{3}} \sqrt{4-x^2} \right] dx = \left[4x - \frac{x^3}{3} + \frac{\left(4-x^2\right)^{3/2}}{3\sqrt{3}} \right]_0^{\sqrt{3}}$

    $= \frac{20\sqrt{3}}{9}$

77. $V = \int_0^1 \int_x^{2-x} \left(x^2 + y^2\right) dy \, dx = \int_0^1 \left[x^2 y + \frac{y^3}{3} \right]_x^{2-x} dx$

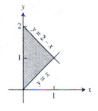

    $= \int_0^1 \left[2x^2 - \frac{7x^3}{3} + \frac{(2-x)^3}{3} \right] dx = \left[\frac{2x^3}{3} - \frac{7x^4}{12} - \frac{(2-x)^4}{12} \right]_0^1$

    $= \left(\frac{2}{3} - \frac{7}{12} - \frac{1}{12} \right) - \left(0 - 0 - \frac{16}{12} \right) = \frac{4}{3}$

79. To maximize the integral, we want the domain to include all points where the integrand is positive and to exclude all points where the integrand is negative. These criteria are met by the points (x, y) such that $4 - x^2 - 2y^2 \geq 0$ or $x^2 + 2y^2 \leq 4$, which is the ellipse $x^2 + 2y^2 = 4$ together with its interior.

81. No, it is not possible. By Fubini's theorem, the two orders of integration must give the same result.

83. $\int_{-b}^b \int_{-b}^b e^{-x^2 - y^2} \, dx \, dy = \int_{-b}^b \int_{-b}^b e^{-y^2} e^{-x^2} \, dx \, dy = \int_{-b}^b e^{-y^2} \left(\int_{-b}^b e^{-x^2} \, dx \right) dy = \left(\int_{-b}^b e^{-x^2} \, dx \right) \left(\int_{-b}^b e^{-y^2} \, dy \right)$

    $= \left(\int_{-b}^b e^{-x^2} \, dx \right)^2 = \left(2 \int_0^b e^{-x^2} \, dx \right)^2 = 4 \left(\int_0^b e^{-x^2} \, dx \right)^2$; taking limits as $b \to \infty$ gives the stated result.

85-87. Example CAS commands:

    Maple:

```
f := (x,y) -> 1/x/y;

q1 := Int( Int (f(x,y) y=1..x), x=1..3 );

evalf(q1);

value(q1);

evalf( value(q1) );
```

89-93.   Example CAS commands:

Maple:

f := (x,y) -> exp(x^2);

c,d:= 0,1;

g1:= y -> 2*y;

g2:= y -> 4;

q5: = Int(Int(f(x,y), x=g1(y)..g2(y)),y=c..d);

value(q5);

plot3d(0, (x=g1(y)..g2(y), y =c..d, color =pink, style =patchnogrid, axes =boxed, orientation =[-90,0]

   scaling =constrained, title ="#89(Section 15.2)");

r5 := Int(Int(f(x,y), y=0..x / 2), x =0..2) + Int(Int(f(x, y=0..1), x =2..4);

value(r5);

value(q5-t5);

85-93.   Example CAS commands:

Mathematica: (functions and bounds will vary)

You can integrate using the built-in integral signs or with the command **Integrate**. In the **Integrate** command, the integration begins with the variable on the right. (In this case, y going from 1 to x).

Clear[x, y, f]

f[x_, y_]:= 1 / (x y)

Integrate[f[x, y], {x, 1, 3}, {y, 1, x}]

To reverse the order of integration, it is best to first plot the region over which the integration extends. This can be done with ImplicitPlot and all bounds involving both x and y can be plotted. A graphics package must be loaded. Remember to use the double equal sign for the equations of the bounding curves.

Clear[x, y, f]

<<Graphics`ImplicitPlot`

ImplicitPlot[{x == 2y,x == 4, y == 0, y == 1},{x,0,4.1}, {y,0,1.1}];

f[x_, y_]:= Exp[x^2]

Integrate[f[x, y], {x, 0, 2}, {y, 0, x/2}] + Integrate[f[x, y], {x, 2, 4}, {y, 0, 1}]

To get a numerical value for the result, use the numerical integrator, **NIntegrate**. Verify that this equals the original.

Integrate[f[x, y], {x, 0, 2}, {y, 0, x/2}] + NIntegrate[f[x, y], {x, 2, 4}, {y, 0, 1}]

NIntegrate[f[x, y], {y, 0, 1}, {x, 2y, 4}]

Another way to show a region is with the FilledPlot command. This assumes that functions are given as y=f(x).

Clear[x, y, f]

<<Graphics`FilledPlot`

FilledPlot[{x^2,9},{x,0,3}, AxesLabels $\rightarrow$ {x,y}];

f[x_, y_]:= x Cos[y^2]

Integrate[f[x, y], [y, 0, 9], {x, 0, Sqrt[y]}]

85. $\int_1^3 \int_1^x \frac{1}{xy}\, dy\, dx \approx 0.603$

87. $\int_0^1 \int_0^1 \tan^{-1} xy\, dy\, dx \approx 0.233$

89. Evaluate the integrals:

$\int_0^1 \int_{2y}^4 e^{x^2}\, dx\, dy$

$= \int_0^2 \int_0^{x/2} e^{x^2}\, dy\, dx + \int_2^4 \int_0^1 e^{x^2}\, dy\, dx$

$= -\frac{1}{4} + \frac{1}{4}\left(e^4 - 2\sqrt{\pi}\, \text{erfi}(2) + 2\sqrt{\pi}\, \text{erfi}(4)\right)$

$\approx 1.1494 \times 10^6$

The following graph was generated using Mathematica.

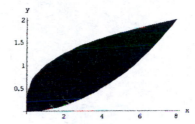

91. Evaluate the integrals:

$\int_0^2 \int_{y^3}^{4\sqrt{2y}} \left(x^2 y - xy^2\right) dx\, dy = \int_0^8 \int_{x^2/32}^{\sqrt[3]{x}} \left(x^2 y - xy^2\right) dy\, dx$

$= \frac{67{,}520}{693} \approx 97.4315$

The following graph was generated using Mathematica.

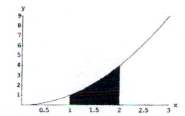

93. Evaluate the integrals:

$\int_1^2 \int_0^{x^2} \frac{1}{x+y}\, dy\, dx$

$= \int_0^1 \int_1^2 \frac{1}{x+y}\, dx\, dy + \int_1^4 \int_{\sqrt{y}}^2 \frac{1}{x+y}\, dx\, dy$

$= -1 + \ln\left(\frac{27}{4}\right) \approx 0.909543$

The following graph was generated using Mathematica.

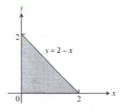

14.3 AREA BY DOUBLE INTEGRATION

1. $\int_0^2 \int_0^{2-x} dy\, dx = \int_0^2 (2-x)\, dx = \left[2x - \frac{x^2}{2}\right]_0^2 = 2$,

   or $\int_0^2 \int_0^{2-y} dx\, dy = \int_0^2 (2-y)\, dy = 2$

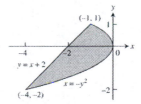

3. $\int_{-2}^1 \int_{y-2}^{-y^2} dx\, dy = \int_{-2}^1 \left(-y^2 - y + 2\right) dy$

   $= \left[-\frac{y^3}{3} - \frac{y^2}{2} + 2y\right]_{-2}^1$

   $= \left(-\frac{1}{3} - \frac{1}{2} + 2\right) - \left(\frac{8}{3} - 2 - 4\right) = \frac{9}{2}$

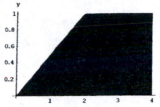

5. $\int_0^{\ln 2} \int_0^{e^x} dy\, dx = \int_0^{\ln 2} e^x\, dx = \left[e^x \right]_0^{\ln 2} = 2 - 1 = 1$

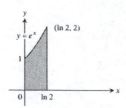

7. $\int_0^1 \int_{y^2}^{2y-y^2} dx\, dy = \int_0^1 \left(2y - 2y^2 \right) dy = \left[y^2 - \frac{2}{3} y^3 \right]_0^1$

$= \frac{1}{3}$

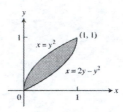

9. $\int_0^2 \int_y^{3y} 1\, dx\, dy = \int_0^2 [x]_y^{3y}\, dy$

$= \int_0^2 (2y)\, dy = \left[y^2 \right]_0^2 = 4$

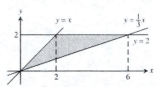

11. $\int_0^1 \int_{x/2}^{2x} 1\, dy\, dx + \int_1^2 \int_{x/2}^{3-x} 1\, dy\, dx$

$= \int_0^1 [y]_{x/2}^{2x}\, dx + \int_1^2 [y]_{x/2}^{3-x}\, dx$

$= \int_0^1 \left(\frac{3}{2} x \right) dx + \int_1^2 \left(3 - \frac{3}{2} x \right) dx$

$= \left[\frac{3}{4} x^2 \right]_0^1 + \left[3x - \frac{3}{4} x^2 \right]_1^2 = \frac{3}{2}$

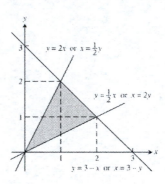

13. $\int_0^6 \int_{y^2/3}^{2y} dx\, dy = \int_0^6 \left(2y - \frac{y^2}{3} \right) dy = \left[y^2 - \frac{y^3}{9} \right]_0^6$

$= 36 - \frac{216}{9} = 12$

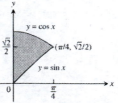

15. $\int_0^{\pi/4} \int_{\sin x}^{\cos x} dy\, dx$

$= \int_0^{\pi/4} (\cos x - \sin x)\, dx = \left[\sin x + \cos x \right]_0^{\pi/4}$

$= \left(\frac{\sqrt{2}}{2} + \frac{\sqrt{2}}{2} \right) - (0 + 1) = \sqrt{2} - 1$

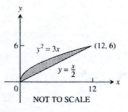

17. $\int_{-1}^0 \int_{-2x}^{1-x} dy\, dx + \int_0^2 \int_{-x/2}^{1-x} dy\, dx$

$= \int_{-1}^0 (1 + x)\, dx + \int_0^2 \left(1 - \frac{x}{2} \right) dx$

$= \left[x + \frac{x^2}{2} \right]_{-1}^0 + \left[x - \frac{x^2}{4} \right]_0^2 = -\left(-1 + \frac{1}{2} \right) + (2 - 1) = \frac{3}{2}$

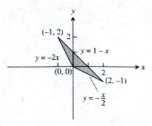

19. (a) average $= \frac{1}{\pi^2} \int_0^\pi \int_0^\pi \sin(x+y)\,dy\,dx = \frac{1}{\pi^2} \int_0^\pi [-\cos(x+y)]_0^\pi\,dx = \frac{1}{\pi^2} \int_0^\pi [-\cos(x+\pi) + \cos x]\,dx$

$= \frac{1}{\pi^2} [-\sin(x+\pi) + \sin x]_0^\pi = \frac{1}{\pi^2} [(-\sin 2\pi + \sin \pi) - (-\sin \pi + \sin 0)] = 0$

(b) average $= \frac{1}{\left(\frac{\pi^2}{2}\right)} \int_0^\pi \int_0^{\pi/2} \sin(x+y)\,dy\,dx = \frac{2}{\pi^2} \int_0^\pi [-\cos(x+y)]_0^{\pi/2}\,dx = \frac{2}{\pi^2} \int_0^\pi \left[-\cos\left(x+\frac{\pi}{2}\right) + \cos x\right]\,dx$

$= \frac{2}{\pi^2} \left[-\sin\left(x+\frac{\pi}{2}\right) + \sin x\right]_0^\pi = \frac{2}{\pi^2} \left[\left(-\sin\frac{3\pi}{2} + \sin \pi\right) - \left(-\sin\frac{\pi}{2} + \sin 0\right)\right] = \frac{4}{\pi^2}$

21. average height $= \frac{1}{4} \int_0^2 \int_0^2 \left(x^2 + y^2\right)\,dy\,dx = \frac{1}{4} \int_0^2 \left[x^2 y + \frac{y^3}{3}\right]_0^2\,dx = \frac{1}{4} \int_0^2 \left(2x^2 + \frac{8}{3}\right)\,dx = \frac{1}{4}\left[\frac{x^3}{3} + \frac{4x}{3}\right]_0^2 = \frac{8}{3}$

23. The region R is shaded in the following figure.

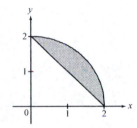

$\int_R dA = \int_0^2 \int_{2-x}^{\sqrt{4-x^2}} 1\,dy\,dx = \int_0^2 \left(\sqrt{4-x^2} - (2-x)\right)\,dx = \left[\frac{x}{2}\sqrt{4-x^2} + 2\sin^{-1}\frac{x}{2} + \frac{x^2}{2} - 2x\right]_0^2 = \pi - 2$, where

we use integration by parts with $u = \sqrt{4-x^2}$ and $dv = 1/2$ to find $\int \sqrt{4-x^2}\,dx$. Geometrically, the region R

25. $\int_{-5}^5 \int_{-2}^0 \frac{10{,}000 e^y}{1+\frac{|x|}{2}}\,dy\,dx = 10{,}000\left(1-e^{-2}\right)\int_{-5}^5 \frac{dx}{1+\frac{|x|}{2}} = 10{,}000\left(1-e^{-2}\right)\left[\int_{-5}^0 \frac{dx}{1-\frac{x}{2}} + \int_0^5 \frac{dx}{1+\frac{x}{2}}\right]$

$= 10{,}000\left(1-e^{-2}\right)\left[-2\ln\left(1-\frac{x}{2}\right)\right]_{-5}^0 + 10{,}000\left(1-e^{-2}\right)\left[2\ln\left(1+\frac{x}{2}\right)\right]_0^5$

$= 10{,}000\left(1-e^{-2}\right)\left[2\ln\left(1+\frac{5}{2}\right)\right] + 10{,}000\left(1-e^{-2}\right)\left[2\ln\left(1+\frac{5}{2}\right)\right] = 40{,}000\left(1-e^{-2}\right)\ln\left(\frac{7}{2}\right) \approx 43{,}329$

27. Let (x_i, y_i) be the location of the weather station in county i for $i = 1, \ldots, 254$. The average temperature in

Texas at time t_0 is approximately $\dfrac{\sum_{i=1}^{254} T(x_i, y_i)\Delta A_i}{A}$, where $T(x_i, y_i)$ is the temperature at time t_0 at the weather

station in county i, ΔA_i is the area of country i, and A is the area of Texas.

29. Since f is continuous on R, if $m \le f(x, y) \le M$, property 3(b) of double integrals gives us

$\iint_R m\,dA \le \iint_R f(x, y)\,dA \le \iint_R M\,dA$ and hence $mA(R) \le \iint_R f(x, y)\,dA \le MA(R)$.

14.4 DOUBLE INTEGRALS IN POLAR FORM

1. $x^2 + y^2 = 9^2 \Rightarrow r = 9 \Rightarrow \frac{\pi}{2} \leq \theta \leq 2\pi, 0 \leq r \leq 9$

3. $y = x \Rightarrow \theta = \frac{\pi}{4}, y = -x \Rightarrow \theta = \frac{3\pi}{4}, y = 1 \Rightarrow r = \csc\theta \Rightarrow \frac{\pi}{4} \leq \theta \leq \frac{3\pi}{4}, 0 \leq r \leq \csc\theta$

5. $x^2 + y^2 = 1^2 \Rightarrow r = 1, x = 2\sqrt{3} \Rightarrow r = 2\sqrt{3}\sec\theta, y = 2 \Rightarrow r = 2\csc\theta;$

   $2\sqrt{3}\sec\theta = 2\csc\theta \Rightarrow \theta = \frac{\pi}{6} \Rightarrow 0 \leq \theta \leq \frac{\pi}{6}, 1 \leq r \leq 2\sqrt{3}\sec\theta; \frac{\pi}{6} \leq \theta \leq \frac{\pi}{2}, 1 \leq r \leq 2\sqrt{3}\csc\theta$

7. $x^2 + y^2 = 2x \Rightarrow r = 2\cos\theta \Rightarrow -\frac{\pi}{2} \leq \theta \leq \frac{\pi}{2}, 0 \leq r \leq 2\cos\theta$

9. $\int_{-1}^{1}\int_{0}^{\sqrt{1-x^2}} dy\,dx = \int_{0}^{\pi}\int_{0}^{1} r\,dr\,d\theta = \frac{1}{2}\int_{0}^{\pi} d\theta = \frac{\pi}{2}$

11. $\int_{0}^{2}\int_{0}^{\sqrt{4-y^2}}\left(x^2 + y^2\right) dx\,dy = \int_{0}^{\pi/2}\int_{0}^{2} r^3\,dr\,d\theta = 4\int_{0}^{\pi/2} d\theta = 2\pi$

13. $\int_{0}^{6}\int_{0}^{y} x\,dx\,dy = \int_{\pi/4}^{\pi/2}\int_{0}^{6\csc\theta} r^2\cos\theta\,dr\,d\theta = 72\int_{\pi/4}^{\pi/2}\cot\theta\csc^2\theta\,d\theta = -36\left[\cot^2\theta\right]_{\pi/4}^{\pi/2} = 36$

15. $\int_{1}^{\sqrt{3}}\int_{1}^{x} dy\,dx = \int_{\pi/6}^{\pi/4}\int_{\csc\theta}^{\sqrt{3}\sec\theta} r\,dr\,d\theta = \int_{\pi/6}^{\pi/4}\left(\frac{3}{2}\sec^2\theta - \frac{1}{2}\csc^2\theta\right) d\theta = \left[\frac{3}{2}\tan\theta + \frac{1}{2}\cot\theta\right]_{\pi/6}^{\pi/4} = 2 - \sqrt{3}$

17. $\int_{-1}^{0}\int_{-\sqrt{1-x^2}}^{0}\frac{2}{1+\sqrt{x^2+y^2}}\,dy\,dx = \int_{\pi}^{3\pi/2}\int_{0}^{1}\frac{2r}{1+r}\,dr\,d\theta = 2\int_{\pi}^{3\pi/2}\int_{0}^{1}\left(1 - \frac{1}{1+r}\right) dr\,d\theta = 2\int_{\pi}^{3\pi/2}(1 - \ln 2)\,d\theta = (1-\ln 2)\pi$

19. $\int_{0}^{\ln 2}\int_{0}^{\sqrt{(\ln 2)^2-y^2}} e^{\sqrt{x^2+y^2}}\,dx\,dy = \int_{0}^{\pi/2}\int_{0}^{\ln 2} re^r\,dr\,d\theta = \int_{0}^{\pi/2}(2\ln 2 - 1)\,d\theta = \frac{\pi}{2}(2\ln 2 - 1)$

21. $\int_{0}^{1}\int_{x}^{\sqrt{2-x^2}}(x+2y)\,dy\,dx = \int_{\pi/4}^{\pi/2}\int_{0}^{\sqrt{2}}(r\cos\theta + 2r\sin\theta)\,r\,dr\,d\theta = \int_{\pi/4}^{\pi/2}\left[\frac{r^3}{3}\cos\theta + \frac{2r^3}{3}\sin\theta\right]_{0}^{\sqrt{2}} d\theta$

   $= \int_{\pi/4}^{\pi/2}\left(\frac{2\sqrt{2}}{3}\cos\theta + \frac{4\sqrt{2}}{3}\sin\theta\right) d\theta = \left[\frac{2\sqrt{2}}{3}\sin\theta - \frac{4\sqrt{2}}{3}\cos\theta\right]_{\pi/4}^{\pi/2} = \frac{2(1+\sqrt{2})}{3}$

23. $\int_{0}^{1}\int_{0}^{\sqrt{1-x^2}} x\,y\,dy\,dx$ or $\int_{0}^{1}\int_{0}^{\sqrt{1-y^2}} x\,y\,dy\,dx$

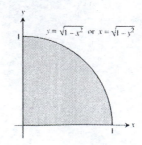

$y = \sqrt{1-x^2}$ or $x = \sqrt{1-y^2}$

25. $\int_0^2 \int_0^x y^2 \left(x^2 + y^2 \right) dy\, dx$ or

$\int_0^2 \int_y^2 y^2 \left(x^2 + y^2 \right) dx\, dy$

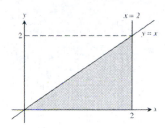

27. $\int_0^{\pi/2} \int_0^{2\sqrt{2-\sin 2\theta}} r\, dr\, d\theta = 2 \int_0^{\pi/2} (2 - \sin 2\theta)\, d\theta = 2(\pi - 1)$

29. $A = 2 \int_0^{\pi/6} \int_0^{12\cos 3\theta} r\, dr\, d\theta = 144 \int_0^{\pi/6} \cos^2 3\theta\, d\theta = 12\pi$

31. $A = \int_0^{\pi/2} \int_0^{1+\sin\theta} r\, dr\, d\theta = \frac{1}{2} \int_0^{\pi/2} \left(\frac{3}{2} + 2\sin\theta - \frac{\cos 2\theta}{2} \right) d\theta = \frac{3\pi}{8} + 1$

33. average $= \frac{4}{\pi a^2} \int_0^{\pi/2} \int_0^a r\sqrt{a^2 - r^2}\, dr\, d\theta = \frac{4}{3\pi a^2} \int_0^{\pi/2} a^3\, d\theta = \frac{2a}{3}$

35. average $= \frac{1}{\pi a^2} \int_{-a}^a \int_{-\sqrt{a^2-x^2}}^{\sqrt{a^2-x^2}} \sqrt{x^2 + y^2}\, dy\, dx = \frac{1}{\pi a^2} \int_0^{2\pi} \int_0^a r^2\, dr\, d\theta = \frac{a}{3\pi} \int_0^{2\pi} d\theta = \frac{2a}{3}$

37. $\int_0^{2\pi} \int_1^{\sqrt{e}} \left(\frac{\ln r^2}{r} \right) r\, dr\, d\theta = \int_0^{2\pi} \int_1^{\sqrt{e}} 2\ln r\, dr\, d\theta = 2 \int_0^{2\pi} \left[r\ln r - r \right]_1^{e^{1/2}} d\theta = 2 \int_0^{2\pi} \sqrt{e} \left[\left(\frac{1}{2} - 1 \right) + 1 \right] d\theta = 2\pi \left(2 - \sqrt{e} \right)$

39. $V = 2 \int_0^{\pi/2} \int_1^{1+\cos\theta} r^2 \cos\theta\, dr\, d\theta = \frac{2}{3} \int_0^{\pi/2} \left(3\cos^2\theta + 3\cos^3\theta + \cos^4\theta \right) d\theta$

$= \frac{2}{3} \left[\frac{15\theta}{8} + \sin 2\theta + 3\sin\theta - \sin^3\theta + \frac{\sin 4\theta}{32} \right]_0^{\pi/2} = \frac{4}{3} + \frac{5\pi}{8}$

41. (a) $I^2 = \int_0^\infty \int_0^\infty e^{-\left(x^2 + y^2 \right)} dx\, dy = \int_0^{\pi/2} \int_0^\infty \left(e^{-r^2} \right) r\, dr\, d\theta = \int_0^{\pi/2} \left[\lim_{b\to\infty} \int_0^b r e^{-r^2}\, dr \right] d\theta$

$= -\frac{1}{2} \int_0^{\pi/2} \lim_{b\to\infty} \left(e^{-b^2} - 1 \right) d\theta = \frac{1}{2} \int_0^{\pi/2} d\theta = \frac{\pi}{4} \Rightarrow I = \frac{\sqrt{\pi}}{2}$

(b) $\lim_{x\to\infty} \int_0^x \frac{2e^{-t^2}}{\sqrt{\pi}}\, dt = \frac{2}{\sqrt{\pi}} \int_0^\infty e^{-t^2}\, dt = \left(\frac{2}{\sqrt{\pi}} \right) \left(\frac{\sqrt{\pi}}{2} \right) = 1$, from part (a)

43. Over the disk $x^2 + y^2 \le \frac{3}{4}$: $\iint_R \frac{1}{1-x^2-y^2}\, dA = \int_0^{2\pi} \int_0^{\sqrt{3}/2} \frac{r}{1-r^2}\, dr\, d\theta = \int_0^{2\pi} \left[-\frac{1}{2}\ln\left(1 - r^2 \right) \right]_0^{\sqrt{3}/2} d\theta$

$= \int_0^{2\pi} \left(-\frac{1}{2}\ln\frac{1}{4} \right) d\theta = (\ln 2) \int_0^{2\pi} d\theta = \pi \ln 4$

Over the disk $x^2 + y^2 \le 1$: $\iint_R \frac{1}{1-x^2-y^2}\, dA = \int_0^{2\pi} \int_0^1 \frac{r}{1-r^2}\, dr\, d\theta = \int_0^{2\pi} \left[\lim_{a\to 1^-} \int_0^a \frac{r}{1-r^2}\, dr \right] d\theta$

$= \int_0^{2\pi} \lim_{a\to 1^-} \left[-\frac{1}{2}\ln\left(1 - a^2 \right) \right] d\theta = 2\pi \cdot \lim_{a\to 1^-} \left[-\frac{1}{2}\ln\left(1 - a^2 \right) \right] = 2\pi \cdot \infty$, so the integral does not exist over

$x^2 + y^2 \le 1$

45. average $= \frac{1}{\pi a^2} \int_0^{2\pi} \int_0^a \left[(r\cos\theta - h)^2 + r^2 \sin^2\theta \right] r \, dr \, d\theta = \frac{1}{\pi a^2} \int_0^{2\pi} \int_0^a \left(r^3 - 2r^2 h \cos\theta + rh^2 \right) dr \, d\theta$

$= \frac{1}{\pi a^2} \int_0^{2\pi} \left(\frac{a^4}{4} - \frac{2a^3 h \cos\theta}{3} + \frac{a^2 h^2}{2} \right) d\theta = \frac{1}{\pi} \int_0^{2\pi} \left(\frac{a^2}{4} - \frac{2ah\cos\theta}{3} + \frac{h^2}{2} \right) d\theta = \frac{1}{\pi} \left[\frac{a^2\theta}{4} - \frac{2ah\sin\theta}{3} + \frac{h^2\theta}{2} \right]_0^{2\pi}$

$= \frac{1}{2}\left(a^2 + 2h^2 \right)$

47. The region R is shaded in the graph below.

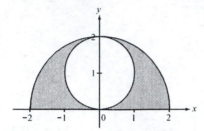

The polar equation of the outer circle is just $r = 2$. The inner circle is $x^2 + (y-1)^2 = 1$ or $x^2 + y^2 = 2y$. This is equivalent to $r^2 = 2r\sin\theta$ or $r = 2\sin\theta$. The integrand is r in polar coordinates, so

$$\iint_R \sqrt{x^2 + y^2} \, dA = \int_0^\pi \int_{2\sin\theta}^2 r \cdot r \, dr \, d\theta$$

$$= \int_0^\pi \left. \frac{r^3}{3} \right|_{2\sin\theta}^2 d\theta = \int_0^\pi \frac{8}{3}\left(1 - \sin^3\theta\right) d\theta$$

Write the integrand as $\frac{8}{3}\left(1 - \sin\theta\left(1 - \cos^2\theta\right)\right)$. The indefinite integral is then $\frac{8}{3}\,\theta + \cos\theta - \frac{1}{3}\cos^3\theta$ and the

definite integral is $\frac{8}{3}\,\theta + \cos\theta - \frac{1}{3}\cos^3\theta \Big|_0^\pi = \frac{8}{9}(3\pi - 4)$

49-51. Example CAS commands:

Maple:

```
f := (x,y) -> y/(x^2+y^2);

    a,b := 0,1;

    f1 := x -> x;

    f2 := x ->1;

    plot3d( f(x,y), y=f1(x)..f2(x), x=a..b, axes=boxed, style=patchnogrid, shading=zhue, orientation=[0,180],
        title="#49(a) (Section 15.4" );                              # (a)

    q1 := eval( x=a, [x=r*cos(theta),y=r*sin(theta)] );              # (b)

    q2 := eval( x=b, [x=r*cos(theta),y=r*sin(theta)] );

    q3 := eval( y=f1(x), [x=r*cos(theta),y=r*sin(theta)] );

    q4 := eval( y=f2(x), [x=r*cos(theta),y=r*sin(theta)] );
```

theta1 := solve(q3, theta);

theta2 := solve(q1, theta);

r1 := 0;

r2 := solve(q4, r);

plot3d(0,r=r1..r2, theta=theta1..theta2, axes=boxed, style=patchnogrid, shading=zhue, orientation=[-90,0],
    title="#49(c) (Section 15.4)");

fP := simplify(eval(f(x,y), [x=r*cos(theta),y=r*sin(theta)]));   # (d)

q5 := Int(Int(fP*r, r=r1..r2), theta=theta1..theta2);

value(q5);

Mathematica: (functions and bounds will vary)

For 49 and 50, begin by drawing the region of integration with the **FilledPlot** command.

Clear[x, y, r, t]

<<Graphics`FilledPlot`

FilledPlot[{x, 1}, {x, 0, 1}, AspectRatio → 1, AxesLabel → {x,y}];

The picture demonstrates that r goes from 0 to the line y=1 or r = 1/Sin[t], while t goes from $\pi/4$ to $\pi/2$.

f:= y / ($x^2 + y^2$)

topolar={x → r Cos[t], y → r Sin[t]};

fp= f/.topolar //Simplify

Integrate[r fp, {t, $\pi/4$, $\pi/2$}, {r, 0, 1/Sin[t]}]

For 51 and 52, drawing the region of integration with the ImplicitPlot command.

Clear[x, y]

<<Graphics`ImplicitPlot`

ImplicitPlot[{x==y, x==2 – y, y==0, y==1}, {x, 0, 2.1}, {y, 0, 1.1}];

The picture shows that as t goes from 0 to $\pi/4$, r goes from 0 to the line x=2 – y. **Solve** will find the bound for
    r.

bdr=Solve[r Cos[t]==2 – r Sin[t], r]//Simplify

f:=Sqrt[x + y]

topolar={x → r Cos[t], y → r Sin[t]};

fp= f/.topolar //Simplify

Integrate[r fp, {t, 0, $\pi/4$}, {r, 0, bdr[[1, 1, 2]]}]

14.5 TRIPLE INTEGRALS IN RECTANGULAR COORDINATES

1. $\int_0^1 \int_0^{1-x} \int_{x+z}^1 F(x,\,y,\,z)\,dy\,dz\,dx = \int_0^1 \int_0^{1-x} \int_{x+z}^1 dy\,dz\,dx = \int_0^1 \int_0^{1-x} (1-x-z)\,dz\,dx$

$= \int_0^1 \left[(1-x) - x(1-x) - \frac{(1-x)^2}{2} \right] dx = \int_0^1 \frac{(1-x)^2}{2}\,dx = \left[-\frac{(1-x)^3}{6} \right]_0^1 = \frac{1}{6}$

3. $\int_0^1 \int_0^{2-2x} \int_0^{3-3x-3y/2} dz\,dy\,dx = \int_0^1 \int_0^{2-2x} \left(3 - 3x - \frac{3}{2}y \right) dy\,dx$

$= \int_0^1 \left[3(1-x) \cdot 2(1-x) - \frac{3}{4} \cdot 4(1-x)^2 \right] dx$

$= 3\int_0^1 (1-x)^2\,dx = \left[-(1-x)^3 \right]_0^1 = 1,$

$\int_0^2 \int_0^{1-y/2} \int_0^{3-3x-3y/2} dz\,dx\,dy, \quad \int_0^1 \int_0^{3-3x} \int_0^{2-2x-2z/3} dy\,dz\,dx,$

$\int_0^3 \int_0^{1-z/3} \int_0^{2-2x-2z/3} dy\,dx\,dz, \quad \int_0^2 \int_0^{3-3y/2} \int_0^{1-y/2-z/3} dx\,dz\,dy,$

$\int_0^3 \int_0^{2-2z/3} \int_0^{1-y/2-z/3} dx\,dy\,dz$

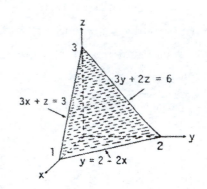

5. $\int_{-2}^2 \int_{-\sqrt{4-x^2}}^{\sqrt{4-x^2}} \int_{x^2+y^2}^{8-x^2-y^2} dz\,dy\,dx = 4\int_0^2 \int_0^{\sqrt{4-x^2}} \int_{x^2+y^2}^{8-x^2-y^2} dz\,dy\,dx$

$= 4\int_0^2 \int_0^{\sqrt{4-x^2}} \left[8 - 2(x^2+y^2) \right] dy\,dx$

$= 8\int_0^2 \int_0^{\sqrt{4-x^2}} \left(4 - x^2 - y^2 \right) dy\,dx = 8\int_0^{\pi/2} \int_0^2 \left(4 - r^2 \right) r\,dr\,d\theta$

$= 8\int_0^{\pi/2} \left[2r^2 - \frac{r^4}{4} \right]_0^2 d\theta = 32\int_0^{\pi/2} d\theta = 32\left(\frac{\pi}{2} \right) = 16\pi,$

$\int_{-2}^2 \int_{-\sqrt{4-y^2}}^{\sqrt{4-y^2}} \int_{x^2+y^2}^{8-x^2-y^2} dz\,dx\,dy,$

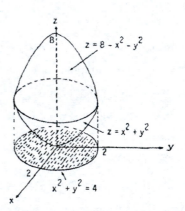

$\int_{-2}^2 \int_{y^2}^4 \int_{-\sqrt{z-y^2}}^{\sqrt{z-y^2}} dx\,dz\,dy + \int_{-2}^2 \int_4^{8-y^2} \int_{-\sqrt{8-z-y^2}}^{\sqrt{8-z-y^2}} dx\,dz\,dy, \quad \int_0^4 \int_{-\sqrt{z}}^{\sqrt{z}} \int_{-\sqrt{z-y^2}}^{\sqrt{z-y^2}} dx\,dy\,dz + \int_4^8 \int_{-\sqrt{8-z}}^{\sqrt{8-z}} \int_{-\sqrt{8-z-y^2}}^{\sqrt{8-z-y^2}} dx\,dy\,dz,$

$\int_{-2}^2 \int_{x^2}^4 \int_{-\sqrt{z-x^2}}^{\sqrt{z-x^2}} dy\,dz\,dx + \int_{-2}^2 \int_4^{8-x^2} \int_{-\sqrt{8-z-x^2}}^{\sqrt{8-z-x^2}} dy\,dz\,dx, \quad \int_0^4 \int_{-\sqrt{z}}^{\sqrt{z}} \int_{-\sqrt{z-x^2}}^{\sqrt{z-x^2}} dy\,dx\,dz + \int_4^8 \int_{-\sqrt{8-z}}^{\sqrt{8-z}} \int_{-\sqrt{8-z-x^2}}^{\sqrt{8-z-x^2}} dy\,dx\,dz$

7. $\int_0^1 \int_0^1 \int_0^1 \left(x^2 + y^2 + z^2 \right) dz\,dy\,dx = \int_0^1 \int_0^1 \left(x^2 + y^2 + \frac{1}{3} \right) dy\,dx = \int_0^1 \left(x^2 + \frac{2}{3} \right) dx = 1$

9. $\int_1^e \int_1^{e^2} \int_1^{e^3} \frac{1}{xyz}\,dx\,dy\,dz = \int_1^e \int_1^{e^2} \left[\frac{\ln x}{yz} \right]_1^{e^3} dy\,dz = \int_1^e \int_1^{e^2} \frac{3}{yz}\,dy\,dz = 3\int_1^e \left[\frac{\ln y}{z} \right]_1^{e^2} dz = \int_1^e \frac{6}{z}\,dz = 6$

11. $\int_0^{\pi/6} \int_0^1 \int_{-2}^3 y \sin z\,dx\,dy\,dz = \int_0^{\pi/6} \int_0^1 5y \sin z\,dy\,dz = \frac{5}{2}\int_0^{\pi/6} \sin z\,dz = \frac{5(2-\sqrt{3})}{4}$

13. $\int_0^3 \int_0^{\sqrt{9-x^2}} \int_0^{\sqrt{9-x^2}} dz\,dy\,dx = \int_0^3 \int_0^{\sqrt{9-x^2}} \sqrt{9-x^2}\,dy\,dx = \int_0^3 \left(9 - x^2 \right) dx = \left[9x - \frac{x^3}{3} \right]_0^3 = 18$

15. $\int_0^1\int_0^{2-x}\int_0^{2-x-y} dz\,dy\,dx = \int_0^1\int_0^{2-x}(2-x-y)\,dy\,dx = \int_0^1\left[(2-x)^2-\frac{1}{2}(2-x)^2\right]dx = \frac{1}{2}\int_0^1(2-x)^2\,dx$

$= \left[-\frac{1}{6}(2-x)^3\right]_0^1 = -\frac{1}{6}+\frac{8}{6} = \frac{7}{6}$

17. $\int_0^\pi\int_0^\pi\int_0^\pi \cos(u+v+w)\,du\,dv\,dw = \int_0^\pi\int_0^\pi[\sin(w+v+\pi)-\sin(w+v)]\,dv\,dw$

$= \int_0^\pi[(-\cos(w+2\pi)+\cos(w+\pi))+(\cos(w+\pi)-\cos w)]\,dw$

$= \left[-\sin(w+2\pi)+\sin(w+\pi)-\sin w+\sin(w+\pi)\right]_0^\pi = 0$

19. $\int_0^{\pi/4}\int_0^{\ln\sec v}\int_{-\infty}^{2t} e^x\,dx\,dt\,dv = \int_0^{\pi/4}\int_0^{\ln\sec v}\lim_{b\to-\infty}\left(e^{2t}-e^b\right)dt\,dv = \int_0^{\pi/4}\int_0^{\ln\sec v}e^{2t}\,dt\,dv$

$= \int_0^{\pi/4}\left(\frac{1}{2}e^{2\ln\sec v}-\frac{1}{2}\right)dv = \int_0^{\pi/4}\left(\frac{\sec^2 v}{2}-\frac{1}{2}\right)dv = \left[\frac{\tan v}{2}-\frac{v}{2}\right]_0^{\pi/4} = \frac{1}{2}-\frac{\pi}{8}$

21. (a) $\int_{-1}^1\int_0^{1-x^2}\int_{x^2}^{1-z} dy\,dz\,dx$ (b) $\int_0^1\int_{-\sqrt{1-z}}^{\sqrt{1-z}}\int_{x^2}^{1-z} dy\,dx\,dz$ (c) $\int_0^1\int_0^{1-z}\int_{-\sqrt{y}}^{\sqrt{y}} dx\,dy\,dz$

(d) $\int_0^1\int_0^{1-y}\int_{-\sqrt{y}}^{\sqrt{y}} dx\,dz\,dy$ (e) $\int_0^1\int_{-\sqrt{y}}^{\sqrt{y}}\int_0^{1-y} dz\,dx\,dy$

23. $V = \int_0^1\int_{-1}^1\int_0^{y^2} dz\,dy\,dx = \int_0^1\int_{-1}^1 y^2\,dy\,dx = \frac{2}{3}\int_0^1 dx = \frac{2}{3}$

25. $V = \int_0^4\int_0^{\sqrt{4-x}}\int_0^{2-y} dz\,dy\,dx = \int_0^4\int_0^{\sqrt{4-x}}(2-y)\,dy\,dx = \int_0^4\left[2\sqrt{4-x}-\left(\frac{4-x}{2}\right)\right]dx = \left[-\frac{4}{3}(4-x)^{3/2}+\frac{1}{4}(4-x)^2\right]_0^4$

$= \frac{4}{3}(4)^{3/2}-\frac{1}{4}(16) = \frac{32}{3}-4 = \frac{20}{3}$

27. $V = \int_0^1\int_0^{2-2x}\int_0^{3-3x-3y/2} dz\,dy\,dx = \int_0^1\int_0^{2-2x}\left(3-3x-\frac{3}{2}y\right)dy\,dx = \int_0^1\left[6(1-x)^2-\frac{3}{4}\cdot4(1-x)^2\right]dx$

$= \int_0^1 3(1-x)^2\,dx = \left[-(1-x)^3\right]_0^1 = 1$

29. $V = 8\int_0^1\int_0^{\sqrt{1-x^2}}\int_0^{\sqrt{1-x^2}} dz\,dy\,dx = 8\int_0^1\int_0^{\sqrt{1-x^2}}\sqrt{1-x^2}\,dy\,dx = 8\int_0^1(1-x^2)\,dx = \frac{16}{3}$

31. $V = \int_0^4\int_0^{\sqrt{16-y^2}/2}\int_0^{4-y} dx\,dz\,dy = \int_0^4\int_0^{\sqrt{16-y^2}/2}(4-y)\,dz\,dy = \int_0^4\frac{\sqrt{16-y^2}}{2}(4-y)\,dy$

$= \int_0^4 2\sqrt{16-y^2}\,dy-\frac{1}{2}\int_0^4 y\sqrt{16-y^2}\,dy = \left[y\sqrt{16-y^2}+16\sin^{-1}\frac{y}{4}\right]_0^4+\left[\frac{1}{6}(16-y^2)^{3/2}\right]_0^4$

$= 16\left(\frac{\pi}{2}\right)-\frac{1}{6}(16)^{3/2} = 8\pi-\frac{32}{3}$

33. $\int_0^2 \int_0^{2-x} \int_{(2-x-y)/2}^{4-2x-2y} dz\, dy\, dx = \int_0^2 \int_0^{2-x} \left(3 - \frac{3x}{2} - \frac{3y}{2}\right) dy\, dx$

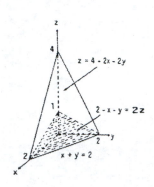

$= \int_0^2 \left[3\left(1 - \frac{x}{2}\right)(2-x) - \frac{3}{4}(2-x)^2\right] dx = \int_0^2 \left[6 - 6x + \frac{3x^2}{2} - \frac{3(2-x)^2}{4}\right] dx$

$= \left[6x - 3x^2 + \frac{x^3}{2} + \frac{(2-x)^3}{4}\right]_0^2 = (12 - 12 + 4 + 0) - \frac{2^3}{4} = 2$

35. $V = 2\int_{-2}^{2} \int_0^{\sqrt{4-x^2}/2} \int_0^{x+2} dz\, dy\, dx = 2\int_{-2}^{2} \int_0^{\sqrt{4-x^2}/2} (x+2)\, dy\, dx = \int_{-2}^{2} (x+2)\sqrt{4-x^2}\, dx$

$= \int_{-2}^{2} 2\sqrt{4-x^2}\, dx + \int_{-2}^{2} x\sqrt{4-x^2}\, dx = \left[x\sqrt{4-x^2} + 4\sin^{-1}\frac{x}{2}\right]_{-2}^{2} + \left[-\frac{1}{3}\left(4-x^2\right)^{3/2}\right]_{-2}^{2} = 4\left(\frac{\pi}{2}\right) - 4\left(-\frac{\pi}{2}\right) = 4\pi$

37. average $= \frac{1}{8}\int_0^2 \int_0^2 \int_0^2 \left(x^2 + 9\right) dz\, dy\, dx = \frac{1}{8}\int_0^2 \int_0^2 \left(2x^2 + 18\right) dy\, dx = \frac{1}{8}\int_0^2 \left(4x^2 + 36\right) dx = \frac{31}{3}$

39. average $= \int_0^1 \int_0^1 \int_0^1 \left(x^2 + y^2 + z^2\right) dz\, dy\, dx = \int_0^1 \int_0^1 \left(x^2 + y^2 + \frac{1}{3}\right) dy\, dx = \int_0^1 \left(x^2 + \frac{2}{3}\right) dx = 1$

41. $\int_0^4 \int_0^1 \int_{2y}^2 \frac{4\cos\left(x^2\right)}{2\sqrt{z}}\, dx\, dy\, dz = \int_0^4 \int_0^2 \int_0^{x/2} \frac{4\cos\left(x^2\right)}{2\sqrt{z}}\, dy\, dx\, dz = \int_0^4 \int_0^2 \frac{x\cos\left(x^2\right)}{\sqrt{z}}\, dx\, dz = \int_0^4 \left(\frac{\sin 4}{2}\right) z^{-1/2}\, dz$

$= \left[(\sin 4) z^{1/2}\right]_0^4 = 2\sin 4$

43. $\int_0^1 \int_{\sqrt[3]{z}}^1 \int_0^{\ln 3} \frac{\pi e^{2x} \sin\left(\pi y^2\right)}{y^2}\, dx\, dy\, dz = \int_0^1 \int_{\sqrt[3]{z}}^1 \frac{4\pi \sin\left(\pi y^2\right)}{y^2}\, dy\, dz = \int_0^1 \int_0^{y^3} \frac{4\pi \sin\left(\pi y^2\right)}{y^2}\, dz\, dy$

$= \int_0^1 4\pi y \sin\left(\pi y^2\right) dy = \left[-2\cos\left(\pi y^2\right)\right]_0^1 = -2(-1) + 2(1) = 4$

45. $\int_0^1 \int_0^{4-a-x^2} \int_a^{4-x^2-y} dz\, dy\, dx = \frac{4}{15} \Rightarrow \int_0^1 \int_0^{4-a-x^2} \left(4 - x^2 - y - a\right) dy\, dx = \frac{4}{15}$

$\Rightarrow \int_0^1 \left[\left(4 - a - x^2\right)^2 - \frac{1}{2}\left(4 - a - x^2\right)^2\right] dx = \frac{4}{15} \Rightarrow \frac{1}{2}\int_0^1 \left(4 - a - x^2\right)^2 dx = \frac{4}{15} \Rightarrow \int_0^1 \left[(4-a)^2 - 2x^2(4-a) + x^4\right] dx$

$= \frac{8}{15} \Rightarrow \left[(4-a)^2 x - \frac{2}{3} x^3 (4-a) + \frac{x^5}{5}\right]_0^1 = \frac{8}{15} \Rightarrow (4-a)^2 - \frac{2}{3}(4-a) + \frac{1}{5} = \frac{8}{15} \Rightarrow 15(4-a)^2 - 10(4-a) - 5 = 0$

$\Rightarrow 3(4-a)^2 - 2(4-a) - 1 = 0 \Rightarrow [3(4-a)+1][(4-a)-1] = 0 \Rightarrow 4 - a = -\frac{1}{3}$ or $4 - a = 1 \Rightarrow a = \frac{13}{3}$ or $a = 3$

47. To minimize the integral, we want the domain to include all points where the integrand is negative and to exclude all points where it is positive. These criteria are met by the points (x, y, z) such that

$4x^2 + 4y^2 + z^2 - 4 \leq 0$ or $4x^2 + 4y^2 + z^2 \leq 4$, which is a solid ellipsoid centered at the origin.

49-51.   Example CAS commands:

Maple:

F := (x,y,z) -> x^2*y^2*z;

q1 := Int(Int(Int(F(x,y,z), y=-sqrt(1-x^2)..sqrt(1-x^2)), x=-1..1), z=0..1);

value(q1);

Mathematica: (functions and bounds will vary)

Clear[f, x, y, z];

$f := x^2 y^2 z$

Integrate[f, {x,−1,1}, {y,−Sqrt[$1-x^2$], Sqrt[$1-x^2$]}, {z, 0, 1}]

N[%]

topolar={x → r Cos[t], y → r Sin[t]};

fp= f/.topolar //Simplify

Integrate[r fp, {t, 0, 2π}, {r, 0, 1},{z, 0, 1}]

N[%]

14.6 MOMENTS AND CENTERS OF MASS

1. $M = \int_0^1 \int_x^{2-x^2} 3\, dy\, dx = 3\int_0^1 \left(2 - x^2 - x\right) dx = \frac{7}{2}$; $M_y = \int_0^1 \int_x^{2-x^2} 3x\, dy\, dx = 3\int_0^1 [xy]_x^{2-x^2}\, dx$

   $= 3\int_0^1 \left(2x - x^3 - x^2\right) dx = \frac{5}{4}$; $M_x = \int_0^1 \int_x^{2-x^2} 3y\, dy\, dx = \frac{3}{2}\int_0^1 \left[y^2\right]_x^{2-x^2} dx = \frac{3}{2}\int_0^1 \left(4 - 5x^2 + x^4\right) dx = \frac{19}{5}$

   $\Rightarrow \bar{x} = \frac{5}{14}$ and $\bar{y} = \frac{38}{35}$

3. $M = \int_0^2 \int_{y^2/2}^{4-y} dx\, dy = \int_0^2 \left(4 - y - \frac{y^2}{2}\right) dy = \frac{14}{3}$; $M_y = \int_0^2 \int_{y^2/2}^{4-y} x\, dx\, dy = \frac{1}{2}\int_0^2 \left[x^2\right]_{y^2/2}^{4-y} dy$

   $= \frac{1}{2}\int_0^2 \left(16 - 8y + y^2 - \frac{y^4}{4}\right) dy = \frac{128}{3}$; $M_x = \int_0^2 \int_{y^2/2}^{4-y} y\, dx\, dy = \int_0^2 \left(4y - y^2 - \frac{y^3}{2}\right) dy = \frac{10}{3} \Rightarrow \bar{x} = \frac{64}{35}$ and $\bar{y} = \frac{5}{7}$

5. $M = \int_0^a \int_0^{\sqrt{a^2-x^2}} dy\, dx = \frac{\pi a^2}{4}$; $M_y = \int_0^a \int_0^{\sqrt{a^2-x^2}} x\, dy\, dx = \int_0^a [xy]_0^{\sqrt{a^2-x^2}}\, dx = \int_0^a x\sqrt{a^2 - x^2}\, dx = \frac{a^3}{3}$

   $\Rightarrow \bar{x} = \bar{y} = \frac{4a}{3\pi}$, by symmetry

7. $I_x = \int_{-2}^2 \int_{-\sqrt{4-x^2}}^{\sqrt{4-x^2}} y^2 dy\, dx = \int_{-2}^2 \left[\frac{y^3}{3}\right]_{-\sqrt{4-x^2}}^{\sqrt{4-x^2}} dx = \frac{2}{3}\int_{-2}^2 \left(4 - x^2\right)^{3/2} dx = 4\pi$; $I_y = 4\pi$, by symmetry;

   $I_o = I_x + I_y = 8\pi$

9. $M = \int_{-\infty}^{0}\int_{0}^{e^x} dy\, dx = \int_{-\infty}^{0} e^x\, dx = \lim_{b\to-\infty}\int_{b}^{0} e^x\, dx = 1 - \lim_{b\to-\infty} e^b = 1;$ $M_y = \int_{-\infty}^{0}\int_{0}^{e^x} x\, dy\, dx = \int_{-\infty}^{0} xe^x dx$

$= \lim_{b\to-\infty}\int_{b}^{0} xe^x dx = \lim_{b\to-\infty}\left[xe^x - e^x\right]_{b}^{0} = -1 - \lim_{b\to-\infty}\left(be^b - e^b\right) = -1;$ $M_x = \int_{-\infty}^{0}\int_{0}^{e^x} y\, dy\, dx = \tfrac{1}{2}\int_{-\infty}^{0} e^{2x}\, dx$

$= \tfrac{1}{2}\lim_{b\to-\infty}\int_{b}^{0} e^{2x} dx = \tfrac{1}{4} \Rightarrow \bar{x} = -1$ and $\bar{y} = \tfrac{1}{4}$

11. $M = \int_{0}^{2}\int_{-y}^{y-y^2}(x+y)\, dx\, dy = \int_{0}^{2}\left[\tfrac{x^2}{2}+xy\right]_{-y}^{y-y^2} dy = \int_{0}^{2}\left(\tfrac{y^4}{2}-2y^3+2y^2\right) dy = \left[\tfrac{y^5}{10}-\tfrac{y^4}{2}+\tfrac{2y^3}{3}\right]_{0}^{2} = \tfrac{8}{15};$

$I_x = \int_{0}^{2}\int_{-y}^{y-y^2} y^2(x+y)\, dx\, dy = \int_{0}^{2}\left[\tfrac{x^2 y^2}{2}+xy^3\right]_{-y}^{y-y^2} dy = \int_{0}^{2}\left(\tfrac{y^6}{2}-2y^5+2y^4\right) dy = \tfrac{64}{105};$

13. $M = \int_{0}^{1}\int_{x}^{2-x}(6x+3y+3)\, dy\, dx = \int_{0}^{1}\left[6xy+\tfrac{3}{2}y^2+3y\right]_{x}^{2-x} dx = \int_{0}^{1}\left(12-12x^2\right) dx = 8;$

$M_y = \int_{0}^{1}\int_{x}^{2-x} x(6x+3y+3)\, dy\, dx = \int_{0}^{1}\left(12x-12x^3\right) dx = 3;$ $M_x = \int_{0}^{1}\int_{x}^{2-x} y(6x+3y+3)\, dy\, dx$

$= \int_{0}^{1}\left(14-6x-6x^2-2x^3\right) dx = \tfrac{17}{2} \Rightarrow \bar{x} = \tfrac{3}{8}$ and $\bar{y} = \tfrac{17}{16}$

15. $M = \int_{0}^{1}\int_{0}^{6}(x+y+1)\, dx\, dy = \int_{0}^{1}(6y+24)\, dy = 27;$ $M_x = \int_{0}^{1}\int_{0}^{6} y(x+y+1)\, dx\, dy = \int_{0}^{1} y(6y+24)\, dy = 14;$

$M_y = \int_{0}^{1}\int_{0}^{6} x(x+y+1)\, dx\, dy = \int_{0}^{1}(18y+90)\, dy = 99 \Rightarrow \bar{x} = \tfrac{11}{3}$ and $\bar{y} = \tfrac{14}{27};$

$I_y = \int_{0}^{1}\int_{0}^{6} x^2(x+y+1)\, dx\, dy = 216\int_{0}^{1}\left(\tfrac{y}{3}+\tfrac{11}{6}\right) dy = 432$

17. $M = \int_{-1}^{1}\int_{0}^{x^2}(7y+1)\, dy\, dx = \int_{-1}^{1}\left(\tfrac{7x^4}{2}+x^2\right) dx = \tfrac{31}{15};$ $M_x = \int_{-1}^{1}\int_{0}^{x^2} y(7y+1)\, dy\, dx = \int_{-1}^{1}\left(\tfrac{7x^6}{3}+\tfrac{x^4}{2}\right) dx = \tfrac{13}{15};$

$M_y = \int_{-1}^{1}\int_{0}^{x^2} x(7y+1)\, dy\, dx = \int_{-1}^{1}\left(\tfrac{7x^5}{2}+x^3\right) dx = 0 \Rightarrow \bar{x} = 0$ and $\bar{y} = \tfrac{13}{31};$

$I_y = \int_{-1}^{1}\int_{0}^{x^2} x^2(7y+1)\, dy\, dx = \int_{-1}^{1}\left(\tfrac{7x^6}{2}+x^4\right) dx = \tfrac{7}{5}$

19. $M = \int_{0}^{1}\int_{-y}^{y}(y+1)\, dx\, dy = \int_{0}^{1}\left(2y^2+2y\right) dy = \tfrac{5}{3};$ $M_x = \int_{0}^{1}\int_{-y}^{y} y(y+1)\, dx\, dy = 2\int_{0}^{1}\left(y^3+y^2\right) dy = \tfrac{7}{6};$

$M_y = \int_{0}^{1}\int_{-y}^{y} x(y+1)\, dx\, dy = \int_{0}^{1} 0\, dy = 0 \Rightarrow \bar{x} = 0$ and $\bar{y} = \tfrac{7}{10};$ $I_x = \int_{0}^{1}\int_{-y}^{y} y^2(y+1)\, dx\, dy$

$= \left(\int_{0}^{1} 2y^4+2y^3\, dy\right) = \tfrac{9}{10};$ $I_y = \int_{0}^{1}\int_{-y}^{y} x^2(y+1)\, dx\, dy = \tfrac{1}{3}\int_{0}^{1}\left(2y^4+2y^3\right) dy = \tfrac{3}{10} \Rightarrow I_o = I_x + I_y = \tfrac{6}{5}$

21. $I_x = \int_{0}^{a}\int_{0}^{b}\int_{0}^{c}\left(y^2+z^2\right) dz\, dy\, dx = \int_{0}^{a}\int_{0}^{b}\left(cy^2+\tfrac{c^3}{3}\right) dy\, dx = \int_{0}^{a}\left(\tfrac{cb^3}{3}+\tfrac{c^3 b}{3}\right) dx = \tfrac{abc(b^2+c^2)}{3} = \tfrac{M}{3}\left(b^2+c^2\right)$ where

$M = abc;$ $I_y = \tfrac{M}{3}\left(a^2+c^2\right)$ and $I_z = \tfrac{M}{3}\left(a^2+b^2\right),$ by symmetry

23. $M = 4\int_0^1\int_0^1\int_{4y^2}^4 dz\, dy\, dx = 4\int_0^1\int_0^1\left(4-4y^2\right)dy\, dx = 16\int_0^1\frac{2}{3}\, dx = \frac{32}{3}$; $M_{xy} = 4\int_0^1\int_0^1\int_{4y^2}^4 z\, dz\, dy\, dx$

$= 2\int_0^1\int_0^1\left(16-16y^4\right)dy\, dx = \frac{128}{5}\int_0^1 dx = \frac{128}{5} \Rightarrow \bar{z} = \frac{12}{5}$, and $\bar{x} = \bar{y} = 0$, by symmetry;

$I_x = 4\int_0^1\int_0^1\int_{4y^2}^4\left(y^2+z^2\right)dz\, dy\, dx = 4\int_0^1\int_0^1\left[\left(4y^2+\frac{64}{3}\right)-\left(4y^4+\frac{64y^6}{3}\right)\right]dy\, dx = 4\int_0^1\frac{1976}{105}\, dx = \frac{7904}{105}$;

$I_y = 4\int_0^1\int_0^1\int_{4y^2}^4\left(x^2+z^2\right)dz\, dy\, dx = 4\int_0^1\int_0^1\left[\left(4x^2+\frac{64}{3}\right)-\left(4x^2y^4+\frac{64y^6}{3}\right)\right]dy\, dx = 4\int_0^1\left(\frac{8}{3}x^2+\frac{128}{7}\right)dx = \frac{4832}{63}$;

$I_z = 4\int_0^1\int_0^1\int_{4y^2}^4\left(x^2+y^2\right)dz\, dy\, dx = 16\int_0^1\int_0^1\left(x^2-x^2y^2+y^2-y^4\right)dy\, dx = 16\int_0^1\left(\frac{2x^2}{3}+\frac{2}{15}\right)dx = \frac{256}{45}$

25. (a) $M = 4\int_0^2\int_0^{\sqrt{4-x^2}}\int_{x^2+y^2}^4 dz\, dy\, dx = 4\int_0^{\pi/2}\int_0^2\int_{r^2}^4 r\, dz\, dr\, d\theta = 4\int_0^{\pi/2}\int_0^2\left(4r-r^3\right)dr\, d\theta = 4\int_0^{\pi/2}4\, d\theta = 8\pi$;

$M_{xy} = \int_0^{2\pi}\int_0^2\int_{r^2}^4 zr\, dz\, dr\, d\theta = \int_0^{2\pi}\int_0^2\frac{r}{2}\left(16-r^4\right)dr\, d\theta = \frac{32}{3}\int_0^{2\pi}d\theta = \frac{64\pi}{3} \Rightarrow \bar{z} = \frac{8}{3}$, and $\bar{x} = \bar{y} = 0$,

by symmetry

(b) $M = 8\pi \Rightarrow 4\pi = \int_0^{2\pi}\int_0^{\sqrt{c}}\int_{r^2}^c r\, dz\, dr\, d\theta = \int_0^{2\pi}\int_0^{\sqrt{c}}\left(cr-r^3\right)dr\, d\theta = \int_0^{2\pi}\frac{c^2}{4}\, d\theta = \frac{c^2\pi}{2} \Rightarrow c^2 = 8 \Rightarrow c = 2\sqrt{2}$,

since $c > 0$

27. The plane $y+2z = 2$ is the top of the wedge $\Rightarrow I_L = \int_{-2}^2\int_{-2}^4\int_{-1}^{(2-y)/2}\left[(y-6)^2+z^2\right]dz\, dy\, dx$

$= \int_{-2}^2\int_{-2}^4\left[\frac{(y-6)^2(4-y)}{2}+\frac{(2-y)^3}{24}+\frac{1}{3}\right]dy\, dx$; let $t = 2-y \Rightarrow I_L = 4\int_{-2}^4\left(\frac{13t^3}{24}+5t^2+16t+\frac{49}{3}\right)dt = 1386$;

$M = \frac{1}{2}(3)(6)(4) = 36$

29. (a) $M = \int_0^2\int_0^{2-x}\int_0^{2-x-y}2x\, dz\, dy\, dx = \int_0^2\int_0^{2-x}\left(4x-2x^2-2xy\right)dy\, dx = \int_0^2\left(x^3-4x^2+4x\right)dx = \frac{4}{3}$

(b) $M_{xy} = \int_0^2\int_0^{2-x}\int_0^{2-x-y}2xz\, dz\, dy\, dx = \int_0^2\int_0^{2-x}x(2-x-y)^2\, dy\, dx = \int_0^2\frac{x(2-x)^3}{3}\, dx = \frac{8}{15}$; $M_{xz} = \frac{8}{15}$ by

symmetry; $M_{yz} = \int_0^2\int_0^{2-x}\int_0^{2-x-y}2x^2\, dz\, dy\, dx = \int_0^2\int_0^{2-x}2x^2(2-x-y)\, dy\, dx = \int_0^2\left(2x-x^2\right)^2 dx = \frac{16}{15}$

$\Rightarrow \bar{x} = \frac{4}{5}$, and $\bar{y} = \bar{z} = \frac{2}{5}$

31. (a) $M = \int_0^1\int_0^1\int_0^1(x+y+z+1)\, dz\, dy\, dx = \int_0^1\int_0^1\left(x+y+\frac{3}{2}\right)dy\, dx = \int_0^1(x+2)\, dx = \frac{5}{2}$

(b) $M_{xy} = \int_0^1\int_0^1\int_0^1 z(x+y+z+1)\, dz\, dy\, dx = \frac{1}{2}\int_0^1\int_0^1\left(x+y+\frac{5}{3}\right)dy\, dx = \frac{1}{2}\int_0^1\left(x+\frac{13}{6}\right)dx = \frac{4}{3}$

$\Rightarrow M_{xy} = M_{yz} = M_{xz} = \frac{4}{3}$, by symmetry $\Rightarrow \bar{x} = \bar{y} = \bar{z} = \frac{8}{15}$

(c) $I_z = \int_0^1\int_0^1\int_0^1\left(x^2+y^2\right)(x+y+z+1)\, dz\, dy\, dx = \int_0^1\int_0^1\left(x^2+y^2\right)\left(x+y+\frac{3}{2}\right)dy\, dx$

$= \int_0^1\left(x^3+2x^2+\frac{1}{3}x+\frac{3}{4}\right)dx = \frac{11}{6} \Rightarrow I_x = I_y = I_z = \frac{11}{6}$, by symmetry

33. $M = \int_{-1}^1\int_{z-1}^{1-z}\int_0^{\sqrt{z}}(2y+5)\, dy\, dx\, dz = \int_0^1\int_{z-1}^{1-z}\left(z+5\sqrt{z}\right)dx\, dz = \int_0^1 2\left(z+5\sqrt{z}\right)(1-z)\, dz$

$= 2\int_0^1\left(5z^{1/2}+z-5z^{3/2}-z^2\right)dz = 2\left[\frac{10}{3}z^{3/2}+\frac{1}{2}z^2-2z^{5/2}-\frac{1}{3}z^3\right]_0^1 = 2\left(\frac{9}{3}-\frac{3}{2}\right) = 3$

14.7 TRIPLE INTEGRALS IN CYLINDRICAL AND SPHERICAL COORDINATES

1. $\int_0^{2\pi}\int_0^1\int_r^{\sqrt{2-r^2}} dz\, r\, dr\, d\theta = \int_0^{2\pi}\int_0^1\left[r\left(2-r^2\right)^{1/2} - r^2 \right] dr\, d\theta = \int_0^{2\pi}\left[-\frac{1}{3}\left(2-r^2\right)^{3/2} - \frac{r^3}{3} \right]_0^1 d\theta$

$= \int_0^{2\pi}\left(\frac{2^{3/2}}{3} - \frac{2}{3} \right) d\theta = \frac{4\pi\left(\sqrt{2}-1\right)}{3}$

3. $\int_0^{2\pi}\int_0^{\theta/(2\pi)}\int_0^{3+24r^3} dz\, r\, dr\, d\theta = \int_0^{2\pi}\int_0^{\theta/(2\pi)}\left(3r + 24r^3 \right) dr\, d\theta = \int_0^{2\pi}\left[\frac{3}{2}r^2 + 6r^4 \right]_0^{\theta/(2\pi)} d\theta$

$= \frac{3}{2}\int_0^{2\pi}\left(\frac{\theta^2}{4\pi^2} + \frac{4\theta^4}{16\pi^4} \right) d\theta = \frac{3}{2}\left[\frac{\theta^3}{12\pi^2} + \frac{\theta^5}{20\pi^4} \right]_0^{2\pi} = \frac{17\pi}{5}$

5. $\int_0^{2\pi}\int_0^1\int_r^{\left(2-r^2\right)^{-1/2}} 3\, dz\, r\, dr\, d\theta = 3\int_0^{2\pi}\int_0^1\left[r\left(2-r^2\right)^{-1/2} - r^2 \right] dr\, d\theta = 3\int_0^{2\pi}\left[-\left(2-r^2\right)^{1/2} - \frac{r^3}{3} \right]_0^1 d\theta$

$= 3\int_0^{2\pi}\left(\sqrt{2} - \frac{4}{3} \right) d\theta = \pi\left(6\sqrt{2} - 8 \right)$

7. $\int_0^{2\pi}\int_0^3\int_0^{z/3} r^3\, dr\, dz\, d\theta = \int_0^{2\pi}\int_0^3 \frac{z^4}{324}\, dz\, d\theta = \int_0^{2\pi}\frac{3}{20}\, d\theta = \frac{3\pi}{10}$

9. $\int_0^1\int_0^{\sqrt{z}}\int_0^{2\pi}\left(r^2\cos^2\theta + z^2 \right) r\, d\theta\, dr\, dz = \int_0^1\int_0^{\sqrt{z}}\left[\frac{r^2\theta}{2} + \frac{r^2\sin 2\theta}{4} + z^2\theta \right]_0^{2\pi} r\, dr\, dz = \int_0^1\int_0^{\sqrt{z}}\left(\pi r^3 + 2\pi r z^2 \right) dr\, dz$

$= \int_0^1\left[\frac{\pi r^4}{4} + \pi r^2 z^2 \right]_0^{\sqrt{z}} dz = \int_0^1\left(\frac{\pi z^2}{4} + \pi z^3 \right) dz = \left[\frac{\pi z^3}{12} + \frac{\pi z^4}{4} \right]_0^1 = \frac{\pi}{3}$

11. (a) $\int_0^{2\pi}\int_0^1\int_0^{\sqrt{4-r^2}} dz\, r\, dr\, d\theta$

(b) $\int_0^{2\pi}\int_0^{\sqrt{3}}\int_0^1 r\, dr\, dz\, d\theta + \int_0^{2\pi}\int_{\sqrt{3}}^2\int_0^{\sqrt{4-z^2}} r\, dr\, dz\, d\theta$

(c) $\int_0^1\int_0^{\sqrt{4-r^2}}\int_0^{2\pi} r\, d\theta\, dz\, dr$

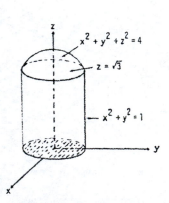

13. $\int_{-\pi/2}^{\pi/2}\int_0^{\cos\theta}\int_0^{3r^2} f(r,\theta,z)\, dz\, r\, dr\, d\theta$

15. $\int_0^{\pi}\int_0^{2\sin\theta}\int_0^{4-r\sin\theta} f(r,\theta,z)\, dz\, r\, dr\, d\theta$

17. $\int_{-\pi/2}^{\pi/2}\int_1^{1+\cos\theta}\int_0^4 f(r,\theta,z)\, dz\, r\, dr\, d\theta$

19. $\int_0^{\pi/4}\int_0^{\sec\theta}\int_0^{2-r\sin\theta} f(r,\theta,z)\, dz\, r\, dr\, d\theta$

21. $\int_0^\pi \int_0^\pi \int_0^{2\sin\phi} \rho^2 \sin\phi \, d\rho \, d\phi \, d\theta = \frac{8}{3} \int_0^\pi \int_0^\pi \sin^4\phi \, d\phi \, d\theta = \frac{8}{3} \int_0^\pi \left(\left[-\frac{\sin^3\phi\cos\phi}{4} \right]_0^\pi + \frac{3}{4} \int_0^\pi \sin^2\phi \, d\phi \right) d\theta$

$= 2 \int_0^\pi \int_0^\pi \sin^2\phi \, d\phi \, d\theta = \int_0^\pi \left[\theta - \frac{\sin 2\theta}{2} \right]_0^\pi d\theta = \int_0^\pi \pi \, d\theta = \pi^2$

23. $\int_0^{2\pi} \int_0^\pi \int_0^{(1-\cos\phi)/2} \rho^2 \sin\phi \, d\rho \, d\phi \, d\theta = \frac{1}{24} \int_0^{2\pi} \int_0^\pi (1-\cos\phi)^3 \sin\phi \, d\phi \, d\theta = \frac{1}{96} \int_0^{2\pi} \left[(1-\cos\phi)^4 \right]_0^\pi d\theta$

$= \frac{1}{96} \int_0^{2\pi} \left(2^4 - 0 \right) d\theta = \frac{16}{96} \int_0^{2\pi} d\theta = \frac{1}{6}(2\pi) = \frac{\pi}{3}$

25. $\int_0^{2\pi} \int_0^{\pi/3} \int_{\sec\phi}^2 3\rho^2 \sin\phi \, d\rho \, d\phi \, d\theta = \int_0^{2\pi} \int_0^{\pi/3} \left(8 - \sec^3\phi \right) \sin\phi \, d\phi \, d\theta = \int_0^{2\pi} \left[-8\cos\phi - \frac{1}{2}\sec^2\phi \right]_0^{\pi/3} d\theta$

$= \int_0^{2\pi} \left[(-4-2) - \left(-8 - \frac{1}{2} \right) \right] d\theta = \frac{5}{2} \int_0^{2\pi} d\theta = 5\pi$

27. $\int_0^2 \int_{-\pi}^0 \int_{\pi/4}^{\pi/2} \rho^3 \sin 2\phi \, d\phi \, d\theta \, d\rho = \int_0^2 \int_{-\pi}^0 \rho^3 \left[-\frac{\cos 2\phi}{2} \right]_{\pi/4}^{\pi/2} d\theta \, d\rho = \int_0^2 \int_{-\pi}^0 \frac{\rho^3}{2} d\theta \, d\rho = \int_0^2 \frac{\rho^2 \pi}{2} d\rho$

$= \left[\frac{\pi\rho^4}{8} \right]_0^2 = 2\pi$

29. $\int_0^1 \int_0^\pi \int_0^{\pi/4} 12\rho \sin^3\phi \, d\phi \, d\theta \, d\rho = \int_0^1 \int_0^\pi \left(12\rho \left[\frac{-\sin^2\phi\cos\phi}{3} \right]_0^{\pi/4} + 8\rho \int_0^{\pi/4} \sin\phi \, d\phi \right) d\theta \, d\rho$

$= \int_0^1 \int_0^\pi \left(-\frac{2\rho}{\sqrt{2}} - 8\rho \left[\cos\phi \right]_0^{\pi/4} \right) d\theta \, d\rho = \int_0^1 \int_0^\pi \left(8\rho - \frac{10\rho}{\sqrt{2}} \right) d\theta \, d\rho = \pi \int_0^1 \left(8\rho - \frac{10\rho}{\sqrt{2}} \right) d\rho = \pi \left[4\rho^2 - \frac{5\rho^2}{\sqrt{2}} \right]_0^1$

$= \frac{(4\sqrt{2}-5)\pi}{\sqrt{2}}$

31. (a) $x^2 + y^2 = 1 \Rightarrow \rho^2 \sin^2\phi = 1$, and $\rho\sin\phi = 1 \Rightarrow \rho = \csc\phi$;

thus $\int_0^{2\pi} \int_0^{\pi/6} \int_0^2 \rho^2 \sin\phi \, d\rho \, d\phi \, d\theta + \int_0^{2\pi} \int_{\pi/6}^{\pi/2} \int_0^{\csc\phi} \rho^2 \sin\phi \, d\rho \, d\phi \, d\theta$

(b) $\int_0^{2\pi} \int_1^2 \int_0^{\sin^{-1}(1/\rho)} \rho^2 \sin\phi \, d\phi \, d\rho \, d\theta + \int_0^{2\pi} \int_0^2 \int_0^{\pi/6} \rho^2 \sin\phi \, d\phi \, d\rho \, d\theta$

33. $V = \int_0^{2\pi} \int_0^{\pi/2} \int_{\cos\phi}^2 \rho^2 \sin\phi \, d\rho \, d\phi \, d\theta = \frac{1}{3} \int_0^{2\pi} \int_0^{\pi/2} \left(8 - \cos^3\phi \right) \sin\phi \, d\phi \, d\theta = \frac{1}{3} \int_0^{2\pi} \left[-8\cos\phi + \frac{\cos^4\phi}{4} \right]_0^{\pi/2} d\theta$

$= \frac{1}{3} \int_0^{2\pi} \left(8 - \frac{1}{4} \right) d\theta = \left(\frac{31}{12} \right)(2\pi) = \frac{31\pi}{6}$

35. $V = \int_0^{2\pi} \int_0^\pi \int_0^{1-\cos\phi} \rho^2 \sin\phi \, d\rho \, d\phi \, d\theta = \frac{1}{3} \int_0^{2\pi} \int_0^\pi (1-\cos\phi)^3 \sin\phi \, d\phi \, d\theta = \frac{1}{3} \int_0^{2\pi} \left[\frac{(1-\cos\phi)^4}{4} \right]_0^\pi d\theta$

$= \frac{1}{12}(2)^4 \int_0^{2\pi} d\theta = \frac{4}{3}(2\pi) = \frac{8\pi}{3}$

37. $V = \int_0^{2\pi} \int_{\pi/4}^{\pi/2} \int_0^{2\cos\phi} \rho^2 \sin\phi \, d\rho \, d\phi \, d\theta = \frac{8}{3} \int_0^{2\pi} \int_{\pi/4}^{\pi/2} \cos^3\phi \sin\phi \, d\phi \, d\theta = \frac{8}{3} \int_0^{2\pi} \left[-\frac{\cos^4\phi}{4} \right]_{\pi/4}^{\pi/2} d\theta$

$= \left(\frac{8}{3}\right)\left(\frac{1}{16}\right) \int_0^{2\pi} d\theta = \frac{1}{6}(2\pi) = \frac{\pi}{3}$

39. (a) $8\int_0^{\pi/2} \int_0^{\pi/2} \int_0^2 \rho^2 \sin\phi \, d\rho \, d\phi \, d\theta$ (b) $8\int_0^{\pi/2} \int_0^2 \int_0^{\sqrt{4-r^2}} dz \, r \, dr \, d\theta$

    (c) $8\int_0^2 \int_0^{\sqrt{4-x^2}} \int_0^{\sqrt{4-x^2-y^2}} dz \, dy \, dx$

41. (a) $V = \int_0^{2\pi} \int_0^{\pi/3} \int_{\sec\phi}^2 \rho^2 \sin\phi \, d\rho \, d\phi \, d\theta$ (b) $V = \int_0^{2\pi} \int_0^{\sqrt{3}} \int_1^{\sqrt{4-r^2}} dz \, r \, dr \, d\theta$

    (c) $V = \int_{-\sqrt{3}}^{\sqrt{3}} \int_{-\sqrt{3-x^2}}^{\sqrt{3-x^2}} \int_1^{\sqrt{4-x^2-y^2}} dz \, dy \, dx$

    (d) $V = \int_0^{2\pi} \int_0^{\sqrt{3}} \left[r\left(4-r^2\right)^{1/2} - r \right] dr \, d\theta = \int_0^{2\pi} \left[-\frac{\left(4-r^2\right)^{3/2}}{3} - \frac{r^2}{2} \right]_0^{\sqrt{3}} d\theta = \int_0^{2\pi} \left(-\frac{1}{3} - \frac{3}{2} + \frac{4^{3/2}}{3} \right) d\theta$

$= \frac{5}{6} \int_0^{2\pi} d\theta = \frac{5\pi}{3}$

43. $V = 4\int_0^{\pi/2} \int_0^1 \int_{r^4-1}^{4-4r^2} dz \, r \, dr \, d\theta = 4\int_0^{\pi/2} \int_0^1 \left(5r - 4r^3 - r^5\right) dr \, d\theta = 4\int_0^{\pi/2} \left(\frac{5}{2} - 1 - \frac{1}{6}\right) d\theta = 4\int_0^{\pi/2} d\theta = \frac{8\pi}{3}$

45. $V = \int_{3\pi/2}^{2\pi} \int_0^{3\cos\theta} \int_0^{-r\sin\theta} dz \, r \, dr \, d\theta = \int_{3\pi/2}^{2\pi} \int_0^{3\cos\theta} \left(-r^2 \sin\theta\right) dr \, d\theta = \int_{3\pi/2}^{2\pi} \left(-9\cos^3\theta\right)(\sin\theta) \, d\theta$

$= \left[\frac{9}{4} \cos^4\theta \right]_{3\pi/2}^{2\pi} = \frac{9}{4} - 0 = \frac{9}{4}$

47. $V = \int_0^{\pi/2} \int_0^{\sin\theta} \int_0^{\sqrt{1-r^2}} dz \, r \, dr \, d\theta = \int_0^{\pi/2} \int_0^{\sin\theta} r\sqrt{1-r^2} \, dr \, d\theta = \int_0^{\pi/2} \left[-\frac{1}{3}\left(1-r^2\right)^{3/2} \right]_0^{\sin\theta} d\theta$

$= -\frac{1}{3} \int_0^{\pi/2} \left[\left(1-\sin^2\theta\right)^{3/2} - 1 \right] d\theta = -\frac{1}{3} \int_0^{\pi/2} \left(\cos^3\theta - 1\right) d\theta = -\frac{1}{3}\left(\left[\frac{\cos^2\theta \sin\theta}{3} \right]_0^{\pi/2} + \frac{2}{3} \int_0^{\pi/2} \cos\theta \, d\theta \right) + \left[\frac{\theta}{3} \right]_0^{\pi/2}$

$= -\frac{2}{9} [\sin\theta]_0^{\pi/2} + \frac{\pi}{6} = \frac{-4+3\pi}{18}$

49. $V = \int_0^{2\pi} \int_{\pi/3}^{2\pi/3} \int_0^a \rho^2 \sin\phi \, d\rho \, d\phi \, d\theta = \int_0^{2\pi} \int_{\pi/3}^{2\pi/3} \frac{a^3}{3} \sin\phi \, d\phi \, d\theta = \frac{a^3}{3} \int_0^{2\pi} [-\cos\phi]_{\pi/3}^{2\pi/3} d\theta = \frac{a^3}{3} \int_0^{2\pi} \left(\frac{1}{2} + \frac{1}{2}\right) d\theta$

$= \frac{2\pi a^3}{3}$

51. $V = \int_0^{2\pi} \int_0^{\pi/3} \int_{\sec\phi}^{2} \rho^2 \sin\phi \, d\rho \, d\phi \, d\theta$

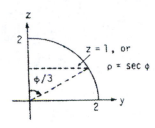

$= \frac{1}{3} \int_0^{2\pi} \int_0^{\pi/3} \left(8\sin\phi - \tan\phi \sec^2\phi \right) d\phi \, d\theta$

$= \frac{1}{3} \int_0^{2\pi} \left[-8\cos\phi - \frac{1}{2}\tan^2\phi \right]_0^{\pi/3} d\theta$

$= \frac{1}{3} \int_0^{2\pi} \left[-4 - \frac{1}{2}(3) + 8 \right] d\theta = \frac{1}{3} \int_0^{2\pi} \frac{5}{2} \, d\theta = \frac{5}{6}(2\pi) = \frac{5\pi}{3}$

53. $V = 4\int_0^{\pi/2} \int_0^1 \int_0^{r^2} dz \, r \, dr \, d\theta = 4\int_0^{\pi/2} \int_0^1 r^3 \, dr \, d\theta = \int_0^{\pi/2} d\theta = \frac{\pi}{2}$

55. $V = 8\int_0^{\pi/2} \int_1^{\sqrt{2}} \int_0^r dz \, r \, dr \, d\theta = 8\int_0^{\pi/2} \int_1^{\sqrt{2}} r^2 \, dr \, d\theta = 8\left(\frac{2\sqrt{2}-1}{3} \right) \int_0^{\pi/2} d\theta = \frac{4\pi(2\sqrt{2}-1)}{3}$

57. $V = \int_0^{2\pi} \int_0^2 \int_0^{4 - r\sin\theta} dz \, r \, dr \, d\theta = \int_0^{2\pi} \int_0^2 \left(4r - r^2\sin\theta \right) dr \, d\theta = 8\int_0^{2\pi} \left(1 - \frac{\sin\theta}{3} \right) d\theta = 16\pi$

59. The paraboloids intersect when $4x^2 + 4y^2 = 5 - x^2 - y^2 \Rightarrow x^2 + y^2 = 1$ and $z = 4$

$\Rightarrow V = 4\int_0^{\pi/2} \int_0^1 \int_{4r^2}^{5 - r^2} dz \, r \, dr \, d\theta = 4\int_0^{\pi/2} \int_0^1 \left(5r - 5r^3 \right) dr \, d\theta = 20\int_0^{\pi/2} \left[\frac{r^2}{2} - \frac{r^4}{4} \right]_0^1 d\theta = 5\int_0^{\pi/2} d\theta = \frac{5\pi}{2}$

61. $V = 8\int_0^{2\pi} \int_0^1 \int_0^{\sqrt{4 - r^2}} dz \, r \, dr \, d\theta = 8\int_0^{2\pi} \int_0^1 r\left(4 - r^2 \right)^{1/2} dr \, d\theta = 8\int_0^{2\pi} \left[-\frac{1}{3}\left(4 - r^2 \right)^{3/2} \right]_0^1 d\theta = -\frac{8}{3} \int_0^{2\pi} \left(3^{3/2} - 8 \right) d\theta$

$= \frac{4\pi(8 - 3\sqrt{3})}{3}$

63. average $= \frac{1}{2\pi} \int_0^{2\pi} \int_0^1 \int_{-1}^1 r^2 dz \, dr \, d\theta = \frac{1}{2\pi} \int_0^{2\pi} \int_0^1 2r^2 dr \, d\theta = \frac{1}{3\pi} \int_0^{2\pi} d\theta = \frac{2}{3}$

65. average $= \frac{1}{\left(\frac{4\pi}{3} \right)} \int_0^{2\pi} \int_0^\pi \int_0^1 \rho^3 \sin\phi \, d\rho \, d\phi \, d\theta = \frac{3}{16\pi} \int_0^{2\pi} \int_0^\pi \sin\phi \, d\phi \, d\theta = \frac{3}{8\pi} \int_0^{2\pi} d\theta = \frac{3}{4}$

67. $M = 4\int_0^{\pi/2} \int_0^1 \int_0^r dz \, r \, dr \, d\theta = 4\int_0^{\pi/2} \int_0^1 r^2 \, dr \, d\theta = \frac{4}{3} \int_0^{\pi/2} d\theta = \frac{2\pi}{3}$; $M_{xy} = \int_0^{2\pi} \int_0^1 \int_0^r z \, dz \, r \, dr \, d\theta$

$= \frac{1}{2} \int_0^{2\pi} \int_0^1 r^3 dr \, d\theta = \frac{1}{8} \int_0^{2\pi} d\theta = \frac{\pi}{4} \Rightarrow \bar{z} = \frac{M_{xy}}{M} = \left(\frac{\pi}{4} \right)\left(\frac{3}{2\pi} \right) = \frac{3}{8}$, and $\bar{x} = \bar{y} = 0$, by symmetry

69. $M = \frac{8\pi}{3}$; $M_{xy} = \int_0^{2\pi} \int_{\pi/3}^{\pi/2} \int_0^2 z\rho^2 \sin\phi \, d\rho \, d\phi \, d\theta = \int_0^{2\pi} \int_{\pi/3}^{\pi/2} \int_0^2 \rho^3 \cos\phi \sin\phi \, d\rho \, d\phi \, d\theta$

$= 4\int_0^{2\pi} \int_{\pi/3}^{\pi/2} \cos\phi \sin\phi \, d\phi \, d\theta = 4\int_0^{2\pi} \left[\frac{\sin^2\phi}{2} \right]_{\pi/3}^{\pi/2} d\theta = 4\int_0^{2\pi} \left(\frac{1}{2} - \frac{3}{8} \right) d\theta = \frac{1}{2} \int_0^{2\pi} d\theta = \pi$

$\Rightarrow \bar{z} = \frac{M_{xy}}{M} = (\pi)\left(\frac{3}{8\pi} \right) = \frac{3}{8}$, and $\bar{x} = \bar{y} = 0$, by symmetry

71. $M = \int_0^{2\pi} \int_0^4 \int_0^{\sqrt{r}} dz\, r\, dr\, d\theta = \int_0^{2\pi} \int_0^4 r^{3/2}\, dr\, d\theta = \frac{64}{5} \int_0^{2\pi} d\theta = \frac{128\pi}{5}; \quad M_{xy} = \int_0^{2\pi} \int_0^4 \int_0^{\sqrt{r}} z\, dz\, r\, dr\, d\theta$

$= \frac{1}{2} \int_0^{2\pi} \int_0^4 r^2\, dr\, d\theta = \frac{32}{3} \int_0^{2\pi} d\theta = \frac{64\pi}{3} \Rightarrow \bar{z} = \frac{M_{xy}}{M} = \frac{5}{6}$, and $\bar{x} = \bar{y} = 0$, by symmetry

73. We orient the cone with its vertex at the origin and axis along the z-axis $\Rightarrow \phi = \frac{\pi}{4}$. We use the x-axis which

is through the vertex and parallel to the base of the cone $\Rightarrow I_x = \int_0^{2\pi} \int_0^1 \int_r^1 \left(r^2 \sin^2\theta + z^2 \right) dz\, r\, dr\, d\theta$

$= \int_0^{2\pi} \int_0^1 \left(r^3 \sin^2\theta - r^4 \sin^2\theta + \frac{r}{3} - \frac{r^4}{3} \right) dr\, d\theta = \int_0^{2\pi} \left(\frac{\sin^2\theta}{20} + \frac{1}{10} \right) d\theta = \left[\frac{\theta}{40} - \frac{\sin 2\theta}{80} + \frac{\theta}{10} \right]_0^{2\pi} = \frac{\pi}{20} + \frac{\pi}{5} = \frac{\pi}{4}$

75. $I_z = \int_0^{2\pi} \int_0^a \int_{\left(\frac{h}{a}\right)r}^h \left(x^2 + y^2 \right) dz\, r\, dr\, d\theta = \int_0^{2\pi} \int_0^a \int_{\frac{hr}{a}}^h r^3\, dz\, dr\, d\theta = \int_0^{2\pi} \int_0^a \left(hr^3 - \frac{hr^4}{a} \right) dr\, d\theta$

$= \int_0^{2\pi} h \left[\frac{r^4}{4} - \frac{r^5}{5a} \right]_0^a d\theta = \int_0^{2\pi} h \left(\frac{a^4}{4} - \frac{a^4}{5a} \right) d\theta = \frac{ha^4}{20} \int_0^{2\pi} d\theta = \frac{\pi h a^4}{10}$

77. (a) $M = \int_0^{2\pi} \int_0^1 \int_r^1 z\, dz\, r\, dr\, d\theta = \frac{1}{2} \int_0^{2\pi} \int_0^1 \left(r - r^3 \right) dr\, d\theta = \frac{1}{8} \int_0^{2\pi} d\theta = \frac{\pi}{4};$

$M_{xy} = \int_0^{2\pi} \int_0^1 \int_r^1 z^2\, dz\, r\, dr\, d\theta = \frac{1}{3} \int_0^{2\pi} \int_0^1 \left(r - r^4 \right) dr\, d\theta = \frac{1}{10} \int_0^{2\pi} d\theta = \frac{\pi}{5} \Rightarrow \bar{z} = \frac{4}{5}$, and $\bar{x} = \bar{y} = 0$, by

symmetry; $I_z = \int_0^{2\pi} \int_0^1 \int_r^1 zr^3\, dz\, dr\, d\theta = \frac{1}{2} \int_0^{2\pi} \int_0^1 \left(r^3 - r^5 \right) dr\, d\theta = \frac{1}{24} \int_0^{2\pi} d\theta = \frac{\pi}{12}$

(b) $M = \int_0^{2\pi} \int_0^1 \int_r^1 z^2\, dz\, r\, dr\, d\theta = \frac{\pi}{5}$ from part (a); $M_{xy} = \int_0^{2\pi} \int_0^1 \int_r^1 z^3\, dz\, r\, dr\, d\theta = \frac{1}{4} \int_0^{2\pi} \int_0^1 \left(r - r^5 \right) dr\, d\theta$

$= \frac{1}{12} \int_0^{2\pi} d\theta = \frac{\pi}{6} \Rightarrow \bar{z} = \frac{5}{6}$, and $\bar{x} = \bar{y} = 0$, by symmetry; $I_z = \int_0^{2\pi} \int_0^1 \int_r^1 z^2 r^3\, dz\, dr\, d\theta$

$= \frac{1}{3} \int_0^{2\pi} \int_0^1 \left(r^3 - r^6 \right) dr\, d\theta = \frac{1}{28} \int_0^{2\pi} d\theta = \frac{\pi}{14}$

79. $M = \int_0^{2\pi} \int_0^a \int_0^{\frac{h}{a}\sqrt{a^2 - r^2}} dz\, r\, dr\, d\theta = \int_0^{2\pi} \int_0^a \frac{h}{a} r\sqrt{a^2 - r^2}\, dr\, d\theta = \frac{h}{a} \int_0^{2\pi} \left[-\frac{1}{3} \left(a^2 - r^2 \right)^{3/2} \right]_0^a d\theta = \frac{h}{a} \int_0^{2\pi} \frac{a^3}{3}\, d\theta$

$= \frac{2ha^2\pi}{3}; \quad M_{xy} = \int_0^{2\pi} \int_0^a \int_0^{\frac{h}{a}\sqrt{a^2 - r^2}} z\, dz\, r\, dr\, d\theta = \frac{h^2}{2a^2} \int_0^{2\pi} \int_0^a \left(a^2 r - r^3 \right) dr\, d\theta = \frac{h^2}{2a^2} \int_0^{2\pi} \left(\frac{a^4}{2} - \frac{a^4}{4} \right) d\theta = \frac{a^2 h^2 \pi}{4}$

$\Rightarrow \bar{z} = \left(\frac{\pi a^2 h^2}{4} \right) \left(\frac{3}{2ha^2\pi} \right) = \frac{3}{8}h$, and $\bar{x} = \bar{y} = 0$, by symmetry

Copyright © 2016 Pearson Education, Inc.

81. The density distribution function is linear so it has the form $\delta(\rho) = k\rho + C$, where ρ is the distance from the center of the planet. Now, $\delta(R) = 0 \Rightarrow kR + C = 0$, and $\delta(\rho) = k\rho - kR$. It remains to determine the constant

$$k: \ M = \int_0^{2\pi}\int_0^\pi\int_0^R (k\rho - kR)\,\rho^2\sin\phi\,d\rho\,d\phi\,d\theta = \int_0^{2\pi}\int_0^\pi \left[k\frac{\rho^4}{4} - kR\frac{\rho^3}{3}\right]_0^R \sin\phi\,d\phi\,d\theta$$

$$= \int_0^{2\pi}\int_0^\pi k\left(\frac{R^4}{4} - \frac{R^4}{3}\right)\sin\phi\,d\phi\,d\theta = \int_0^{2\pi}\left(-\frac{k}{12}R^4\left[-\cos\phi\right]_0^\pi\right)d\theta = \int_0^{2\pi}\left(-\frac{k}{6}R^4\right)d\theta = -\frac{k\pi R^4}{3} \Rightarrow k = -\frac{3M}{\pi R^4}$$

$$\Rightarrow \delta(\rho) = -\frac{3M}{\pi R^4}\rho + \frac{3M}{\pi R^4}R. \ \text{At the center of the planet } \rho = 0 \Rightarrow \delta(0) = \left(\frac{3M}{\pi R^4}\right)R = \frac{3M}{\pi R^3}.$$

14.8 SUBSTITUTIONS IN MULTIPLE INTEGRALS

1. (a) $x - y = u$ and $2x + y = v \Rightarrow 3x = u + v$ and $y = x - u \Rightarrow x = \frac{1}{3}(u + v)$ and $y = \frac{1}{3}(-2u + v)$;

$$\frac{\partial(x, y)}{\partial(u, v)} = \begin{vmatrix} \frac{1}{3} & \frac{1}{3} \\ -\frac{2}{3} & \frac{1}{3} \end{vmatrix} = \frac{1}{9} + \frac{2}{9} = \frac{1}{3}$$

   (b) The line segment $y = x$ from $(0, 0)$ to $(1, 1)$ is
   $x - y = 0 \Rightarrow u = 0$; the line segment $y = -2x$
   from $(0, 0)$ to $(1, -2)$ is $2x + y = 0 \Rightarrow v = 0$;
   the line segment $x = 1$ from $(1, 1)$ to $(1, -2)$ is
   $(x - y) + (2x + y) = 3 \Rightarrow u + v = 3$. The
   transformed region is sketched at the right.

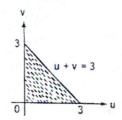

3. (a) $3x + 2y = u$ and $x + 4y = v \Rightarrow -5x = -2u + v$ and $y = \frac{1}{2}(u - 3x) \Rightarrow x = \frac{1}{5}(2u - v)$ and $y = \frac{1}{10}(3v - u)$;

$$\frac{\partial(x, y)}{\partial(u, v)} = \begin{vmatrix} \frac{2}{5} & -\frac{1}{5} \\ -\frac{1}{10} & \frac{3}{10} \end{vmatrix} = \frac{6}{50} - \frac{1}{50} = \frac{1}{10}$$

   (b) The x-axis $y = 0 \Rightarrow u = 3v$;
   the y-axis $x = 0 \Rightarrow v = 2u$;
   the line $x + y = 1 \Rightarrow \frac{1}{5}(2u - v) + \frac{1}{10}(3v - u) = 1$
   $\Rightarrow 2(2u - v) + (3v - u) = 10 \Rightarrow 3u + v = 10$. The
   transformed region is sketched at the right.

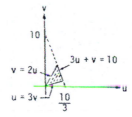

5. $\int_0^4\int_{y/2}^{(y/2)+1}\left(x - \frac{y}{2}\right)dx\,dy = \int_0^4\left[\frac{x^2}{2} - \frac{xy}{2}\right]_{\frac{y}{2}}^{\frac{y}{2}+1}dy = \frac{1}{2}\int_0^4\left[\left(\frac{y}{2}+1\right)^2 - \left(\frac{y}{2}\right)^2 - \left(\frac{y}{2}+1\right)y + \left(\frac{y}{2}\right)y\right]dy$

$= \frac{1}{2}\int_0^4(y + 1 - y)\,dy = \frac{1}{2}\int_0^4 dy = \frac{1}{2}(4) = 2$

7. $\iint_R\left(3x^2 + 14xy + 8y^2\right)dx\,dy$

$= \iint_R(3x + 2y)(x + 4y)\,dx\,dy$

$= \iint_G uv\left|\frac{\partial(x, y)}{\partial(u, v)}\right|du\,dv = \frac{1}{10}\iint_G uv\,du\,dv;$

We find the boundaries of G from the boundaries
of R, shown in the accompanying figure:

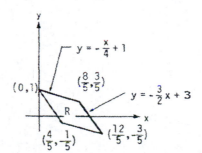

xy-equations for the boundary of R	Corresponding uv-equations for the boundary of G	Simplified uv-equations
$y = -\frac{3}{2}x + 1$	$\frac{1}{10}(3v - u) = -\frac{3}{10}(2u - v) + 1$	$u = 2$
$y = -\frac{3}{2}x + 3$	$\frac{1}{10}(3v - u) = -\frac{3}{10}(2u - v) + 3$	$u = 6$
$y = -\frac{1}{4}x$	$\frac{1}{10}(3v - u) = -\frac{1}{20}(2u - v)$	$v = 0$
$y = -\frac{1}{4}x + 1$	$\frac{1}{10}(3v - u) = -\frac{1}{20}(2u - v) + 1$	$v = 4$

$$\Rightarrow \frac{1}{10}\iint\limits_{G} uv \, du \, dv = \frac{1}{10}\int_2^6\int_0^4 uv \, dv \, du = \frac{1}{10}\int_2^6 u\left[\frac{v^2}{2}\right]_0^4 du = \frac{4}{5}\int_2^6 u \, du = \left(\frac{4}{5}\right)\left[\frac{u^2}{2}\right]_2^6 = \left(\frac{4}{5}\right)(18 - 2) = \frac{64}{5}$$

9. $x = \frac{u}{v}$ and $y = uv \Rightarrow \frac{y}{x} = v^2$ and $xy = u^2$; $\frac{\partial(x, y)}{\partial(u, v)} = J(u, v) = \begin{vmatrix} v^{-1} & -uv^{-2} \\ v & u \end{vmatrix} = v^{-1}u + v^{-1}u = \frac{2u}{v}$;

$y = x \Rightarrow uv = \frac{u}{v} \Rightarrow v = 1$, and $y = 4x \Rightarrow v = 2$; $xy = 1 \Rightarrow u = 1$, and $xy = 9 \Rightarrow u = 3$; thus

$$\iint\limits_{R}\left(\sqrt{\frac{y}{x}} + \sqrt{xy}\right) dx \, dy = \int_1^3\int_1^2 (v + u)\left(\frac{2u}{v}\right) dv \, du = \int_1^3\int_1^2\left(2u + \frac{2u^2}{v}\right) dv \, du = \int_1^3\left[2uv + 2u^2 \ln v\right]_1^2 du$$

$$= \int_1^3\left(2u + 2u^2 \ln 2\right) du = \left[u^2 + \frac{2}{3}u^2 \ln 2\right]_1^3 = 8 + \frac{2}{3}(26)(\ln 2) = 8 + \frac{52}{3}(\ln 2)$$

11. $x = ar \cos\theta$ and $y = ar \sin\theta \Rightarrow \frac{\partial(x, y)}{\partial(r, \theta)} = J(r, \theta) = \begin{vmatrix} a\cos\theta & -ar\sin\theta \\ b\sin\theta & br\cos\theta \end{vmatrix} = abr\cos^2\theta + abr\sin^2\theta = abr$;

$$I_0 = \iint\limits_{R}\left(x^2 + y^2\right) dA = \int_0^{2\pi}\int_0^1 r^2\left(a^2\cos^2\theta + b^2\sin^2\theta\right)|J(r, \theta)| dr \, d\theta$$

$$= \int_0^{2\pi}\int_0^1 abr^3\left(a^2\cos^2\theta + b^2\sin^2\theta\right) dr \, d\theta = \frac{ab}{4}\left(\int_0^{2\pi} a^2\cos^2\theta + b^2\sin^2\theta\right) d\theta$$

$$= \frac{ab}{4}\left[\frac{a^2\theta}{2} + \frac{a^2\sin 2\theta}{4} + \frac{b^2\theta}{2} - \frac{b^2\sin 2\theta}{4}\right]_0^{2\pi} = \frac{ab\pi\left(a^2 + b^2\right)}{4}$$

13. The region of integration R in the xy-plane is sketched in the figure at the right. The boundaries of the image G are obtained as follows, with G sketched at the right:

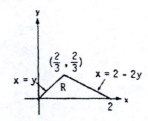

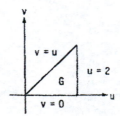

xy-equations for the boundary of R	Corresponding uv-equations for the boundary of G	Simplified uv-equations
$x = y$	$\frac{1}{3}(u+2v) = \frac{1}{3}(u-v)$	$v = 0$
$x = 2 - 2y$	$\frac{1}{3}(u+2v) = 2 - \frac{2}{3}(u-v)$	$u = 2$
$y = 0$	$0 = \frac{1}{3}(u-v)$	$v = u$

Also, from Exercise 2, $\dfrac{\partial(x,\,y)}{\partial(u,\,v)} = J(u,\,v) = -\frac{1}{3} \Rightarrow \displaystyle\int_0^{2/3}\int_y^{2-2y}(x+2y)e^{(y-x)}\,dx\,dy = \int_0^2\int_0^u ue^{-v}\left|-\frac{1}{3}\right|\,dv\,du$

$= \frac{1}{3}\displaystyle\int_0^2 u\left[-e^{-v}\right]_0^u\,du = \frac{1}{3}\int_0^2 u\left(1 - e^{-u}\right)\,du = \frac{1}{3}\left[u\left(u + e^{-u}\right) - \frac{u^2}{2} + e^{-u}\right]_0^2 = \frac{1}{3}\left[2\left(2 + e^{-2}\right) - 2 + e^{-2} - 1\right]$

$= \frac{1}{3}\left(3e^{-2} + 1\right) \approx 0.4687$

15. $x = \frac{u}{v}$ and $y = uv \Rightarrow \frac{y}{x} = v^2$ and $xy = u^2$; $\dfrac{\partial(x,\,y)}{\partial(u,\,v)} = J(u,\,v) = \begin{vmatrix} v^{-1} & -uv^{-2} \\ v & u \end{vmatrix} = v^{-1}u + v^{-1}u = \frac{2u}{v}$;

$y = x \Rightarrow uv = \frac{u}{v} \Rightarrow v = 1$, and $y = 4x \Rightarrow v = 2$; $xy = 1 \Rightarrow u = 1$, and $xy = 4 \Rightarrow u = 2$; thus

$\displaystyle\int_1^2\int_{1/y}^y\left(x^2 + y^2\right)dx\,dy + \int_2^4\int_{y/4}^{4/y}\left(x^2 + y^2\right)dx\,dy = \int_1^2\int_1^2\left(\frac{u^2}{v^2} + u^2v^2\right)\left(\frac{2u}{v}\right)du\,dv = \int_1^2\int_1^2\left(\frac{2u^3}{v^3} + 2u^3v\right)du\,dv$

$= \displaystyle\int_1^2\left[\frac{u^4}{2v^3} + \frac{1}{2}u^4v\right]_1^2\,dv = \int_1^2\left(\frac{15}{2v^3} + \frac{15v}{2}\right)dv = \left[-\frac{15}{4v^2} + \frac{15v^2}{4}\right]_1^2 = \frac{225}{16}$

17. (a) $x = u\cos v$ and $y = u\sin v \Rightarrow \dfrac{\partial(x,\,y)}{\partial(u,\,v)} = \begin{vmatrix} \cos v & -u\sin v \\ \sin v & u\cos v \end{vmatrix} = u\cos^2 v + u\sin^2 v = u$

(b) $x = u\sin v$ and $y = u\cos v \Rightarrow \dfrac{\partial(x,\,y)}{\partial(u,\,v)} = \begin{vmatrix} \sin v & u\cos v \\ \cos v & -u\sin v \end{vmatrix} = -u\sin^2 v - u\cos^2 v = -u$

19. $\begin{vmatrix} \sin\phi\cos\theta & \rho\cos\phi\cos\theta & -\rho\sin\phi\sin\theta \\ \sin\phi\sin\theta & \rho\cos\phi\sin\theta & \rho\sin\phi\cos\theta \\ \cos\phi & -\rho\sin\phi & 0 \end{vmatrix}$

$= (\cos\phi)\begin{vmatrix} \rho\cos\phi\cos\theta & -\rho\sin\phi\sin\theta \\ \rho\cos\phi\sin\theta & \rho\sin\phi\cos\theta \end{vmatrix} + (\rho\sin\phi)\begin{vmatrix} \sin\phi\cos\theta & -\rho\sin\phi\sin\theta \\ \sin\phi\sin\theta & \rho\sin\phi\cos\theta \end{vmatrix}$

$= \left(\rho^2\cos\phi\right)\left(\sin\phi\cos\phi\cos^2\theta + \sin\phi\cos\phi\sin^2\theta\right) + \left(\rho^2\sin\phi\right)\left(\sin^2\phi\cos^2\theta + \sin^2\phi\sin^2\theta\right)$

$= \rho^2\sin\phi\cos^2\phi + \rho^2\sin^3\phi = \left(\rho^2\sin\phi\right)\left(\cos^2\phi + \sin^2\phi\right) = \rho^2\sin\phi$

21. $\displaystyle\int_0^3\int_0^4\int_{y/2}^{1+(y/2)}\left(\frac{2x-y}{2} + \frac{z}{3}\right)dx\,dy\,dz = \int_0^3\int_0^4\left[\frac{x^2}{2} - \frac{xy}{2} + \frac{xz}{3}\right]_{y/2}^{1+(y/2)}dy\,dz = \int_0^3\int_0^4\left[\frac{1}{2}(y+1) - \frac{y}{2} + \frac{z}{3}\right]dy\,dz$

$= \displaystyle\int_0^3\left[\frac{(y+1)^2}{4} - \frac{y^2}{4} + \frac{yz}{3}\right]_0^4\,dz = \int_0^3\left(\frac{9}{4} + \frac{4z}{3} - \frac{1}{4}\right)dz = \int_0^3\left(2 + \frac{4z}{3}\right)dz = \left[2z + \frac{2z^2}{3}\right]_0^3 = 12$

23. $J(u, v, w) = \begin{vmatrix} a & 0 & 0 \\ 0 & b & 0 \\ 0 & 0 & c \end{vmatrix} = abc;$ for R and G as in Exercise 22, $\iiint\limits_{R} |\ xyz\ |\ dx\ dy\ dz \iiint\limits_{G} a^2 b^2 c^2 uvw\ dw\ dv\ du$

$= 8a^2 b^2 c^2 \int_0^{\pi/2} \int_0^{\pi/2} \int_0^1 (\rho \sin\phi \cos\theta)(\rho \sin\phi \sin\theta)(\rho \cos\phi)\left(\rho^2 \sin\phi\right) d\rho\ d\phi\ d\theta$

$= \frac{4a^2 b^2 c^2}{3} \int_0^{\pi/2} \int_0^{\pi/2} \sin\theta \cos\theta \sin^3\phi \cos\phi\ d\phi\ d\theta = \frac{a^2 b^2 c^2}{3} \int_0^{\pi/2} \sin\theta \cos\theta\ d\theta = \frac{a^2 b^2 c^2}{6}$

25. The first moment about the xy-coordinate plane for the semi-ellipsoid, $\frac{x^2}{a^2} + \frac{y^2}{b^2} + \frac{z^2}{c^2} = 1$ using the

transformation in Exercise 23 is, $M_{xy} = \iiint\limits_{D} z\ dz\ dy\ dx = \iiint\limits_{G} cw\ |J(u, v, w)|\ du\ dv\ dw$

$= abc^2 \iiint\limits_{G} w\ du\ dv\ dw = \left(abc^2\right) \cdot \left(M_{xy} \text{ of the hemisphere } x^2 + y^2 + z^2 = 1, z \geq 0\right) = \frac{abc^2 \pi}{4};$

the mass of the semi-ellipsoid is $\frac{2abc\pi}{3} \Rightarrow \overline{z} = \left(\frac{abc^2\pi}{4}\right)\left(\frac{3}{2abc\pi}\right) = \frac{3}{8}c$

27. The region R is shaded in the graph below.

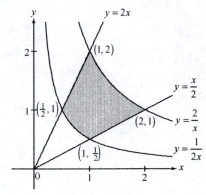

Solving explicitly for the transformation that gives x and y in terms of u and v yields a complicated

expression for $\frac{\partial(x, y)}{\partial(u, v)}$. However, its reciprocal, $\frac{\partial(u, v)}{\partial(x, y)}$ is relatively easy to compute.

Since $u(x, y) = xy$ and $v(x, y) = y/x$, $J(x, y) = \begin{vmatrix} y & x \\ -\dfrac{y}{x^2} & \dfrac{1}{x} \end{vmatrix} = 2\dfrac{y}{x} = 2v$. Thus $J(u, v) = 1/2v$. In the uv-plane

the region corresponding to R is $G: \dfrac{1}{2} \leq u \leq 2$, $\dfrac{1}{2} \leq v \leq 2$. Thus v is positive and $\left|J(u, v)\right| = 1/2v$.

$dA = \int_R \int = \int_{1/2}^2 \int_{1/2}^2 \frac{1}{2v}\ du\ dv = \int_{1/2}^2 \frac{\ln u}{2} \Big|_{1/2}^2\ dv = \int_{1/2}^2 \ln 2\ dv = \frac{3}{2}\ln 2$

CHAPTER 14 PRACTICE EXERCISES

1. $\displaystyle\int_1^{10}\int_0^{1/y} ye^{xy}\,dx\,dy = \int_1^{10}\left[e^{xy}\right]_0^{1/y}dy$

$= \displaystyle\int_1^{10}(e-1)\,dy = 9e-9$

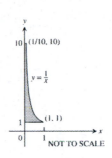

NOT TO SCALE

3. $\displaystyle\int_0^{3/2}\int_{-\sqrt{9-4t^2}}^{\sqrt{9-4t^2}} t\,ds\,dt = \int_0^{3/2}\left[ts\right]_{-\sqrt{9-4t^2}}^{\sqrt{9-4t^2}}dt$

$= \displaystyle\int_0^{3/2} 2t\sqrt{9-4t^2}\,dt = \left[-\frac{1}{6}\left(9-4t^2\right)^{3/2}\right]_0^{3/2}$

$= -\dfrac{1}{6}\left(0^{3/2}-9^{3/2}\right) = \dfrac{27}{6} = \dfrac{9}{2}$

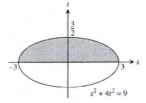

5. $\displaystyle\int_{-2}^0\int_{2x+4}^{4-x^2} dy\,dx = \int_{-2}^0\left(-x^2-2x\right)dx$

$= \left[-\dfrac{x^3}{3}-x^2\right]_{-2}^0 = -\left(\dfrac{8}{3}-4\right) = \dfrac{4}{3}$

$\displaystyle\int_0^4\int_{-\sqrt{4-y}}^{(y-4)/2} dx\,dy = \int_0^4\left(\dfrac{y-4}{2}+\sqrt{4-y}\right)dy$

$= \left[\dfrac{y^2}{2}-2y-\dfrac{2}{3}(4-y)^{3/2}\right]_0^4$

$= 4-8+\dfrac{2}{3}\cdot 4^{3/2} = -4+\dfrac{16}{3} = \dfrac{4}{3}$

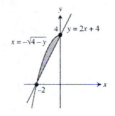

7. $\displaystyle\int_{-3}^3\int_0^{(1/2)\sqrt{9-x^2}} y\,dy\,dx = \int_{-3}^3\left[\dfrac{y^2}{2}\right]_0^{(1/2)\sqrt{9-x^2}}dx$

$= \displaystyle\int_{-3}^3 \dfrac{1}{8}\left(9-x^2\right)dx = \left[\dfrac{9x}{8}-\dfrac{x^3}{24}\right]_{-3}^3$

$= \left(\dfrac{27}{8}-\dfrac{27}{24}\right)-\left(-\dfrac{27}{8}+\dfrac{27}{24}\right) = \dfrac{27}{6} = \dfrac{9}{2}$

$\displaystyle\int_0^{3/2}\int_{-\sqrt{9-4y^2}}^{\sqrt{9-4y^2}} y\,dx\,dy = \int_0^{3/2} 2y\sqrt{9-4y^2}\,dy$

$= \left[-\dfrac{1}{4}\cdot\dfrac{2}{3}\left(9-4y^2\right)^{3/2}\right]_0^{3/2} = \dfrac{1}{6}\cdot 9^{3/2} = \dfrac{27}{6} = \dfrac{9}{2}$

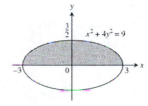

9. $\displaystyle\int_0^1\int_{2y}^2 4\cos\left(x^2\right)dx\,dy = \int_0^2\int_0^{x/2} 4\cos\left(x^2\right)dy\,dx = \int_0^2 2x\cos\left(x^2\right)dx = \left[\sin\left(x^2\right)\right]_0^2 = \sin 4$

11. $\displaystyle\int_0^8\int_{\sqrt[3]{x}}^2 \dfrac{1}{y^4+1}\,dy\,dx = \int_0^2\int_0^{y^3}\dfrac{1}{y^4+1}\,dx\,dy = \dfrac{1}{4}\int_0^2\dfrac{4y^3}{y^4+1}\,dy = \dfrac{\ln 17}{4}$

13. $A = \displaystyle\int_{-2}^0\int_{2x+4}^{4-x^2} dy\,dx = \int_{-2}^0\left(-x^2-2x\right)dx = \dfrac{4}{3}$

532 Chapter 14 Multiple Integrals

15. $V = \int_0^1 \int_x^{2-x} \left(x^2 + y^2\right) dy\, dx = \int_0^1 \left[x^2 y + \frac{y^3}{3} \right]_x^{2-x} dx = \int_0^1 \left[2x^2 + \frac{(2-x)^3}{3} - \frac{7x^3}{3} \right] dx = \left[\frac{2x^3}{3} - \frac{(2-x)^4}{12} - \frac{7x^4}{12} \right]_0^1$

$= \left(\frac{2}{3} - \frac{1}{12} - \frac{7}{12} \right) + \frac{2^4}{12} = \frac{4}{3}$

17. average value $= \int_0^1 \int_0^1 xy\, dy\, dx = \int_0^1 \left[\frac{xy^2}{2} \right]_0^1 dx = \int_0^1 \frac{x}{2}\, dx = \frac{1}{4}$

19. $\int_{-1}^1 \int_{-\sqrt{1-x^2}}^{\sqrt{1-x^2}} \frac{2}{\left(1+x^2+y^2\right)^2} dy\, dx = \int_0^{2\pi} \int_0^1 \frac{2r}{\left(1+r^2\right)^2} dr\, d\theta = \int_0^{2\pi} \left[-\frac{1}{1+r^2} \right]_0^1 d\theta = \frac{1}{2} \int_0^{2\pi} d\theta = \pi$

21. $\left(x^2 + y^2\right)^2 - \left(x^2 - y^2\right) = 0 \Rightarrow r^4 - r^2 \cos 2\theta = 0 \Rightarrow r^2 = \cos 2\theta$ so the integral is $\int_{-\pi/4}^{\pi/4} \int_0^{\sqrt{\cos 2\theta}} \frac{r}{\left(1+r^2\right)^2} dr\, d\theta$

$= \int_{-\pi/4}^{\pi/4} \left[-\frac{1}{2\left(1+r^2\right)} \right]_0^{\sqrt{\cos 2\theta}} d\theta = \frac{1}{2} \int_{-\pi/4}^{\pi/4} \left(1 - \frac{1}{1+\cos 2\theta}\right) d\theta = \frac{1}{2} \int_{-\pi/4}^{\pi/4} \left(1 - \frac{1}{2\cos^2 \theta}\right) d\theta = \frac{1}{2} \int_{-\pi/4}^{\pi/4} \left(1 - \frac{\sec^2 \theta}{2}\right) d\theta$

$= \frac{1}{2} \left[\theta - \frac{\tan \theta}{2} \right]_{-\pi/4}^{\pi/4} = \frac{\pi-2}{4}$

23. $\int_0^\pi \int_0^\pi \int_0^\pi \cos(x+y+z)\, dx\, dy\, dz = \int_0^\pi \int_0^\pi \left[\sin(z+y+\pi) - \sin(z+y) \right] dy\, dz$

$= \int_0^\pi \left[-\cos(z+2\pi) + \cos(z+\pi) - \cos z + \cos(z+\pi) \right] dz = 0$

25. $\int_0^1 \int_0^{x^2} \int_0^{x+y} (2x - y - z)\, dz\, dy\, dx = \int_0^1 \int_0^{x^2} \left(\frac{3x^2}{2} - \frac{3y^2}{2} \right) dy\, dx = \int_0^1 \left(\frac{3x^4}{2} - \frac{x^6}{2} \right) dx = \frac{8}{35}$

27. $V = 2\int_0^{\pi/2} \int_{-\cos y}^0 \int_0^{-2x} dz\, dx\, dy = 2\int_0^{\pi/2} \int_{-\cos y}^0 (-2x)\, dx\, dy = 2\int_0^{\pi/2} \cos^2 y\, dy = 2\left[\frac{y}{2} + \frac{\sin 2y}{4} \right]_0^{\pi/2} = \frac{\pi}{2}$

29. average $= \frac{1}{3} \int_0^1 \int_0^3 \int_0^1 30xz\sqrt{x^2+y}\, dz\, dy\, dx = \frac{1}{3} \int_0^1 \int_0^3 15x\sqrt{x^2+y}\, dy\, dx = \frac{1}{3} \int_0^3 \int_0^1 15x\sqrt{x^2+y}\, dx\, dy$

$= \frac{1}{3} \int_0^3 \left[5\left(x^2 + y\right)^{3/2} \right]_0^1 dy = \frac{1}{3} \int_0^3 \left[5(1+y)^{3/2} - 5y^{3/2} \right] dy = \frac{1}{3} \left[2(1+y)^{5/2} - 2y^{5/2} \right]_0^3 = \frac{1}{3} \left[2(4)^{5/2} - 2(3)^{5/2} - 2 \right]$

$= \frac{1}{3} \left[2\left(31 - 3^{5/2}\right) \right]$

31. (a) $\int_{-\sqrt{2}}^{\sqrt{2}} \int_{-\sqrt{2-y^2}}^{\sqrt{2-y^2}} \int_{\sqrt{x^2+y^2}}^{\sqrt{4-x^2-y^2}} 3\, dz\, dx\, dy$

(b) $\int_0^{2\pi} \int_0^{\pi/4} \int_0^2 3\rho^2 \sin \phi\, d\rho\, d\phi\, d\theta$

(c) $\int_0^{2\pi} \int_0^{\sqrt{2}} \int_r^{\sqrt{4-r^2}} 3\, dz\, r\, dr\, d\theta = 3\int_0^{2\pi} \int_0^{\sqrt{2}} \left[r\left(4-r^2\right)^{1/2} - r^2 \right] dr\, d\theta = 3\int_0^{2\pi} \left[-\frac{1}{3}\left(4-r^2\right)^{3/2} - \frac{r^3}{3} \right]_0^{\sqrt{2}} d\theta$

$= \int_0^{2\pi} \left(-2^{3/2} - 2^{3/2} + 4^{3/2} \right) d\theta = \left(8 - 4\sqrt{2}\right) \int_0^{2\pi} d\theta = 2\pi\left(8 - 4\sqrt{2}\right)$

Copyright © 2016 Pearson Education, Inc.

33. (a) $\int_0^{2\pi}\int_0^{\pi/4}\int_0^{\sec\phi}\rho^2\sin\phi\,d\rho\,d\phi\,d\theta$

(b) $\int_0^{2\pi}\int_0^{\pi/4}\int_0^{\sec\phi}\rho^2\sin\phi\,d\rho\,d\phi\,d\theta=\frac{1}{3}\int_0^{2\pi}\int_0^{\pi/4}(\sec\phi)(\sec\phi\tan\phi)\,d\phi\,d\theta=\frac{1}{3}\int_0^{2\pi}\left[\frac{1}{2}\tan^2\phi\right]_0^{\pi/4}d\theta$

$=\frac{1}{6}\int_0^{2\pi}d\theta=\frac{\pi}{3}$

35. $\int_0^1\int_{\sqrt{1-x^2}}^{\sqrt{3-x^2}}\int_1^{\sqrt{4-x^2-y^2}}z^2yx\,dz\,dy\,dx+\int_1^{\sqrt{3}}\int_0^{\sqrt{3-x^2}}\int_1^{\sqrt{4-x^2-y^2}}z^2yx\,dz\,dy\,dx$

37. (a) $V=\int_0^{2\pi}\int_0^2\int_2^{\sqrt{8-r^2}}dz\,r\,dr\,d\theta=\int_0^{2\pi}\int_0^2\left(r\sqrt{8-r^2}-2r\right)dr\,d\theta=\int_0^{2\pi}\left[-\frac{1}{3}(8-r^2)^{3/2}-r^2\right]_0^2 d\theta$

$=\int_0^{2\pi}\left[-\frac{1}{3}(4)^{3/2}-4+\frac{1}{3}(8)^{3/2}\right]d\theta=\int_0^{2\pi}\frac{4}{3}\left(-2-3+2\sqrt{8}\right)d\theta=\frac{4}{3}\left(4\sqrt{2}-5\right)\int_0^{2\pi}d\theta=\frac{8\pi(4\sqrt{2}-5)}{3}$

(b) $V=\int_0^{2\pi}\int_0^{\pi/4}\int_{2\sec\phi}^{\sqrt{8}}\rho^2\sin\phi\,d\rho\,d\phi\,d\theta=\frac{8}{3}\int_0^{2\pi}\int_0^{\pi/4}\left(2\sqrt{2}\sin\phi-\sec^3\phi\sin\phi\right)d\phi\,d\theta$

$=\frac{8}{3}\int_0^{2\pi}\int_0^{\pi/4}\left(2\sqrt{2}\sin\phi-\tan\phi\sec^2\phi\right)d\phi\,d\theta=\frac{8}{3}\int_0^{2\pi}\left[-2\sqrt{2}\cos\phi-\frac{1}{2}\tan^2\phi\right]_0^{\pi/4}d\theta$

$=\frac{8}{3}\int_0^{2\pi}\left(-2-\frac{1}{2}+2\sqrt{2}\right)d\theta=\frac{8}{3}\int_0^{2\pi}\left(\frac{-5+4\sqrt{2}}{2}\right)d\theta=\frac{8\pi(4\sqrt{2}-5)}{3}$

39. With the centers of the spheres at the origin, $I_z=\int_0^{2\pi}\int_0^{\pi}\int_a^b\delta(\rho\sin\phi)^2\left(\rho^2\sin\phi\right)d\rho\,d\phi\,d\theta$

$=\frac{\delta(b^5-a^5)}{5}\int_0^{2\pi}\int_0^{\pi}\sin^3\phi\,d\phi\,d\theta=\frac{\delta(b^5-a^5)}{5}\int_0^{2\pi}\int_0^{\pi}\left(\sin\phi-\cos^2\phi\sin\phi\right)d\phi\,d\theta$

$=\frac{\delta(b^5-a^5)}{5}\int_0^{2\pi}\left[-\cos\phi+\frac{\cos^3\phi}{3}\right]_0^{\pi}d\theta=\frac{4\delta(b^5-a^5)}{15}\int_0^{2\pi}d\theta=\frac{8\pi\delta(b^5-a^5)}{15}$

41. $M=\int_1^2\int_{2/x}^2 dy\,dx=\int_1^2\left(2-\frac{2}{x}\right)dx=2-\ln 4;\quad M_y=\int_1^2\int_{2/x}^2 x\,dy\,dx=\int_1^2 x\left(2-\frac{2}{x}\right)dx=1;$

$M_x=\int_1^2\int_{2/x}^2 y\,dy\,dx=\int_1^2\left(2-\frac{2}{x^2}\right)dx=1\Rightarrow\bar{x}=\bar{y}=\frac{1}{2-\ln 4}$

43. $I_o=\int_0^2\int_{2x}^4(x^2+y^2)(3)\,dy\,dx=3\int_0^2\left(4x^2+\frac{64}{3}-\frac{14x^3}{3}\right)dx=104$

45. $M=\delta\int_0^3\int_0^{2x/3}dy\,dx=\delta\int_0^3\frac{2x}{3}dx=3\delta;\quad I_x=\delta\int_0^3\int_0^{2x/3}y^2\,dy\,dx=\frac{8\delta}{81}\int_0^3 x^3\,dx=\left(\frac{8\delta}{81}\right)\left(\frac{3^4}{4}\right)=2\delta$

47. $M=\int_{-1}^1\int_{-1}^1\left(x^2+y^2+\frac{1}{3}\right)dy\,dx=\int_{-1}^1\left(2x^2+\frac{4}{3}\right)dx=4;\quad M_x=\int_{-1}^1\int_{-1}^1 y\left(x^2+y^2+\frac{1}{3}\right)dy\,dx=\int_{-1}^1 0\,dx=0;$

$M_y=\int_{-1}^1\int_{-1}^1 x\left(x^2+y^2+\frac{1}{3}\right)dy\,dx=\int_{-1}^1\left(2x^3+\frac{4}{3}x\right)dx=0$

49. $M=\int_{-\pi/3}^{\pi/3}\int_0^3 r\,dr\,d\theta=\frac{9}{2}\int_{-\pi/3}^{\pi/3}d\theta=3\pi;\quad M_y=\int_{-\pi/3}^{\pi/3}\int_0^3 r^2\cos\theta\,dr\,d\theta=9\int_{-\pi/3}^{\pi/3}\cos\theta\,d\theta=9\sqrt{3}\Rightarrow\bar{x}=\frac{3\sqrt{3}}{\pi}$, and $\bar{y}=0$ by symmetry

51. (a) $M = 2\int_0^{\pi/2}\int_1^{1+\cos\theta} r\,dr\,d\theta$

$= \int_0^{\pi/2}\left(2\cos\theta + \frac{1+\cos2\theta}{2}\right)d\theta = \frac{8+\pi}{4}$;

$M_y = \int_{-\pi/2}^{\pi/2}\int_1^{1+\cos\theta}(r\cos\theta)\,r\,dr\,d\theta$

$= \int_{-\pi/2}^{\pi/2}\left(\cos^2\theta + \cos^3\theta + \frac{\cos^4\theta}{3}\right)d\theta$

$= \frac{32+15\pi}{24} \Rightarrow \bar{x} = \frac{15\pi+32}{6\pi+48}$, and $\bar{y} = 0$ by symmetry

(b)

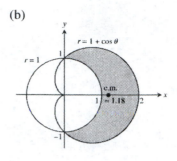

53. $x = u + y$ and $y = v \Rightarrow x = u+v$ and $y = v$

$\Rightarrow J(u,v) = \begin{vmatrix} 1 & 1 \\ 0 & 1 \end{vmatrix} = 1$; the boundary of the image G

is obtained from the boundary of R as follows:

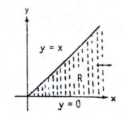

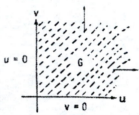

xy-equations for the boundary of R	Corresponding uv-equations for the boundary of G	Simplified uv-equations
$y = x$	$v = u + v$	$u = 0$
$y = 0$	$v = 0$	$v = 0$

$\Rightarrow \int_0^\infty\int_0^x e^{-sx}f(x-y,y)\,dy\,dx = \int_0^\infty\int_0^\infty e^{-s(u+v)}f(u,v)\,du\,dv$

CHAPTER 14 ADDITIONAL AND ADVANCED EXERCISES

1. (a) $V = \int_{-3}^2\int_x^{6-x^2} x^2\,dy\,dx$ (b) $V = \int_{-3}^2\int_x^{6-x^2}\int_0^{x^2} dz\,dy\,dx$

(c) $V = \int_{-3}^2\int_x^{6-x^2} x^2\,dy\,dx = \int_{-3}^2\int_x^{6-x^2}\left(6x^2 - x^4 - x^3\right)dx = \left[2x^3 - \frac{x^5}{5} - \frac{x^4}{4}\right]_{-3}^2 = \frac{125}{4}$

3. Using cylindrical coordinates, $V = \int_0^{2\pi}\int_0^1\int_0^{2-r(\cos\theta+\sin\theta)} dz\,r\,dr\,d\theta = \int_0^{2\pi}\int_0^1\left(2r - r^2\cos\theta - r^2\sin\theta\right)dr\,d\theta$

$= \int_0^{2\pi}\left(1 - \frac{1}{3}\cos\theta - \frac{1}{3}\sin\theta\right)d\theta = \left[\theta - \frac{1}{3}\sin\theta + \frac{1}{3}\cos\theta\right]_0^{2\pi} = 2\pi$

5. The surfaces intersect when $3 - x^2 - y^2 = 2x^2 + 2y^2 \Rightarrow x^2 + y^2 = 1$. Thus the volume is

$V = 4\int_0^1\int_0^{\sqrt{1-x^2}}\int_{2x^2+2y^2}^{3-x^2-y^2} dz\,dy\,dx = 4\int_0^{\pi/2}\int_0^1\int_{2r^2}^{3-r^2} dz\,r\,dr\,d\theta = 4\int_0^{\pi/2}\int_0^1\left(3r - 3r^3\right)dr\,d\theta = 3\int_0^{\pi/2} d\theta = \frac{3\pi}{2}$

7. (a) The radius of the hole is 1, and the radius of the sphere is 2.

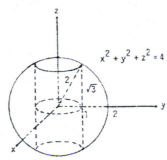

(b) $V = 2\int_0^{2\pi}\int_0^{\sqrt{3}}\int_1^{\sqrt{4-z^2}} r\, dr\, dz\, d\theta = \int_0^{2\pi}\int_0^{\sqrt{3}}\left(3-z^2\right)dz\, d\theta = 2\sqrt{3}\int_0^{2\pi} d\theta = 4\sqrt{3}\pi$

9. The surfaces intersect when $x^2 + y^2 = \frac{x^2+y^2+1}{2} \Rightarrow x^2 + y^2 = 1$. Thus the volume in cylindrical coordinates is

$V = 4\int_0^{\pi/2}\int_0^1\int_{r^2}^{(r^2+1)/2} dz\, r\, dr\, d\theta = 4\int_0^{\pi/2}\int_0^1\left(\frac{r}{2}-\frac{r^3}{2}\right)dr\, d\theta = 4\int_0^{\pi/2}\left[\frac{r^2}{4}-\frac{r^4}{8}\right]_0^1 d\theta = \frac{1}{2}\int_0^{\pi/2} d\theta = \frac{\pi}{4}$

11. $\int_0^\infty \frac{e^{-ax}-e^{-bx}}{x}\, dx = \int_0^\infty \int_a^b e^{-xy}\, dy\, dx = \int_a^b \int_0^\infty e^{-xy}\, dx\, dy = \int_a^b\left(\lim_{t\to\infty}\int_0^t e^{-xy}\, dx\right)dy = \int_a^b \lim_{t\to\infty}\left[-\frac{e^{-xy}}{y}\right]_0^t dy$

$= \int_a^b \lim_{t\to\infty}\left(\frac{1}{y}-\frac{e^{-yt}}{y}\right)dy = \int_a^b \frac{1}{y}\, dy = \left[\ln y\right]_a^b = \ln\left(\frac{b}{a}\right)$

13. $\int_0^x \int_0^u e^{m(x-t)} f(t)\, dt\, du = \int_0^x \int_t^x e^{m(x-t)} f(t)\, du\, dt = \int_0^x (x-t)e^{m(x-t)} f(t)\, dt$; also

$\int_0^x \int_0^v \int_0^u e^{m(x-t)} f(t)\, dt\, du\, dv = \int_0^x \int_t^x \int_t^v e^{m(x-t)} f(t)\, du\, dv\, dt = \int_0^x \int_t^x (v-t)e^{m(x-t)} f(t)\, dv\, dt$

$= \int_0^x \left[\frac{1}{2}(v-t)^2 e^{m(x-t)} f(t)\right]_t^x dt = \int_0^x \frac{(x-t)^2}{2} e^{m(x-t)} f(t)\, dt$

15. $I_o(a) = \int_0^a \int_0^{x/a^2}\left(x^2+y^2\right)dy\, dx = \int_0^a\left[x^2 y+\frac{y^3}{3}\right]_0^{x/a^2} dx = \int_0^a\left(\frac{x^3}{a^2}+\frac{x^3}{3a^6}\right)dx = \left[\frac{x^4}{4a^2}+\frac{x^4}{12a^6}\right]_0^a = \frac{a^2}{4}+\frac{1}{12}a^{-2}$;

$I_o'(a) = \frac{1}{2}a-\frac{1}{6}a^{-3} = 0 \Rightarrow a^4 = \frac{1}{3} \Rightarrow a = \sqrt[4]{\frac{1}{3}} = \frac{1}{\sqrt[4]{3}}$. Since $I_o''(a) = \frac{1}{2}+\frac{1}{2}a^{-4} > 0$, the value of a does provide a

minimum for the polar moment of inertia $I_o(a)$.

17. $M = \int_{-\theta}^{\theta}\int_{b\sec\theta}^{a} r\, dr\, d\theta = \int_{-\theta}^{\theta}\left(\frac{a^2}{2}-\frac{b^2}{2}\sec^2\theta\right)d\theta$

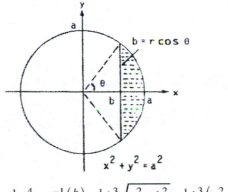

$= a^2\theta-b^2\tan\theta = a^2\cos^{-1}\left(\frac{b}{a}\right)-b^2\left(\frac{\sqrt{a^2-b^2}}{b}\right)$

$= a^2\cos^{-1}\left(\frac{b}{a}\right)-b\sqrt{a^2-b^2}$;

$I_o = \int_{-\theta}^{\theta}\int_{b\sec\theta}^{a} r^3\, dr\, d\theta = \frac{1}{4}\int_{-\theta}^{\theta}\left(a^4+b^4\sec^4\theta\right)d\theta$

$= \frac{1}{4}\int_{-\theta}^{\theta}\left[a^4+b^4\left(1+\tan^2\theta\right)\left(\sec^2\theta\right)\right]d\theta$

$= \frac{1}{4}\left[a^4\theta-b^4\tan\theta-\frac{b^4\tan^3\theta}{3}\right]_{-\theta}^{\theta} = \frac{a^4\theta}{2}-\frac{b^4\tan\theta}{2}-\frac{b^4\tan^3\theta}{6} = \frac{1}{2}a^4\cos^{-1}\left(\frac{b}{a}\right)-\frac{1}{2}b^3\sqrt{a^2-b^2}-\frac{1}{6}b^3\left(a^2-b^2\right)^{3/2}$

19. $= \left[\frac{1}{2ab}e^{b^2x^2}\right]_0^a + \left[\frac{1}{2ba}e^{a^2y^2}\right]_0^b = \frac{1}{2ab}\left(e^{b^2a^2}-1\right)+\frac{1}{2ab}\left(e^{a^2b^2}-1\right) = \frac{1}{ab}\left(e^{a^2b^2}-1\right)$

21. (a) (i) Fubini's Theorem

      (ii) Treating $G(y)$ as a constant

      (iii) Algebraic rearrangement

      (iv) The definite integral is a constant number

  (b) $\int_0^{\ln 2} \int_0^{\pi/2} e^x \cos y \, dy \, dx = \left(\int_0^{\ln 2} e^x dx \right) \left(\int_0^{\pi/2} \cos y \, dy \right) = \left(e^{\ln 2} - e^0 \right) \left(\sin \frac{\pi}{2} - \sin 0 \right) = (1)(1) = 1$

  (c) $\int_1^2 \int_{-1}^1 \frac{x}{y^2} \, dx \, dy = \left(\int_1^2 \frac{1}{y^2} \, dy \right) \left(\int_{-1}^1 x \, dx \right) = \left[-\frac{1}{y} \right]_1^2 \left[\frac{x^2}{2} \right]_{-1}^1 = \left(-\frac{1}{2} + 1 \right) \left(\frac{1}{2} - \frac{1}{2} \right) = 0$

23. (a) $I^2 = \int_0^\infty \int_0^\infty e^{-(x^2+y^2)} dx \, dy = \int_0^{\pi/2} \int_0^\infty \left(e^{-r^2} \right) r \, dr \, d\theta = \int_0^{\pi/2} \left[\lim_{b \to \infty} \int_0^b r e^{-r^2} dr \right] d\theta$

    $= -\frac{1}{2} \int_0^{\pi/2} \lim_{b \to \infty} \left(e^{-b^2} - 1 \right) d\theta = \frac{1}{2} \int_0^{\pi/2} d\theta = \frac{\pi}{4} \Rightarrow I = \frac{\sqrt{\pi}}{2}$

  (b) $\Gamma\left(\frac{1}{2} \right) = \int_0^\infty t^{-1/2} e^{-t} dt = \int_0^\infty \left(y^2 \right)^{-1/2} e^{-y^2} (2y) \, dy = 2 \int_0^\infty e^{-y^2} dy = 2 \left(\frac{\sqrt{\pi}}{2} \right) = \sqrt{\pi}$, where $y = \sqrt{t}$

25. For a height h in the bowl the volume of water is $V = \int_{-\sqrt{h}}^{\sqrt{h}} \int_{-\sqrt{h-x^2}}^{\sqrt{h-x^2}} \int_{x^2+y^2}^h dz \, dy \, dx$

  $= \int_{-\sqrt{h}}^{\sqrt{h}} \int_{-\sqrt{h-x^2}}^{\sqrt{h-x^2}} \left(h - x^2 - y^2 \right) dy \, dx = \int_0^{2\pi} \int_0^{\sqrt{h}} \left(h - r^2 \right) r \, dr \, d\theta = \int_0^{2\pi} \left[\frac{hr^2}{2} - \frac{r^4}{4} \right]_0^{\sqrt{h}} d\theta = \int_0^{2\pi} \frac{h^2}{4} \, d\theta = \frac{h^2 \pi}{2}$.

Since the top of the bowl has area 10π, then we calibrate the bowl by comparing it to a right circular cylinder whose cross sectional area is 10π from $z = 0$ to $z = 10$. If such a cylinder contains $\frac{h^2 \pi}{2}$ cubic inches of water to a depth w then we have $10\pi w = \frac{h^2 \pi}{2} \Rightarrow w = \frac{h^2}{20}$. So for 1 inch of rain, $w = 1$ and $h = \sqrt{20}$; for 3 inches of rain, $w = 3$ and $h = \sqrt{60}$.

27. The cylinder is given by $x^2 + y^2 = 1$ from $z = 1$ to $\infty \Rightarrow \iiint\limits_D z \left(r^2 + z^2 \right)^{-5/2} dV$

  $= \int_0^{2x} \int_0^1 \int_1^\infty \frac{z}{\left(r^2 + z^2 \right)^{5/2}} \, dz \, r \, dr \, d\theta = \lim_{a \to \infty} \int_0^{2\pi} \int_0^1 \int_1^a \frac{rz}{\left(r^2 + z^2 \right)^{5/2}} \, dz \, dr \, d\theta = \lim_{a \to \infty} \int_0^{2\pi} \int_0^1 \left[\left(-\frac{1}{3} \right) \frac{r}{\left(r^2 + z^2 \right)^{3/2}} \right]_1^a dr \, d\theta$

  $= \lim_{a \to \infty} \int_0^{2\pi} \int_0^1 \left[\left(-\frac{1}{3} \right) \frac{r}{\left(r^2 + a^2 \right)^{3/2}} + \left(\frac{1}{3} \right) \frac{r}{\left(r^2 + 1 \right)^{3/2}} \right] dr \, d\theta = \lim_{a \to \infty} \int_0^{2\pi} \left[\frac{1}{3} \left(r^2 + a^2 \right)^{-1/2} - \frac{1}{3} \left(r^2 + 1 \right)^{-1/2} \right]_0^1 d\theta$

  $= \lim_{a \to \infty} \int_0^{2\pi} \left[\frac{1}{3} \left(1 + a^2 \right)^{-1/2} - \frac{1}{3} \left(2^{-1/2} \right) - \frac{1}{3} \left(a^2 \right)^{-1/2} + \frac{1}{3} \right] d\theta = \lim_{a \to \infty} 2\pi \left[\frac{1}{3} \left(1 + a^2 \right)^{-1/2} - \frac{1}{3} \left(\frac{\sqrt{2}}{2} \right) - \frac{1}{3} \left(\frac{1}{a} \right) + \frac{1}{3} \right]$

  $= 2\pi \left[\frac{1}{3} - \left(\frac{1}{3} \right) \frac{\sqrt{2}}{2} \right]$.

CHAPTER 15   INTEGRALS AND VECTOR FIELDS

15.1 LINE INTEGRALS

1. $\mathbf{r} = t\mathbf{i} + (1-t)\mathbf{j} \Rightarrow x = t$ and $y = 1-t \Rightarrow y = 1-x \Rightarrow (c)$

3. $\mathbf{r} = (2\cos t)\mathbf{i} + (2\sin t)\mathbf{j} \Rightarrow x = 2\cos t$ and $y = 2\sin t \Rightarrow x^2 + y^2 = 4 \Rightarrow (g)$

5. $\mathbf{r} = t\mathbf{i} + t\mathbf{j} + t\mathbf{k} \Rightarrow x = t, \ y = t,$ and $z = t \Rightarrow (d)$

7. $\mathbf{r} = \left(t^2 - 1\right)\mathbf{j} + 2t\mathbf{k} \Rightarrow y = t^2 - 1$ and $z = 2t \Rightarrow y = \frac{z^2}{4} - 1 \Rightarrow (f)$

9. $\mathbf{r}(t) = t\mathbf{i} + (1-t)\mathbf{j}, \ 0 \le t \le 1 \Rightarrow \frac{d\mathbf{r}}{dt} = \mathbf{i} - \mathbf{j} \Rightarrow \left|\frac{d\mathbf{r}}{dt}\right| = \sqrt{2}\mathbf{j}; \ x = t$ and $y = 1-t \Rightarrow x + y = t + (1-t) = 1$

$\Rightarrow \int_C f(x, y, z)\, ds = \int_0^1 f(t, 1-t, 0)\left|\frac{d\mathbf{r}}{dt}\right| dt = \int_0^1 (1)\left(\sqrt{2}\right) dt = \left[\sqrt{2}\, t\right]_0^1 = \sqrt{2}$

11. $\mathbf{r}(t) = 2t\mathbf{i} + t\mathbf{j} + (2-2t)\mathbf{k}, \ 0 \le t \le 1 \Rightarrow \frac{d\mathbf{r}}{dt} = 2\mathbf{i} + \mathbf{j} - 2\mathbf{k} \Rightarrow \left|\frac{d\mathbf{r}}{dt}\right| = \sqrt{4+1+4} = 3; \ xy + y + z = (2t)t + t + (2-2t)$

$\Rightarrow \int_C f(x, y, z)\, ds = \int_0^1 \left(2t^2 - t + 2\right) 3\, dt = 3\left[\frac{2}{3}t^3 - \frac{1}{2}t^2 + 2t\right]_0^1 = 3\left(\frac{2}{3} - \frac{1}{2} + 2\right) = \frac{13}{2}$

13. $\mathbf{r}(t) = (\mathbf{i} + 2\mathbf{j} + 3\mathbf{k}) + t(-\mathbf{i} - 3\mathbf{j} - 2\mathbf{k}) = (1-t)\mathbf{i} + (2-3t)\mathbf{j} + (3-2t)\mathbf{k}, \ 0 \le t \le 1 \Rightarrow \frac{d\mathbf{r}}{dt} = -\mathbf{i} - 3\mathbf{j} - 2\mathbf{k}$

$\Rightarrow \left|\frac{d\mathbf{r}}{dt}\right| = \sqrt{1+9+4} = \sqrt{14}; \ x + y + z = (1-t) + (2-3t) + (3-2t) = 6 - 6t \Rightarrow \int_C f(x, y, z)\, ds$

$= \int_0^1 (6 - 6t)\sqrt{14}\, dt = 6\sqrt{14}\left[t - \frac{t^2}{2}\right]_0^1 = \left(6\sqrt{14}\right)\left(\frac{1}{2}\right) = 3\sqrt{14}$

15. $C_1 : \mathbf{r}(t) = t\mathbf{i} + t^2\mathbf{j}, \ 0 \le t \le 1 \Rightarrow \frac{d\mathbf{r}}{dt} = \mathbf{i} + 2t\mathbf{j} \Rightarrow \left|\frac{d\mathbf{r}}{dt}\right| = \sqrt{1+4t^2}; \ x + \sqrt{y} - z^2 = t + \sqrt{t^2} - 0 = t + |t| = 2t$ since

$t \ge 0 \Rightarrow \int_{C_1} f(x, y, z)\, ds = \int_0^1 2t\sqrt{1+4t^2}\, dt = \left[\frac{1}{6}\left(1+4t^2\right)^{3/2}\right]_0^1 = \frac{1}{6}(5)^{3/2} - \frac{1}{6} = \frac{1}{6}\left(5\sqrt{5} - 1\right);$

$C_2 : \mathbf{r}(t) = \mathbf{i} + \mathbf{j} + t\mathbf{k}, \ 0 \le t \le 1 \Rightarrow \frac{d\mathbf{r}}{dt} = \mathbf{k} \Rightarrow \left|\frac{d\mathbf{r}}{dt}\right| = 1; \ x + \sqrt{y} - z^2 = 1 + \sqrt{1} - t^2 = 2 - t^2$

$\Rightarrow \int_{C_2} f(x, y, z)\, ds = \int_0^1 \left(2 - t^2\right)(1)dt = \left[2t - \frac{1}{3}t^3\right]_0^1 = 2 - \frac{1}{3} = \frac{5}{3};$

therefore $\int_C f(x, y, z)\, ds = \int_{C_1} f(x, y, z)\, ds + \int_{C_2} f(x, y, z)\, ds = \frac{5}{6}\sqrt{5} + \frac{3}{2}$

17. $\mathbf{r}(t) = t\mathbf{i} + t\mathbf{j} + t\mathbf{k}, \ 0 < a \le t \le b \Rightarrow \frac{d\mathbf{r}}{dt} = \mathbf{i} + \mathbf{j} + \mathbf{k} \Rightarrow \left|\frac{d\mathbf{r}}{dt}\right| = \sqrt{3}; \ \frac{x+y+z}{x^2+y^2+z^2} = \frac{t+t+t}{t^2+t^2+t^2} = \frac{1}{t}$

$\Rightarrow \int_C f(x, y, z)\, ds = \int_a^b \left(\frac{1}{t}\right)\sqrt{3}\, dt = \left[\sqrt{3} \ln|t|\right]_a^b = \sqrt{3} \ln\left(\frac{b}{a}\right),$ since $0 < a \le b$

19. (a) $\mathbf{r}(t) = t\mathbf{i} + \frac{1}{2}t\mathbf{j}, \ 0 \le t \le 4 \Rightarrow \frac{d\mathbf{r}}{dt} = \mathbf{i} + \frac{1}{2}\mathbf{j} \Rightarrow \left|\frac{d\mathbf{r}}{dt}\right| = \frac{\sqrt{5}}{2} \Rightarrow \int_C x\, ds = \int_0^4 t\frac{\sqrt{5}}{2}\, dt = \frac{\sqrt{5}}{2}\int_0^4 t\, dt = \left[\frac{\sqrt{5}}{4}t^2\right]_0^4 = 4\sqrt{5}$

(b) $\mathbf{r}(t) = t\mathbf{i} + t^2\mathbf{j}, 0 \le t \le 2 \Rightarrow \frac{d\mathbf{r}}{dt} = \mathbf{i} + 2t\mathbf{j} \Rightarrow \left|\frac{d\mathbf{r}}{dt}\right| = \sqrt{1+4t^2} \Rightarrow \int_C x \, ds = \int_0^2 t\sqrt{1+4t^2} \, dt$

$\qquad = \left[\frac{1}{12}\left(1+4t^2\right)^{3/2}\right]_0^2 = \frac{17\sqrt{17}-1}{12}$

21. $\mathbf{r}(t) = 4t\mathbf{i} - 3t\mathbf{j}, -1 \le t \le 2 \Rightarrow \frac{d\mathbf{r}}{dt} = 4\mathbf{i} - 3\mathbf{j} \Rightarrow \left|\frac{d\mathbf{r}}{dt}\right| = 5 \Rightarrow \int_C ye^{x^2} \, ds = \int_{-1}^2 (-3t)\, e^{(4t)^2} \cdot 5 \, dt$

$\qquad = -15\int_{-1}^2 t\, e^{16t^2} \, dt = \left[-\frac{15}{32} e^{16t^2}\right]_{-1}^2 = -\frac{15}{32} e^{64} + \frac{15}{32} e^{16} = \frac{15}{32}\left(e^{16} - e^{64}\right)$

23. $\mathbf{r}(t) = t^2\mathbf{i} + t^3\mathbf{j}, 1 \le t \le 2 \Rightarrow \frac{d\mathbf{r}}{dt} = 2t\mathbf{i} + 3t^2\mathbf{j} \Rightarrow \left|\frac{d\mathbf{r}}{dt}\right| = \sqrt{(2t)^2 + \left(3t^2\right)^2} = t\sqrt{4+9t^2}$

$\qquad \Rightarrow \int_C \frac{x^2}{y^{4/3}} \, ds = \int_1^2 \frac{\left(t^2\right)^2}{\left(t^3\right)^{4/3}} \cdot t\sqrt{4+9t^2} \, dt = \int_1^2 t\sqrt{4+9t^2} \, dt = \left[\frac{1}{27}\left(4+9t^2\right)^{3/2}\right]_1^2 = \frac{80\sqrt{10}-13\sqrt{13}}{27}$

25. $C_1 : \mathbf{r}(t) = t\mathbf{i} + t^2\mathbf{j}, 0 \le t \le 1 \Rightarrow \frac{d\mathbf{r}}{dt} = \mathbf{i} + 2t\mathbf{j} \Rightarrow \left|\frac{d\mathbf{r}}{dt}\right| = \sqrt{1+4t^2}; C_2 : \mathbf{r}(t) = (1-t)\mathbf{i} + (1-t)\mathbf{j}, 0 \le t \le 1$

$\qquad \Rightarrow \frac{d\mathbf{r}}{dt} = -\mathbf{i} - \mathbf{j} \Rightarrow \left|\frac{d\mathbf{r}}{dt}\right| = \sqrt{2} \Rightarrow \int_C \left(x + \sqrt{y}\right) ds = \int_{C_1} \left(x + \sqrt{y}\right) ds + \int_{C_2} \left(x + \sqrt{y}\right) ds$

$\qquad = \int_0^1 \left(t + \sqrt{t^2}\right)\sqrt{1+4t^2} \, dt + \int_0^1 \left((1-t) + \sqrt{1-t}\right)\sqrt{2} \, dt = \int_0^1 2t\sqrt{1+4t^2} \, dt + \int_0^1 \left(1-t + \sqrt{1-t}\right)\sqrt{2} \, dt$

$\qquad = \left[\frac{1}{6}\left(1+4t^2\right)^{3/2}\right]_0^1 + \sqrt{2}\left[t - \frac{1}{2}t^2 - \frac{2}{3}(1-t)^{3/2}\right]_0^1 = \frac{5\sqrt{5}-1}{6} + \frac{7\sqrt{2}}{6} = \frac{5\sqrt{5}+7\sqrt{2}-1}{6}$

27. $\mathbf{r}(x) = x\mathbf{i} + y\mathbf{j} = x\mathbf{i} + \frac{x^2}{2}\mathbf{j}, 0 \le x \le 2 \Rightarrow \frac{d\mathbf{r}}{dx} = \mathbf{i} + x\mathbf{j} \Rightarrow \left|\frac{d\mathbf{r}}{dx}\right| = \sqrt{1+x^2}; f(x, y) = f\left(x, \frac{x^2}{2}\right) = \frac{x^3}{\left(\frac{x^2}{2}\right)} = 2x$

$\qquad \Rightarrow \int_C f \, ds = \int_0^2 (2x)\sqrt{1+x^2} \, dx = \left[\frac{2}{3}\left(1+x^2\right)^{3/2}\right]_0^2 = \frac{2}{3}\left(5^{3/2} - 1\right) = \frac{10\sqrt{5}-2}{3}$

29. $\mathbf{r}(t) = (2\cos t)\mathbf{i} + (2\sin t)\mathbf{j}, 0 \le t \le \frac{\pi}{2} \Rightarrow \frac{d\mathbf{r}}{dt} = (-2\sin t)\mathbf{i} + (2\cos t)\mathbf{j} \Rightarrow \left|\frac{d\mathbf{r}}{dt}\right| = 2; f(x, y) = f(2\cos t, 2\sin t)$

$\qquad = 2\cos t + 2\sin t \Rightarrow \int_C f \, ds = \int_0^{\pi/2} (2\cos t + 2\sin t)(2) \, dt = \left[4\sin t - 4\cos t\right]_0^{\pi/2} = 4 - (-4) = 8$

31. $y = x^2, 0 \le x \le 2 \Rightarrow \mathbf{r}(t) = t\mathbf{i} + t^2\mathbf{j}, 0 \le t \le 2 \Rightarrow \frac{d\mathbf{r}}{dt} = \mathbf{i} + 2t\mathbf{j} \Rightarrow \left|\frac{d\mathbf{r}}{dt}\right| = \sqrt{1+4t^2} \Rightarrow A = \int_C f(x, y) \, ds$

$\qquad = \int_C \left(x + \sqrt{y}\right) ds = \int_0^2 \left(t + \sqrt{t^2}\right)\sqrt{1+4t^2} \, dt = \int_0^2 2t\sqrt{1+4t^2} \, dt = \left[\frac{1}{6}\left(1+4t^2\right)^{3/2}\right]_0^2 = \frac{17\sqrt{17}-1}{6}$

33. $\mathbf{r}(t) = \left(t^2 - 1\right)\mathbf{j} + 2t\mathbf{k}, 0 \le t \le 1 \Rightarrow \frac{d\mathbf{r}}{dt} = 2t\mathbf{j} + 2\mathbf{k} \Rightarrow \left|\frac{d\mathbf{r}}{dt}\right| = 2\sqrt{t^2+1}; M = \int_C \delta(x, y, z) \, ds = \int_0^1 \delta(t)\left(2\sqrt{t^2+1}\right) dt$

$\qquad = \int_0^1 \left(\frac{3}{2}t\right)\left(2\sqrt{t^2+1}\right) dt = \left[\left(t^2+1\right)^{3/2}\right]_0^1 = 2^{3/2} - 1 = 2\sqrt{2} - 1$

35. $\mathbf{r}(t) = \sqrt{2}t\mathbf{i} + \sqrt{2}t\mathbf{j} + \left(4 - t^2\right)\mathbf{k}, \; 0 \le t \le 1 \Rightarrow \frac{d\mathbf{r}}{dt} = \sqrt{2}\mathbf{i} + \sqrt{2}\mathbf{j} - 2t\mathbf{k} \Rightarrow \left|\frac{d\mathbf{r}}{dt}\right| = \sqrt{2 + 2 + 4t^2} = 2\sqrt{1 + t^2}$;

   (a) $M = \int_C \delta \, ds = \int_0^1 (3t)\left(2\sqrt{1 + t^2}\right) dt = \left[2\left(1 + t^2\right)^{3/2}\right]_0^1 = 2\left(2^{3/2} - 1\right) = 4\sqrt{2} - 2$

   (b) $M = \int_C \delta \, ds = \int_0^1 (1)\left(2\sqrt{1 + t^2}\right) dt = \left[t\sqrt{1 + t^2} + \ln\left(t + \sqrt{1 + t^2}\right)\right]_0^1 = \left[\sqrt{2} + \ln\left(1 + \sqrt{2}\right)\right] - (0 + \ln 1)$

   $= \sqrt{2} + \ln\left(1 + \sqrt{2}\right)$

37. Let $x = a \cos t$ and $y = a \sin t$, $0 \le t \le 2\pi$. Then $\frac{dx}{dt} = -a \sin t$, $\frac{dy}{dt} = a \cos t$, $\frac{dz}{dt} = 0$

   $\Rightarrow \sqrt{\left(\frac{dx}{dt}\right)^2 + \left(\frac{dy}{dt}\right)^2 + \left(\frac{dz}{dt}\right)^2} \, dt = a \, dt; \; I_z = \int_C \left(x^2 + y^2\right) \delta \, ds = \int_0^{2\pi} \left(a^2 \sin^2 t + a^2 \cos^2 t\right) a\delta \, dt$

   $= \int_0^{2\pi} a^3 \delta \, dt = 2\pi\delta a^3$.

39. $\mathbf{r}(t) = (\cos t)\mathbf{i} + (\sin t)\mathbf{j} + t\mathbf{k}, \; 0 \le t \le 2\pi \Rightarrow \frac{d\mathbf{r}}{dt} = (-\sin t)\mathbf{i} + (\cos t)\mathbf{j} + \mathbf{k} \Rightarrow \left|\frac{d\mathbf{r}}{dt}\right| = \sqrt{\sin^2 t + \cos^2 t + 1} = \sqrt{2}$;

   (a) $I_z = \int_C \left(x^2 + y^2\right) \delta \, ds = \int_0^{2\pi} \left(\cos^2 t + \sin^2 t\right) \delta\sqrt{2} \, dt = 2\pi\delta\sqrt{2}$

   (b) $I_z = \int_C \left(x^2 + y^2\right) \delta \, ds = \int_0^{4\pi} \delta\sqrt{2} \, dt = 4\pi\delta\sqrt{2}$

41. $\delta(x, y, z) = 2 - z$ and $\mathbf{r}(t) = (\cos t)\mathbf{j} + (\sin t)\mathbf{k}, \; 0 \le t \le \pi \Rightarrow M = 2\pi - 2$ as found in Example 3 of the text; also

   $\left|\frac{d\mathbf{r}}{dt}\right| = 1; \; I_x = \int_C \left(y^2 + z^2\right) \delta \, ds = \int_0^\pi \left(\cos^2 t + \sin^2 t\right)(2 - \sin t) \, dt = \int_0^\pi (2 - \sin t) \, dt = 2\pi - 2$

43-45. Example CAS commands:

   Maple:

```
f := (x,y,z) -> sqrt(1+30*x^2+10*y);
g := t -> t;
h := t -> t^2;
k := t -> 3*t^2;
a,b := 0.2;
ds := ( D(g)^2 + D(h)^2 + D(k)^2)^(1/2):     #(a)
'ds'=ds(t)*'dt';
F := f(g,h,k):                               #(b)
'F(t)'=F(t);
Int( f, s=C..NULL ) = Int( simplify(F(t)*ds(t)), t=a..b);   #(c)
``= value(rhs(%));
```

<u>Mathematica:</u> (functions and domains may vary)

```
Clear[x, y, z, r, t, f]
f[x_,y_,z_]:=Sqrt(1+30x² +10y]
{a,b} = {0,2};
x[t_]:=t
y[t_]:=t²
z[t_]:=3t²
r[t_]:={x[t], y[t], z[t]}
v[t_]:= D[r[t],t]
mag[vector_]:=Sqrt[vector.vector]
Integrate[f[x(t),y(t),z[t]] mag[v[t]], {t, a, b}]
N[%]
```

15.2 VECTOR FIELDS AND LINE INTEGRALS: WORK, CIRCULATION, AND FLUX

1. $f(x, y, z) = \left(x^2 + y^2 + z^2\right)^{-1/2} \Rightarrow \frac{\partial f}{\partial x} = -\frac{1}{2}\left(x^2 + y^2 + z^2\right)^{-3/2}(2x) = -x\left(x^2 + y^2 + z^2\right)^{-3/2}$; similarly,

$\frac{\partial f}{\partial y} = -y\left(x^2 + y^2 + z^2\right)^{-3/2}$ and $\frac{\partial f}{\partial z} = -z\left(x^2 + y^2 + z^2\right)^{-3/2} \Rightarrow \nabla f = \frac{-x\mathbf{i}-y\mathbf{j}-z\mathbf{k}}{\left(x^2+y^2+z^2\right)^{3/2}}$

3. $g(x, y, z) = e^z - \ln\left(x^2 + y^2\right) \Rightarrow \frac{\partial g}{\partial x} = -\frac{2x}{x^2+y^2}, \frac{\partial g}{\partial y} = -\frac{2y}{x^2+y^2}$ and $\frac{\partial g}{\partial z} = e^z \Rightarrow \nabla g = \left(\frac{-2x}{x^2+y^2}\right)\mathbf{i} - \left(\frac{2y}{x^2+y^2}\right)\mathbf{j} + e^z\mathbf{k}$

5. $|\mathbf{F}|$ inversely proportional to the square of the distance from (x, y) to the origin $\Rightarrow \sqrt{\left(M(x, y)\right)^2 + \left(N(x, y)\right)^2}$

$= \frac{k}{x^2+y^2}, k > 0$; $\mathbf{F}$ points toward the origin $\Rightarrow \mathbf{F}$ is in the direction of $\mathbf{n} = \frac{-x}{\sqrt{x^2+y^2}}\mathbf{i} - \frac{y}{\sqrt{x^2+y^2}}\mathbf{j} \Rightarrow \mathbf{F} = a\mathbf{n}$, for

some constant $a > 0$. Then $M(x, y) = \frac{-ax}{\sqrt{x^2+y^2}}$ and $N(x, y) = \frac{-ay}{\sqrt{x^2+y^2}} \Rightarrow \sqrt{\left(M(x, y)\right)^2 + \left(N(x, y)\right)^2} = a$

$\Rightarrow a = \frac{k}{x^2+y^2} \Rightarrow \mathbf{F} = \frac{-kx}{\left(x^2+y^2\right)^{3/2}}\mathbf{i} - \frac{ky}{\left(x^2+y^2\right)^{3/2}}\mathbf{j}$, for any constant $k > 0$

7. Substitute the parametric representations for $\mathbf{r}(t) = x(t)\mathbf{i} + y(t)\mathbf{j} + z(t)\mathbf{k}$ representing each path into the vector field $\mathbf{F}$, and calculate $\int_C \mathbf{F} \cdot \frac{d\mathbf{r}}{dt} dt$.

(a) $\mathbf{F} = 3t\mathbf{i} + 2t\mathbf{j} + 4t\mathbf{k}$ and $\frac{d\mathbf{r}}{dt} = \mathbf{i} + \mathbf{j} + \mathbf{k} \Rightarrow \mathbf{F} \cdot \frac{d\mathbf{r}}{dt} = 9t \Rightarrow \int_0^1 9t \, dt = \frac{9}{2}$

(b) $\mathbf{F} = 3t^2\mathbf{i} + 2t\mathbf{j} + 4t^4\mathbf{k}$ and $\frac{d\mathbf{r}}{dt} = \mathbf{i} + 2t\mathbf{j} + 4t^3\mathbf{k} \Rightarrow \mathbf{F} \cdot \frac{d\mathbf{r}}{dt} = 7t^2 + 16t^7 \Rightarrow \int_0^1 \left(7t^2 + 16t^7\right) dt = \left[\frac{7}{3}t^3 + 2t^8\right]_0^1$

$= \frac{7}{3} + 2 = \frac{13}{3}$

(c) $\mathbf{r}_1 = t\mathbf{i} + t\mathbf{j}$ and $\mathbf{r}_2 = \mathbf{i} + \mathbf{j} + t\mathbf{k}$; $\mathbf{F}_1 = 3t\mathbf{i} + 2t\mathbf{j}$ and $\frac{d\mathbf{r}_1}{dt} = \mathbf{i} + \mathbf{j} \Rightarrow \mathbf{F}_1 \cdot \frac{d\mathbf{r}_1}{dt} = 5t \Rightarrow \int_0^1 5t \, dt = \frac{5}{2}$;

$\mathbf{F}_2 = 3\mathbf{i} + 2\mathbf{j} + 4t\mathbf{k}$ and $\frac{d\mathbf{r}_2}{dt} = \mathbf{k} \Rightarrow \mathbf{F}_2 \cdot \frac{d\mathbf{r}_2}{dt} = 4t \Rightarrow \int_0^1 4t \, dt = 2 \Rightarrow \frac{5}{2} + 2 = \frac{9}{2}$

9. Substitute the parametric representation for $\mathbf{r}(t) = x(t)\mathbf{i} + y(t)\mathbf{j} + z(t)\mathbf{k}$ representing each path into the vector field $\mathbf{F}$, and calculate $\int_C \mathbf{F} \cdot \frac{d\mathbf{r}}{dt}\, dt$.

   (a) $\mathbf{F} = \sqrt{t}\,\mathbf{i} - 2t\mathbf{j} + \sqrt{t}\,\mathbf{k}$ and $\frac{d\mathbf{r}}{dt} = \mathbf{i} + \mathbf{j} + \mathbf{k} \Rightarrow \mathbf{F} \cdot \frac{d\mathbf{r}}{dt} = 2\sqrt{t} - 2t \Rightarrow \int_0^1 \left(2\sqrt{t} - 2t\right) dt = \left[\frac{4}{3}t^{3/2} - t^2\right]_0^1 = \frac{1}{3}$

   (b) $\mathbf{F} = t^2\mathbf{i} - 2t\mathbf{j} + t\mathbf{k}$ and $\frac{d\mathbf{r}}{dt} = \mathbf{i} + 2t\mathbf{j} + 4t^3\mathbf{k} \Rightarrow \mathbf{F} \cdot \frac{d\mathbf{r}}{dt} = 4t^4 - 3t^2 \Rightarrow \int_0^1\left(4t^4 - 3t^2\right) dt = \left[\frac{4}{5}t^5 - t^3\right]_0^1 = -\frac{1}{5}$

   (c) $\mathbf{r}_1 = t\mathbf{i} + t\mathbf{j}$ and $\mathbf{r}_2 = \mathbf{i} + \mathbf{j} + t\mathbf{k}$; $\mathbf{F}_1 = -2t\mathbf{j} + \sqrt{t}\,\mathbf{k}$ and $\frac{d\mathbf{r}_1}{dt} = \mathbf{i} + \mathbf{j} \Rightarrow \mathbf{F}_1 \cdot \frac{d\mathbf{r}_1}{dt} = -2t \Rightarrow \int_0^1 -2t\, dt = -1;$

   $\mathbf{F}_2 = \sqrt{t}\,\mathbf{i} - 2\mathbf{j} + \mathbf{k}$ and $\frac{d\mathbf{r}_2}{dt} = \mathbf{k} \Rightarrow \mathbf{F}_2 \cdot \frac{d\mathbf{r}_2}{dt} = 1 \Rightarrow \int_0^1 dt = 1 \Rightarrow -1 + 1 = 0$

11. Substitute the parametric representation for $\mathbf{r}(t) = x(t)\mathbf{i} + y(t)\mathbf{j} + z(t)\mathbf{k}$ representing each path into the vector field $\mathbf{F}$, and calculate $\int_C \mathbf{F} \cdot \frac{d\mathbf{r}}{dt}\, dt$.

   (a) $\mathbf{F} = \left(3t^2 - 3t\right)\mathbf{i} + 3t\mathbf{j} + \mathbf{k}$ and $\frac{d\mathbf{r}}{dt} = \mathbf{i} + \mathbf{j} + \mathbf{k} \Rightarrow \mathbf{F} \cdot \frac{d\mathbf{r}}{dt} = 3t^2 + 1 \Rightarrow \int_0^1\left(3t^2 + 1\right) dt = \left[t^3 + t\right]_0^1 = 2$

   (b) $\mathbf{F} = \left(3t^2 - 3t\right)\mathbf{i} + 3t^4\mathbf{j} + \mathbf{k}$ and $\frac{d\mathbf{r}}{dt} = \mathbf{i} + 2t\mathbf{j} + 4t^3\mathbf{k} \Rightarrow \mathbf{F} \cdot \frac{d\mathbf{r}}{dt} = 6t^5 + 4t^3 + 3t^2 - 3t$

   $\Rightarrow \int_0^1\left(6t^5 + 4t^3 + 3t^2 - 3t\right) dt = \left[t^6 + t^4 + t^3 - \frac{3}{2}t^2\right]_0^1 = \frac{3}{2}$

   (c) $\mathbf{r}_1 = t\mathbf{i} + t\mathbf{j}$ and $\mathbf{r}_2 = \mathbf{i} + \mathbf{j} + t\mathbf{k}$; $\mathbf{F}_1 = \left(3t^2 - 3t\right)\mathbf{i} + \mathbf{k}$ and $\frac{d\mathbf{r}_1}{dt} = \mathbf{i} + \mathbf{j} \Rightarrow \mathbf{F}_1 \cdot \frac{d\mathbf{r}_1}{dt} = 3t^2 - 3t$

   $\Rightarrow \int_0^1\left(3t^2 - 3t\right) dt = \left[t^3 - \frac{3}{2}t^2\right]_0^1 = -\frac{1}{2}; \mathbf{F}_2 = 3t\mathbf{j} + \mathbf{k}$ and $\frac{d\mathbf{r}_2}{dt} = \mathbf{k} \Rightarrow \mathbf{F}_2 \cdot \frac{d\mathbf{r}_2}{dt} = 1 \Rightarrow \int_0^1 dt = 1 \Rightarrow -\frac{1}{2} + 1 = \frac{1}{2}$

13. $x = t,\ y = 2t + 1,\ 0 \le t \le 3 \Rightarrow dx = dt \Rightarrow \int_C (x - y)\, dx = \int_0^3 (t - (2t + 1))\, dt = \int_0^3 (-t - 1)\, dt = \left[-\frac{1}{2}t^2 - t\right]_0^3 = -\frac{15}{2}$

15. $C_1 : x = t,\ y = 0,\ 0 \le t \le 3 \Rightarrow dy = 0;\ C_2 : x = 3,\ y = t,\ 0 \le t \le 3 \Rightarrow dy = dt \Rightarrow \int_C \left(x^2 + y^2\right) dy$

   $= \int_{C_1}\left(x^2 + y^2\right) dx + \int_{C_2}\left(x^2 + y^2\right) dx = \int_0^3\left(t^2 + 0^2\right) \cdot 0 + \int_0^3\left(3^2 + t^2\right) dt = \int_0^3\left(9 + t^2\right) dt = \left[9t + \frac{1}{3}t^3\right]_0^3 = 36$

17. $\mathbf{r}(t) = t\mathbf{i} - \mathbf{j} + t^2\mathbf{k},\ 0 \le t \le 1 \Rightarrow dx = dt,\ dy = 0,\ dz = 2t\, dt$

   (a) $\int_C (x + y - z)\, dx = \int_0^1\left(t - 1 - t^2\right) dt = \left[\frac{1}{2}t^2 - t - \frac{1}{3}t^3\right]_0^1 = -\frac{5}{6}$

   (b) $\int_C (x + y - z)\, dy = \int_0^1\left(t - 1 - t^2\right) \cdot 0 = 0$

   (c) $\int_C (x + y - z)\, dz = \int_0^1\left(t - 1 - t^2\right) 2t\, dt = \int_0^1\left(2t^2 - 2t - 2t^3\right) dt = \left[\frac{2}{3}t^3 - t^2 - \frac{1}{2}t^4\right]_0^1 = -\frac{5}{6}$

19. $\mathbf{r} = t\mathbf{i} + t^2\mathbf{j} + t\mathbf{k},\ 0 \le t \le 1$, and $\mathbf{F} = xy\mathbf{i} + y\mathbf{j} - yz\mathbf{k} \Rightarrow \mathbf{F} = t^3\mathbf{i} + t^2\mathbf{j} - t^3\mathbf{k}$ and $\frac{d\mathbf{r}}{dt} = \mathbf{i} + 2t\mathbf{j} + \mathbf{k}$

   $\Rightarrow \mathbf{F} \cdot \frac{d\mathbf{r}}{dt} = 2t^3 \Rightarrow \text{work} = \int_0^1 2t^3\, dt = \frac{1}{2}$

21. $\mathbf{r} = (\sin t)\mathbf{i} + (\cos t)\mathbf{j} + t\mathbf{k},\ 0 \le t \le 2\pi$, and $\mathbf{F} = z\mathbf{i} + x\mathbf{j} + y\mathbf{k} \Rightarrow \mathbf{F} = t\mathbf{i} + (\sin t)\mathbf{j} + (\cos t)\mathbf{k}$ and

   $\frac{d\mathbf{r}}{dt} = (\cos t)\mathbf{i} - (\sin t)\mathbf{j} + \mathbf{k} \Rightarrow \mathbf{F} \cdot \frac{d\mathbf{r}}{dt} = t\cos t - \sin^2 t + \cos t \Rightarrow \text{work} = \int_0^{2\pi}\left(t\cos t - \sin^2 t + \cos t\right) dt$

   $= \left[\cos t + t\sin t - \frac{t}{2} + \frac{\sin 2t}{4} + \sin t\right]_0^{2\pi} = -\pi$

23. $x = t$ and $y = x^2 = t^2 \Rightarrow \mathbf{r} = t\mathbf{i} + t^2\mathbf{j}, -1 \le t \le 2$, and $\mathbf{F} = xy\mathbf{i} + (x+y)\mathbf{j} \Rightarrow \mathbf{F} = t^3\mathbf{i} + \left(t + t^2\right)\mathbf{j}$ and

$\frac{d\mathbf{r}}{dt} = \mathbf{i} + 2t\mathbf{j} \Rightarrow \mathbf{F} \cdot \frac{d\mathbf{r}}{dt} = t^3 + \left(2t^2 + 2t^3\right) = 3t^3 + 2t^2 \Rightarrow \int_C xy\,dx + (x+y)\,dy = \int_C \mathbf{F} \cdot \frac{d\mathbf{r}}{dt}\,dt = \int_{-1}^{2}\left(3t^3 + 2t^2\right)dt$

$= \left[\frac{3}{4}t^4 + \frac{2}{3}t^3\right]_{-1}^{2} = \left(12 + \frac{16}{3}\right) - \left(\frac{3}{4} - \frac{2}{3}\right) = \frac{45}{4} + \frac{18}{3} = \frac{69}{4}$

25. $\mathbf{r} = x\mathbf{i} + y\mathbf{j} = y^2\mathbf{i} + y\mathbf{j}, 2 \ge y \ge -1$, and $\mathbf{F} = x^2\mathbf{i} - y\mathbf{j} = y^4\mathbf{i} - y\mathbf{j} \Rightarrow \frac{d\mathbf{r}}{dy} = 2y\mathbf{i} + \mathbf{j}$ and $\mathbf{F} \cdot \frac{d\mathbf{r}}{dy} = 2y^5 - y$

$\Rightarrow \int_C \mathbf{F} \cdot \mathbf{T}\,ds = \int_2^{-1} \mathbf{F} \cdot \frac{d\mathbf{r}}{dy}\,dy = \int_2^{-1}\left(2y^5 - y\right)dy = \left[\frac{1}{3}y^6 - \frac{1}{2}y^2\right]_2^{-1} = \left(\frac{1}{3} - \frac{1}{2}\right) - \left(\frac{64}{3} - \frac{4}{2}\right) = \frac{3}{2} - \frac{63}{3} = -\frac{39}{2}$

27. $\mathbf{r} = (\mathbf{i} + \mathbf{j}) + t(\mathbf{i} + 2\mathbf{j}) = (1+t)\mathbf{i} + (1+2t)\mathbf{j}, 0 \le t \le 1$, and $\mathbf{F} = xy\mathbf{i} + (y-x)\mathbf{j} \Rightarrow \mathbf{F} = \left(1 + 3t + 2t^2\right)\mathbf{i} + t\mathbf{j}$ and

$\frac{d\mathbf{r}}{dt} = \mathbf{i} + 2\mathbf{j} \Rightarrow \mathbf{F} \cdot \frac{d\mathbf{r}}{dt} = 1 + 5t + 2t^2 \Rightarrow \text{work} = \int_C \mathbf{F} \cdot \frac{d\mathbf{r}}{dt}\,dt = \int_0^1\left(1 + 5t + 2t^2\right)dt = \left[t + \frac{5}{2}t^2 + \frac{2}{3}t^3\right]_0^1 = \frac{25}{6}$

29. (a) $\mathbf{r} = (\cos t)\mathbf{i} + (\sin t)\mathbf{j}, 0 \le t \le 2\pi$, $\mathbf{F}_1 = x\mathbf{i} + y\mathbf{j}$, and $\mathbf{F}_2 = -y\mathbf{i} + x\mathbf{j} \Rightarrow \frac{d\mathbf{r}}{dt} = (-\sin t)\mathbf{i} + (\cos t)\mathbf{j}$,

$\mathbf{F}_1 = (\cos t)\mathbf{i} + (\sin t)\mathbf{j}$, and $\mathbf{F}_2 = (-\sin t)\mathbf{i} + (\cos t)\mathbf{j} \Rightarrow \mathbf{F}_1 \cdot \frac{d\mathbf{r}}{dt} = 0$ and $\mathbf{F}_2 \cdot \frac{d\mathbf{r}}{dt} = \sin^2 t + \cos^2 t = 1$

$\Rightarrow \text{Circ}_1 = \int_0^{2\pi} 0\,dt = 0$ and $\text{Circ}_2 = \int_0^{2\pi} dt = 2\pi$; $\mathbf{n} = (\cos t)\mathbf{i} + (\sin t)\mathbf{j} \Rightarrow \mathbf{F}_1 \cdot \mathbf{n} = \cos^2 t + \sin^2 t = 1$ and

$\mathbf{F}_2 \cdot \mathbf{n} = 0 \Rightarrow \text{Flux}_1 = \int_0^{2\pi} dt = 2\pi$ and $\text{Flux}_2 = \int_0^{2\pi} 0\,dt = 0$

(b) $\mathbf{r} = (\cos t)\mathbf{i} + (4\sin t)\mathbf{j}, 0 \le t \le 2\pi \Rightarrow \frac{d\mathbf{r}}{dt} = (-\sin t)\mathbf{i} + (4\cos t)\mathbf{j}$, $\mathbf{F}_1 = (\cos t)\mathbf{i} + (4\sin t)\mathbf{j}$, and

$\mathbf{F}_2 = (-4\sin t)\mathbf{i} + (\cos t)\mathbf{j} \Rightarrow \mathbf{F}_1 \cdot \frac{d\mathbf{r}}{dt} = 15\sin t \cos t$ and $\mathbf{F}_2 \cdot \frac{d\mathbf{r}}{dt} = 4 \Rightarrow \text{Circ}_1 = \int_0^{2\pi} 15\sin t \cos t\,dt$

$= \left[\frac{15}{2}\sin^2 t\right]_0^{2\pi} = 0$ and $\text{Circ}_2 = \int_0^{2\pi} 4\,dt = 8\pi$; $\mathbf{n} = \left(\frac{4}{\sqrt{17}}\cos t\right)\mathbf{i} + \left(\frac{1}{\sqrt{17}}\sin t\right)\mathbf{j} \Rightarrow \mathbf{F}_1 \cdot \mathbf{n}$

$= \frac{4}{\sqrt{17}}\cos^2 t + \frac{4}{\sqrt{17}}\sin^2 t$ and $\mathbf{F}_2 \cdot \mathbf{n} = -\frac{15}{\sqrt{17}}\sin t \cos t \Rightarrow \text{Flux}_1 = \int_0^{2\pi}(\mathbf{F}_1 \cdot \mathbf{n})\,|\mathbf{v}|\,dt = \int_0^{2\pi}\left(\frac{4}{\sqrt{17}}\right)\sqrt{17}\,dt$

$= 8\pi$ and $\text{Flux}_2 = \int_0^{2\pi}(\mathbf{F}_2 \cdot \mathbf{n})\,|\mathbf{v}|\,dt = \int_0^{2\pi}\left(-\frac{15}{\sqrt{17}}\sin t \cos t\right)\sqrt{17}\,dt = \left[-\frac{15}{2}\sin^2 t\right]_0^{2\pi} = 0$

31. $\mathbf{F}_1 = (a\cos t)\mathbf{i} + (a\sin t)\mathbf{j}, \frac{d\mathbf{r}_1}{dt} = (-a\sin t)\mathbf{i} + (a\cos t)\mathbf{j} \Rightarrow \mathbf{F}_1 \cdot \frac{d\mathbf{r}_1}{dt} = 0 \Rightarrow \text{Circ}_1 = 0$; $M_1 = a\cos t$,

$N_1 = a\sin t, dx = -a\sin t\,dt, dy = a\cos t\,dt \Rightarrow \text{Flux}_1 = \int_C M_1\,dy - N_1\,dx = \int_0^{\pi}\left(a^2\cos^2 t + a^2\sin^2 t\right)dt$

$= \int_0^{\pi} a^2\,dt = a^2\pi$;

$\mathbf{F}_2 = t\mathbf{i}, \frac{d\mathbf{r}_2}{dt} = \mathbf{i} \Rightarrow \mathbf{F}_2 \cdot \frac{d\mathbf{r}_2}{dt} = t \Rightarrow \text{Circ}_2 = \int_{-a}^{a} t\,dt = 0$; $M_2 = t, N_2 = 0, dx = dt, dy = 0$

$\Rightarrow \text{Flux}_2 = \int_C M_2\,dy - N_2\,dx = \int_{-a}^{a} 0\,dt = 0$;

therefore, $\text{Circ} = \text{Circ}_1 + \text{Circ}_2 = 0$ and $\text{Flux} = \text{Flux}_1 + \text{Flux}_2 = a^2\pi$

33. $\mathbf{F}_1 = (-a \sin t)\mathbf{i} + (a \cos t)\mathbf{j}, \frac{d\mathbf{r}_1}{dt} = (-a \sin t)\mathbf{i} + (a \cos t)\mathbf{j} \Rightarrow \mathbf{F}_1 \cdot \frac{d\mathbf{r}_1}{dt} = a^2 \sin^2 t + a^2 \cos^2 t = a^2$

$\Rightarrow \text{Circ}_1 = \int_0^\pi a^2 \, dt = a^2 \pi; \ M_1 = -a \sin t, \ N_1 = a \cos t, \ dx = -a \sin t \, dt, \ dy = a \cos t \, dt$

$\Rightarrow \text{Flux}_1 = \int_C M_1 \, dy - N_1 \, dx = \int_0^\pi \left(-a^2 \sin t \cos t + a^2 \sin t \cos t \right) dt = 0; \ \mathbf{F}_2 = t\mathbf{j}, \frac{d\mathbf{r}_2}{dt} = \mathbf{i} \Rightarrow \mathbf{F}_2 \cdot \frac{d\mathbf{r}_2}{dt} = 0$

$\Rightarrow \text{Circ}_2 = 0; \ M_2 = 0, \ N_2 = t, \ dx = dt, \ dy = 0 \Rightarrow \text{Flux}_2 = \int_C M_2 \, dy - N_2 \, dx = \int_{-a}^a -t \, dt = 0; \ \text{therefore,}$

$\text{Circ} = \text{Circ}_1 + \text{Circ}_2 = a^2 \pi$ and $\text{Flux} = \text{Flux}_1 + \text{Flux}_2 = 0$

35. (a) $\mathbf{r} = (\cos t)\mathbf{i} + (\sin t)\mathbf{j}, 0 \le t \le \pi$, and $\mathbf{F} = (x+y)\mathbf{i} - \left(x^2 + y^2\right)\mathbf{j} \Rightarrow \frac{d\mathbf{r}}{dt} = (-\sin t)\mathbf{i} + (\cos t)\mathbf{j}$ and

$\mathbf{F} = (\cos t + \sin t)\mathbf{i} - \left(\cos^2 t + \sin^2 t\right)\mathbf{j} \Rightarrow \mathbf{F} \cdot \frac{d\mathbf{r}}{dt} = -\sin t \cos t - \sin^2 t - \cos t \Rightarrow \int_C \mathbf{F} \cdot \mathbf{T} \, ds$

$= \int_0^\pi \left(-\sin t \cos t - \sin^2 t - \cos t \right) dt = \left[-\frac{1}{2} \sin^2 t - \frac{t}{2} + \frac{\sin 2t}{4} - \sin t \right]_0^\pi = -\frac{\pi}{2}$

(b) $\mathbf{r} = (1-2t)\mathbf{i}, 0 \le t \le 1$, and $\mathbf{F} = (x+y)\mathbf{i} - \left(x^2 + y^2\right)\mathbf{j} \Rightarrow \frac{d\mathbf{r}}{dt} = -2\mathbf{i}$ and $\mathbf{F} = (1-2t)\mathbf{i} - (1-2t)^2 \mathbf{j} \Rightarrow$

$\mathbf{F} \cdot \frac{d\mathbf{r}}{dt} = 4t - 2 \Rightarrow \int_C \mathbf{F} \cdot \mathbf{T} \, ds = \int_0^1 (4t - 2) \, dt = \left[2t^2 - 2t \right]_0^1 = 0$

(c) $\mathbf{r}_1 = (1-t)\mathbf{i} - t\mathbf{j}, 0 \le t \le 1$, and $\mathbf{F} = (x+y)\mathbf{i} - \left(x^2 + y^2\right)\mathbf{j} \Rightarrow \frac{d\mathbf{r}_1}{dt} = -\mathbf{i} - \mathbf{j}$ and $\mathbf{F} = (1-2t)\mathbf{i} - \left(1 - 2t + 2t^2\right)\mathbf{j}$

$\Rightarrow \mathbf{F} \cdot \frac{d\mathbf{r}_1}{dt} = (2t - 1) + \left(1 - 2t + 2t^2\right) = 2t^2 \Rightarrow \text{Flow}_1 = \int_{C_1} \mathbf{F} \cdot \frac{d\mathbf{r}_1}{dt} = \int_0^1 2t^2 \, dt = \frac{2}{3}; \ \mathbf{r}_2 = -t\mathbf{i} + (t-1)\mathbf{j},$

$0 \le t \le 1$, and $\mathbf{F} = (x+y)\mathbf{i} - \left(x^2 + y^2\right)\mathbf{j} \Rightarrow \frac{d\mathbf{r}_2}{dt} = -\mathbf{i} + \mathbf{j}$ and $\mathbf{F} = -\mathbf{i} - \left(t^2 + t^2 - 2t + 1\right)\mathbf{j}$

$= -\mathbf{i} - \left(2t^2 - 2t + 1\right)\mathbf{j} \Rightarrow \mathbf{F} \cdot \frac{d\mathbf{r}_2}{dt} = 1 - \left(2t^2 - 2t + 1\right) = 2t - 2t^2 \Rightarrow \text{Flow}_2 = \int_{C_2} \mathbf{F} \cdot \frac{d\mathbf{r}_2}{dt} = \int_0^1 \left(2t - 2t^2 \right) dt$

$= \left[t^2 - \frac{2}{3}t^3 \right]_0^1 = \frac{1}{3} \Rightarrow \text{Flow} = \text{Flow}_1 + \text{Flow}_2 = \frac{2}{3} + \frac{1}{3} = 1$

37. (a) $y = 2x, 0 \le x \le 2 \Rightarrow r(t) = t\mathbf{i} + 2t\mathbf{j}, 0 \le t \le 2 \Rightarrow \frac{d\mathbf{r}}{dt} = \mathbf{i} + 2t\mathbf{j} \Rightarrow \mathbf{F} \cdot \frac{d\mathbf{r}}{dt} = \left((2t)^2 \mathbf{i} + 2(t)(2t)\mathbf{j} \right) \cdot (\mathbf{i} + 2\mathbf{j})$

$= 4t^2 + 8t^2 = 12t^2 \Rightarrow \text{Flow} = \int_C \mathbf{F} \cdot \frac{d\mathbf{r}}{dt} \, dt = \int_0^2 12t^2 \, dt = \left[4t^3 \right]_0^2 = 32$

(b) $y = x^2, 0 \le x \le 2 \Rightarrow \mathbf{r}(t) = t\mathbf{i} + t^2\mathbf{j}, \ 0 \le t \le 2 \Rightarrow \frac{d\mathbf{r}}{dt} = \mathbf{i} + 2t\mathbf{j} \Rightarrow \mathbf{F} \cdot \frac{d\mathbf{r}}{dt} = \left(\left(t^2\right)^2 \mathbf{i} + 2(t)\left(t^2\right)\mathbf{j} \right) \cdot (\mathbf{i} + 2t\mathbf{j})$

$= t^4 + 4t^4 = 5t^4 \Rightarrow \text{Flow} = \int_c \mathbf{F} \cdot \frac{d\mathbf{r}}{dt} \, dt = \int_0^2 5t^4 \, dt = \left[t^5 \right]_0^2 = 32$

(c) answers will vary, one possible path is $y = \frac{1}{2}x^3, 0 \le x \le 2 \Rightarrow \mathbf{r}(t) = t\mathbf{i} + \frac{1}{2}t^3\mathbf{j}, \ 0 \le t \le 2 \Rightarrow \frac{d\mathbf{r}}{dt} = \mathbf{i} + 3t^2\mathbf{j}$

$\Rightarrow \mathbf{F} \cdot \frac{d\mathbf{r}}{dt} = \left(\left(\frac{1}{2}t^3\right)^2 \mathbf{i} + 2(t)\left(\frac{1}{2}t^3\right)\mathbf{j} \right) \cdot (\mathbf{i} + 3t^2\mathbf{j}) = \frac{1}{4}t^6 + \frac{3}{2}t^6 = \frac{7}{4}t^6 \Rightarrow \text{Flow} = \int_c \mathbf{F} \cdot \frac{d\mathbf{r}}{dt} \, dt = \int_0^2 \frac{7}{4}t^6 \, dt$

$= \left[\frac{1}{4}t^7 \right]_0^2 = 32$

544   Chapter 15   Integrals and Vector Fields

39. $\mathbf{F} = -\dfrac{y}{\sqrt{x^2+y^2}}\mathbf{i} + \dfrac{x}{\sqrt{x^2+y^2}}\mathbf{j}$ on $x^2+y^2=4$;

at $(2,0)$, $\mathbf{F}=\mathbf{j}$; at $(0,2)$, $\mathbf{F}=-\mathbf{i}$;

at $(-2,0)$, $\mathbf{F}=-\mathbf{j}$; at $(0,-2)$, $\mathbf{F}=\mathbf{i}$;

at $\left(\sqrt{2},\sqrt{2}\right)$, $\mathbf{F}=-\dfrac{\sqrt{3}}{2}\mathbf{i}+\dfrac{1}{2}\mathbf{j}$;

at $\left(\sqrt{2},-\sqrt{2}\right)$, $\mathbf{F}=\dfrac{\sqrt{3}}{2}\mathbf{i}+\dfrac{1}{2}\mathbf{j}$;

at $\left(-\sqrt{2},\sqrt{2}\right)$, $\mathbf{F}=-\dfrac{\sqrt{3}}{2}\mathbf{i}-\dfrac{1}{2}\mathbf{j}$;

at $\left(-\sqrt{2},-\sqrt{2}\right)$, $\mathbf{F}=\dfrac{\sqrt{3}}{2}\mathbf{i}-\dfrac{1}{2}\mathbf{j}$

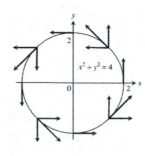

41. (a) $\mathbf{G}=P(x,y)\mathbf{i}+Q(x,y)\mathbf{j}$ is to have a magnitude $\sqrt{a^2+b^2}$ and to be tangent to $x^2+y^2=a^2+b^2$ in a
counterclockwise direction. Thus $x^2+y^2=a^2+b^2 \Rightarrow 2x+2yy'=0 \Rightarrow y'=-\frac{x}{y}$ is the slope of the
tangent line at any point on the circle $\Rightarrow y'=-\frac{a}{b}$ at (a,b). Let $\mathbf{v}=-b\mathbf{i}+a\mathbf{j} \Rightarrow |\mathbf{v}|=\sqrt{a^2+b^2}$, with $\mathbf{v}$ in a
counterclockwise direction and tangent to the circle. Then let $P(x,y)=-y$ and $Q(x,y)=x$
$\Rightarrow \mathbf{G}=-y\mathbf{i}+x\mathbf{j} \Rightarrow$ for (a,b) on $x^2+y^2=a^2+b^2$ we have $\mathbf{G}=-b\mathbf{i}+a\mathbf{j}$ and $|\mathbf{G}|=\sqrt{a^2+b^2}$.

(b) $\mathbf{G}=\left(\sqrt{x^2+y^2}\right)\mathbf{F}=\left(\sqrt{a^2+b^2}\right)\mathbf{F}$.

43. The slope of the line through (x,y) and the origin is $\frac{y}{x} \Rightarrow \mathbf{v}=x\mathbf{i}+y\mathbf{j}$ is a vector parallel to that line and
pointing away from the origin $\Rightarrow \mathbf{F}=-\dfrac{x\mathbf{i}+y\mathbf{j}}{\sqrt{x^2+y^2}}$ is the unit vector pointing toward the origin.

45. Yes. The work and area have the same numerical value because work $=\displaystyle\int_C \mathbf{F}\cdot d\mathbf{r}=\int_C y\mathbf{i}\cdot d\mathbf{r}$

$=\displaystyle\int_b^a [f(t)\mathbf{i}]\cdot\left[\mathbf{i}+\frac{df}{dt}\mathbf{j}\right]dt$ [On the path, y equals $f(t)$]

$=\displaystyle\int_a^b f(t)\,dt$ = Area under the curve   [because $f(t)>0$]

47. $\mathbf{F}=-4t^3\mathbf{i}+8t^2\mathbf{j}+2\mathbf{k}$ and $\frac{d\mathbf{r}}{dt}=\mathbf{i}+2t\mathbf{j} \Rightarrow \mathbf{F}\cdot\frac{d\mathbf{r}}{dt}=12t^3 \Rightarrow$ Flow $=\displaystyle\int_0^2 12t^3 dt=\Big[3t^4\Big]_0^2=48$

49. $\mathbf{F}=(\cos t-\sin t)\mathbf{i}+(\cos t)\mathbf{k}$ and $\frac{d\mathbf{r}}{dt}=(-\sin t)\mathbf{i}+(\cos t)\mathbf{k} \Rightarrow \mathbf{F}\cdot\frac{d\mathbf{r}}{dt}=-\sin t\cos t+1$

$\Rightarrow$ Flow $=\displaystyle\int_0^\pi(-\sin t\cos t+1)\,dt=\Big[\tfrac{1}{2}\cos^2 t+t\Big]_0^\pi=\left(\tfrac{1}{2}+\pi\right)-\left(\tfrac{1}{2}+0\right)=\pi$

51. $C_1:\mathbf{r}=(\cos t)\mathbf{i}+(\sin t)\mathbf{j}+t\mathbf{k}, 0\le t\le\frac{\pi}{2} \Rightarrow \mathbf{F}=(2\cos t)\mathbf{i}+2t\mathbf{j}+(2\sin t)\mathbf{k}$ and $\frac{d\mathbf{r}}{dt}=(-\sin t)\mathbf{i}+(\cos t)\mathbf{j}+\mathbf{k}$

$\Rightarrow \mathbf{F}\cdot\frac{d\mathbf{r}}{dt}=-2\cos t\sin t+2t\cos t+2\sin t=-\sin 2t+2t\cos t+2\sin t$

$\Rightarrow$ Flow$_1=\displaystyle\int_0^{\pi/2}(-\sin 2t+2t\cos t+2\sin t\,)\,dt=\Big[\tfrac{1}{2}\cos 2t+2t\sin t+2\cos t-2\cos t\Big]_0^{\pi/2}=-1+\pi$;

$C_2:\mathbf{r}=\mathbf{j}+\frac{\pi}{2}(1-t)\mathbf{k}, 0\le t\le 1 \Rightarrow \mathbf{F}=\pi(1-t)\mathbf{j}+2\mathbf{k}$ and $\frac{d\mathbf{r}}{dt}=-\frac{\pi}{2}\mathbf{k} \Rightarrow \mathbf{F}\cdot\frac{d\mathbf{r}}{dt}=-\pi$

Copyright © 2016 Pearson Education, Inc.

$$\Rightarrow \text{Flow}_2 = \int_0^1 -\pi \, dt = \left[-\pi t\right]_0^1 = -\pi;$$

$C_3: \mathbf{r} = t\mathbf{i} + (1-t)\mathbf{j}, \, 0 \le t \le 1 \Rightarrow \mathbf{F} = 2t\mathbf{i} + 2(1-t)\mathbf{k}$ and $\frac{d\mathbf{r}}{dt} = \mathbf{i} - \mathbf{j} \Rightarrow \mathbf{F} \cdot \frac{d\mathbf{r}}{dt} = 2t$

$$\Rightarrow \text{Flow}_3 = \int_0^1 2t \, dt = \left[t^2\right]_0^1 = 1 \Rightarrow \text{Circulation} = (-1+\pi) - \pi + 1 = 0$$

53. Let $x = t$ be the parameter $\Rightarrow y = x^2 = t^2$ and $z = x = t \Rightarrow \mathbf{r} = t\mathbf{i} + t^2\mathbf{j} + t\mathbf{k}, \, 0 \le t \le 1$ from $(0, 0, 0)$ to $(1, 1, 1)$

$\Rightarrow \frac{d\mathbf{r}}{dt} = \mathbf{i} + 2t\mathbf{j} + \mathbf{k}$ and $\mathbf{F} = xy\mathbf{i} + y\mathbf{j} - yz\mathbf{k} = t^3\mathbf{i} + t^2\mathbf{j} - t^3\mathbf{k} \Rightarrow \mathbf{F} \cdot \frac{d\mathbf{r}}{dt} = t^3 + 2t^3 - t^3 = 2t^3 \Rightarrow \text{Flow} = \int_0^1 2t^3 \, dt = \frac{1}{2}$

55-59.   Example CAS commands:

Maple:

```
with( LinearAlgebra );#55
F:=r-><r[1]*r[2]^6|3*r[1]*(r[1]*r[2]^5+2>;
r:=t-><2*cos(t)|sin(t)>;
a,b:=0.2*Pi;
dr:=map(diff,r(t),t):                    # (a)
F(r(t));                                 # (b)
q1:=simplify( F(r(t)) . dr ) assuming t::real;    # (c)
q2:=Int( q1, t=a..b );
value( q2 );
```

Mathematica: (functions and bounds will vary):
Exercises 55 and 56 use vectors in 2 dimensions

```
Clear[x, y, t, f, r, v]
f[x_,y_]:={x y^6, 3x(x y^5 + 2)}
{a, b} = {0, 2π};
x[t_]:= 2 Cos[t]
y[t_]:= Sin[t]
r[t_]:= {x[t], y[t]}
v[t_]:= r'{t}
integrand = f[x[t], y[t]]. v[t]//Simplify
Integrate[integrand, (t, a, b)]
N[%]
```

If the integration takes too long or cannot be done, use NIntegrate to integrate numerically. This is suggested for Exercises 57 - 60 that use vectors in 3 dimensions. Be certain to leave spaces between variables to be multiplied.

```
Clear[x, y, z, t, f, r, v]
f[x_,y_, z_]:=[y + y z Cos[x, y, z], x^2 + x z Cos[x, y, z], z + x y Cos[x y z]}
[a, b] = {0, 2π};
x[t_]:= 2 Cos[t]
y[t_]:= 3 Sin[t]
z[t_]:= 1
r[t_]:= {x[t], y[t], z[t]}
v[t_]:= r'[t]
integrand = f[x[t], y[t], z[t]] · v[t]/Simplify
NIntegrate[integrand (t, a, b}]
```

15.3 PATH INDEPENDENCE, POTENTIAL FUNCTIONS, AND CONSERVATIVE FIELDS

1. $\frac{\partial P}{\partial y} = x = \frac{\partial N}{\partial z}, \frac{\partial M}{\partial z} = y = \frac{\partial P}{\partial x}, \frac{\partial N}{\partial x} = z = \frac{\partial M}{\partial y} \Rightarrow$ Conservative

3. $\frac{\partial P}{\partial y} = -1 \neq 1 = \frac{\partial N}{\partial z} \Rightarrow$ Not Conservative

5. $\frac{\partial N}{\partial x} = 0 \neq 1 = \frac{\partial M}{\partial y} \Rightarrow$ Not Conservative

7. $\frac{\partial f}{\partial x} = 2x \Rightarrow f(x, y, z) = x^2 + g(y, z) \Rightarrow \frac{\partial f}{\partial y} = \frac{\partial g}{\partial y} = 3y \Rightarrow g(y, z) = \frac{3y^2}{2} + h(z) \Rightarrow f(x, y, z) = x^2 + \frac{3y^2}{2} + h(z)$

    $\Rightarrow \frac{\partial f}{\partial z} = h'(z) = 4z \Rightarrow h(z) = 2z^2 + C \Rightarrow f(x, y, z) = x^2 + \frac{3y^2}{2} + 2z^2 + C$

9. $\frac{\partial f}{\partial x} = e^{y+2z} \Rightarrow f(x, y, z) = xe^{y+2z} + g(y, z) \Rightarrow \frac{\partial f}{\partial y} = xe^{y+2z} + \frac{\partial g}{\partial y} = xe^{y+2z} \Rightarrow \frac{\partial g}{\partial y} = 0 \Rightarrow f(x, y, z)$

    $= xe^{y+2z} + h(z) \Rightarrow \frac{\partial f}{\partial z} = 2xe^{y+2z} + h'(z) = 2xe^{y+2z} \Rightarrow h'(z) = 0 \Rightarrow h(z) = C \Rightarrow f(x, y, z) = xe^{y+2z} + C$

11. $\frac{\partial f}{\partial z} = \frac{z}{y^2+z^2} \Rightarrow f(x, y, z) = \frac{1}{2}\ln\left(y^2 + z^2\right) + g(x, y) \Rightarrow \frac{\partial f}{\partial x} = \frac{\partial g}{\partial x} = \ln x + \sec^2(x+y) \Rightarrow g(x, y)$

    $= (x\ln x - x) + \tan(x+y) + h(y) \Rightarrow f(x, y, z) = \frac{1}{2}\ln\left(y^2 + z^2\right) + (x\ln x - x) + \tan(x+y) + h(y)$

    $\Rightarrow \frac{\partial f}{\partial y} = \frac{y}{y^2+z^2} + \sec^2(x+y) + h'(y) = \sec^2(x+y) + \frac{y}{y^2+z^2} \Rightarrow h'(y) = 0 \Rightarrow h(y) = C \Rightarrow f(x, y, z)$

    $= \frac{1}{2}\ln\left(y^2 + z^2\right) + (x\ln x - x) + \tan(x+y) + C$

13. Let $\mathbf{F}(x, y, z) = 2x\mathbf{i} + 2y\mathbf{j} + 2z\mathbf{k} \Rightarrow \frac{\partial P}{\partial y} = 0 = \frac{\partial N}{\partial z}, \frac{\partial M}{\partial z} = 0 = \frac{\partial P}{\partial x}, \frac{\partial N}{\partial x} = 0 = \frac{\partial M}{\partial y} \Rightarrow M\,dx + N\,dy + P\,dz$ is exact;

    $\frac{\partial f}{\partial x} = 2x \Rightarrow f(x, y, z) = x^2 + g(y, z) \Rightarrow \frac{\partial f}{\partial y} = \frac{\partial g}{\partial y} = 2y \Rightarrow g(y, z) = y^2 + h(z) \Rightarrow f(x, y, z) = x^2 + y^2 = h(z)$

    $\Rightarrow \frac{\partial f}{\partial z} = h'(z) = 2z \Rightarrow h(z) = z^2 + C \Rightarrow f(x, y, z) = x^2 + y^2 + z^2 + C \Rightarrow \int_{(0,0,0)}^{(2,3,-6)} 2x\,dx + 2y\,dy + 2z\,dz$

    $= f(2, 3, -6) - f(0, 0, 0) = 2^2 + 3^2 + (-6)^2 = 49$

15. Let $\mathbf{F}(x, y, z) = 2xy\mathbf{i} + \left(x^2 - z^2\right)\mathbf{j} - 2yz\mathbf{k} \Rightarrow \frac{\partial P}{\partial y} = -2z = \frac{\partial N}{\partial z}, \frac{\partial M}{\partial z} = 0 = \frac{\partial P}{\partial x}, \frac{\partial N}{\partial x} = 2x = \frac{\partial M}{\partial y}$

    $\Rightarrow M\,dx + N\,dy + P\,dz$ is exact; $\frac{\partial f}{\partial x} = 2xy \Rightarrow f(x, y, z) = x^2y + g(y, z) \Rightarrow \frac{\partial f}{\partial y} = x^2 + \frac{\partial g}{\partial y} = x^2 - z^2 \Rightarrow \frac{\partial g}{\partial y} = -z^2$

    $\Rightarrow g(y, z) = -yz^2 + h(z) \Rightarrow f(x, y, z) = x^2y - yz^2 + h(z) \Rightarrow \frac{\partial f}{\partial z} = -2yz + h'(z) = -2yz \Rightarrow h'(z) = 0 \Rightarrow h(z) = C$

    $\Rightarrow f(x, y, z) = x^2y - yz^2 + C \Rightarrow \int_{(0,0,0)}^{(1,2,3)} 2xy\,dx + \left(x^2 - z^2\right)dy - 2yz\,dz = f(1, 2, 3) - f(0, 0, 0) = 2 - 2(3)^2$

    $= -16$

17. Let $\mathbf{F}(x, y, z) = (\sin y \cos x)\mathbf{i} + (\cos y \sin x)\mathbf{j} + \mathbf{k} \Rightarrow \frac{\partial P}{\partial y} = 0 = \frac{\partial N}{\partial z}, \frac{\partial M}{\partial z} = 0 = \frac{\partial P}{\partial x}, \frac{\partial N}{\partial x} = \cos y \cos x = \frac{\partial M}{\partial y}$

$\Rightarrow M\,dx + N\,dy + P\,dz$ is exact; $\frac{\partial f}{\partial x} = \sin y \cos x \Rightarrow f(x, y, z) = \sin y \sin x + g(y, z) \Rightarrow \frac{\partial f}{\partial y} = \cos y \sin x + \frac{\partial g}{\partial y}$

$= \cos y \sin x \Rightarrow \frac{\partial g}{\partial y} = 0 \Rightarrow g(y, z) = h(z) \Rightarrow f(x, y, z) = \sin y \sin x + h(z) \Rightarrow \frac{\partial f}{\partial z} = h'(z) = 1 \Rightarrow h(z) = z + C$

$\Rightarrow f(x, y, z) = \sin y \sin x + z + C \Rightarrow \int_{(1, 0, 0)}^{(0, 1, 1)} \sin y \cos x\,dx + \cos y \sin x\,dy + dz = f(0, 1, 1) - f(1, 0, 0)$

$= (0 + 1) - (0 + 0) = 1$

19. Let $\mathbf{F}(x, y, z) = 3x^2\mathbf{i} + \left(\frac{z^2}{y}\right)\mathbf{j} + (2z \ln y)\mathbf{k} \Rightarrow \frac{\partial P}{\partial y} = \frac{2z}{y} = \frac{\partial N}{\partial z}, \frac{\partial M}{\partial z} = 0 = \frac{\partial P}{\partial x}, \frac{\partial N}{\partial x} = 0 = \frac{\partial M}{\partial y}$

$\Rightarrow M\,dx + N\,dy + P\,dz$ is exact; $\frac{\partial f}{\partial x} = 3x^2 \Rightarrow f(x, y, z) = x^3 + g(y, z) \Rightarrow \frac{\partial f}{\partial y} = \frac{\partial g}{\partial y} = \frac{z^2}{y} \Rightarrow g(y, z) = z^2 \ln y + h(z)$

$\Rightarrow f(x, y, z) = x^3 + z^2 \ln y + h(z) \Rightarrow \frac{\partial f}{\partial z} = 2z \ln y + h'(z) = 2z \ln y \Rightarrow h'(z) = 0 \Rightarrow h(z) = C$

$\Rightarrow f(x, y, z) = x^3 + z^2 \ln y + C \Rightarrow \int_{(1, 1, 1)}^{(1, 2, 3)} 3x^2\,dx + \frac{z^2}{y}\,dy + 2z \ln y\,dz = f(1, 2, 3) - f(1, 1, 1)$

$= (1 + 9 \ln 2 + C) - (1 + 0 + C) = 9 \ln 2$

21. Let $\mathbf{F}(x, y, z) = \left(\frac{1}{y}\right)\mathbf{i} + \left(\frac{1}{z} - \frac{x}{y^2}\right)\mathbf{j} - \left(\frac{y}{z^2}\right)\mathbf{k} \Rightarrow \frac{\partial P}{\partial y} = -\frac{1}{z^2} = \frac{\partial N}{\partial z}, \frac{\partial M}{\partial z} = 0 = \frac{\partial P}{\partial x}, \frac{\partial N}{\partial x} = -\frac{1}{y^2} = \frac{\partial M}{\partial y}$

$\Rightarrow M\,dx + N\,dy + P\,dz$ is exact; $\frac{\partial f}{\partial x} = \frac{1}{y} \Rightarrow f(x, y, z) = \frac{x}{y} + g(y, z) \Rightarrow \frac{\partial f}{\partial y} = -\frac{x}{y^2} + \frac{\partial g}{\partial y} = \frac{1}{z} - \frac{x}{y^2}$

$\Rightarrow \frac{\partial g}{\partial y} = \frac{1}{z} \Rightarrow g(y, z) = \frac{y}{z} + h(z) \Rightarrow f(x, y, z) = \frac{x}{y} + \frac{y}{z} + h(z) \Rightarrow \frac{\partial f}{\partial z} = -\frac{y}{z^2} + h'(z) = -\frac{y}{z^2} \Rightarrow h'(z) = 0 \Rightarrow h(z) = C$

$\Rightarrow f(x, y, z) = \frac{x}{y} + \frac{y}{z} + C \Rightarrow \int_{(1, 1, 1)}^{(2, 2, 2)} \frac{1}{y}\,dx + \left(\frac{1}{z} - \frac{x}{y^2}\right)dy - \frac{y}{z^2}\,dz = f(2, 2, 2) - f(1, 1, 1) = \left(\frac{2}{2} + \frac{2}{2} + C\right) - \left(\frac{1}{1} + \frac{1}{1} + C\right)$

$= 0$

23. $\mathbf{r} = (\mathbf{i} + \mathbf{j} + \mathbf{k}) + t(\mathbf{i} + 2\mathbf{j} - 2\mathbf{k}) = (1 + t)\mathbf{i} + (1 + 2t)\mathbf{j} + (1 - 2t)\mathbf{k}, 0 \le t \le 1 \Rightarrow dx = dt, dy = 2\,dt, dz = -2\,dt$

$\Rightarrow \int_{(1, 1, 1)}^{(2, 3, -1)} y\,dx + x\,dy + 4\,dz = \int_0^1 (2t + 1)\,dt + (t + 1)(2\,dt) + 4(-2)\,dt = \int_0^1 (4t - 5)\,dt = \left[2t^2 - 5t\right]_0^1 = -3$

25. $\frac{\partial P}{\partial y} = 0 = \frac{\partial N}{\partial z}, \frac{\partial M}{\partial z} = 2z = \frac{\partial P}{\partial x}, \frac{\partial N}{\partial x} = 0 = \frac{\partial M}{\partial y} \Rightarrow M\,dx + N\,dy + P\,dz$ is exact $\Rightarrow \mathbf{F}$ is conservative

$\Rightarrow$ path independence

27. $\frac{\partial P}{\partial y} = 0 = \frac{\partial N}{\partial z}, \frac{\partial M}{\partial z} = 0 = \frac{\partial P}{\partial x}, \frac{\partial N}{\partial x} = -\frac{2x}{y^2} = \frac{\partial M}{\partial y} \Rightarrow \mathbf{F}$ is conservative $\Rightarrow$ there exists an f so that $\mathbf{F} = \nabla f$;

$\frac{\partial f}{\partial x} = \frac{2x}{y} \Rightarrow f(x, y) = \frac{x^2}{y} + g(y) \Rightarrow \frac{\partial f}{\partial y} = -\frac{x^2}{y^2} + g'(y) = \frac{1 - x^2}{y^2} \Rightarrow g'(y) = \frac{1}{y^2} \Rightarrow g(y) = -\frac{1}{y} + C$

$\Rightarrow f(x, y) = \frac{x^2}{y} - \frac{1}{y} + C \Rightarrow \mathbf{F} = \nabla\left(\frac{x^2 - 1}{y}\right)$

29. $\frac{\partial P}{\partial y} = 0 = \frac{\partial N}{\partial z}, \frac{\partial M}{\partial y} = 0 = \frac{\partial P}{\partial x}, \frac{\partial N}{\partial x} = 1 = \frac{\partial M}{\partial y} \Rightarrow \mathbf{F}$ is conservative $\Rightarrow$ there exists an f so that $\mathbf{F} = \nabla f$;

$\frac{\partial f}{\partial x} = x^2 + y \Rightarrow f(x, y, z) = \frac{1}{3}x^3 + xy + g(y, z) \Rightarrow \frac{\partial f}{\partial y} = x + \frac{\partial g}{\partial y} = y^2 + x \Rightarrow \frac{\partial g}{\partial y} = y^2 \Rightarrow g(y, z) = \frac{1}{3}y^3 + h(z)$

$\Rightarrow f(x, y, z) = \frac{1}{3}x^3 + xy + \frac{1}{3}y^3 + h(z) \Rightarrow \frac{\partial f}{\partial z} = h'(z) = ze^z \Rightarrow h(z) = ze^z - e^z + C$

$\Rightarrow f(x, y, z) = \frac{1}{3}x^3 + xy + \frac{1}{3}y^3 + ze^z - e^z + C \Rightarrow \mathbf{F} = \nabla\left(\frac{1}{3}x^3 + xy + \frac{1}{3}y^3 + ze^z - e^z\right)$

(a) work $= \int_A^B \mathbf{F} \cdot \frac{d\mathbf{r}}{dt}\, dt = \int_A^B \mathbf{F} \cdot d\mathbf{r} = \left[\frac{1}{3}x^3 + xy + \frac{1}{3}y^3 + ze^z - e^z\right]_{(1,0,0)}^{(1,0,1)} = \left(\frac{1}{3} + 0 + 0 + e - e\right) - \left(\frac{1}{3} + 0 + 0 - 1\right) = 1$

(b) work $= \int_A^B \mathbf{F} \cdot d\mathbf{r} = \left[\frac{1}{3}x^3 + xy + \frac{1}{3}y^3 + ze^z - e^z\right]_{(1,0,0)}^{(1,0,1)} = 1$

(c) work $= \int_A^B \mathbf{F} \cdot d\mathbf{r} = \left[\frac{1}{3}x^3 + xy + \frac{1}{3}y^3 + ze^z - e^z\right]_{(1,0,0)}^{(1,0,1)} = 1$

Note: Since $\mathbf{F}$ is conservative, $\int_A^B \mathbf{F} \cdot d\mathbf{r}$ is independent of the path from $(1, 0, 0)$ to $(1, 0, 1)$.

31. (a) $\mathbf{F} = \nabla\left(x^3 y^2\right) \Rightarrow \mathbf{F} = 3x^2 y^2 \mathbf{i} + 2x^3 y \mathbf{j}$; let C_1 be the path from $(-1, 1)$ to $(0, 0) \Rightarrow x = t - 1$ and

$y = -t + 1, 0 \le t \le 1 \Rightarrow \mathbf{F} = 3(t-1)^2(-t+1)^2 \mathbf{i} + 2(t-1)^3(-t+1)\mathbf{j} = 3(t-1)^4 \mathbf{i} - 2(t-1)^4 \mathbf{j}$ and

$\mathbf{r}_1 = (t-1)\mathbf{i} + (-t+1)\mathbf{j} \Rightarrow d\mathbf{r}_1 = dt\,\mathbf{i} - dt\,\mathbf{j} \Rightarrow \int_{C_1} \mathbf{F} \cdot d\mathbf{r}_1 = \int_0^1 \left[3(t-1)^4 + 2(t-1)^4\right] dt$

$= \int_0^1 5(t-1)^4 dt = \left[(t-1)^5\right]_0^1 = 1$; let C_2 be the path from $(0, 0)$ to $(1, 1) \Rightarrow x = t$ and $y = t$,

$0 \le t \le 1 \Rightarrow \mathbf{F} = 3t^4 \mathbf{i} + 2t^4 \mathbf{j}$ and $\mathbf{r}_2 = t\mathbf{i} + t\mathbf{j} \Rightarrow d\mathbf{r}_2 = dt\mathbf{i} + dt\mathbf{j} \Rightarrow \int_{C_2} \mathbf{F} \cdot d\mathbf{r}_2 = \int_0^1 \left(3t^4 + 2t^4\right) dt$

$= \int_0^1 5t^4 dt = 1 \Rightarrow \int_C \mathbf{F} \cdot d\mathbf{r} = \int_{C_1} \mathbf{F} \cdot d\mathbf{r}_1 + \int_{C_2} \mathbf{F} \cdot d\mathbf{r}_2 = 2$

(b) Since $f(x, y) = x^3 y^2$ is a potential function for $\mathbf{F}, \int_{(-1,1)}^{(1,1)} \mathbf{F} \cdot d\mathbf{r} = f(1, 1) - f(-1, 1) = 2$

33. (a) If the differential form is exact, then $\frac{\partial P}{\partial y} = \frac{\partial N}{\partial z} \Rightarrow 2ay = cy$ for all $y \Rightarrow 2a = c, \frac{\partial M}{\partial z} = \frac{\partial P}{\partial x} \Rightarrow 2cx = 2cx$ for all

x, and $\frac{\partial N}{\partial x} = \frac{\partial M}{\partial y} \Rightarrow by = 2ay$ for all $y \Rightarrow b = 2a$ and $c = 2a$

(b) $\mathbf{F} = \nabla f \Rightarrow$ the differential form with $a = 1$ in part (a) is exact $\Rightarrow b = 2$ and $c = 2$

35. The path will not matter; the work along any path will be the same because the field is conservative.

37. Let the coordinates of points A and B be (x_A, y_A, z_A) and (x_B, y_B, z_B), respectively. The force
$\mathbf{F} = a\mathbf{i} + b\mathbf{j} + c\mathbf{k}$ is conservative because all the partial derivatives of M, N, and P are zero. Therefore, the
potential function is $f(x, y, z) = ax + by + cz + C$, and the work done by the force in moving a particle along
any path from A to B is $f(B) - f(A) = f(x_B, y_B, z_B) - f(x_A, y_A, z_A)$

$= (ax_B + by_B + cz_B + C) - (ax_A + by_A + cz_A + C) = a(x_B - x_A) + b(y_B - y_A) + c(z_B - z_A) = \mathbf{F} \cdot \overrightarrow{BA}$

15.4 GREEN'S THEOREM IN THE PLANE

1. $M = -y = -a \sin t$, $N = x = a \cos t$, $dx = -a \sin t \, dt$, $dy = a \cos t \, dt \Rightarrow \frac{\partial M}{\partial y} = 0$, $\frac{\partial M}{\partial y} = -1$, $\frac{\partial N}{\partial x} = 1$, and $\frac{\partial N}{\partial y} = 0$;

   Equation (3): $\oint_C M \, dy - N \, dx = \int_0^{2\pi} [(-a \sin t)(a \cos t) - (a \cos t)(-a \sin t)] \, dt = \int_0^{2\pi} 0 \, dt = 0$;

   $\iint_R \left(\frac{\partial M}{\partial x} + \frac{\partial N}{\partial y} \right) dx \, dy = \iint_R 0 \, dx \, dy = 0$, Flux

   Equation (4): $\oint_C M \, dx + N \, dy = \int_0^{2\pi} [(-a \sin t)(-a \sin t) - (a \cos t)(a \cos t)] \, dt = \int_0^{2\pi} a^2 \, dt = 2\pi a^2$;

   $\iint_R \left(\frac{\partial N}{\partial x} - \frac{\partial M}{\partial y} \right) dx \, dy = \int_{-a}^a \int_{-c}^{\sqrt{a^2 - x^2}} 2 \, dy \, dx = \int_{-a}^a 4\sqrt{a^2 - x^2} \, dx = 4 \left[\frac{x}{2} \sqrt{a^2 - x^2} + \frac{a^2}{2} \sin^{-1} \frac{x}{a} \right]_{-a}^a$

   $= 2a^2 \left(\frac{\pi}{2} + \frac{\pi}{2} \right) = 2a^2 \pi$, Circulation

3. $M = 2x = 2a \cos t$, $N = -3y = -3a \sin t$, $dx = -a \sin t \, dt$, $dy = a \cos t \, dt \Rightarrow \frac{\partial M}{\partial x} = 2$, $\frac{\partial M}{\partial y} = 0$, $\frac{\partial N}{\partial x} = 0$, and

   $\frac{\partial N}{\partial y} = -3$;

   Equation (3): $\oint_C M \, dy - N \, dx = \int_0^{2\pi} [(2a \cos t)(a \cos t) + (3a \sin t)(-a \sin t)] \, dt$

   $= \int_0^{2\pi} \left(2a^2 \cos^2 t - 3a^2 \sin^2 t \right) dt = 2a^2 \left[\frac{t}{2} + \frac{\sin 2t}{4} \right]_0^{2\pi} - 3a^2 \left[\frac{t}{2} - \frac{\sin 2t}{4} \right]_0^{2\pi} = 2\pi a^2 - 3\pi a^2 = -\pi a^2$;

   $\iint_R \left(\frac{\partial M}{\partial x} + \frac{\partial N}{\partial y} \right) = \iint_R -1 \, dx \, dy = \int_0^{2\pi} \int_0^a -r \, dr \, d\theta = \int_0^{2\pi} -\frac{a^2}{2} \, d\theta = -\pi a^2$, Flux

   Equation (4): $\oint_C M \, dx + N \, dy = \int_0^{2\pi} [(2a \cos t)(-a \sin t) + (-3a \sin t)(a \cos t)] \, dt$

   $= \int_0^{2\pi} \left(-2a^2 \sin t \cos t - 3a^2 \sin t \cos t \right) dt = -5a^2 \left[\frac{1}{2} \sin^2 t \right]_0^{2\pi} = 0$; $\iint_R 0 \, dx \, dy = 0$, Circulation

5. $M = x - y$, $N = y - x \Rightarrow \frac{\partial M}{\partial x} = 1$, $\frac{\partial M}{\partial y} = -1$, $\frac{\partial N}{\partial x} = -1$, $\frac{\partial N}{\partial y} = 1 \Rightarrow$ Flux $= \iint_R 2 \, dx \, dy = \int_0^1 \int_0^1 2 \, dx \, dy = 2$;

   Circ $= \iint_R [-1 - (-1)] \, dx \, dy = 0$

7. $M = y^2 - x^2$, $N = x^2 + y^2 \Rightarrow \frac{\partial M}{\partial x} = -2x$, $\frac{\partial M}{\partial y} = 2y$, $\frac{\partial N}{\partial x} = 2x$, $\frac{\partial N}{\partial y} = 2y \Rightarrow$ Flux $= \iint_R (-2x + 2y) \, dx \, dy$

   $= \int_0^3 \int_0^x (-2x + 2y) \, dy \, dx = \int_0^3 \left(-2x^2 + x^2 \right) dx = \left[-\frac{1}{3} x^3 \right]_0^3 = -9$; Circ $= \iint_R (2x - 2y) \, dx \, dy$

   $= \int_0^3 \int_0^x (2x - 2y) \, dy \, dx = \int_0^3 x^2 \, dx = 9$

9. $M = xy + y^2$, $N = x - y \Rightarrow \frac{\partial M}{\partial x} = y$, $\frac{\partial M}{\partial y} = x + 2y$, $\frac{\partial N}{\partial x} = 1$, $\frac{\partial N}{\partial y} = -1 \Rightarrow$ Flux $= \iint_R (y + (-1)) \, dy \, dx$

   $= \int_0^1 \int_{x^2}^{\sqrt{x}} (y - 1) \, dy \, dx = \int_0^1 \left(\frac{1}{2} x - \sqrt{x} - \frac{1}{2} x^4 + x^2 \right) dx = -\frac{11}{60}$; Circ $= \iint_R (1 - (x + 2y)) \, dy \, dx$

   $= \int_0^1 \int_{x^2}^{\sqrt{x}} (1 - x - 2y) \, dy \, dx = \int_0^1 \left(\sqrt{x} - x^{3/2} - x - x^2 + x^3 + x^4 \right) dx = -\frac{7}{60}$

11.　$M = x^3 y^2$, $N = \frac{1}{2} x^4 y \Rightarrow \frac{\partial M}{\partial x} = 3x^2 y^2$, $\frac{\partial M}{\partial y} = 2x^3 y$, $\frac{\partial N}{\partial x} = 2x^3 y$, $\frac{\partial N}{\partial y} = \frac{1}{2} x^4 \Rightarrow$ Flux $= \iint\limits_R \left(3x^2 y^2 + \frac{1}{2} x^4 \right) dy\, dx$

$\quad = \int_0^2 \int_{x^2-x}^x \left(3x^2 y^2 + \frac{1}{2} x^4 \right) dy\, dx = \int_0^2 \left(3x^5 - \frac{7}{2} x^6 + 3x^7 - x^8 \right) dx = \frac{64}{9}$; Circ $= \iint\limits_R \left(2x^3 y - 2x^3 y \right) dy\, dx = 0$

13.　$M = x + e^x \sin\, y$, $N = x + e^x \cos\, y \Rightarrow \frac{\partial M}{\partial x} = 1 + e^x \sin\, y$, $\frac{\partial M}{\partial y} = e^x \cos\, y$, $\frac{\partial N}{\partial x} = 1 + e^x \cos\, y$, $\frac{\partial N}{\partial y} = -e^x \sin\, y$

$\quad \Rightarrow$ Flux $= \iint\limits_R dx\, dy = \int_{-\pi/4}^{\pi/4} \int_0^{\sqrt{\cos 2\theta}} r\, dr\, d\theta = \int_{-\pi/4}^{\pi/4} \left(\frac{1}{2} \cos 2\theta \right) d\theta = \left[\frac{1}{4} \sin 2\theta \right]_{-\pi/4}^{\pi/4} = \frac{1}{2}$;

$\quad$ Circ $= \iint\limits_R \left(1 + e^x \cos\, y - e^x \cos\, y \right) dx\, dy = \iint\limits_R dx\, dy = \int_{-\pi/4}^{\pi/4} \int_0^{\sqrt{\cos 2\theta}} r\, dr\, d\theta = \int_{-\pi/4}^{\pi/4} \left(\frac{1}{2} \cos 2\theta \right) d\theta = \frac{1}{2}$

15.　$M = xy$, $N = y^2 \Rightarrow \frac{\partial M}{\partial x} = y$, $\frac{\partial M}{\partial y} = x$, $\frac{\partial N}{\partial x} = 0$, $\frac{\partial N}{\partial y} = 2y \Rightarrow$ Flux $= \iint\limits_R (y + 2y)\, dy\, dx = \int_0^1 \int_{x^2}^x 3y\, dy\, dx$

$\quad = \int_0^1 \left(\frac{3x^2}{2} - \frac{3x^4}{2} \right) dx = \frac{1}{5}$; Circ $= \iint\limits_R -x\, dy\, dx = \int_0^1 \int_{x^2}^x -x\, dy\, dx = \int_0^1 \left(-x^2 + x^3 \right) dx = -\frac{1}{12}$

17.　$M = 3xy - \frac{x}{1+y^2}$, $N = e^x + \tan^{-1} y \Rightarrow \frac{\partial M}{\partial x} = 3y - \frac{1}{1+y^2}$, $\frac{\partial N}{\partial y} = \frac{1}{1+y^2}$

$\quad \Rightarrow$ Flux $= \iint\limits_R \left(3y - \frac{1}{1+y^2} + \frac{1}{1+y^2} \right) dx\, dy = \iint\limits_R 3y\, dx\, dy = \int_0^{2\pi} \int_0^{a(1+\cos\theta)} (3r \sin\theta) r\, dr\, d\theta$

$\quad = \int_0^{2\pi} a^3 (1+\cos\theta)^3 (\sin\theta)\, d\theta = \left[-\frac{a^3}{4} (1+\cos\theta)^4 \right]_0^{2\pi} = -4a^3 - \left(-4a^3 \right) = 0$

19.　$M = 2xy^3$, $N = 4x^2 y^2 \Rightarrow \frac{\partial M}{\partial y} = 6xy^2$, $\frac{\partial N}{\partial x} = 8xy^2 \Rightarrow$ work $= \oint_C 2xy^3 dx + 4x^2 y^2 dy = \iint\limits_R \left(8xy^2 - 6xy^2 \right) dx\, dy$

$\quad = \int_0^1 \int_0^{x^3} 2xy^2 dy\, dx = \int_0^1 \frac{2}{3} x^{10}\, dx = \frac{2}{33}$

21.　$M = y^2$, $N = x^2 \Rightarrow \frac{\partial M}{\partial y} = 2y$, $\frac{\partial N}{\partial x} = 2x \Rightarrow \oint_C y^2 dx + x^2 dy = \iint\limits_R (2x - 2y)\, dy\, dx$

$\quad = \int_0^1 \int_0^{1-x} (2x - 2y)\, dy\, dx = \int_0^1 \left(-3x^2 + 4x - 1 \right) dx = \left[-x^3 + 2x^2 - x \right]_0^1 = -1 + 2 - 1 = 0$

23.　$M = 6y + x$, $N = y + 2x \Rightarrow \frac{\partial M}{\partial y} = 6$, $\frac{\partial N}{\partial x} = 2 \Rightarrow \oint_C (6y + x)\, dx + (y + 2x)\, dy = \iint\limits_R (2 - 6)\, dy\, dx$

$\quad = -4(\text{Area of the circle}) = -16\pi$

25.　$M = x = a \cos\, t$, $N = y = a \sin\, t \Rightarrow dx = -a \sin\, t\, dt$, $dy = a \cos\, t\, dt \Rightarrow$ Area $= \frac{1}{2} \oint_C x\, dy - y\, dx$

$\quad = \frac{1}{2} \int_0^{2\pi} \left(a^2 \cos^2 t + a^2 \sin^2 t \right) dt = \frac{1}{2} \int_0^{2\pi} a^2\, dt = \pi a^2$

27. $M = x = \cos^3 t,\ N = y = \sin^3 t \Rightarrow dx = -3\cos^2 t \sin t\ dt,\ dy = 3\sin^2 t \cos t\ dt \Rightarrow \text{Area} = \frac{1}{2}\oint_C x\ dy - y\ dx$

$= \frac{1}{2}\int_0^{2\pi}\left(3\sin^2 t \cos^2 t\right)\left(\cos^2 t + \sin^2 t\right) dt = \frac{1}{2}\int_0^{2\pi}\left(3\sin^2 t \cos^2 t\right) dt = \frac{3}{8}\int_0^{2\pi}\sin^2 2t\ dt = \frac{3}{16}\int_0^{4\pi}\sin^2 u\ du$

$= \frac{3}{16}\left[\frac{u}{2} - \frac{\sin 2u}{4}\right]_0^{4\pi} = \frac{3}{8}\pi$

29. (a) $M = f(x),\ N = g(y) \Rightarrow \frac{\partial M}{\partial y} = 0, \frac{\partial N}{\partial x} = 0 \Rightarrow \oint_C f(x)\ dx + g(y)\ dy = \iint_R \left(\frac{\partial N}{\partial x} - \frac{\partial M}{\partial y}\right) dx\ dy = \iint_R 0\ dx\ dy = 0$

(b) $M = ky,\ N = hx \Rightarrow \frac{\partial M}{\partial y} = k, \frac{\partial N}{\partial x} = h \Rightarrow \oint_C ky\ dx + hx\ dy = \iint_R \left(\frac{\partial N}{\partial x} - \frac{\partial M}{\partial y}\right) dx\ dy$

$= \iint_R (h-k)\ dx\ dy = (h-k)(\text{Area of the region})$

31. The integral is 0 for any simple closed plane curve C. The reasoning: By the tangential form of Green's

Theorem, with $M = 4x^3 y$ and $N = x^4, \oint_C 4x^3 y\ dx + x^4 dy = \iint_R \left[\frac{\partial}{\partial x}\left(x^4\right) - \frac{\partial}{\partial y}\left(4x^3 y\right)\right] dx\ dy$

$= \iint_R \underbrace{\left(4x^3 - 4x^3\right)}_{0} dx\ dy = 0.$

33. Let $M = x$ and $N = 0 \Rightarrow \frac{\partial M}{\partial x} = 1$ and $\frac{\partial N}{\partial y} = 0 \Rightarrow \oint_C M\ dy - N\ dx = \iint_R \left(\frac{\partial M}{\partial x} + \frac{\partial N}{\partial y}\right) dx\ dy \Rightarrow \oint_C x\ dy$

$\iint_R (1+0)\ dx\ dy \Rightarrow \text{Area of } R = \iint_R dx\ dy = \oint_C x\ dy;$ similarly, $M = y$ and $N = 0 \Rightarrow \frac{\partial M}{\partial y} = 1$ and $\frac{\partial N}{\partial x} = 0$

$\Rightarrow \oint_C M\ dx + N\ dy = \iint_R \left(\frac{\partial N}{\partial x} + \frac{\partial M}{\partial y}\right) dy\ dx \Rightarrow \oint_C y\ dx = \iint_R (0-1)\ dy\ dx \Rightarrow -\oint_C y\ dx = \iint_R dx\ dy = \text{Area of } R$

35. Let $\delta(x, y) = 1 \Rightarrow \bar{x} = \frac{M_y}{M} = \frac{\iint_R x\ \delta(x, y)\ dA}{\iint_R \delta(x, y)\ dA} = \frac{\iint_R x\ dA}{\iint_R dA} = \frac{\iint_R x\ dA}{A} \Rightarrow A\bar{x} = \iint_R x\ dA = \iint_R (x+0)\ dx\ dy$

$= \oint_C \frac{x^2}{2}\ dy, A\bar{x} = \iint_R x\ dA = \iint_R (0+x)\ dx\ dy = -\oint_C xy\ dx,$ and $A\bar{x} = \iint_R x\ dA = \iint_R \left(\frac{2}{3}x + \frac{1}{3}x\right) dx\ dy$

$= \oint_C \frac{1}{3}x^2 dy - \frac{1}{3}xy\ dx \Rightarrow \frac{1}{2}\oint_C x^2 dy = -\oint_C xy\ dx = \frac{1}{3}\oint_C x^2 dy - xy\ dx = A\bar{x}$

37. $M = \frac{\partial f}{\partial y},\ N = -\frac{\partial f}{\partial x} \Rightarrow \frac{\partial M}{\partial y} = \frac{\partial^2 f}{\partial y^2}, \frac{\partial N}{\partial x} = -\frac{\partial^2 f}{\partial x^2} \Rightarrow \oint_C \frac{\partial f}{\partial y}\ dx - \frac{\partial f}{\partial x}\ dy = \iint_R \left(-\frac{\partial^2 f}{\partial x^2} - \frac{\partial^2 f}{\partial y^2}\right) dx\ dy = 0$ for such curves C

39. (a) $\nabla f = \left(\frac{2x}{x^2+y^2}\right)\mathbf{i} + \left(\frac{2y}{x^2+y^2}\right)\mathbf{j} \Rightarrow M = \frac{2x}{x^2+y^2},\ N = \frac{2y}{x^2+y^2};$ since M, N are discontinuous at $(0, 0)$, we

compute $\int_C \nabla f \cdot \mathbf{n}\ ds$ directly since Green's Theorem does not apply. Let $x = a\cos t,\ y = a\sin t$

$\Rightarrow dx = -a\sin t\ dt,\ dy = a\cos t\ dt,\ M = \frac{2}{a}\cos t,\ N = \frac{2}{a}\sin t, 0 \le t \le 2\pi,$ so $\int_C \nabla f \cdot \mathbf{n}\ ds = \int_C M\ dy - N\ dx$

$= \int_0^{2\pi}\left[\left(\frac{2}{a}\cos t\right)(a\cos t) - \left(\frac{2}{a}\sin t\right)(-a\sin t)\right] dt = \int_0^{2\pi} 2\left(\cos^2 t + \sin^2 t\right) dt = 4\pi.$ Note that this holds for any $a > 0$, so $\int_C \nabla f \cdot \mathbf{n}\, ds = 4\pi$ for any circle C centered at $(0,0)$ traversed counterclockwise and $\int_C \nabla f \cdot \mathbf{n}\, ds = -4\pi$ if C is traversed clockwise.

(b) If K does not enclose the point $(0,0)$ we may apply Green's Theorem: $\int_C \nabla f \cdot \mathbf{n}\, ds = \int_C M\, dy - N\, dx$

$$= \iint\limits_R \left(\frac{\partial M}{\partial x} + \frac{\partial N}{\partial y}\right) dx\, dy = \iint\limits_R \left(\frac{2\left(y^2 - x^2\right)}{\left(x^2 + y^2\right)^2} + \frac{2\left(x^2 - y^2\right)}{\left(x^2 + y^2\right)^2}\right) dx\, dy = \iint\limits_R 0\, dx\, dy = 0.$$ If K does enclose the point $(0,0)$

we proceed as follows:

Choose a small enough so that the circle C centered at $(0,0)$ of radius a lies entirely within K. Green's

Theorem applies to the region R that lies between K and C. Thus, as before, $0 = \iint\limits_R \left(\frac{\partial M}{\partial x} + \frac{\partial N}{\partial y}\right) dx\, dy$

$$= \int_K M\, dy - N\, dx + \int_C M\, dy - N\, dx$$ where K is traversed counterclockwise and C is traversed clockwise.

Hence by part (a) $0 = \left[\int_K M\, dy - N\, dx\right] - 4\pi \Rightarrow 4\pi = \int_K M\, dy - N\, dx = \int_K \nabla f \cdot \mathbf{n}\, ds.$ We have shown:

$$\int_K \nabla f \cdot \mathbf{n}\, ds = \begin{cases} 0 & \text{if } (0,0) \text{ lies inside } K \\ 4\pi & \text{if } (0,0) \text{ lies outside } K \end{cases}$$

41. $\int_{g_1(y)}^{g_2(y)} \frac{\partial N}{\partial x} dx\, dy = N\left(g_2(y), y\right) - N\left(g_1(y), y\right) \Rightarrow \int_c^d \int_{g_1(y)}^{g_2(y)} \left(\frac{\partial N}{\partial x} dx\right) dy = \int_c^d \left[N\left(g_2(y), y\right) - N\left(g_1(y), y\right)\right] dy$

$$= \int_c^d N\left(g_2(y), y\right) dy - \int_c^d N\left(g_1(y), y\right) dy = \int_c^d N\left(g_2(y), y\right) dy + \int_d^c N\left(g_1(y), y\right) dy = \int_{C_2} N\, dy + \int_{C_1} N\, dy$$

$$= \oint_C dy \Rightarrow \oint_C N\, dy = \iint\limits_R \frac{\partial N}{\partial x} dx\, dy$$

43-45. Example CAS commands:

Maple:

       with(plots);#43

       M:= (x,y) -> 2*x-y;

       N:= (x,y) -> x+3*y;

       C:= x^2+4*y^2= 4;

       implicitplot(C, x =-2..2, y = 2..2, scaling=constrained, title = "#43(a)(Section 16.4)");

       curlF_k:=D[1](N) - D[2](M):   #(b)

       'curlF_k' = curlF_k(x,y);

       top,bot:= solve(C, y);   #(c)

       left,right:= -2,2;

       q1:= Int(Int(curlF_k(x,y),y=bot..top), x =left..right);

       value(q1);

Mathematica: (functions and bounds will vary)

The **ImplicitPlot** command will be useful for 43 and 44, but is not needed for 45 and 46. In 46, the equation of the line from $(0, 4)$ to $(2, 0)$ must be determined first.

       Clear[x, y, f]

       <<Graphics`ImplicitPlot`

f[x_, y_]:= {2x − y, x + 3y}

curve= $x^2 + 4y^2$ ==4

ImplicitPlot[curve, [x, −3,3},{y, −2,2}, AspectRatio → Automatic, AxesLabel → {x, y}];

ybonds= Solve[curve, y]

{y1, y2]=y/.bounds;

integrand:=D[f(x,y][[2]],x] − D[f[x,y]],y]/Simplify

Integrate[integrand, {x, −2,2},{y, y1, y2}]

N[%]

Bounds for y are determined differently in 45 and 46. In 46, note equation of the line from (0, 4) to (2, 0).

Clear[x, y, f]

f[x_, y_]:= {x Exp[y], $4x^2$Log[y]}

ybound = 4 − 2x

Plot[{0, ybound}, {x, 0, 2.1}, AspectRatio → Automatic, AxesLabel → {x,y}];

integrand:=D[f[x, y][[2]],x] − D[f[x,y][[1]],y]//Simplify

Integrate[integrand, [x, 0, 2], {y, 0, ybound }]

N[%]

15.5 SURFACES AND AREA

1. In cylindrical coordinates, let $x = r\cos\theta$, $y = r\sin\theta$, $z = \left(\sqrt{x^2 + y^2}\right)^2 = r^2$.

   Then $\mathbf{r}(r, \theta) = (r\cos\theta)\mathbf{i} + (r\sin\theta)\mathbf{j} + r^2\mathbf{k}$, $0 \le r \le 2, 0 \le \theta \le 2\pi$.

3. In cylindrical coordinates, let $x = r\cos\theta$, $y = r\sin\theta$, $z = \frac{\sqrt{x^2+y^2}}{2} \Rightarrow z = \frac{r}{2}$.

   Then $\mathbf{r}(r, \theta) = (r\cos\theta)\mathbf{i} + (r\sin\theta)\mathbf{j} + \left(\frac{r}{2}\right)\mathbf{k}$. For $0 \le z \le 3, 0 \le \frac{r}{2} \le 3 \Rightarrow 0 \le r \le 6$; to get only the first octant,

   let $0 \le \theta \le \frac{\pi}{2}$.

5. In cylindrical coordinates, let $x = r\cos\theta$, $y = r\sin\theta$; since $x^2 + y^2 = r^2 \Rightarrow z^2 = 9 - \left(x^2 + y^2\right) = 9 - r^2$

   $\Rightarrow z = \sqrt{9 - r^2}, z \ge 0$. Then $\mathbf{r}(r, \theta) = (r\cos\theta)\mathbf{i} + (r\sin\theta)\mathbf{j} + \sqrt{9 - r^2}\mathbf{k}$. Let $0 \le \theta \le 2\pi$. For the domain of r:

   $z = \sqrt{x^2 + y^2}$ and $x^2 + y^2 + z^2 = 9 \Rightarrow x^2 + y^2 + \left(\sqrt{x^2 + y^2}\right)^2 = 9 \Rightarrow 2\left(x^2 + y^2\right) = 9 \Rightarrow 2r^2 = 9$

   $\Rightarrow r = \frac{3}{\sqrt{2}} \Rightarrow 0 \le r \le \frac{3}{\sqrt{2}}$.

7. In spherical coordinates, $x = \rho\sin\phi\cos\theta$, $y = \rho\sin\phi\sin\theta$, $\rho = \sqrt{x^2 + y^2 + z^2} \Rightarrow \rho^2 = 3 \Rightarrow \rho = \sqrt{3}$

   $\Rightarrow z = \sqrt{3}\cos\phi$ for the sphere; $z = \frac{\sqrt{3}}{2} = \sqrt{3}\cos\phi \Rightarrow \cos\phi = \frac{1}{2} \Rightarrow \phi = \frac{\pi}{3}; z = -\frac{\sqrt{3}}{2} \Rightarrow -\frac{\sqrt{3}}{2} = \sqrt{3}\cos\phi$

   $\Rightarrow \cos\phi = -\frac{1}{2} \Rightarrow \phi = \frac{2\pi}{3}$. Then $\mathbf{r}(\phi, \theta) = \left(\sqrt{3}\sin\phi\cos\theta\right)\mathbf{i} + \left(\sqrt{3}\sin\phi\sin\theta\right)\mathbf{j} + \left(\sqrt{3}\cos\phi\right)\mathbf{k}$,

   $\frac{\pi}{3} \le \phi \le \frac{2\pi}{3}$ and $0 \le \theta \le 2\pi$.

9. Since $z = 4 - y^2$, we can let $\mathbf{r}$ be a function of x and $y \Rightarrow \mathbf{r}(x, y) = x\mathbf{i} + y\mathbf{j} + \left(4 - y^2\right)\mathbf{k}$. Then $z = 0$

$\Rightarrow 0 = 4 - y^2 \Rightarrow y = \pm 2$. Thus, let $-2 \leq y \leq 2$ and $0 \leq x \leq 2$.

11. When $x = 0$, let $y^2 + z^2 = 9$ be the circular section in the yz-plane. Use polar coordinates in the yz-plane $\Rightarrow y = 3 \cos \theta$ and $z = 3 \sin \theta$. Thus let $x = u$ and $\theta = v \Rightarrow \mathbf{r}(u, v) = u\mathbf{i} + (3 \cos v)\mathbf{j} + (3 \sin v)\mathbf{k}$ where $0 \leq u \leq 3$, and $0 \leq v \leq 2\pi$.

13. (a) $x + y + z = 1 \Rightarrow z = 1 - x - y$. In cylindrical coordinates, let $x = r \cos \theta$ and $y = r \sin \theta$
$\Rightarrow z = 1 - r \cos \theta - r \sin \theta \Rightarrow \mathbf{r}(r, \theta) = (r \cos \theta)\mathbf{i} + (r \sin \theta)\mathbf{j} + (1 - r \cos \theta - r \sin \theta)\mathbf{k}, 0 \leq \theta \leq 2\pi$
and $0 \leq r \leq 3$.

   (b) In a fashion similar to cylindrical coordinates, but working in the yz-plane instead of the xy-plane, let

   $y = u \cos v, z = u \sin v$ where $u = \sqrt{y^2 + z^2}$ and v is the angle formed by (x, y, z), $(x, 0, 0)$, and $(x, y, 0)$ with $(x, 0, 0)$ as vertex. Since $x + y + z = 1 \Rightarrow x = 1 - y - z \Rightarrow x = 1 - u \cos v - u \sin v$, then $\mathbf{r}$ is a function of u and $v \Rightarrow \mathbf{r}(u, v) = (1 - u \cos v - u \sin v)\mathbf{i} + (u \cos v)\mathbf{j} + (u \sin v)\mathbf{k}, 0 \leq u \leq 3$ and $0 \leq v \leq 2\pi$.

15. Let $x = w \cos v$ and $z = w \sin v$. Then $(x - 2)^2 + z^2 = 4 \Rightarrow x^2 - 4x + z^2 = 0$

$\Rightarrow w^2 \cos^2 v - 4w \cos v + w^2 \sin^2 v = 0 \Rightarrow w^2 - 4w \cos v = 0 \Rightarrow w = 0$ or $w - 4 \cos v = 0 \Rightarrow w = 0$ or $w = 4 \cos v$. Now $w = 0 \Rightarrow x = 0$ and $y = 0$, which is a line not a cylinder. Therefore, let

$w = 4 \cos v \Rightarrow x = (4 \cos v)(\cos v) = 4\cos^2 v$ and $z = 4 \cos v \sin v$. Finally, let $y = u$. Then

$\mathbf{r}(u, v) = \left(4 \cos^2 v\right)\mathbf{i} + u\mathbf{j} + (4 \cos v \sin v)\mathbf{k}, -\frac{\pi}{2} \leq v \leq \frac{\pi}{2}$ and $0 \leq u \leq 3$.

17. Let $x = r \cos \theta$ and $y = r \sin \theta$. Then $\mathbf{r}(r, \theta) = (r \cos \theta)\mathbf{i} + (r \sin \theta)\mathbf{j} + \left(\frac{2 - r \sin \theta}{2}\right)\mathbf{k}$, $0 \leq r \leq 1$ and $0 \leq \theta \leq 2\pi$

$\Rightarrow \mathbf{r}_r = (\cos \theta)\mathbf{i} + (\sin \theta)\mathbf{j} - \left(\frac{\sin \theta}{2}\right)\mathbf{k}$ and $\mathbf{r}_\theta = (-r \sin \theta)\mathbf{i} + (r \cos \theta)\mathbf{j} - \left(\frac{r \cos \theta}{2}\right)\mathbf{k}$

$$\Rightarrow \mathbf{r}_r \times \mathbf{r}_\theta = \begin{vmatrix} \mathbf{i} & \mathbf{j} & \mathbf{k} \\ \cos \theta & \sin \theta & -\frac{\sin \theta}{2} \\ -r \sin \theta & r \cos \theta & -\frac{r \cos \theta}{2} \end{vmatrix}$$

$= \left(\frac{-r \sin \theta \cos \theta}{2} + \frac{(\sin \theta)(r \cos \theta)}{2}\right)\mathbf{i} + \left(\frac{r \sin^2 \theta}{2} + \frac{r \cos^2 \theta}{2}\right)\mathbf{j} + \left(r \cos^2 \theta + r \sin^2 \theta\right)\mathbf{k} = \frac{r}{2}\mathbf{j} + r\mathbf{k}$

$\Rightarrow |\mathbf{r}_r \times \mathbf{r}_\theta| = \sqrt{\frac{r^2}{4} + r^2} = \frac{\sqrt{5}r}{2} \Rightarrow A = \int_0^{2\pi} \int_0^1 \frac{\sqrt{5}r}{2} \, dr \, d\theta = \int_0^{2\pi} \left[\frac{\sqrt{5}r^2}{4}\right]_0^1 d\theta = \int_0^{2\pi} \frac{\sqrt{5}}{4} \, d\theta = \frac{\pi\sqrt{5}}{2}$

19. Let $x = r \cos \theta$ and $y = r \sin \theta \Rightarrow z = 2\sqrt{x^2 + y^2} = 2r, 1 \leq r \leq 3$ and $0 \leq \theta \leq 2\pi$. Then
$\mathbf{r}(r, \theta) = (r \cos \theta)\mathbf{i} + (r \sin \theta)\mathbf{j} + 2r\mathbf{k} \Rightarrow \mathbf{r}_r = (\cos \theta)\mathbf{i} + (\sin \theta)\mathbf{j} + 2\mathbf{k}$ and $\mathbf{r}_\theta = (-r \sin \theta)\mathbf{i} + (r \cos \theta)\mathbf{j}$

$$\Rightarrow \mathbf{r}_r \times \mathbf{r}_\theta = \begin{vmatrix} \mathbf{i} & \mathbf{j} & \mathbf{k} \\ \cos \theta & \sin \theta & 2 \\ -r \sin \theta & r \cos \theta & 0 \end{vmatrix} = (-2r \cos \theta)\mathbf{i} - (2r \sin \theta)\mathbf{j} + \left(r \cos^2 \theta + r \sin^2 \theta\right)\mathbf{k}$$

$\Rightarrow (-2r \cos \theta)\mathbf{i} - (2r \sin \theta)\mathbf{j} + r\mathbf{k} \Rightarrow |\mathbf{r}_r \times \mathbf{r}_\theta| = \sqrt{4r^2 \cos^2 \theta + 4r^2 \sin^2 \theta + r^2} = \sqrt{5r^2} = r\sqrt{5}$

$\Rightarrow A = \int_0^{2\pi} \int_1^3 r\sqrt{5} \, dr \, d\theta = \int_0^{2\pi} \left[\frac{r^2\sqrt{5}}{2}\right]_1^3 d\theta = \int_0^{2\pi} 4\sqrt{5} \, d\theta = 8\pi\sqrt{5}$

21. Let $x = r\cos\theta$ and $y = r\sin\theta \Rightarrow r^2 = x^2 + y^2 = 1, 1 \le z \le 4$ and $0 \le \theta \le 2\pi$. Then

$\mathbf{r}(z, \theta) = (\cos\theta)\mathbf{i} + (\sin\theta)\mathbf{j} + z\mathbf{k} \Rightarrow \mathbf{r}_z = \mathbf{k}$ and $\mathbf{r}_\theta = (-\sin\theta)\mathbf{i} + (\cos\theta)\mathbf{j}$

$\Rightarrow \mathbf{r}_\theta \times \mathbf{r}_z = \begin{vmatrix} \mathbf{i} & \mathbf{j} & \mathbf{k} \\ -\sin\theta & \cos\theta & 0 \\ 0 & 0 & 1 \end{vmatrix} = (\cos\theta)\mathbf{i} + (\sin\theta)\mathbf{j} \Rightarrow |\mathbf{r}_\theta \times \mathbf{r}_z| = \sqrt{\cos^2\theta + \sin^2\theta} = 1$

$\Rightarrow A = \int_0^{2\pi}\int_1^4 1\, dr\, d\theta = \int_0^{2\pi} 3\, d\theta = 6\pi$

23. $z = 2 - x^2 - y^2$ and $z = \sqrt{x^2 + y^2} \Rightarrow z = 2 - z^2 \Rightarrow z^2 + z - 2 = 0 \Rightarrow z = -2$ or $z = 1$. Since $z = \sqrt{x^2 + y^2} \ge 0$,

we get $z = 1$ where the cone intersects the paraboloid. When $x = 0$ and $y = 0, z = 2 \Rightarrow$ the vertex of the

paraboloid is $(0, 0, 2)$. Therefore, z ranges from 1 to 2 on the "cap" $\Rightarrow r$ ranges from 1 (when $x^2 + y^2 = 1$) to

0 (when $x = 0$ and $y = 0$ at the vertex). Let $x = r\cos\theta$, $y = r\sin\theta$, and $z = 2 - r^2$. Then

$\mathbf{r}(r, \theta) = (r\cos\theta)\mathbf{i} + (r\sin\theta)\mathbf{j} + \left(2 - r^2\right)\mathbf{k}, \; 0 \le r \le 1, 0 \le \theta \le 2\pi \Rightarrow \mathbf{r}_r = (\cos\theta)\mathbf{i} + (\sin\theta)\mathbf{j} - 2r\mathbf{k}$ and

$\mathbf{r}_\theta = (-r\sin\theta)\mathbf{i} + (r\cos\theta)\mathbf{j} \Rightarrow \mathbf{r}_r \times \mathbf{r}_\theta = \begin{vmatrix} \mathbf{i} & \mathbf{j} & \mathbf{k} \\ \cos\theta & \sin\theta & -2r \\ -r\sin\theta & r\cos\theta & 0 \end{vmatrix}$

$= \left(2r^2\cos\theta\right)\mathbf{i} + \left(2r^2\sin\theta\right)\mathbf{j} + r\mathbf{k} \Rightarrow |\mathbf{r}_r \times \mathbf{r}_\theta| = \sqrt{4r^4\cos^2\theta + 4r^4\sin^2\theta + r^2} = r\sqrt{4r^2 + 1}$

$\Rightarrow A = \int_0^{2\pi}\int_0^1 r\sqrt{4r^2 + 1}\, dr\, d\theta = \int_0^{2\pi}\left[\frac{1}{12}\left(4r^2 + 1\right)^{3/2}\right]_0^1 d\theta = \int_0^{2\pi}\left(\frac{5\sqrt{5}-1}{12}\right)d\theta = \frac{\pi}{6}\left(5\sqrt{5} - 1\right)$

25. Let $x = \rho\sin\phi\cos\theta$, $y = \rho\sin\phi\sin\theta$, and $z = \rho\cos\phi \Rightarrow \rho = \sqrt{x^2 + y^2 + z^2} = \sqrt{2}$ on the sphere. Next,

$x^2 + y^2 + z^2 = 2$ and $z = \sqrt{x^2 + y^2} \Rightarrow z^2 + z^2 = 2 \Rightarrow z^2 = 1 \Rightarrow z = 1$ since $z \ge 0 \Rightarrow \phi = \frac{\pi}{4}$. For the lower

portion of the sphere cut by the cone, we get $\phi = \pi$. Then

$\mathbf{r}(\phi, \theta) = \left(\sqrt{2}\sin\phi\cos\theta\right)\mathbf{i} + \left(\sqrt{2}\sin\phi\sin\theta\right)\mathbf{j} + \left(\sqrt{2}\cos\phi\right)\mathbf{k}, \; \frac{\pi}{4} \le \phi \le \pi, 0 \le \theta \le 2\pi$

$\Rightarrow \mathbf{r}_\phi = \left(\sqrt{2}\sin\phi\cos\theta\right)\mathbf{i} + \left(\sqrt{2}\cos\phi\sin\theta\right)\mathbf{j} - \left(\sqrt{2}\sin\phi\right)\mathbf{k}$ and $\mathbf{r}_\theta = \left(-\sqrt{2}\sin\phi\sin\theta\right)\mathbf{i} + \left(\sqrt{2}\sin\phi\cos\theta\right)\mathbf{j}$

$\Rightarrow \mathbf{r}_\phi \times \mathbf{r}_\theta = \begin{vmatrix} \mathbf{i} & \mathbf{j} & \mathbf{k} \\ \sqrt{2}\cos\phi\cos\theta & \sqrt{2}\cos\phi\sin\theta & -\sqrt{2}\sin\phi \\ -\sqrt{2}\sin\phi\sin\theta & \sqrt{2}\sin\phi\cos\theta & 0 \end{vmatrix} = \left(2\sin^2\phi\cos\theta\right)\mathbf{i} + \left(2\sin^2\phi\sin\theta\right)\mathbf{j} + (2\sin\phi\cos\phi)\mathbf{k}$

$\Rightarrow |\mathbf{r}_\phi \times \mathbf{r}_\theta| = \sqrt{4\sin^4\phi\cos^2\theta + 4\sin^4\phi\sin^2\theta + 4\sin^2\phi\cos^2\phi} = \sqrt{4\sin^2\phi} = 2|\sin\phi| = 2\sin\phi$

$\Rightarrow A = \int_0^{2\pi}\int_{\pi/4}^{\pi} 2\sin\phi\, d\phi\, d\theta = \int_0^{2\pi}\left(2 + \sqrt{2}\right)d\theta = \left(4 + 2\sqrt{2}\right)\pi$

27. The parametrization $\mathbf{r}(r, \theta) = (r\cos\theta)\mathbf{i} + (r\sin\theta)\mathbf{j} + r\mathbf{k}$

at $P_0 = \left(\sqrt{2}, \sqrt{2}, 2\right) \Rightarrow \theta = \frac{\pi}{4}, r = 2,$

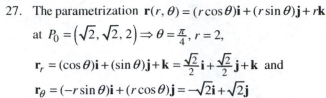

$\mathbf{r}_r = (\cos\theta)\mathbf{i} + (\sin\theta)\mathbf{j} + \mathbf{k} = \frac{\sqrt{2}}{2}\mathbf{i} + \frac{\sqrt{2}}{2}\mathbf{j} + \mathbf{k}$ and

$\mathbf{r}_\theta = (-r\sin\theta)\mathbf{i} + (r\cos\theta)\mathbf{j} = -\sqrt{2}\mathbf{i} + \sqrt{2}\mathbf{j}$

$\Rightarrow \mathbf{r}_r \times \mathbf{r}_\theta = \begin{vmatrix} \mathbf{i} & \mathbf{j} & \mathbf{k} \\ \sqrt{2}/2 & \sqrt{2}/2 & 1 \\ -\sqrt{2} & \sqrt{2} & 0 \end{vmatrix} = -\sqrt{2}\mathbf{i} - \sqrt{2}\mathbf{j} + 2\mathbf{k}$

$\Rightarrow$ the tangent plane is $0 = \left(-\sqrt{2}\mathbf{i} - \sqrt{2}\mathbf{j} + 2\mathbf{k}\right) \cdot \left[\left(x - \sqrt{2}\right)\mathbf{i} + \left(y - \sqrt{2}\right)\mathbf{j} + (z - 2)\mathbf{k}\right] \Rightarrow \sqrt{2}x + \sqrt{2}y - 2z = 0$, or

$x + y - \sqrt{2}z = 0$. The parametrization $\mathbf{r}(r, \theta) \Rightarrow x = r\cos\theta, y = r\sin\theta$ and $z = r \Rightarrow x^2 + y^2 = r^2 = z^2$

$\Rightarrow$ the surface is $z = \sqrt{x^2 + y^2}$.

29. The parametrization

$\mathbf{r}(\theta, z) = (3\sin 2\theta)\mathbf{i} + \left(6\sin^2\theta\right)\mathbf{j} + z\mathbf{k}$ at

$P_0 = \left(\frac{3\sqrt{3}}{2}, \frac{9}{2}, 0\right) \Rightarrow \theta = \frac{\pi}{3}$ and $z = 0$. Then

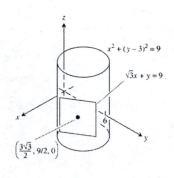

$\mathbf{r}_\theta = (6\cos 2\theta)\mathbf{i} + (12\sin\theta\cos\theta)\mathbf{j} = -3\mathbf{i} + 3\sqrt{3}\mathbf{j}$ and

$\mathbf{r}_z = \mathbf{k}$ at $P_0 \Rightarrow \mathbf{r}_\theta \times \mathbf{r}_z \begin{vmatrix} \mathbf{i} & \mathbf{j} & \mathbf{k} \\ -3 & 3\sqrt{3} & 0 \\ 0 & 0 & 1 \end{vmatrix} = 3\sqrt{3}\mathbf{i} + 3\mathbf{j}$

$\Rightarrow$ the tangent plane is $\left(3\sqrt{3}\mathbf{i} + 3\mathbf{j}\right) \cdot \left[\left(x - \frac{3\sqrt{3}}{2}\right)\mathbf{i} + \left(y - \frac{9}{2}\right)\mathbf{j} + (z - 0)\mathbf{k}\right] = 0 \Rightarrow \sqrt{3}x + y = 9$. The parametrization

$\Rightarrow x = 3\sin 2\theta$ and $y = 6\sin^2\theta \Rightarrow x^2 + y^2 = 9\sin^2 2\theta + \left(6\sin^2\theta\right)^2$

$= 9\left(4\sin^2\theta\cos^2\theta\right) + 36\sin^4\theta = 6\left(6\sin^2\theta\right) = 6y \Rightarrow x^2 + y^2 - 6y + 9 = 9 \Rightarrow x^2 + (y - 3)^2 = 9$

31. (a) An arbitrary point on the circle C is $(x, z) = (R + r\cos u, r\sin u) \Rightarrow (x, y, z)$ is on the torus with

$x = (R + r\cos u)\cos v, y = (R + r\cos u)\sin v$, and $z = r\sin u, 0 \le u \le 2\pi, 0 \le v \le 2\pi$

(b) $\mathbf{r}_u = (-r\sin u\cos v)\mathbf{i} - (r\sin u\sin v)\mathbf{j} + (r\cos u)\mathbf{k}$ and $\mathbf{r}_v = \left(-(R + r\cos u)\sin v\right)\mathbf{i} + \left((R + r\cos u)\cos v\right)\mathbf{j}$

$\Rightarrow \mathbf{r}_u \times \mathbf{r}_v = \begin{vmatrix} \mathbf{i} & \mathbf{j} & \mathbf{k} \\ -r\sin u\cos v & -r\sin u\sin v & r\cos u \\ -(R + r\cos u)\sin v & (R + r\cos u)\cos v & 0 \end{vmatrix}$

$= -(R + r\cos u)(r\cos v\cos u)\mathbf{i} - (R + r\cos u)(r\sin v\cos u)\mathbf{j} + (-r\sin u)(R + r\cos u)\mathbf{k}$

$\Rightarrow |\mathbf{r}_u \times \mathbf{r}_v|^2 = (R + r\cos u)^2 \left(r^2\cos^2 v\cos^2 u + r^2\sin^2 v\cos^2 u + r^2\sin^2 u\right) \Rightarrow |\mathbf{r}_u \times \mathbf{r}_v| = r(R + r\cos u)$

$\Rightarrow A = \int_0^{2\pi} \int_0^{2\pi} \left(rR + r^2\cos u\right) du\,dv = \int_0^{2\pi} 2\pi rR\,dv = 4\pi^2 rR$

33. (a) Let $w^2 + \frac{z^2}{c^2} = 1$ where $w = \cos\phi$ and $\frac{z}{c} = \sin\phi \Rightarrow \frac{x^2}{a^2} + \frac{y^2}{b^2} = \cos^2\phi \Rightarrow \frac{x}{a} = \cos\phi\cos\theta$ and $\frac{y}{b} = \cos\phi\sin\theta$

$\Rightarrow x = a\cos\theta\cos\phi,\ y = b\sin\theta\cos\phi,$ and $z = c\sin\phi$

$\Rightarrow \mathbf{r}(\theta,\phi) = (a\cos\theta\cos\phi)\mathbf{i} + (b\sin\theta\cos\phi)\mathbf{j} + (c\sin\phi)\mathbf{k}$

(b) $\mathbf{r}_\theta = (-a\sin\theta\cos\phi)\mathbf{i} + (b\cos\theta\cos\phi)\mathbf{j}$ and $\mathbf{r}_\phi = (-a\cos\theta\sin\phi)\mathbf{i} - (b\sin\theta\sin\phi)\mathbf{j} + (c\cos\phi)\mathbf{k}$

$$\Rightarrow \mathbf{r}_\theta \times \mathbf{r}_\phi = \begin{vmatrix} \mathbf{i} & \mathbf{j} & \mathbf{k} \\ -a\sin\theta\cos\phi & b\cos\theta\cos\phi & 0 \\ -a\cos\theta\sin\phi & -b\sin\theta\sin\phi & c\cos\phi \end{vmatrix}$$

$= \left(bc\cos\theta\cos^2\phi\right)\mathbf{i} + \left(ac\sin\theta\cos^2\phi\right)\mathbf{j} + (ab\sin\phi\cos\phi)\mathbf{k}$

$\Rightarrow |\mathbf{r}_\theta \times \mathbf{r}_\phi|^2 = b^2c^2\cos^2\theta\cos^4\phi + a^2c^2\sin^2\theta\cos^4\phi + a^2b^2\sin^2\phi\cos^2\phi,$ and the result follows.

$A = \int_0^{2\pi}\int_0^{2\pi} |\mathbf{r}_\theta\times\mathbf{r}_\phi|d\phi\,d\theta = \int_0^{2\pi}\int_0^\pi \left[a^2b^2\sin^2\phi\cos^2\phi + b^2c^2\cos^2\theta\cos^4\phi + a^2c^2\sin^2\theta\cos^4\phi\right]^{1/2}d\phi\,d\theta$

35. $\mathbf{r}(\theta,u) = (5\cosh u\cos\theta)\mathbf{i} + (5\cosh u\sin\theta)\mathbf{j} + (5\sinh u)\mathbf{k} \Rightarrow \mathbf{r}_\theta = (-5\cosh u\sin\theta)\mathbf{i} + (5\cosh u\cos\theta)\mathbf{j}$ and

$$\mathbf{r}_u = (5\sinh u\cos\theta)\mathbf{i} + (5\sinh u\sin\theta)\mathbf{j} + (5\cosh u)\mathbf{k} \Rightarrow \mathbf{r}_\theta\times\mathbf{r}_u = \begin{vmatrix} \mathbf{i} & \mathbf{j} & \mathbf{k} \\ -5\cosh u\sin\theta & 5\cosh u\cos\theta & 0 \\ 5\sinh u\cos\theta & 5\sinh u\sin\theta & 5\cosh u \end{vmatrix}$$

$= \left(25\cosh^2 u\cos\theta\right)\mathbf{i} + \left(25\cosh^2 u\sin\theta\right)\mathbf{j} - (25\cosh u\sinh u)\mathbf{k}.$ At the point $(x_0, y_0, 0)$, where $x_0^2 + y_0^2 = 25$

we have $5\sinh u = 0 \Rightarrow u = 0$ and $x_0 = 25\cos\theta,\ y_0 = 25\sin\theta \Rightarrow$ the tangent plane is

$5(x_0\mathbf{i} + y_0\mathbf{j})\cdot[(x-x_0)\mathbf{i} + (y-y_0)\mathbf{j} + z\mathbf{k}] = 0 \Rightarrow x_0 x - x_0^2 + y_0 y - y_0^2 = 0 \Rightarrow x_0 x + y_0 y = 25$

37. $\mathbf{p} = \mathbf{k},\ \nabla f = 2x\mathbf{i} + 2y\mathbf{j} - \mathbf{k} \Rightarrow |\nabla f| = \sqrt{(2x)^2 + (2y)^2 + (-1)^2} = \sqrt{4x^2 + 4y^2 + 1}$ and $|\nabla f\cdot\mathbf{p}| = 1;$

$z = 2 \Rightarrow x^2 + y^2 = 2;$ thus $S = \iint_R \frac{|\nabla f|}{|\nabla f\cdot\mathbf{p}|}dA = \iint_R \sqrt{4x^2 + 4y^2 + 1}\,dx\,dy = \iint_R \sqrt{4r^2\cos^2\theta + 4r^2\sin^2\theta + 1}\ r\,dr\,d\theta$

$= \int_0^{2\pi}\int_0^{\sqrt2}\sqrt{4r^2+1}\ r\,dr\,d\theta = \int_0^{2\pi}\left[\frac{1}{12}(4r^2+1)^{3/2}\right]_0^{\sqrt2}d\theta = \int_0^{2\pi}\frac{13}{6}d\theta = \frac{13}{3}\pi$

39. $\mathbf{p} = \mathbf{k},\ \nabla f = \mathbf{i} + 2\mathbf{j} + 2\mathbf{k} \Rightarrow |\nabla f| = 3$ and $|\nabla f\cdot\mathbf{p}| = 2;\ x = y^2$ and $x = 2 - y^2$ intersect at $(1,1)$ and $(1,-1)$

$\Rightarrow S = \iint_R \frac{|\nabla f|}{|\nabla f\cdot\mathbf{p}|}dA = \iint_R \frac{3}{2}dx\,dy = \int_{-1}^1\int_{y^2}^{2-y^2}\frac{3}{2}dx\,dy = \int_{-1}^1 (3-3y^2)dy = 4$

41. $\mathbf{p} = \mathbf{k},\ \nabla f = 2x\mathbf{i} - 2\mathbf{j} - 2\mathbf{k} \Rightarrow |\nabla f| = \sqrt{(2x)^2 + (-2)^2 + (-2)^2} = \sqrt{4x^2 + 8} = 2\sqrt{x^2 + 2}$ and $|\nabla f\cdot\mathbf{p}| = 2$

$\Rightarrow S = \iint_R \frac{|\nabla f|}{|\nabla f\cdot\mathbf{p}|}dA = \iint_R \frac{2\sqrt{x^2+2}}{2}dx\,dy = \int_0^2\int_0^{3x}\sqrt{x^2+2}\,dy\,dx = \int_0^2 3x\sqrt{x^2+2}\,dx = \left[(x^2+2)^{3/2}\right]_0^2 = 6\sqrt6 - 2\sqrt2$

43. $\mathbf{p} = \mathbf{k},\ \nabla f = c\mathbf{i} - \mathbf{k} \Rightarrow |\nabla f| = \sqrt{c^2 + 1^2}$ and $|\nabla f\cdot\mathbf{p}| = 1 \Rightarrow S = \iint_R \frac{|\nabla f|}{|\nabla f\cdot\mathbf{p}|}dA = \iint_R \sqrt{c^2+1}\,dx\,dy$

$= \int_0^{2\pi}\int_0^1 \sqrt{c^2+1}\ r\,dr\,d\theta = \int_0^{2\pi}\frac{\sqrt{c^2+1}}{2}d\theta = \pi\sqrt{c^2+1}$

45. $\mathbf{p} = \mathbf{i}, \nabla f = \mathbf{i} + 2y\mathbf{j} + 2z\mathbf{k} \Rightarrow |\nabla f| = \sqrt{1^2 + (2y)^2 + (2z)^2} = \sqrt{1 + 4y^2 + 4z^2}$ and $|\nabla f \cdot \mathbf{p}| = 1;\ 1 \le y^2 + z^2 \le 4$

$\Rightarrow S = \iint\limits_R \frac{|\nabla f|}{|\nabla f \cdot \mathbf{p}|}\, dA = \iint\limits_R \sqrt{1 + 4y^2 + 4z^2}\, dy\, dz = \int_0^{2\pi}\int_1^2 \sqrt{1 + 4r^2\cos^2\theta + 4r^2\sin^2\theta}\ r\, dr\, d\theta$

$= \int_0^{2\pi}\int_1^2 \sqrt{1 + 4r^2}\ r\, dr\, d\theta = \int_0^{2\pi}\left[\frac{1}{12}(1 + 4r^2)^{3/2}\right]_1^2 d\theta = \int_0^{2\pi}\frac{1}{12}\left(17\sqrt{17} - 5\sqrt{5}\right)d\theta = \frac{\pi}{6}\left(17\sqrt{17} - 5\sqrt{5}\right)$

47. $\mathbf{p} = \mathbf{k}, \nabla f = \left(2x - \frac{2}{x}\right)\mathbf{i} + \sqrt{15}\,\mathbf{j} - \mathbf{k} \Rightarrow |\nabla f| = \sqrt{\left(2x - \frac{2}{x}\right)^2 + \left(\sqrt{15}\right)^2 + (-1)^2} = \sqrt{4x^2 + 8 + \frac{4}{x^2}} = \sqrt{\left(2x + \frac{2}{x}\right)^2}$

$= 2x + \frac{2}{x}$, on $1 \le x \le 2$ and $|\nabla f \cdot \mathbf{p}| = 1 \Rightarrow S = \iint\limits_R \frac{|\nabla f|}{|\nabla f \cdot \mathbf{p}|}\, dA = \iint\limits_R \left(2x + 2x^{-1}\right)dx\, dy = \int_0^1\int_1^2\left(2x + 2x^{-1}\right)dx\, dy$

$= \int_0^1\left[x^2 + 2\ln x\right]_1^2 dy = \int_0^1 (3 + 2\ln 2)\, dy = 3 + 2\ln 2$

49. $f_x(x, y) = 2x, f_y(x, y) = 2y \Rightarrow \sqrt{f_x^2 + f_y^2 + 1} = \sqrt{4x^2 + 4y^2 + 1} \Rightarrow \text{Area} = \iint\limits_R \sqrt{4x^2 + 4y^2 + 1}\, dx\, dy$

$= \int_0^{2\pi}\int_0^{\sqrt{3}} \sqrt{4r^2 + 1}\ r\, dr\, d\theta = \frac{\pi}{6}\left(13\sqrt{13} - 1\right)$

51. $f_x(x, y) = \frac{x}{\sqrt{x^2 + y^2}}, f_y(x, y) = \frac{x}{\sqrt{x^2 + y^2}} \Rightarrow \sqrt{f_x^2 + f_y^2 + 1} = \sqrt{\frac{x^2}{x^2 + y^2} + \frac{y^2}{x^2 + y^2} + 1} = \sqrt{2} \Rightarrow \text{Area} = \iint\limits_{R_{xy}} \sqrt{2}\, dx\, dy$

$= \sqrt{2}$ (Area between the ellipse and the circle) $= \sqrt{2}(6\pi - \pi) = 5\pi\sqrt{2}$

53. $y = \frac{2}{3}z^{3/2} \Rightarrow f_x(x, z) = 0, f_z(x, z) = z^{1/2} \Rightarrow \sqrt{f_x^2 + f_z^2 + 1} = \sqrt{z + 1};\ y = \frac{16}{3} \Rightarrow \frac{16}{3} = \frac{2}{3}z^{3/2} \Rightarrow z = 4$

$\Rightarrow \text{Area} = \int_0^4\int_0^1 \sqrt{z + 1}\, dx\, dz = \int_0^4 \sqrt{z + 1}\, dz = \frac{2}{3}\left(5\sqrt{5} - 1\right)$

55. $\mathbf{r}(x, y) = x\mathbf{i} + y\mathbf{j} + f(x, y)\mathbf{k} \Rightarrow \mathbf{r}_x(x, y) = \mathbf{i} + f_x(x, y)\mathbf{k}, \mathbf{r}_y(x, y) = \mathbf{j} + f_y(x, y)\mathbf{k}$

$\Rightarrow \mathbf{r}_x \times \mathbf{r}_y = \begin{vmatrix} \mathbf{i} & \mathbf{j} & \mathbf{k} \\ 1 & 0 & f_x(x, y) \\ 0 & 1 & f_y(x, y) \end{vmatrix} = -f_x(x, y)\mathbf{i} - f_y(x, y)\mathbf{j} + \mathbf{k}$

$\Rightarrow |\mathbf{r}_x \times \mathbf{r}_y| = \sqrt{\left(-f_x(x, y)\right)^2 + \left(-f_y(x, y)\right)^2 + 1^2} = \sqrt{f_x(x, y)^2 + f_y(x, y)^2 + 1}$

$\Rightarrow d\sigma = \sqrt{f_x(x, y)^2 + f_y(x, y)^2 + 1}\, dA$

15.6 SURFACE INTEGRALS

1. Let the parametrization be $\mathbf{r}(x, z) = x\mathbf{i} + x^2\mathbf{j} + z\mathbf{k} \Rightarrow \mathbf{r}_x = \mathbf{i} + 2x\mathbf{j}$ and $\mathbf{r}_z = \mathbf{k} \Rightarrow \mathbf{r}_x \times \mathbf{r}_z = \begin{vmatrix} \mathbf{i} & \mathbf{j} & \mathbf{k} \\ 1 & 2x & 0 \\ 0 & 0 & 1 \end{vmatrix}$

$= 2x\mathbf{i} + \mathbf{j} \Rightarrow |\mathbf{r}_x \times \mathbf{r}_z| = \sqrt{4x^2 + 1} \Rightarrow \iint_S G(x, y, z)\, d\sigma = \int_0^3 \int_0^2 x\sqrt{4x^2 + 1}\, dx\, dz = \int_0^3 \left[\frac{1}{12}(4x^2 + 1)^{3/2} \right]_0^2 dz$

$= \int_0^3 \frac{1}{12}\left(17\sqrt{17} - 1\right) dz = \frac{17\sqrt{17} - 1}{4}$

3. Let the parametrization be $\mathbf{r}(\phi, \theta) = (\sin\phi\cos\theta)\mathbf{i} + (\sin\phi\sin\theta)\mathbf{j} + (\cos\phi)\mathbf{k}$ (spherical coordinates with $\rho = 1$ on the sphere), $0 \le \phi \le \pi, 0 \le \theta \le 2\pi \Rightarrow \mathbf{r}_\phi = (\cos\phi\cos\theta)\mathbf{i} + (\cos\phi\sin\theta)\mathbf{j} - (\sin\phi)\mathbf{k}$ and

$\mathbf{r}_\theta = (-\sin\phi\sin\theta)\mathbf{i} + (\sin\phi\cos\theta)\mathbf{j} \Rightarrow \mathbf{r}_\phi \times \mathbf{r}_\theta = \begin{vmatrix} \mathbf{i} & \mathbf{j} & \mathbf{k} \\ \cos\phi\cos\theta & \cos\phi\sin\theta & -\sin\phi \\ -\sin\phi\sin\theta & \sin\phi\cos\theta & 0 \end{vmatrix}$

$= \left(\sin^2\phi\cos\theta\right)\mathbf{i} + \left(\sin^2\phi\sin\theta\right)\mathbf{j} + (\sin\phi\cos\phi)\mathbf{k} \Rightarrow |\mathbf{r}_\phi \times \mathbf{r}_\theta| = \sqrt{\sin^4\phi\cos^2\theta + \sin^4\phi\sin^2\theta + \sin^2\phi\cos^2\phi}$

$= \sin\phi; x = \sin\phi\cos\theta \Rightarrow G(x, y, z) = \cos^2\theta\sin^2\phi \Rightarrow \iint_S G(x, y, z)\, d\sigma = \int_0^{2\pi} \int_0^\pi \left(\cos^2\theta\sin^2\phi\right)(\sin\phi)\, d\phi\, d\theta$

$= \int_0^{2\pi} \int_0^\pi \left(\cos^2\theta\right)\left(1 - \cos^2\phi\right)(\sin\phi)\, d\phi\, d\theta; \begin{bmatrix} u = \cos\phi \\ du = -\sin\phi\, d\phi \end{bmatrix} \to \int_0^{2\pi} \int_1^{-1} \left(\cos^2\theta\right)\left(u^2 - 1\right) du\, d\theta$

$= \int_0^{2\pi} \left(\cos^2\theta\right)\left[\frac{u^3}{3} - u\right]_1^{-1} d\theta = \frac{4}{3}\int_0^{2\pi} \cos^2\theta\, d\theta = \frac{4}{3}\left[\frac{\theta}{2} + \frac{\sin 2\theta}{4}\right]_0^{2\pi} = \frac{4\pi}{3}$

5. Let the parametrization be $\mathbf{r}(x, y) = x\mathbf{i} + y\mathbf{j} + (4 - x - y)\mathbf{k} \Rightarrow \mathbf{r}_x = \mathbf{i} - \mathbf{k}$ and $\mathbf{r}_y = \mathbf{j} - \mathbf{k} \Rightarrow \mathbf{r}_x \times \mathbf{r}_y = \begin{vmatrix} \mathbf{i} & \mathbf{j} & \mathbf{k} \\ 1 & 0 & -1 \\ 0 & 1 & -1 \end{vmatrix}$

$= \mathbf{i} + \mathbf{j} + \mathbf{k} \Rightarrow |\mathbf{r}_x \times \mathbf{r}_y| = \sqrt{3} \Rightarrow \iint_S F(x, y, z)\, d\sigma = \int_0^1 \int_0^1 (4 - x - y)\sqrt{3}\, dy\, dx = \int_0^1 \sqrt{3}\left[4y - xy - \frac{y^2}{2}\right]_0^1 dx$

$= \int_0^1 \sqrt{3}\left(\frac{7}{2} - x\right) dx = \sqrt{3}\left[\frac{7}{2}x - \frac{x^2}{2}\right]_0^1 = 3\sqrt{3}$

7. Let the parametrization be $\mathbf{r}(r, \theta) = (r\cos\theta)\mathbf{i} + (r\sin\theta)\mathbf{j} + (1 - r^2)\mathbf{k}, \ 0 \le r \le 1$ (since $0 \le z \le 1$) and $0 \le \theta \le 2\pi \Rightarrow \mathbf{r}_r = (\cos\theta)\mathbf{i} + (\sin\theta)\mathbf{j} - 2r\mathbf{k}$ and $\mathbf{r}_\theta = (-r\sin\theta)\mathbf{i} + (r\cos\theta)\mathbf{j}$

$\Rightarrow \mathbf{r}_r \times \mathbf{r}_\theta = \begin{vmatrix} \mathbf{i} & \mathbf{j} & \mathbf{k} \\ \cos\theta & \sin\theta & -2r \\ -r\sin\theta & r\cos\theta & 0 \end{vmatrix} = \left(2r^2\cos\theta\right)\mathbf{i} + \left(2r^2\sin\theta\right)\mathbf{j} + r\mathbf{k} \Rightarrow |\mathbf{r}_r \times \mathbf{r}_\theta|$

$= \sqrt{\left(2r^2\cos\theta\right)^2 + \left(2r^2\sin\theta\right)^2 + r^2} = r\sqrt{1 + 4r^2}; z = 1 - r^2$ and $x = r\cos\theta \Rightarrow H(x, y\, z) = \left(r^2\cos^2\theta\right)\sqrt{1 + 4r^2}$

$\Rightarrow \iint_S H(x, y, z)\, d\sigma = \int_0^{2\pi} \int_0^1 \left(r^2\cos^2\theta\right)\left(\sqrt{1 + 4r^2}\right)\left(r\sqrt{1 + 4r^2}\right) dr\, d\theta = \int_0^{2\pi} \int_0^1 r^3\left(1 + 4r^2\right)\cos^2\theta\, dr\, d\theta = \frac{11\pi}{12}$

9. The bottom face S of the cube is in the xy-plane $\Rightarrow z = 0 \Rightarrow G(x, y, 0) = x + y$ and $f(x, y, z) = z = 0 \Rightarrow \mathbf{p} = \mathbf{k}$

and $\nabla f = \mathbf{k} \Rightarrow |\nabla f| = 1$ and $|\nabla f \cdot \mathbf{p}| = 1 \Rightarrow d\sigma = dx\, dy \Rightarrow \iint\limits_S G\, d\sigma = \iint\limits_R (x + y)\, dx\, dy = \int_0^a \int_0^a (x + y)\, dx\, dy$

$= \int_0^a \left(\frac{a^2}{2} + ay \right) dy = a^3$. Because of symmetry, we also get a^3 over the face of the cube in the xz-plane and

a^3 over the face of the cube in the yz-plane. Next, on the top of the cube, $G(x, y, z) = G(x, y, a) = x + y + a$

and $f(x, y, z) = z = a \Rightarrow \mathbf{p} = \mathbf{k}$ and $\nabla f = \mathbf{k} \Rightarrow |\nabla f| = 1$ and $|\nabla f \cdot \mathbf{p}| = 1 \Rightarrow d\sigma = dx\, dy$

$\Rightarrow \iint\limits_S G\, d\sigma = \iint\limits_R (x + y + a)\, dx\, dy = \int_0^a \int_0^a (x + y + a)\, dx\, dy \int_0^a \int_0^a (x + y)\, dx\, dy + \int_0^a \int_0^a a\, dx\, dy = 2a^3$. Because of

symmetry, the integral is also $2a^3$ over each of the other two faces. Therefore,

$\iint\limits_{\text{cube}} (x + y + z)\, d\sigma = 3\left(a^3 + 2a^3 \right) = 9a^3$.

11. On the faces in the coordinate planes, $G(x, y, z) = 0 \Rightarrow$ the integral over these faces is 0.

On the face $x = a$, we have $f(x, y, z) = x = a$ and $G(x, y, z) = G(a, y, z) = ayz \Rightarrow \mathbf{p} = \mathbf{i}$ and

$\nabla f = \mathbf{i} \Rightarrow |\nabla f| = 1$ and $|\nabla f \cdot \mathbf{p}| = 1 \Rightarrow d\sigma = dy\, dz \Rightarrow \iint\limits_S G\, d\sigma = \iint\limits_S ayz\, d\sigma = \int_0^c \int_0^b ayz\, dy\, dz = \frac{ab^2c^2}{4}$.

On the face $y = b$, we have $f(x, y, z) = y = b$ and $G(x, y, z) = G(x, b, z) = bxz \Rightarrow \mathbf{p} = \mathbf{j}$ and

$\nabla f = \mathbf{j} \Rightarrow |\nabla f| = 1$ and $|\nabla f \cdot \mathbf{p}| = 1 \Rightarrow d\sigma = dx\, dz \Rightarrow \iint\limits_S G\, d\sigma = \iint\limits_S bxz\, d\sigma = \int_0^c \int_0^b bxz\, dx\, dz = \frac{a^2bc^2}{4}$.

On the face $z = c$, we have $f(x, y, z) = z = c$ and $G(x, y, z) = G(x, y, c) = cxy \Rightarrow \mathbf{p} = \mathbf{k}$ and

$\nabla f = \mathbf{k} \Rightarrow |\nabla f| = 1$ and $|\nabla f \cdot \mathbf{p}| = 1 \Rightarrow d\sigma = dy\, dx \Rightarrow \iint\limits_S G\, d\sigma = \iint\limits_S cxy\, d\sigma = \int_0^b \int_0^a cxy\, dx\, dy = \frac{a^2b^2c}{4}$.

Therefore, $\iint\limits_S G(x, y, z)\, d\sigma = \frac{abc(ab + ac + bc)}{4}$.

13. $f(x, y, z) = 2x + 2y + z = 2 \Rightarrow \nabla f = 2\mathbf{i} + 2\mathbf{j} + \mathbf{k}$ and $G(x, y, z) = x + y + (2 - 2x - 2y) = 2 - x - y \Rightarrow \mathbf{p} = \mathbf{k}$,

$|\nabla f| = 3$ and $|\nabla f \cdot \mathbf{p}| = 1 \Rightarrow d\sigma = 3\, dy\, dx$; $z = 0 \Rightarrow 2x + 2y = 2 \Rightarrow y = 1 - x \Rightarrow \iint\limits_S G\, d\sigma = \iint\limits_S (2 - x - y)\, d\sigma$

$= 3\int_0^1 \int_0^{1-x} (2 - x - y)\, dy\, dx = 3\int_0^1 \left[(2 - x)(1 - x) - \frac{1}{2}(1 - x)^2 \right] dx = 3\int_0^1 \left(\frac{3}{2} - 2x + \frac{x^2}{2} \right) dx = 2$

15. $f(x, y, z) = x + y^2 - z = 0 \Rightarrow \nabla f = \mathbf{i} + 2y\mathbf{j} - \mathbf{k} \Rightarrow |\nabla f| = \sqrt{4y^2 + 2} = \sqrt{2}\sqrt{2y^2 + 1}$ and

$\mathbf{p} = \mathbf{k} \Rightarrow |\nabla f \cdot \mathbf{p}| = 1 \Rightarrow d\sigma = \frac{\sqrt{2}\sqrt{2y^2+1}}{1}\, dx\, dy \Rightarrow \iint\limits_S G\, d\sigma = \int_0^1 \int_0^y \left(x + y^2 - x \right) \sqrt{2}\sqrt{2y^2 + 1}\, dx\, dy$

$= \sqrt{2}\int_0^1 \int_0^y y^2 \sqrt{2y^2 + 1}\, dx\, dy = \sqrt{2}\int_0^1 y^3 \sqrt{2y^2 + 1}\, dy = \frac{6\sqrt{6} + \sqrt{2}}{30}$

17. $f(x, y, z) = 2x + y + z = 2 \Rightarrow \nabla f = 2\mathbf{i} + \mathbf{j} + \mathbf{k} \Rightarrow |\nabla f| = \sqrt{6}$ and $\mathbf{p} = \mathbf{k} \Rightarrow |\nabla f \cdot \mathbf{p}| = 1 \Rightarrow d\sigma = \frac{\sqrt{6}}{1}\, dy\, dx$

$\Rightarrow \iint\limits_S G\, d\sigma = \int_0^1 \int_{1-x}^{2-2x} xy(2 - 2x - y)\sqrt{6}\, dy\, dx = \sqrt{6}\int_0^1 \int_{1-x}^{2-2x} \left(2xy - 2x^2 y - xy^2 \right) dy\, dx$

$= \sqrt{6}\int_0^1 \left(\frac{2}{3}x - 2x^2 + 2x^3 - \frac{2}{3}x^4 \right) dx = \frac{\sqrt{6}}{30}$

19. Let the parametrization be $\mathbf{r}(x, y) = x\mathbf{i} + y\mathbf{j} + \left(4 - y^2\right)\mathbf{k}, 0 \leq x \leq 1, -2 \leq y \leq 2; z = 0 \Rightarrow 0 = 4 - y^2$

$$\Rightarrow y = \pm 2; \mathbf{r}_x = \mathbf{i} \text{ and } \mathbf{r}_y = \mathbf{j} - 2y\mathbf{k} \Rightarrow \mathbf{r}_x \times \mathbf{r}_y = \begin{vmatrix} \mathbf{i} & \mathbf{j} & \mathbf{k} \\ 1 & 0 & 0 \\ 0 & 1 & -2y \end{vmatrix} = 2y\mathbf{j} + \mathbf{k} \Rightarrow \mathbf{F} \cdot \mathbf{n} \, d\sigma = \mathbf{F} \cdot \frac{\mathbf{r}_x \times \mathbf{r}_y}{|\mathbf{r}_x \times \mathbf{r}_y|} |\mathbf{r}_x \times \mathbf{r}_y| \, dy \, dx$$

$$= (2xy - 3z) \, dy \, dx = \left[2xy - 3\left(4 - y^2\right)\right] dy \, dx \Rightarrow \iint_S \mathbf{F} \cdot \mathbf{n} \, d\sigma = \int_0^1 \int_{-2}^2 \left(2xy + 3y^2 - 12\right) dy \, dx$$

$$= \int_0^1 \left[xy^2 + y^3 - 12y\right]_{-2}^2 dx = \int_0^1 (-32) \, dx = -32$$

21. Let the parametrization be $\mathbf{r}(\phi, \theta) = (a \sin \phi \cos \theta)\mathbf{i} + (a \sin \phi \sin \theta)\mathbf{j} + (a \cos \phi)\mathbf{k}$ (spherical coordinates with

$\rho = a, a \geq 0$, on the sphere), $0 \leq \phi \leq \frac{\pi}{2}$ (for the first octant), $0 \leq \theta \leq \frac{\pi}{2}$ (for the first octant)

$$\Rightarrow \mathbf{r}_\phi = (a \cos \phi \cos \theta)\mathbf{i} + (a \cos \phi \sin \theta)\mathbf{j} - (a \sin \phi)\mathbf{k} \text{ and } \mathbf{r}_\theta = (-a \sin \phi \sin \theta)\mathbf{i} + (a \sin \phi \cos \theta)\mathbf{j}$$

$$\Rightarrow \mathbf{r}_\phi \times \mathbf{r}_\theta = \begin{vmatrix} \mathbf{i} & \mathbf{j} & \mathbf{k} \\ a \cos \phi \cos \theta & a \cos \phi \sin \theta & -a \sin \phi \\ -a \sin \phi \sin \theta & a \sin \phi \cos \theta & 0 \end{vmatrix} = \left(a^2 \sin^2 \phi \cos \theta\right)\mathbf{i} + \left(a^2 \sin^2 \phi \sin \theta\right)\mathbf{j} + \left(a^2 \sin \phi \cos \phi\right)\mathbf{k}$$

$$\Rightarrow \mathbf{F} \cdot \mathbf{n} \, d\sigma = \mathbf{F} \cdot \frac{\mathbf{r}_\phi \times \mathbf{r}_\theta}{|\mathbf{r}_\phi \times \mathbf{r}_\theta|} |\mathbf{r}_\phi \times \mathbf{r}_\theta| \, d\theta \, d\phi = a^3 \cos^2 \phi \sin \phi \, d\theta \, d\phi \text{ since } \mathbf{F} = z\mathbf{k} = (a \cos \phi)\mathbf{k}$$

$$\Rightarrow \iint_S \mathbf{F} \cdot \mathbf{n} \, d\sigma = \int_0^{\pi/2} \int_0^{\pi/2} a^3 \cos^2 \phi \sin \phi \, d\phi \, d\theta = \frac{\pi a^3}{6}$$

23. Let the parametrization be $\mathbf{r}(x, y) = x\mathbf{i} + y\mathbf{j} + (2a - x - y)\mathbf{k}, 0 \leq x \leq a, 0 \leq y \leq a \Rightarrow \mathbf{r}_x = \mathbf{i} - \mathbf{k} \text{ and } \mathbf{r}_y = \mathbf{j} - \mathbf{k}$

$$\Rightarrow \mathbf{r}_x \times \mathbf{r}_y = \begin{vmatrix} \mathbf{i} & \mathbf{j} & \mathbf{k} \\ 1 & 0 & -1 \\ 0 & 1 & -1 \end{vmatrix} = \mathbf{i} + \mathbf{j} + \mathbf{k} \Rightarrow \mathbf{F} \cdot \mathbf{n} \, d\sigma = \mathbf{F} \cdot \frac{\mathbf{r}_x \times \mathbf{r}_y}{|\mathbf{r}_x \times \mathbf{r}_y|} |\mathbf{r}_x \times \mathbf{r}_y| \, dy \, dx$$

$$= [2xy + 2y(2a - x - y) + 2x(2a - x - y)] \, dy \, dx \text{ since } \mathbf{F} = 2xy\mathbf{i} + 2yz\mathbf{j} + 2xz\mathbf{k}$$

$$= 2xy\mathbf{i} + 2y(2a - x - y)\mathbf{j} + 2x(2a - x - y)\mathbf{k} \Rightarrow \iint_S \mathbf{F} \cdot \mathbf{n} \, d\sigma$$

$$= \int_0^a \int_0^a [2xy + 2y(2a - x - y) + 2x(2a - x - y)] \, dy \, dx = \int_0^a \int_0^a \left(4ay - 2y^2 + 4ax - 2x^2 - 2xy\right) dy \, dx$$

$$= \int_0^a \left(\tfrac{4}{3}a^3 + 3a^2 x - 2ax^2\right) dx = \left(\tfrac{4}{3} + \tfrac{3}{2} - \tfrac{2}{3}\right)a^4 = \frac{13a^4}{6}$$

25. Let the parametrization be $\mathbf{r}(r, \theta) = (r \cos \theta)\mathbf{i} + (r \sin \theta)\mathbf{j} + r\mathbf{k}, 0 \leq r \leq 1$ (since $0 \leq z \leq 1$) and $0 \leq \theta \leq 2\pi$

$$\Rightarrow \mathbf{r}_r = (\cos \theta)\mathbf{i} + (\sin \theta)\mathbf{j} + \mathbf{k} \text{ and } \mathbf{r}_\theta = (-r \sin \theta)\mathbf{i} + (r \cos \theta)\mathbf{j} \Rightarrow \mathbf{r}_\theta \times \mathbf{r}_r = \begin{vmatrix} \mathbf{i} & \mathbf{j} & \mathbf{k} \\ -r \sin \theta & r \cos \theta & 0 \\ \cos \theta & \sin \theta & 1 \end{vmatrix}$$

$$= (r \cos \theta)\mathbf{i} + (r \sin \theta)\mathbf{j} - r\mathbf{k} \Rightarrow \mathbf{F} \cdot \mathbf{n} \, d\sigma = \mathbf{F} \cdot \frac{\mathbf{r}_\theta \times \mathbf{r}_r}{|\mathbf{r}_\theta \times \mathbf{r}_r|} |\mathbf{r}_\theta \times \mathbf{r}_r| \, d\theta \, dr = \left(r^3 \sin \theta \cos^2 \theta + r^2\right) d\theta \, dr \text{ since}$$

$$\mathbf{F} = \left(r^2 \sin \theta \cos \theta\right)\mathbf{i} - r\mathbf{k} \Rightarrow \iint_S \mathbf{F} \cdot \mathbf{n} \, d\sigma = \int_0^{2\pi} \int_0^1 \left(r^3 \sin \theta \cos^2 \theta + r^2\right) dr \, d\theta = \int_0^{2\pi} \left(\tfrac{1}{4} \sin \theta \cos^2 \theta + \tfrac{1}{3}\right) d\theta$$

$$= \left[-\tfrac{1}{12} \cos^3 \theta + \tfrac{\theta}{3}\right]_0^{2\pi} = \frac{2\pi}{3}$$

27. Let the parametrization be $\mathbf{r}(r, \theta) = (r\cos\theta)\mathbf{i} + (r\sin\theta)\mathbf{j} + r\mathbf{k}, 1 \le r \le 2$ (since $1 \le z \le 2$) and $0 \le \theta \le 2\pi$

$$\Rightarrow \mathbf{r}_r = (\cos\theta)\mathbf{i} + (\sin\theta)\mathbf{j} + \mathbf{k} \text{ and } \mathbf{r}_\theta = (-r\sin\theta)\mathbf{i} + (r\cos\theta)\mathbf{j} \Rightarrow \mathbf{r}_\theta \times \mathbf{r}_r = \begin{vmatrix} \mathbf{i} & \mathbf{j} & \mathbf{k} \\ -r\sin\theta & r\cos\theta & 0 \\ \cos\theta & \sin\theta & 1 \end{vmatrix}$$

$$= (r\cos\theta)\mathbf{i} + (r\sin\theta)\mathbf{j} - r\mathbf{k} \Rightarrow \mathbf{F} \cdot \mathbf{n} \, d\sigma = \mathbf{F} \cdot \frac{\mathbf{r}_\theta \times \mathbf{r}_r}{|\mathbf{r}_\theta \times \mathbf{r}_r|} |\mathbf{r}_\theta \times \mathbf{r}_r| \, d\theta \, dr = \left(-r^2\cos^2\theta - r^2\sin^2\theta - r^3\right) d\theta \, dr$$

$$= \left(-r^2 - r^3\right) d\theta \, dr \text{ since } \mathbf{F} = (-r\cos\theta)\mathbf{i} - (r\sin\theta)\mathbf{j} + r^2\mathbf{k} \Rightarrow \iint\limits_S \mathbf{F} \cdot \mathbf{n} \, d\sigma = \int_0^{2\pi}\int_1^2 \left(-r^2 - r^3\right) dr \, d\theta = -\frac{73\pi}{6}$$

29. $g(x, y, z) = z, \mathbf{p} = \mathbf{k} \Rightarrow \nabla g = \mathbf{k} \Rightarrow |\nabla g| = 1$ and $|\nabla g \cdot \mathbf{p}| = 1 \Rightarrow \text{Flux} = \iint\limits_S \mathbf{F} \cdot \mathbf{n} \, d\sigma = \iint\limits_R (\mathbf{F} \cdot \mathbf{k}) \, dA$

$$= \int_0^2 \int_0^3 3 \, dy \, dx = 18$$

31. $\nabla g = 2x\mathbf{i} + 2y\mathbf{j} + 2z\mathbf{k} \Rightarrow |\nabla g| = \sqrt{4x^2 + 4y^2 + 4z^2} = 2a; \mathbf{n} = \frac{2x\mathbf{i} + 2y\mathbf{j} + 2z\mathbf{k}}{2\sqrt{x^2 + y^2 + z^2}} = \frac{x\mathbf{i} + y\mathbf{j} + z\mathbf{k}}{a} \Rightarrow \mathbf{F} \cdot \mathbf{n} = \frac{z^2}{a};$

$$|\nabla g \cdot \mathbf{k}| = 2z \Rightarrow d\sigma = \frac{2a}{2z} dA \Rightarrow \text{Flux} = \iint\limits_S \left(\frac{z^2}{a}\right)\left(\frac{a}{z}\right) dA = \iint\limits_S z \, dA = \iint\limits_S \sqrt{a^2 - (x^2 + y^2)} \, dx \, dy$$

$$= \int_0^{\pi/2} \int_0^a \sqrt{a^2 - r^2} \, r \, dr \, d\theta = \frac{\pi a^3}{6}$$

33. From Exercise 31, $\mathbf{n} = \frac{x\mathbf{i} + y\mathbf{j} + z\mathbf{k}}{a}$ and $d\sigma = \frac{a}{z} dA \Rightarrow \mathbf{F} \cdot \mathbf{n} = \frac{xy}{a} - \frac{xy}{a} + \frac{z}{a} = \frac{z}{a} \Rightarrow \text{Flux} = \iint\limits_R \left(\frac{z}{a}\right)\left(\frac{a}{z}\right) dA = \iint\limits_R 1 \, dA = \frac{\pi a^2}{4}$

35. From Exercise 31, $\mathbf{n} = \frac{x\mathbf{i} + y\mathbf{j} + z\mathbf{k}}{a}$ and $d\sigma = \frac{a}{z} dA \Rightarrow \mathbf{F} \cdot \mathbf{n} = \frac{x^2}{a} + \frac{y^2}{a} + \frac{z^2}{a} = a \Rightarrow \text{Flux} = \iint\limits_R a\left(\frac{a}{z}\right) dA = \iint\limits_R \frac{a^2}{z} dA$

$$= \iint\limits_R \frac{a^2}{\sqrt{a^2 - (x^2 + y^2)}} dA = \int_0^{\pi/2} \int_0^a \frac{a^2}{\sqrt{a^2 - r^2}} r \, dr \, d\theta = \int_0^{\pi/2} a^2 \left[-\sqrt{a^2 - r^2}\right]_0^a d\theta = \frac{\pi a^3}{2}$$

37. $g(x, y, z) = y^2 + z = 4 \Rightarrow \nabla g = 2y\mathbf{j} + \mathbf{k} \Rightarrow |\nabla g| = \sqrt{4y^2 + 1} \Rightarrow \mathbf{n} = \frac{2y\mathbf{j} + \mathbf{k}}{\sqrt{4y^2 + 1}} \Rightarrow \mathbf{F} \cdot \mathbf{n} = \frac{2xy - 3z}{\sqrt{4y^2 + 1}};$

$$\mathbf{p} = \mathbf{k} \Rightarrow |\nabla g \cdot \mathbf{p}| = 1 \Rightarrow d\sigma = \sqrt{4y^2 + 1} \, dA \Rightarrow \text{Flux} = \iint\limits_R \left(\frac{2xy - 3z}{\sqrt{4y^2 + 1}}\right)\sqrt{4y^2 + 1} \, dA = \iint\limits_R (2xy - 3z) \, dA;$$

$$z = 0 \text{ and } z = 4 - y^2 \Rightarrow y^2 = 4 \Rightarrow \text{Flux} = \iint\limits_R \left[2xy - 3\left(4 - y^2\right)\right] dA = \int_0^1 \int_{-2}^2 \left(2xy - 12 + 3y^2\right) dy \, dx$$

$$= \int_0^1 \left[xy^2 - 12y + y^3\right]_{-2}^2 dx = \int_0^1 (-32) \, dx = -32$$

39. $g(x, y, z) = y - e^x = 0 \Rightarrow \nabla g = -e^x\mathbf{i} + \mathbf{j} \Rightarrow |\nabla g| = \sqrt{e^{2x} + 1} \Rightarrow \mathbf{n} = \frac{e^x\mathbf{i} - \mathbf{j}}{\sqrt{e^{2x} + 1}} \Rightarrow \mathbf{F} \cdot \mathbf{n} = \frac{-2e^x - 2y}{\sqrt{e^{2x} + 1}};$

$$\mathbf{p} = \mathbf{i} \Rightarrow |\nabla g \cdot \mathbf{p}| = e^x \Rightarrow d\sigma = \frac{\sqrt{e^{2x} + 1}}{e^x} dA \Rightarrow \text{Flux} = \iint\limits_R \left(\frac{-2e^x - 2y}{\sqrt{e^{2x} + 1}}\right)\left(\frac{\sqrt{e^{2x} + 1}}{e^x}\right) dA = \iint\limits_R \frac{-2e^x - 2e^x}{e^x} dA$$

$$= \iint\limits_R (-4) \, dA = \int_0^1 \int_1^2 (-4) \, dy \, dz = -4$$

41. On the face $z = a$: $g(x, y, z) = z \Rightarrow \nabla g = \mathbf{k} \Rightarrow |\nabla g| = 1$; $\mathbf{n} = \mathbf{k} \Rightarrow \mathbf{F} \cdot \mathbf{n} = 2xz = 2ax$ since $z = a$; $d\sigma = dx\,dy$

$\Rightarrow \text{Flux} = \iint\limits_{R} 2ax\,dx\,dy = \int_0^a \int_0^a 2ax\,dx\,dy = a^4$.

On the face $z = 0$: $g(x, y, z) = z \Rightarrow \nabla g = \mathbf{k} \Rightarrow |\nabla g| = 1$; $\mathbf{n} = -\mathbf{k} \Rightarrow \mathbf{F} \cdot \mathbf{n} = -2xz = 0$ since $z = 0$;

$d\sigma = dx\,dy \Rightarrow \text{Flux} = \iint\limits_{R} 0\,dx\,dy = 0$.

On the face $x = a$: $g(x, y, z) = x \Rightarrow \nabla g = \mathbf{i} \Rightarrow |\nabla g| = 1$; $\mathbf{n} = \mathbf{i} \Rightarrow \mathbf{F} \cdot \mathbf{n} = 2xy = 2ay$ since $x = a$;

$d\sigma = dy\,dz \Rightarrow \text{Flux} = \int_0^a \int_0^a 2ay\,dy\,dz = a^4$.

On the face $x = 0$: $g(x, y, z) = x \Rightarrow \nabla g = \mathbf{i} \Rightarrow |\nabla g| = 1$; $\mathbf{n} = -\mathbf{i} \Rightarrow \mathbf{F} \cdot \mathbf{n} = -2xy = 0$ since $x = 0 \Rightarrow \text{Flux} = 0$.

On the face $y = a$: $g(x, y, z) = y \Rightarrow \nabla g = \mathbf{j} \Rightarrow |\nabla g| = 1$; $\mathbf{n} = \mathbf{j} \Rightarrow \mathbf{F} \cdot \mathbf{n} = 2yz = 2az$ since $y = a$;

$d\sigma = dz\,dx \Rightarrow \text{Flux} = \int_0^a \int_0^a 2az\,dz\,dx = a^4$.

On the face $y = 0$: $g(x, y, z) = y \Rightarrow \nabla g = \mathbf{j} \Rightarrow |\nabla g| = 1$; $\mathbf{n} = -\mathbf{j} \Rightarrow \mathbf{F} \cdot \mathbf{n} = -2yz = 0$ since $y = 0 \Rightarrow \text{Flux} = 0$.

Therefore, Total Flux $= 3a^4$.

43. $\nabla f = 2x\mathbf{i} + 2y\mathbf{j} + 2z\mathbf{k} \Rightarrow |\nabla f| = \sqrt{4x^2 + 4y^2 + 4z^2} = 2a$; $\mathbf{p} = \mathbf{k} \Rightarrow |\nabla f \cdot \mathbf{p}| = 2z$ since $z \geq 0 \Rightarrow d\sigma = \frac{2a}{2z}dA$

$= \frac{a}{z}dA$; $M = \iint\limits_{S} \delta\,d\sigma = \frac{\delta}{8}$ (surface area of sphere) $= \frac{\delta \pi a^2}{2}$; $M_{xy} = \iint\limits_{S} z\delta\,d\sigma = \delta \iint\limits_{R} z\left(\frac{a}{z}\right)dA = a\delta \iint\limits_{R} dA$

$= a\delta \int_0^{\pi/2} \int_0^a r\,dr\,d\theta = \frac{\delta \pi a^3}{4} \Rightarrow \bar{z} = \frac{M_{xy}}{M} = \left(\frac{\delta \pi a^3}{4}\right)\left(\frac{2}{\delta \pi a^2}\right) = \frac{a}{2}$. Because of symmetry, $\bar{x} = \bar{y} = \frac{a}{2}$

$\Rightarrow$ the centroid is $\left(\frac{a}{2}, \frac{a}{2}, \frac{a}{2}\right)$.

45. Because of symmetry, $\bar{x} = \bar{y} = 0$; $M = \iint\limits_{S} \delta\,d\sigma = \delta \iint\limits_{S} d\sigma = (\text{Area of } S)\delta = 3\pi\sqrt{2}\delta$; $\nabla f = 2x\mathbf{i} + 2y\mathbf{j} - 2z\mathbf{k}$

$\Rightarrow |\nabla f| = \sqrt{4x^2 + 4y^2 + 4z^2} = 2\sqrt{x^2 + y^2 + z^2}$; $\mathbf{p} = \mathbf{k} \Rightarrow |\nabla f \cdot \mathbf{p}| = 2z \Rightarrow d\sigma = \frac{2\sqrt{x^2+y^2+z^2}}{2z}dA$

$= \frac{\sqrt{x^2+y^2+\left(x^2+y^2\right)}}{z}dA = \frac{\sqrt{2}\sqrt{x^2+y^2}}{z}dA \Rightarrow M_{xy} = \delta \iint\limits_{R} z\left(\frac{\sqrt{2}\sqrt{x^2+y^2}}{z}\right)dA = \delta \iint\limits_{R} \sqrt{2}\sqrt{x^2+y^2}\,dA$

$= \delta \int_0^{2\pi} \int_1^2 \sqrt{2}r^2\,dr\,d\theta = \frac{14\pi\sqrt{2}}{3}\delta \Rightarrow \bar{z} = \frac{\left(\frac{14\pi\sqrt{2}}{3}\delta\right)}{3\pi\sqrt{2}\delta} = \frac{14}{9} \Rightarrow (\bar{x}, \bar{y}, \bar{z}) = \left(0, 0, \frac{14}{9}\right)$. Next, $I_z = \iint\limits_{S} \left(x^2 + y^2\right)\delta\,d\sigma$

$= \iint\limits_{R} (x^2 + y^2)\left(\frac{\sqrt{2}\sqrt{x^2+y^2}}{z}\right)\delta\,dA = \delta\sqrt{2} \iint\limits_{R} \left(x^2 + y^2\right)dA = \delta\sqrt{2} \int_0^{2\pi} \int_1^2 r^3\,dr\,d\theta = \frac{15\pi\sqrt{2}}{2}\delta \Rightarrow R_z = \sqrt{\frac{I_z}{M}} = \frac{\sqrt{10}}{2}$

15.7 STOKES' THEOREM

1. curl $\mathbf{F} = \nabla \times \mathbf{F} = \begin{vmatrix} \mathbf{i} & \mathbf{j} & \mathbf{k} \\ \frac{\partial}{\partial x} & \frac{\partial}{\partial y} & \frac{\partial}{\partial z} \\ x^2 & 2x & z^2 \end{vmatrix} = 0\mathbf{i} + 0\mathbf{j} + (2 - 0)\mathbf{k} = 2\mathbf{k}$ and $\mathbf{n} = \mathbf{k} \Rightarrow$ curl $\mathbf{F} \cdot \mathbf{n} = 2 \Rightarrow d\sigma = dx\,dy$

$\Rightarrow \oint_C \mathbf{F} \cdot d\mathbf{r} = \iint\limits_{R} 2\,dA = 2(\text{Area of the ellipse}) = 4\pi$

3. $\text{curl } \mathbf{F} = \nabla \times \mathbf{F} = \begin{vmatrix} \mathbf{i} & \mathbf{j} & \mathbf{k} \\ \frac{\partial}{\partial x} & \frac{\partial}{\partial y} & \frac{\partial}{\partial z} \\ y & xz & x^2 \end{vmatrix} = -x\mathbf{i} - 2x\mathbf{j} + (z-1)\mathbf{k}$ and $\mathbf{n} = \frac{\mathbf{i}+\mathbf{j}+\mathbf{k}}{\sqrt{3}} \Rightarrow \text{curl } \mathbf{F} \cdot \mathbf{n} = \frac{1}{\sqrt{3}}(-x-2x+z-1)$

$\Rightarrow d\sigma = \frac{\sqrt{3}}{1} dA \Rightarrow \oint_C \mathbf{F} \cdot d\mathbf{r} = \iint_R \frac{1}{\sqrt{3}}(-3x+z-1)\sqrt{3}\, dA = \int_0^1 \int_0^{1-x} [-3x+(1-x-y)-1]\, dy\, dx$

$= \int_0^1 \int_0^{1-x} (-4x-y)\, dy\, dx = \int_0^1 -\left[4x(1-x)+\frac{1}{2}(1-x)^2\right] dx = -\int_0^1 \left(\frac{1}{2}+3x-\frac{7}{2}x^2\right) dx = -\frac{5}{6}$

5. $\text{curl } \mathbf{F} = \nabla \times \mathbf{F} = \begin{vmatrix} \mathbf{i} & \mathbf{j} & \mathbf{k} \\ \frac{\partial}{\partial x} & \frac{\partial}{\partial y} & \frac{\partial}{\partial z} \\ y^2+z^2 & x^2+y^2 & x^2+y^2 \end{vmatrix} = 2y\mathbf{i} + (2z-2x)\mathbf{j} + (2x-2y)\mathbf{k}$ and $\mathbf{n} = \mathbf{k} \Rightarrow \text{curl } \mathbf{F} \cdot \mathbf{n} = 2x-2y$

$\Rightarrow d\sigma = dx\, dy \Rightarrow \oint_C \mathbf{F} \cdot d\mathbf{r} = \int_{-1}^1 \int_{-1}^1 (2x-2y)\, dx\, dy = \int_{-1}^1 \left[x^2-2xy\right]_{-1}^1 dy = \int_{-1}^1 -4y\, dy = 0$

7. $x = 3\cos t$ and $y = 2\sin t \Rightarrow \mathbf{F} = (2\sin t)\mathbf{i} + \left(9\cos^2 t\right)\mathbf{j} + \left(9\cos^2 t + 16\sin^4 t\right)\sin e^{\sqrt{(6\sin t\cos t)(0)}}\mathbf{k}$ at the base of

the shell; $\mathbf{r} = (3\cos t)\mathbf{i} + (2\sin t)\mathbf{j} \Rightarrow d\mathbf{r} = (-3\sin t)\mathbf{i} + (2\cos t)\mathbf{j} \Rightarrow \mathbf{F} \cdot \frac{d\mathbf{r}}{dt} = -6\sin^2 t + 18\cos^3 t$

$\Rightarrow \iint_S \nabla \times \mathbf{F} \cdot \mathbf{n}\, d\sigma = \int_0^{2\pi} \left(-6\sin^2 t + 18\cos^3 t\right) dt = \left[-3t + \frac{3}{2}\sin 2t + 6(\sin t)\left(\cos^2 t + 2\right)\right]_0^{2\pi} = -6\pi$

9. Flux of $\nabla \times \mathbf{F} = \iint_S \nabla \times \mathbf{F} \cdot \mathbf{n}\, d\sigma = \oint_C \mathbf{F} \cdot d\mathbf{r}$, so let C be parametrized by $\mathbf{r} = (a\cos t)\mathbf{i} + (a\sin t)\mathbf{j}, 0 \le t \le 2\pi$

$\Rightarrow \frac{d\mathbf{r}}{dt} = (-a\sin t)\mathbf{i} + (a\cos t)\mathbf{j} \Rightarrow \mathbf{F} \cdot \frac{d\mathbf{r}}{dt} = ay\sin t + ax\cos t = a^2\sin^2 t + a^2\cos^2 t = a^2$

$\Rightarrow$ Flux of $\nabla \times \mathbf{F} = \oint_C \mathbf{F} \cdot d\mathbf{r} = \int_0^{2\pi} a^2\, dt = 2\pi a^2$

11. For the upper hemisphere with $z \ge 0$, the boundary C is the unit circle of radius 1 centered at the origin in the xy-plane. An outward normal on the upper hemisphere corresponds to counterclockwise circulation around the boundary, so the boundary can be parametrized as $\mathbf{r}(\theta) = (\cos\theta)\mathbf{i} + (\sin\theta)\mathbf{j} + 0\mathbf{k}$, with $0 \le \theta \le 2\pi$. Thus

$d\mathbf{r} = (-\sin\theta\, d\theta)\mathbf{i} + (\cos\theta\, d\theta)\mathbf{j}$. For the field $\mathbf{A} = (y + \sqrt{z})\mathbf{i} + e^{xyz}\mathbf{j} + (\cos xz)\mathbf{k}$, the flux of $\mathbf{F} = \nabla \times \mathbf{A}$ across the upper hemisphere is, by Stokes' Theorem, equal to the circulation of $\mathbf{A}$ on the boundary. Since $z = 0$ and $y = \sin\theta$ on the boundary, the field $\mathbf{A}$ on the boundary is $(\sin\theta)\mathbf{i} + \mathbf{j} + \mathbf{k}$. The circulation of $\mathbf{A}$ on C is

$\oint_C \mathbf{A} \cdot d\mathbf{r} = \oint_C ((\sin\theta)\mathbf{i} + \mathbf{j} + \mathbf{k}) \cdot ((-\sin\theta\, d\theta)\mathbf{i} + (\cos\theta\, d\theta)\mathbf{j}) = \int_0^{2\pi} \left(\cos\theta - \sin^2\theta\right) d\theta$

$= \int_0^{2\pi} \left(\cos\theta + \frac{1}{2}(\cos 2\theta - 1)\right) d\theta = -\pi$

13. $\nabla \times \mathbf{F} = \begin{vmatrix} \mathbf{i} & \mathbf{j} & \mathbf{k} \\ \frac{\partial}{\partial x} & \frac{\partial}{\partial y} & \frac{\partial}{\partial z} \\ 2z & 3x & 5y \end{vmatrix} = 5\mathbf{i} + 2\mathbf{j} + 3\mathbf{k};\ \mathbf{r}_r = (\cos\theta)\mathbf{i} + (\sin\theta)\mathbf{j} - 2r\mathbf{k}$ and $\mathbf{r}_\theta = (-r\sin\theta)\mathbf{i} + (r\cos\theta)\mathbf{j}$

$\Rightarrow \mathbf{r}_r \times \mathbf{r}_\theta = \begin{vmatrix} \mathbf{i} & \mathbf{j} & \mathbf{k} \\ \cos\theta & \sin\theta & -2r \\ -r\sin\theta & r\cos\theta & 0 \end{vmatrix} = \left(2r^2\cos\theta\right)\mathbf{i} + \left(2r^2\sin\theta\right)\mathbf{j} + r\mathbf{k};\ \mathbf{n} = \frac{\mathbf{r}_r \times \mathbf{r}_\theta}{|\mathbf{r}_r \times \mathbf{r}_\theta|}$ and $d\sigma = |\mathbf{r}_r \times \mathbf{r}_\theta|\, dr\, d\theta$

$\Rightarrow \nabla \times \mathbf{F} \cdot \mathbf{n}\, d\sigma = (\nabla \times \mathbf{F}) \cdot (\mathbf{r}_r \times \mathbf{r}_\theta)\, dr\, d\theta = \left(10r^2\cos\theta + 4r^2\sin\theta + 3r\right)dr\, d\theta \Rightarrow \iint\limits_S \nabla \times \mathbf{F} \cdot \mathbf{n}\, d\sigma$

$= \int_0^{2\pi}\int_0^2 \left(10r^2\cos\theta + 4r^2\sin\theta + 3r\right)dr\, d\theta = \int_0^{2\pi}\left[\frac{10}{3}r^3\cos\theta + \frac{4}{3}r^3\sin\theta + \frac{3}{2}r^2\right]_0^2 d\theta$

$= \int_0^{2\pi}\left(\frac{80}{3}\cos\theta + \frac{32}{3}\sin\theta + 6\right)d\theta = 6(2\pi) = 12\pi$

15. $\nabla \times \mathbf{F} = \begin{vmatrix} \mathbf{i} & \mathbf{j} & \mathbf{k} \\ \frac{\partial}{\partial x} & \frac{\partial}{\partial y} & \frac{\partial}{\partial z} \\ x^2 y & 2y^3 z & 3z \end{vmatrix} = -2y^3\mathbf{i} + 0\mathbf{j} - x^2\mathbf{k};\ \mathbf{r}_r \times \mathbf{r}_\theta = \begin{vmatrix} \mathbf{i} & \mathbf{j} & \mathbf{k} \\ \cos\theta & \sin\theta & 1 \\ -r\sin\theta & r\cos\theta & 0 \end{vmatrix} = (-r\cos\theta)\mathbf{i} - (r\sin\theta)\mathbf{j} + r\mathbf{k}$ and

$\nabla \times \mathbf{F} \cdot \mathbf{n}\, d\sigma = (\nabla \times \mathbf{F}) \cdot (\mathbf{r}_r \times \mathbf{r}_\theta)\, dr\, d\theta$ (see Exercise 13 above) $\Rightarrow \iint\limits_S \nabla \times \mathbf{F} \cdot \mathbf{n}\, d\sigma = \iint\limits_R \left(2ry^3\cos\theta - rx^2\right)dr\, d\theta$

$= \int_0^{2\pi}\int_0^1 \left(2r^4\sin^3\theta\cos\theta - r^3\cos^2\theta\right)dr\, d\theta = \int_0^{2\pi}\left(\frac{2}{5}\sin^3\theta\cos\theta - \frac{1}{4}\cos^2\theta\right)d\theta = \left[\frac{1}{10}\sin^4\theta - \frac{1}{4}\left(\frac{\theta}{2} + \frac{\sin 2\theta}{4}\right)\right]_0^{2\pi}$

$= -\frac{\pi}{4}$

17. $\nabla \times \mathbf{F} = \begin{vmatrix} \mathbf{i} & \mathbf{j} & \mathbf{k} \\ \frac{\partial}{\partial x} & \frac{\partial}{\partial y} & \frac{\partial}{\partial z} \\ 3y & 5-2x & z^2-2 \end{vmatrix} = 0\mathbf{i} + 0\mathbf{j} - 5\mathbf{k};\ \mathbf{r}_\phi \times \mathbf{r}_\theta = \begin{vmatrix} \mathbf{i} & \mathbf{j} & \mathbf{k} \\ \sqrt{3}\cos\phi\cos\theta & \sqrt{3}\cos\phi\sin\theta & -\sqrt{3}\sin\phi \\ -\sqrt{3}\sin\phi\sin\theta & \sqrt{3}\sin\phi\cos\theta & 0 \end{vmatrix}$

$= \left(3\sin^2\phi\cos\theta\right)\mathbf{i} + \left(3\sin^2\phi\sin\theta\right)\mathbf{j} + (3\sin\phi\cos\phi)\mathbf{k};\ \nabla \times \mathbf{F} \cdot \mathbf{n}\, d\sigma = (\nabla \times \mathbf{F}) \cdot (\mathbf{r}_\phi \times \mathbf{r}_\theta)\, d\phi\, d\theta$ (see Exercise 13

above) $\Rightarrow \iint\limits_S \nabla \times \mathbf{F} \cdot \mathbf{n}\, d\sigma = \int_0^{2\pi}\int_0^{\pi/2} (-15\cos\phi\sin\phi)\, d\phi\, d\theta = \int_0^{2\pi}\left[\frac{15}{2}\cos^2\phi\right]_0^{\pi/2} d\theta = \int_0^{2\pi} -\frac{15}{2}\, d\theta = -15\pi$

19. We first compute the circulation of $\mathbf{F} = y\mathbf{i} - x\mathbf{j} + x^2\mathbf{k}$ on the curve C given by

$\mathbf{r}(t) = (2\cos t)\mathbf{i} + (2\sin t)\mathbf{j} + (3 - 2\cos^3 t)\mathbf{k}$ for $0 \le t \le 2\pi$. On C $\mathbf{F} = (2\sin t)\mathbf{i} - (2\cos t)\mathbf{j} + (4\cos^2 t)\mathbf{k}$, and

$d\mathbf{r} = (-2\sin t\, dt)\mathbf{i} + (2\cos t\, dt)\mathbf{j} - (6\sin t\cos^2 t\, dt)\mathbf{k}$.

$\oint_C \mathbf{F} \cdot d\mathbf{r} = \oint_C ((2\sin t)\mathbf{i} - (2\cos t)\mathbf{j} + (4\cos^2 t)\mathbf{k}) \cdot ((-2\sin t\, dt)\mathbf{i} + (2\cos t\, dt)\mathbf{j} - (6\sin t\cos^2 t\, dt)\mathbf{k})$

$= -4\int_0^{2\pi}(\sin^2 t + \cos^2 t + 6\sin t\cos^4 t)\, dt = -4\int_0^{2\pi}(1 + 6\sin t\cos^4 t)\, dt = -8\pi.$

Now we find the flux of $\nabla \times \mathbf{F}$ across the surface S. Note that counterclockwise circulation on C corresponds to inward normals on the cylindrical portion of S and upward normals on the base disk.

For the field $\mathbf{F} = y\mathbf{i} - x\mathbf{j} + x^2\mathbf{k}$, $\nabla \times \mathbf{F} = (-2x)\mathbf{j} - 2\mathbf{k}$.

On the base disk, the unit upward normal is $\mathbf{k}$ so $\nabla \times \mathbf{F} \cdot \mathbf{n} = ((-2x)\mathbf{j} - 2\mathbf{k}) \cdot \mathbf{k} = -2$. The integral of the constant -2 over a disk of area 4π is -8π, so to verify Stokes' Theorem in this case it remains to show that the flux across the cylindrical portion of S is 0.

We'll reuse the parameter t and parametrize the cylinder by $\mathbf{s}(t,z) = (2\cos t)\mathbf{i} + (2\sin t)\mathbf{j} + z\mathbf{k}$ with

$0 \le z \le 3 - 2\cos^3 t$. An inward unit normal is $(-\cos t)\mathbf{i} + (-\sin t)\mathbf{j}$ and the area element is $d\sigma = 2\,dz\,dt$. On the cylinder the field $\nabla \times \mathbf{F} = (-4\cos t)\mathbf{j} - 2\mathbf{k}$. Thus

$$\iint_S \nabla \times \mathbf{F} \cdot \mathbf{n}\,d\sigma = \iint_S ((-4\cos t)\mathbf{j} - 2\mathbf{k}) \cdot ((-\cos t)\mathbf{i} + (-\sin t)\mathbf{j})\,d\sigma$$

$$= \int_0^{2\pi} \int_0^{3-2\cos^3 t} (4\sin t \cos t)\,2\,dz\,dt = 8 \int_0^{2\pi} (\sin t \cos t)(3 - 2\cos^3 t)\,dt$$

$$= 8 \int_0^{2\pi} (3\sin t \cos t - 2\sin t \cos^4 t)\,dt = 8 \left[\frac{3}{2}\sin^2 t + \frac{2}{5}\cos^5 t \right]_0^{2\pi} = 0$$

21. (a) $\mathbf{F} = 2x\mathbf{i} + 2y\mathbf{j} + 2z\mathbf{k} \Rightarrow \text{curl } \mathbf{F} = \mathbf{0} \Rightarrow \oint_C \mathbf{F} \cdot d\mathbf{r} = \iint_S \nabla \times \mathbf{F} \cdot \mathbf{n}\,d\sigma = \iint_S 0\,d\sigma = 0$

   (b) Let $f(x, y, z) = x^2 y^2 z^3 \Rightarrow \nabla \times \mathbf{F} = \nabla \times \nabla f = \mathbf{0} \Rightarrow \text{curl } \mathbf{F} = \mathbf{0} \Rightarrow \oint_C \mathbf{F} \cdot d\mathbf{r} = \iint_S \nabla \times \mathbf{F} \cdot \mathbf{n}\,d\sigma = \iint_S 0\,d\sigma = 0$

   (c) $\mathbf{F} = \nabla \times (x\mathbf{i} + y\mathbf{j} + z\mathbf{k}) = \mathbf{0} \Rightarrow \nabla \times \mathbf{F} = \mathbf{0} \Rightarrow \oint_C \mathbf{F} \cdot d\mathbf{r} = \iint_S \nabla \times \mathbf{F} \cdot \mathbf{n}\,d\sigma = \iint_S 0\,d\sigma = 0$

   (d) $\mathbf{F} = \nabla f \Rightarrow \nabla \times \mathbf{F} = \nabla \times \nabla f = \mathbf{0} \Rightarrow \oint_C \mathbf{F} \cdot d\mathbf{r} = \iint_S \nabla \times \mathbf{F} \cdot \mathbf{n}\,d\sigma = \iint_S 0\,d\sigma = 0$

23. Let $\mathbf{F} = 2y\mathbf{i} + 3z\mathbf{j} - x\mathbf{k} \Rightarrow \nabla \times \mathbf{F} = \begin{vmatrix} \mathbf{i} & \mathbf{j} & \mathbf{k} \\ \frac{\partial}{\partial x} & \frac{\partial}{\partial y} & \frac{\partial}{\partial z} \\ 2y & 3z & -x \end{vmatrix} = -3\mathbf{i} + \mathbf{j} - 2\mathbf{k}; \; \mathbf{n} = \frac{2\mathbf{i} + 2\mathbf{j} + \mathbf{k}}{3} \Rightarrow \nabla \times \mathbf{F} \cdot \mathbf{n} = -2$

   $\Rightarrow \oint_C 2y\,dx + 3z\,dy - x\,dz = \oint_C \mathbf{F} \cdot d\mathbf{r} = \iint_S \nabla \times \mathbf{F} \cdot \mathbf{n}\,d\sigma = \iint_S -2\,d\sigma = -2\iint_S d\sigma$, where $\iint_S d\sigma$ is the area of the region enclosed by C on the plane S: $2x + 2y + z = 2$

25. Suppose $\mathbf{F} = M\mathbf{i} + N\mathbf{j} + P\mathbf{k}$ exists such that $\nabla \times \mathbf{F} = \left(\frac{\partial P}{\partial y} - \frac{\partial N}{\partial z} \right)\mathbf{i} + \left(\frac{\partial M}{\partial z} - \frac{\partial P}{\partial x} \right)\mathbf{j} + \left(\frac{\partial N}{\partial x} - \frac{\partial M}{\partial y} \right)\mathbf{k} = x\mathbf{i} + y\mathbf{j} + z\mathbf{k}$.

   Then $\frac{\partial}{\partial x}\left(\frac{\partial P}{\partial y} - \frac{\partial N}{\partial z} \right) = \frac{\partial}{\partial x}(x) \Rightarrow \frac{\partial^2 P}{\partial x \partial y} - \frac{\partial^2 N}{\partial x \partial y} = 1$.

   Likewise, $\frac{\partial}{\partial y}\left(\frac{\partial M}{\partial z} - \frac{\partial P}{\partial x} \right) = \frac{\partial}{\partial y}(y) \Rightarrow \frac{\partial^2 M}{\partial y \partial z} - \frac{\partial^2 P}{\partial y \partial x} = 1$ and $\frac{\partial}{\partial z}\left(\frac{\partial N}{\partial x} - \frac{\partial M}{\partial y} \right) = \frac{\partial}{\partial z}(z) \Rightarrow \frac{\partial^2 N}{\partial z \partial x} - \frac{\partial^2 M}{\partial z \partial y} = 1$.

   Summing the calculated equations $\Rightarrow \left(\frac{\partial^2 P}{\partial x \partial y} - \frac{\partial^2 P}{\partial y \partial x} \right) + \left(\frac{\partial^2 N}{\partial z \partial x} - \frac{\partial^2 N}{\partial x \partial z} \right) + \left(\frac{\partial^2 M}{\partial y \partial z} - \frac{\partial^2 M}{\partial z \partial y} \right) = 3$ or $0 = 3$

   (assuming the second mixed partials are equal). This result is a contradiction, so there is no field $\mathbf{F}$ such that curl $\mathbf{F} = x\mathbf{i} + y\mathbf{j} + z\mathbf{k}$.

27. $r = \sqrt{x^2 + y^2} \Rightarrow r^4 = \left(x^2 + y^2 \right)^2 \Rightarrow \mathbf{F} = \nabla\left(r^4 \right) = 4x\left(x^2 + y^2 \right)\mathbf{i} + 4y\left(x^2 + y^2 \right)\mathbf{j} = M\mathbf{i} + N\mathbf{j}$

   $\Rightarrow \oint_C \nabla\left(r^4 \right) \cdot \mathbf{n}\,ds = \oint_C \mathbf{F} \cdot \mathbf{n}\,ds = \oint_C M\,dy - N\,dx = \iint_R \left(\frac{\partial M}{\partial x} + \frac{\partial N}{\partial y} \right)\,dx\,dy$

   $= \iint_R \left[4\left(x^2 + y^2 \right) + 8x^2 + 4\left(x^2 + y^2 \right) + 8y^2 \right]dA = \iint_R 16\left(x^2 + y^2 \right)dA = 16\iint_R x^2\,dA + 16 \iint_R y^2\,dA$

   $= 16 I_y + 16 I_x$.

15.8 THE DIVERGENCE THEOREM AND A UNIFIED THEORY

1. $\mathbf{F} = \dfrac{-y\mathbf{i}+x\mathbf{j}}{\sqrt{x^2+y^2}} \Rightarrow \operatorname{div}\mathbf{F} = \dfrac{xy-xy}{(x^2+y^2)^{3/2}} = 0$

3. $\mathbf{F} = -\dfrac{GM\,(x\mathbf{i}+y\mathbf{j}+z\mathbf{k})}{(x^2+y^2+z^2)^{3/2}}$

$\Rightarrow \operatorname{div}\mathbf{F} = -GM\left[\dfrac{\left(x^2+y^2+z^2\right)^{3/2}-3x^2\left(x^2+y^2+z^2\right)^{1/2}}{\left(x^2+y^2+z^2\right)^3}\right] -GM\left[\dfrac{\left(x^2+y^2+z^2\right)^{3/2}-3y^2\left(x^2+y^2+z^2\right)^{1/2}}{\left(x^2+y^2+z^2\right)^3}\right]$

$\qquad -GM\left[\dfrac{\left(x^2+y^2+z^2\right)^{3/2}-3z^2\left(x^2+y^2+z^2\right)^{1/2}}{\left(x^2+y^2+z^2\right)^3}\right] = -GM\left[\dfrac{3\left(x^2+y^2+z^2\right)^2-3\left(x^2+y^2+z^2\right)\left(x^2+y^2+z^2\right)}{\left(x^2+y^2+z^2\right)^{7/2}}\right] = 0$

5. $\dfrac{\partial}{\partial x}(y-x) = -1, \dfrac{\partial}{\partial y}(z-y) = -1, \dfrac{\partial}{\partial z}(y-x) = 0 \Rightarrow \nabla\cdot\mathbf{F} = -2 \Rightarrow \text{Flux} = \int_{-1}^{1}\int_{-1}^{1}\int_{-1}^{1} -2\,dx\,dy\,dz = -2\left(2^3\right) = -16$

7. $\dfrac{\partial}{\partial x}(y) = 0, \dfrac{\partial}{\partial y}(xy) = x, \dfrac{\partial}{\partial z}(-z) = -1 \Rightarrow \nabla\cdot\mathbf{F} = x-1;\ z = x^2+y^2 \Rightarrow z = r^2$ in cylindrical coordinates

$\Rightarrow \text{Flux} = \iiint_D (x-1)dz\,dy\,dx = \int_0^{2\pi}\int_0^2\int_0^{r^2}(r\cos\theta-1)\,dz\,r\,dr\,d\theta = \int_0^{2\pi}\int_0^2\left(r^3\cos\theta-r^2\right)r\,dr\,d\theta$

$= \int_0^{2\pi}\left[\dfrac{r^5}{5}\cos\theta-\dfrac{r^4}{4}\right]_0^2 d\theta = \int_0^{2\pi}\left(\dfrac{32}{5}\cos\theta-4\right)d\theta = \left[\dfrac{32}{5}\sin\theta-4\theta\right]_0^{2\pi} = -8\pi$

9. $\dfrac{\partial}{\partial x}\left(x^2\right) = 2x, \dfrac{\partial}{\partial y}(-2xy) = -2x, \dfrac{\partial}{\partial z}(3xz) = 3x \Rightarrow \text{Flux} = \iiint_D 3x\,dx\,dy\,dz$

$= \int_0^{\pi/2}\int_0^{\pi/2}\int_0^2(3\rho\sin\phi\cos\theta)\left(\rho^2\sin\phi\right)d\rho\,d\phi\,d\theta = \int_0^{\pi/2}\int_0^{\pi/2}12\sin^2\phi\cos\theta\,d\phi\,d\theta = \int_0^{\pi/2}3\pi\cos\theta\,d\theta = 3\pi$

11. $\dfrac{\partial}{\partial x}(2xz) = 2z, \dfrac{\partial}{\partial y}(-xy) = -x, \dfrac{\partial}{\partial z}\left(-z^2\right) = -2z \Rightarrow \nabla\cdot\mathbf{F} = -x \Rightarrow \text{Flux} = \iiint_D -x\,dV$

$= \int_0^2\int_0^{\sqrt{16-4x^2}}\int_0^{4-y}(-x)\,dz\,dy\,dx = \int_0^2\int_0^{\sqrt{16-4x^2}}(xy-4x)\,dy\,dx = \int_0^2\left[\dfrac{1}{2}x\left(16-4x^2\right)-4x\sqrt{16-4x^2}\right]dx$

$= \left[4x^2-\dfrac{1}{2}x^4+\dfrac{1}{3}\left(16-4x^2\right)^{3/2}\right]_0^2 = -\dfrac{40}{3}$

13. Let $\rho = \sqrt{x^2+y^2+z^2}$. Then $\dfrac{\partial p}{\partial x} = \dfrac{x}{\rho}, \dfrac{\partial\rho}{\partial y} = \dfrac{y}{\rho}, \dfrac{\partial p}{\partial z} = \dfrac{z}{\rho} \Rightarrow \dfrac{\partial}{\partial x}(\rho x) = \left(\dfrac{\partial\rho}{\partial x}\right)x+\rho = \dfrac{x^2}{\rho}+\rho,$

$\dfrac{\partial}{\partial y}(\rho y) = \left(\dfrac{\partial\rho}{\partial y}\right)y+\rho = \dfrac{y^2}{\rho}+\rho, \dfrac{\partial}{\partial z}(\rho z) = \left(\dfrac{\partial\rho}{\partial z}\right)z+\rho = \dfrac{z^2}{\rho}+\rho \Rightarrow \nabla\cdot\mathbf{F} = \dfrac{x^2+y^2+z^2}{\rho}+3\rho = 4\rho,$ since

$\rho = \sqrt{x^2+y^2+z^2} \Rightarrow \text{Flux} = \iiint_D 4\rho\,dV = \int_0^{2\pi}\int_0^{\pi}\int_1^{\sqrt{2}}(4\rho)\left(\rho^2\sin\phi\right)d\rho\,d\phi\,d\theta = \int_0^{2\pi}\int_0^{\pi}3\sin\phi\,d\phi\,d\theta$

$= \int_0^{2\pi}6\,d\theta = 12\pi$

15. $\frac{\partial}{\partial x}\left(5x^3+12xy^2\right)=15x^2+12y^2, \frac{\partial}{\partial y}\left(y^3+e^y\sin z\right)=3y^2+e^y\sin z, \frac{\partial}{\partial z}\left(5z^3+e^y\cos z\right)=15z^2-e^y\sin z$

$\Rightarrow \nabla\cdot\mathbf{F}=15x^2+15y^2+15z^2=15\rho^2 \Rightarrow \text{Flux}=\iiint_D 15\rho^2 dV=\int_0^{2\pi}\int_0^{\pi}\int_1^{\sqrt{2}}\left(15\rho^2\right)\left(\rho^2\sin\phi\right)d\rho\,d\phi\,d\theta$

$=\int_0^{2\pi}\int_0^{\pi}\left(12\sqrt{2}-3\right)\sin\phi\,d\phi\,d\theta=\int_0^{2\pi}\left(24\sqrt{2}-6\right)d\theta=\left(48\sqrt{2}-12\right)\pi$

17. (a) $\frac{\partial}{\partial x}(x)=1, \frac{\partial}{\partial y}(y)=1, \frac{\partial}{\partial z}(z)=1 \Rightarrow \nabla\cdot\mathbf{F}=3 \Rightarrow \text{Flux}=\iiint_D 3\,dV=3\iiint_D dV=3$ (Volume of the solid)

(b) If $\mathbf{F}$ is orthogonal to $\mathbf{n}$ at every point of S, then $\mathbf{F}\cdot\mathbf{n}=0$ everywhere $\Rightarrow \text{Flux}=\iint_S \mathbf{F}\cdot\mathbf{n}\,d\sigma=0$. But the

flux is 3 (Volume of the solid) $\neq 0$, so $\mathbf{F}$ is not orthogonal to $\mathbf{n}$ at every point.

19. For the field $\mathbf{F}=(y\cos 2x)\mathbf{i}+(y^2\sin 2x)\mathbf{j}+(x^2y+x)\mathbf{k}$, $\nabla\cdot\mathbf{F}=-2y\sin 2x+2y\sin 2x+1=1$. If $\mathbf{F}$ were the curl of a field $\mathbf{A}$ whose component functions have continuous second partial derivatives, then we would have div $\mathbf{F}=\text{div(curl }\mathbf{A})=\nabla\cdot(\nabla\times\mathbf{A})=0$. Since div $\mathbf{F}=1$, $\mathbf{F}$ is not the curl of such a field.

21. The integral's value never exceeds the surface area of S. Since $|\mathbf{F}|\le 1$, we have $|\mathbf{F}\cdot\mathbf{n}|=|\mathbf{F}||\mathbf{n}|\le (1)(1)=1$ and

$\iiint_D \nabla\cdot\mathbf{F}\,d\sigma=\iint_S \mathbf{F}\cdot\mathbf{n}\,d\sigma$ [Divergence Theorem]

$\le \iint_S |\mathbf{F}\cdot\mathbf{n}|\,d\sigma$ [A property of integrals]

$\le \iint_S (1)d\sigma$ $\left[|\mathbf{F}\cdot\mathbf{n}|\le 1\right]$

$= \text{Area of } S.$

23. By the Divergence Theorem, the net outward flux of the field $\mathbf{F}=xy\mathbf{i}+(\sin xz+y^2)\mathbf{j}+(e^{xy^2}+x)\mathbf{k}$ over the surface S will be equal to the integral of $\nabla\cdot\mathbf{F}=y+2y=3y$ over the region D bounded by S. We will

integrate using the area in the zx-plane bounded by $z=0$ and $z=1-x^2$ as the base. The y height at any point (x,z) will be $2-z$. Thus the integral of div $\mathbf{F}$ over D is

$\int_{-1}^{1}\int_0^{1-x^2}\int_0^{2-z} 3y\,dy\,dz\,dx=\int_{-1}^{1}\int_0^{1-x^2}\frac{3}{2}y^2\Big|_0^{2-z}\,dz\,dx=\int_{-1}^{1}\int_0^{1-x^2}\frac{3}{2}(2-z)^2\,dz\,dx$

$=\int_{-1}^{1}-\frac{1}{2}(2-z)^3\Big|_0^{1-x^2}\,dx=\int_{-1}^{1}4-\frac{1}{2}(x^2+1)^3\,dx=\frac{7}{2}x-\frac{1}{3}x^3-\frac{3}{10}x^5-\frac{1}{14}x^7\Big|_{-1}^{1}=\frac{184}{35}$

25. (a) $\text{div}(g\mathbf{F})=\nabla\cdot g\mathbf{F}=\frac{\partial}{\partial x}(gM)+\frac{\partial}{\partial y}(gN)+\frac{\partial}{\partial z}(gP)=\left(g\frac{\partial M}{\partial x}+M\frac{\partial g}{\partial x}\right)+\left(g\frac{\partial N}{\partial y}+N\frac{\partial g}{\partial y}\right)+\left(g\frac{\partial P}{\partial z}+P\frac{\partial g}{\partial z}\right)$

$=\left(M\frac{\partial g}{\partial x}+N\frac{\partial g}{\partial y}+P\frac{\partial g}{\partial z}\right)+g\left(\frac{\partial M}{\partial x}+\frac{\partial N}{\partial y}+\frac{\partial P}{\partial z}\right)=g\nabla\cdot\mathbf{F}+\nabla g\cdot\mathbf{F}$

(b) $\nabla\times(g\mathbf{F})=\left[\frac{\partial}{\partial y}(gP)-\frac{\partial}{\partial z}(gN)\right]\mathbf{i}+\left[\frac{\partial}{\partial z}(gM)-\frac{\partial}{\partial x}(gP)\right]\mathbf{j}+\left[\frac{\partial}{\partial x}(gN)-\frac{\partial}{\partial y}(gM)\right]\mathbf{k}$

$=\left(P\frac{\partial g}{\partial y}+g\frac{\partial P}{\partial y}-N\frac{\partial g}{\partial z}-g\frac{\partial N}{\partial z}\right)\mathbf{i}+\left(M\frac{\partial g}{\partial z}+g\frac{\partial M}{\partial z}-P\frac{\partial g}{\partial x}-g\frac{\partial P}{\partial x}\right)\mathbf{j}+\left(N\frac{\partial g}{\partial x}+g\frac{\partial N}{\partial x}-M\frac{\partial g}{\partial y}-g\frac{\partial M}{\partial y}\right)\mathbf{k}$

$=\left(P\frac{\partial g}{\partial y}-N\frac{\partial g}{\partial z}\right)\mathbf{i}+\left(g\frac{\partial P}{\partial y}-g\frac{\partial N}{\partial z}\right)\mathbf{i}+\left(M\frac{\partial g}{\partial z}-P\frac{\partial g}{\partial x}\right)\mathbf{j}+\left(g\frac{\partial M}{\partial z}-g\frac{\partial P}{\partial x}\right)\mathbf{j}+\left(N\frac{\partial g}{\partial x}-M\frac{\partial g}{\partial y}\right)\mathbf{k}+\left(g\frac{\partial N}{\partial x}-g\frac{\partial M}{\partial y}\right)\mathbf{k}$

$=g\nabla\times\mathbf{F}+\nabla g\times\mathbf{F}$

27. Let $\mathbf{F}_1=M_1\mathbf{i}+N_1\mathbf{j}+P_1\mathbf{k}$ and $\mathbf{F}_2=M_2\mathbf{i}+N_2\mathbf{j}+P_2\mathbf{k}$.

(a) $\mathbf{F}_1\times\mathbf{F}_2=(N_1P_2-P_1N_2)\mathbf{i}+(P_1M_2-M_1P_2)\mathbf{j}+(M_1N_2-N_1M_2)\mathbf{k}$

$\Rightarrow\nabla\times(\mathbf{F}_1\times\mathbf{F}_2)=\left[\frac{\partial}{\partial y}(M_1N_2-N_1M_2)-\frac{\partial}{\partial z}(P_1M_2-M_1P_2)\right]\mathbf{i}+\left[\frac{\partial}{\partial z}(N_1P_2-P_1N_2)-\frac{\partial}{\partial x}(M_1N_2-N_1M_2)\right]\mathbf{j}$

$+\left[\frac{\partial}{\partial x}(P_1M_2-M_1P_2)-\frac{\partial}{\partial y}(N_1P_2-P_1N_2)\right]\mathbf{k}$

consider the **i**-component only: $\frac{\partial}{\partial y}(M_1N_2-N_1M_2)-\frac{\partial}{\partial z}(P_1M_2-M_1P_2)$

$=N_2\frac{\partial M_1}{\partial y}+M_1\frac{\partial N_2}{\partial y}-M_2\frac{\partial N_1}{\partial y}-N_1\frac{\partial M_2}{\partial y}-M_2\frac{\partial P_1}{\partial z}-P_1\frac{\partial M_2}{\partial z}+P_2\frac{\partial M_1}{\partial z}+M_1\frac{\partial P_2}{\partial z}$

$=\left(N_2\frac{\partial M_1}{\partial y}+P_2\frac{\partial M_1}{\partial z}\right)-\left(N_1\frac{\partial M_2}{\partial y}+P_1\frac{\partial M_2}{\partial z}\right)+\left(\frac{\partial N_2}{\partial y}+\frac{\partial P_2}{\partial z}\right)M_1-\left(\frac{\partial N_1}{\partial y}+\frac{\partial P_1}{\partial z}\right)M_2$

$=\left(M_2\frac{\partial M_1}{\partial x}+N_2\frac{\partial M_1}{\partial y}+P_2\frac{\partial M_1}{\partial z}\right)-\left(M_1\frac{\partial M_2}{\partial x}+N_1\frac{\partial M_2}{\partial y}+P_1\frac{\partial M_2}{\partial z}\right)+\left(\frac{\partial M_2}{\partial x}+\frac{\partial N_2}{\partial y}+\frac{\partial P_2}{\partial z}\right)M_1-\left(\frac{\partial M_1}{\partial x}+\frac{\partial N_1}{\partial y}+\frac{\partial P_1}{\partial z}\right)M_2$

Now, **i**-comp of $(\mathbf{F}_2\cdot\nabla)\mathbf{F}_1=\left(M_2\frac{\partial}{\partial x}+N_2\frac{\partial}{\partial y}+P_2\frac{\partial}{\partial z}\right)M_1=\left(M_2\frac{\partial M_1}{\partial x}+N_2\frac{\partial M_1}{\partial y}+P_2\frac{\partial M_1}{\partial z}\right)$;

likewise, **i**-comp of $(\mathbf{F}_1\cdot\nabla)\mathbf{F}_2=\left(M_1\frac{\partial M_2}{\partial x}+N_1\frac{\partial M_2}{\partial y}+P_1\frac{\partial M_2}{\partial z}\right)$;

i comp of $(\nabla\cdot\mathbf{F}_2)\mathbf{F}_1=\left(\frac{\partial M_2}{\partial x}+\frac{\partial N_2}{\partial y}+\frac{\partial P_2}{\partial z}\right)M_1$ and **i**-comp of $(\nabla\cdot\mathbf{F}_1)\mathbf{F}_2=\left(\frac{\partial M_1}{\partial x}+\frac{\partial N_1}{\partial y}+\frac{\partial P_1}{\partial z}\right)M_2$.

Similar results hold for the **j** and **k** components of $\nabla\times(\mathbf{F}_1\times\mathbf{F}_2)$. In summary, since the corresponding components are equal, we have the result $\nabla\times(\mathbf{F}_1\times\mathbf{F}_2)=(\mathbf{F}_2\cdot\nabla)\mathbf{F}_1-(\mathbf{F}_1\cdot\nabla)\mathbf{F}_2+(\nabla\cdot\mathbf{F}_2)\mathbf{F}_1-(\nabla\cdot\mathbf{F}_1)\mathbf{F}_2$

(b) Here again we consider only the **i**-component of each expression. Thus, the **i**-comp of $\nabla(\mathbf{F}_1\cdot\mathbf{F}_2)$

$=\frac{\partial}{\partial x}(M_1M_2+N_1N_2+P_1P_2)=\left(M_1\frac{\partial M_2}{\partial x}+M_2\frac{\partial M_1}{\partial x}+N_1\frac{\partial N_2}{\partial x}+N_2\frac{\partial N_1}{\partial x}+P_1\frac{\partial P_2}{\partial x}+P_2\frac{\partial P_1}{\partial x}\right)$

i-comp of $(\mathbf{F}_1\cdot\nabla)\mathbf{F}_2=\left(M_1\frac{\partial M_2}{\partial x}+N_1\frac{\partial M_2}{\partial y}+P_1\frac{\partial M_2}{\partial z}\right)$,

i-comp of $(\mathbf{F}_2\cdot\nabla)\mathbf{F}_1=\left(M_2\frac{\partial M_1}{\partial x}+N_2\frac{\partial M_1}{\partial y}+P_2\frac{\partial M_1}{\partial z}\right)$,

i-comp of $\mathbf{F}_1\times(\nabla\times\mathbf{F}_2)=N_1\left(\frac{\partial N_2}{\partial x}-\frac{\partial M_2}{\partial y}\right)-P_1\left(\frac{\partial M_2}{\partial z}-\frac{\partial P_2}{\partial x}\right)$, and

i-comp of $\mathbf{F}_2\times(\nabla\times\mathbf{F}_1)=N_2\left(\frac{\partial N_1}{\partial x}-\frac{\partial M_1}{\partial y}\right)-P_2\left(\frac{\partial M_1}{\partial z}-\frac{\partial P_1}{\partial x}\right)$.

Since corresponding components are equal, we see that
$\nabla(\mathbf{F}_1\cdot\mathbf{F}_2)=(\mathbf{F}_1\cdot\nabla)\mathbf{F}_2+(\mathbf{F}_2\cdot\nabla)\mathbf{F}_1+\mathbf{F}_1\times(\nabla\times\mathbf{F}_2)+\mathbf{F}_2\times(\nabla\times\mathbf{F}_1)$, as claimed.

29. $\iint\limits_{S} f\nabla g\cdot\mathbf{n}\,d\sigma = \iiint\limits_{D} \nabla\cdot f\nabla g\,dV = \iiint\limits_{D} \nabla\cdot\left(f\frac{\partial g}{\partial x}\mathbf{i} + f\frac{\partial g}{\partial y}\mathbf{j} + f\frac{\partial g}{\partial z}\mathbf{k}\right)dV$

$= \iiint\limits_{D}\left(f\frac{\partial^2 g}{\partial x^2} + \frac{\partial f}{\partial x}\frac{\partial g}{\partial x} + f\frac{\partial^2 g}{\partial y^2} + \frac{\partial f}{\partial y}\frac{\partial g}{\partial y} + f\frac{\partial^2 g}{\partial z^2} + \frac{\partial f}{\partial z}\frac{\partial g}{\partial z}\right)dV$

$= \iiint\limits_{D}\left[f\left(\frac{\partial^2 g}{\partial x^2} + \frac{\partial^2 g}{\partial y^2} + \frac{\partial^2 g}{\partial z^2}\right) + \left(\frac{\partial f}{\partial x}\frac{\partial g}{\partial x} + \frac{\partial f}{\partial y}\frac{\partial g}{\partial y} + \frac{\partial f}{\partial z}\frac{\partial g}{\partial z}\right)\right]dV = \iiint\limits_{D}\left(f\nabla^2 g + \nabla f\cdot\nabla g\right)dV$

CHAPTER 15 PRACTICE EXERCISES

1. Path 1 : $\mathbf{r} = t\mathbf{i} + t\mathbf{j} + t\mathbf{k} \Rightarrow x = t,\, y = t,\, z = t,\, 0\le t\le 1 \Rightarrow f\left(g(t), h(t), k(t)\right) = 3 - 3t^2$ and $\frac{dx}{dt} = 1, \frac{dy}{dt} = 1,$

$\frac{dz}{dt} = 1 \Rightarrow \sqrt{\left(\frac{dx}{dt}\right)^2 + \left(\frac{dy}{dt}\right)^2 + \left(\frac{dz}{dt}\right)^2}\,dt = \sqrt{3}\,dt \Rightarrow \int_C f(x, y, z)\,ds = \int_0^1 \sqrt{3}\left(3 - 3t^2\right)dt = 2\sqrt{3}$

Path 2 : $\mathbf{r}_1 = t\mathbf{i} + t\mathbf{j},\, 0\le t\le 1 \Rightarrow x = t,\, y = t,\, z = 0 \Rightarrow f\left(g(t), h(t), k(t)\right) = 2t - 3t^2 + 3$ and $\frac{dx}{dt} = 1, \frac{dy}{dt} = 1,$

$\frac{dz}{dt} = 0 \Rightarrow \sqrt{\left(\frac{dx}{dt}\right)^2 + \left(\frac{dy}{dt}\right)^2 + \left(\frac{dz}{dt}\right)^2}\,dt = \sqrt{2}\,dt \Rightarrow \int_{C_1} f(x, y, z)\,ds = \int_0^1 \sqrt{2}\left(2t - 3t^2 + 3\right)dt = 3\sqrt{2};$

$\mathbf{r}_2 = \mathbf{i} + \mathbf{j} + t\mathbf{k} \Rightarrow x = 1,\, y = 1,\, z = t \Rightarrow f\left(g(t), h(t), k(t)\right) = 2 - 2t$ and $\frac{dx}{dt} = 0, \frac{dy}{dt} = 0, \frac{dz}{dt} = 1$

$\Rightarrow \sqrt{\left(\frac{dx}{dt}\right)^2 + \left(\frac{dy}{dt}\right)^2 + \left(\frac{dz}{dt}\right)^2}\,dt = dt \Rightarrow \int_{C_2} f(x, y, z)\,ds = \int_0^1 (2 - 2t)\,dt = 1$

$\Rightarrow \int_C f(x, y, z)\,ds = \int_{C_1} f(x, y, z)\,ds + \int_{C_2} f(x, y, z) = 3\sqrt{2} + 1$

3. $\mathbf{r} = (a\cos t)\mathbf{j} + (a\sin t)\mathbf{k} \Rightarrow x = 0,\, y = a\cos t,\, z = a\sin t \Rightarrow f\left(g(t), h(t), k(t)\right) = \sqrt{a^2\sin^2 t} = a\,|\sin t|$ and

$\frac{dx}{dt} = 0, \frac{dy}{dt} = -a\sin t, \frac{dz}{dt} = a\cos t \Rightarrow \sqrt{\left(\frac{dx}{dt}\right)^2 + \left(\frac{dy}{dt}\right)^2 + \left(\frac{dz}{dt}\right)^2}\,dt = a\,dt$

$\Rightarrow \int_c f(x, y, z)\,ds = \int_0^{2\pi} a^2|\sin t|\,dt = \int_0^{\pi} a^2\sin t\,dt + \int_{\pi}^{2\pi}\left(-a^2\sin t\right)dt = 4a^2$

5. $\frac{\partial P}{\partial y} = -\frac{1}{2}\left(x + y + z\right)^{-3/2} = \frac{\partial N}{\partial z},\ \frac{\partial M}{\partial z} = -\frac{1}{2}\left(x + y + z\right)^{-3/2} = \frac{\partial P}{\partial x},\ \frac{\partial N}{\partial x} = -\frac{1}{2}\left(x + y + z\right)^{-3/2} = \frac{\partial M}{\partial y}$

$\Rightarrow M\,dx + N\,dy + P\,dz$ is exact; $\frac{\partial f}{\partial x} = \frac{1}{\sqrt{x+y+z}} \Rightarrow f(x, y, z) = 2\sqrt{x + y + z} + g(y, z) \Rightarrow \frac{\partial f}{\partial y} = \frac{1}{\sqrt{x+y+z}} + \frac{\partial g}{\partial y}$

$= \frac{1}{\sqrt{x+y+z}} \Rightarrow \frac{\partial g}{\partial y} = 0 \Rightarrow g(y, z) = h(z) \Rightarrow f(x, y, z) = 2\sqrt{x + y + z} + h(z) \Rightarrow \frac{\partial f}{\partial z} = \frac{1}{\sqrt{x+y+z}} + h'(z)$

$= \frac{1}{\sqrt{x+y+z}} \Rightarrow h'(x) = 0 \Rightarrow h(z) = C \Rightarrow f(x, y, z) = 2\sqrt{x + y + z} + C \Rightarrow \int_{(-1, 1, 1)}^{(4, -3, 0)} \frac{dx + dy + dz}{\sqrt{x+y+z}}$

$= f(4, -3, 0) - f(-1, 1, 1) = 2\sqrt{1} - 2\sqrt{1} = 0$

7. $\frac{\partial M}{\partial z} = -y\cos z \ne y\cos z = \frac{\partial P}{\partial x} \Rightarrow \mathbf{F}$ is not conservative; $\mathbf{r} = (2\cos t)\mathbf{i} + (2\sin t)\mathbf{j} - \mathbf{k},\, 0\le t\le 2\pi$

$\Rightarrow d\mathbf{r} = (-2\sin t)\mathbf{i} - (2\cos t)\mathbf{j} \Rightarrow \int_C \mathbf{F}\cdot d\mathbf{r} = \int_0^{2\pi}\left[-(-2\sin t)(\sin(-1))(-2\sin t) + (2\cos t)(\sin(-1))(-2\cos t)\right]dt$

$= 4\sin(1)\int_0^{2\pi}\left(\sin^2 t + \cos^2 t\right)dt = 8\pi\sin(1)$

9. Let $M = 8x \sin y$ and $N = -8y \cos x \Rightarrow \frac{\partial M}{\partial y} = 8x \cos y$ and $\frac{\partial N}{\partial x} = 8y \sin x \Rightarrow \int_C 8x \sin y \, dx - 8y \cos x \, dy$

$= \iint\limits_R (8y \sin x - 8x \cos y) \, dy \, dx = \int_0^{\pi/2} \int_0^{\pi/2} (8y \sin x - 8x \cos y) \, dy \, dx = \int_0^{\pi/2} \left(\pi^2 \sin x - 8x \right) dx = -\pi^2 + \pi^2 = 0$

11. Let $z = 1 - x - y \Rightarrow f_x(x, y) = -1$ and $f_y(x, y) = -1 \Rightarrow \sqrt{f_x^2 + f_y^2 + 1} = \sqrt{3} \Rightarrow$ Surface Area $= \iint\limits_R \sqrt{3} \, dx \, dy$

$= \sqrt{3}$ (Area of the circular region in the xy-plane) $= \pi\sqrt{3}$

13. $\nabla f = 2x\mathbf{i} + 2y\mathbf{j} + 2z\mathbf{k}, \mathbf{p} = \mathbf{k} \Rightarrow |\nabla f| = \sqrt{4x^2 + 4y^2 + 4z^2} = 2\sqrt{x^2 + y^2 + z^2} = 2$ and $|\nabla f \cdot \mathbf{p}| = |2z| = 2z$ since

$z \geq 0 \Rightarrow$ Surface Area $= \iint\limits_R \frac{2}{2z} \, dA = \iint\limits_R \frac{1}{z} \, dA = \iint\limits_R \frac{1}{\sqrt{1-x^2-y^2}} \, dx \, dy = \int_0^{2\pi} \int_0^{1/\sqrt{2}} \frac{1}{\sqrt{1-r^2}} r \, dr \, d\theta$

$= \int_0^{2\pi} \left[-\sqrt{1-r^2} \right]_0^{1/\sqrt{2}} d\theta = \int_0^{2\pi} \left(1 - \frac{1}{\sqrt{2}} \right) d\theta = 2\pi \left(1 - \frac{1}{\sqrt{2}} \right)$

15. $f(x, y, z) = \frac{x}{a} + \frac{y}{b} + \frac{z}{c} = 1 \Rightarrow \nabla f = \left(\frac{1}{a} \right)\mathbf{i} + \left(\frac{1}{b} \right)\mathbf{j} + \left(\frac{1}{c} \right)\mathbf{k} \Rightarrow |\nabla f| = \sqrt{\frac{1}{a^2} + \frac{1}{b^2} + \frac{1}{c^2}}$ and $\mathbf{p} = \mathbf{k} \Rightarrow |\nabla f \cdot \mathbf{p}| = \frac{1}{c}$ since

$c > 0 \Rightarrow$ Surface Area $= \iint\limits_R \frac{\sqrt{\frac{1}{a^2} + \frac{1}{b^2} + \frac{1}{c^2}}}{\left(\frac{1}{c} \right)} \, dA = c\sqrt{\frac{1}{a^2} + \frac{1}{b^2} + \frac{1}{c^2}} \iint\limits_R dA = \frac{1}{2} abc\sqrt{\frac{1}{a^2} + \frac{1}{b^2} + \frac{1}{c^2}}$, since the area of the

triangular region R is $\frac{1}{2} ab$. To check this result, let $\mathbf{v} = a\mathbf{i} + c\mathbf{k}$ and $\mathbf{w} = -a\mathbf{i} + b\mathbf{j}$; the area can be found by

computing $\frac{1}{2}|\mathbf{v} \times \mathbf{w}|$.

17. $\nabla f = 2y\mathbf{j} + 2z\mathbf{k}, \mathbf{p} = \mathbf{k} \Rightarrow |\nabla f| = \sqrt{4y^2 + 4z^2} = 2\sqrt{y^2 + z^2} = 10$ and $|\nabla f \cdot \mathbf{p}| = 2z$ since $z \geq 0$

$\Rightarrow d\sigma = \frac{10}{2z} \, dx \, dy = \frac{5}{z} \, dx \, dy \Rightarrow \iint\limits_S g(x, y, z) \, d\sigma = \iint\limits_R \left(x^4 y \right) \left(y^2 + z^2 \right) \left(\frac{5}{z} \right) dx \, dy$

$= \iint\limits_R \left(x^4 y \right) (25) \left(\frac{5}{\sqrt{25-y^2}} \right) dx \, dy = \int_0^4 \int_0^1 \frac{125 y}{\sqrt{25-y^2}} x^4 \, dx \, dy = \int_0^4 \frac{25 y}{\sqrt{25-y^2}} \, dy = 50$

19. A possible parametrization is $\mathbf{r}(\phi, \theta) = (6 \sin \phi \cos \theta)\mathbf{i} + (6 \sin \phi \sin \theta)\mathbf{j} + (6 \cos \phi)\mathbf{k}$ (spherical coordinates); now

$\rho = 6$ and $z = -3 \Rightarrow -3 = 6 \cos \phi \Rightarrow \cos \phi = -\frac{1}{2} \Rightarrow \phi = \frac{2\pi}{3}$ and $z = 3\sqrt{3} \Rightarrow 3\sqrt{3} = 6 \cos \phi$

$\Rightarrow \cos \phi = \frac{\sqrt{3}}{2} \Rightarrow \phi = \frac{\pi}{6} \Rightarrow \frac{\pi}{6} \leq \phi \leq \frac{2\pi}{3}$; also $0 \leq \theta \leq 2\pi$

21. A possible parametrization is $\mathbf{r}(r, \theta) = (r \cos \theta)\mathbf{i} + (r \sin \theta)\mathbf{j} + (1+r)\mathbf{k}$ (cylindrical coordinates);

now $r = \sqrt{x^2 + y^2} \Rightarrow z = 1 + r$ and $1 \leq z \leq 3 \Rightarrow 1 \leq 1 + r \leq 3 \Rightarrow 0 \leq r \leq 2$; also $0 \leq \theta \leq 2\pi$

23. Let $x = u \cos v$ and $z = u \sin v$, where $u = \sqrt{x^2 + z^2}$ and v is the angle in the xz-plane with the x-axis

$\Rightarrow \mathbf{r}(u, v) = (u \cos v)\mathbf{i} + 2u^2\mathbf{j} + (u \sin v)\mathbf{k}$ is a possible parametrization; $0 \leq y \leq 2 \Rightarrow 2u^2 \leq 2 \Rightarrow u^2 \leq 1$

$\Rightarrow 0 \leq u \leq 1$ since $u \geq 0$; also, for just the upper half of the paraboloid, $0 \leq v \leq \pi$

25. $\mathbf{r}_u = \mathbf{i} + \mathbf{j}, \mathbf{r}_v = \mathbf{i} - \mathbf{j} + \mathbf{k} \Rightarrow \mathbf{r}_u \times \mathbf{r}_v = \begin{vmatrix} \mathbf{i} & \mathbf{j} & \mathbf{k} \\ 1 & 1 & 0 \\ 1 & -1 & 1 \end{vmatrix} = \mathbf{i} - \mathbf{j} - 2\mathbf{k} \Rightarrow |\mathbf{r}_u \times \mathbf{r}_v| = \sqrt{6}$

$\Rightarrow$ Surface Area $= \displaystyle\iint_{R_{uv}} |\mathbf{r}_u \times \mathbf{r}_v| \, du \, dv = \int_0^1 \int_0^1 \sqrt{6} \, du \, dv = \sqrt{6}$

27. $\mathbf{r}_r = (\cos\theta)\mathbf{i} + (\sin\theta)\mathbf{j}, \mathbf{r}_\theta = (-r\sin\theta)\mathbf{i} + (r\cos\theta)\mathbf{j} + \mathbf{k} \Rightarrow \mathbf{r}_r \times \mathbf{r}_\theta = \begin{vmatrix} \mathbf{i} & \mathbf{j} & \mathbf{k} \\ \cos\theta & \sin\theta & 0 \\ -r\sin\theta & r\cos\theta & 1 \end{vmatrix}$

$= (\sin\theta)\mathbf{i} - (\cos\theta)\mathbf{j} + r\mathbf{k} \Rightarrow |\mathbf{r}_r \times \mathbf{r}_\theta| = \sqrt{\sin^2\theta + \cos^2\theta + r^2} = \sqrt{1 + r^2} \Rightarrow$ Surface Area $= \displaystyle\iint_{R_{r\theta}} |\mathbf{r}_r \times \mathbf{r}_\theta| dr \, d\theta$

$= \int_0^{2\pi} \int_0^1 \sqrt{1+r^2} \, dr \, d\theta = \int_0^{2\pi} \left[\frac{r}{2}\sqrt{1+r^2} + \frac{1}{2}\ln\left(r + \sqrt{1+r^2}\right) \right]_0^1 d\theta = \int_0^{2\pi} \left[\frac{1}{2}\sqrt{2} + \frac{1}{2}\ln\left(1+\sqrt{2}\right) \right] d\theta$

$= \pi\left[\sqrt{2} + \ln\left(1+\sqrt{2}\right) \right]$

29. $\frac{\partial P}{\partial y} = 0 = \frac{\partial N}{\partial z}, \frac{\partial M}{\partial z} = 0 = \frac{\partial P}{\partial x}, \frac{\partial N}{\partial x} = 0 = \frac{\partial M}{\partial y} \Rightarrow$ Conservative

31. $\frac{\partial P}{\partial y} = 0 \neq ye^z = \frac{\partial N}{\partial z} \Rightarrow$ Not Conservative

33. $\frac{\partial f}{\partial x} = 2 \Rightarrow f(x, y, z) = 2x + g(y, z) \Rightarrow \frac{\partial f}{\partial y} = \frac{\partial g}{\partial y} = 2y + z \Rightarrow g(y, z) = y^2 + zy + h(z) \Rightarrow f(x, y, z)$

$= 2x + y^2 + zy + h(z) \Rightarrow \frac{\partial f}{\partial z} = y + h'(z) = y + 1 \Rightarrow h'(z) = 1 \Rightarrow h(z) = z + C \Rightarrow f(x, y, z) = 2x + y^2 + zy + z + C$

35. Over Path 1: $\mathbf{r} = t\mathbf{i} + t\mathbf{j} + t\mathbf{k}, 0 \le t \le 1 \Rightarrow x = t, y = t, z = t$ and $d\mathbf{r} = (\mathbf{i} + \mathbf{j} + \mathbf{k}) \, dt \Rightarrow \mathbf{F} = 2t^2\mathbf{i} + \mathbf{j} + t^2\mathbf{k}$

$\Rightarrow \mathbf{F} \cdot d\mathbf{r} = \left(3t^2 + 1\right) dt \Rightarrow$ Work $= \int_0^1 \left(3t^2 + 1\right) dt = 2;$

Over Path 2: $\mathbf{r}_1 = t\mathbf{i} + t\mathbf{j}, 0 \le t \le 1 \Rightarrow x = t, y = t, z = 0$ and $d\mathbf{r}_1 = (\mathbf{i} + \mathbf{j}) dt \Rightarrow \mathbf{F}_1 = 2t^2\mathbf{i} + \mathbf{j} + t^2\mathbf{k}$

$\Rightarrow \mathbf{F}_1 \cdot d\mathbf{r}_1 = \left(2t^2 + 1\right) dt \Rightarrow$ Work$_1 = \int_0^1 \left(2t^2 + 1\right) dt = \frac{5}{3}; \mathbf{r}_2 = \mathbf{i} + \mathbf{j} + t\mathbf{k}, 0 \le t \le 1 \Rightarrow x = 1, y = 1, z = t$ and

$d\mathbf{r}_2 = \mathbf{k} \, dt \Rightarrow \mathbf{F}_2 = 2\mathbf{i} + \mathbf{j} + \mathbf{k} \Rightarrow \mathbf{F}_2 \cdot d\mathbf{r}_2 = dt \Rightarrow$ Work$_2 = \int_0^1 dt = 1 \Rightarrow$ Work $=$ Work$_1 +$ Work$_2 = \frac{5}{3} + 1 = \frac{8}{3}$

37. (a) $\mathbf{r} = \left(e^t \cos t\right)\mathbf{i} + \left(e^t \sin t\right)\mathbf{j} \Rightarrow x = e^t \cos t, y = e^t \sin t$ from $(1, 0)$ to $\left(e^{2\pi}, 0\right) \Rightarrow 0 \le t \le 2\pi$

$\Rightarrow \frac{d\mathbf{r}}{dt} = \left(e^t \cos t - e^t \sin t\right)\mathbf{i} + \left(e^t \sin t + e^t \cos t\right)\mathbf{j}$ and $\mathbf{F} = \frac{x\mathbf{i} + y\mathbf{j}}{\left(x^2 + y^2\right)^{3/2}} = \frac{\left(e^t \cos t\right)\mathbf{i} + \left(e^t \sin t\right)\mathbf{j}}{\left(e^{2t}\cos^2 t + e^{2t}\sin^2 t\right)^{3/2}}$

$= \left(\frac{\cos t}{e^{2t}}\right)\mathbf{i} + \left(\frac{\sin t}{e^{2t}}\right)\mathbf{j} \Rightarrow \mathbf{F} \cdot \frac{d\mathbf{r}}{dt} = \left(\frac{\cos^2 t}{e^t} - \frac{\sin t \cos t}{e^t} + \frac{\sin^2 t}{e^t} + \frac{\sin t \cos t}{e^t}\right) = e^{-t} \Rightarrow$ Work $= \int_0^{2\pi} e^{-t} dt = 1 - e^{-2\pi}$

(b) $\mathbf{F} = \dfrac{x\mathbf{i}+y\mathbf{j}}{\left(x^2+y^2\right)^{3/2}} \Rightarrow \dfrac{\partial f}{\partial x} = \dfrac{x}{\left(x^2+y^2\right)^{3/2}} \Rightarrow f(x,y,z) = -\left(x^2+y^2\right)^{-1/2} + g(y,z) \Rightarrow \dfrac{\partial f}{\partial y} = \dfrac{y}{\left(x^2+y^2\right)^{3/2}} + \dfrac{\partial g}{\partial y}$

$= \dfrac{y}{(x^2+y^2)^{3/2}} \Rightarrow g(y,z) = C \Rightarrow f(x,y,z) = -(x^2+y^2)^{-1/2}$ is a potential function for $\mathbf{F}$

$\Rightarrow \int_C \mathbf{F} \cdot d\mathbf{r} = f\left(e^{2\pi},0\right) - f(1,0) = 1 - e^{-2\pi}$

39. $\nabla \times \mathbf{F} = \begin{vmatrix} \mathbf{i} & \mathbf{j} & \mathbf{k} \\ \dfrac{\partial}{\partial x} & \dfrac{\partial}{\partial y} & \dfrac{\partial}{\partial z} \\ y^2 & -y & 3z^2 \end{vmatrix} = -2y\mathbf{k};$ unit normal to the plane is $\mathbf{n} = \dfrac{2\mathbf{i}+6\mathbf{j}-3\mathbf{k}}{\sqrt{4+36+9}} = \dfrac{2}{7}\mathbf{i} + \dfrac{6}{7}\mathbf{j} - \dfrac{3}{7}\mathbf{k}$

$\Rightarrow \nabla \times \mathbf{F} \cdot \mathbf{n} = \dfrac{6}{7}y; \mathbf{p} = \mathbf{k}$ and $f(x,y,z) = 2x+6y-3z \Rightarrow |\nabla f \cdot \mathbf{p}| = 3 \Rightarrow d\sigma = \dfrac{|\nabla f|}{|\nabla f \cdot \mathbf{p}|}dA = \dfrac{7}{3}dA$

$\Rightarrow \oint_C \mathbf{F} \cdot d\mathbf{r} = \iint_R \dfrac{6}{7}y\,d\sigma = \iint_R \left(\dfrac{6}{7}y\right)\left(\dfrac{7}{3}dA\right) = \iint_R 2y\,dA = \int_0^{2\pi}\int_0^1 2r\sin\theta\, r\,dr\,d\theta = \int_0^{2\pi}\dfrac{2}{3}\sin\theta\,d\theta = 0$

41. (a) $\mathbf{r} = \sqrt{2}t\mathbf{i} + \sqrt{2}t\mathbf{j} + \left(4-t^2\right)\mathbf{k}, 0 \le t \le 1 \Rightarrow x = \sqrt{2}t, y = \sqrt{2}t, z = 4-t^2 \Rightarrow \dfrac{dx}{dt} = \sqrt{2}, \dfrac{dy}{dt} = \sqrt{2}, \dfrac{dz}{dt} = -2t$

$\Rightarrow \sqrt{\left(\dfrac{dx}{dt}\right)^2 + \left(\dfrac{dy}{dt}\right)^2 + \left(\dfrac{dz}{dt}\right)^2}\,dt = \sqrt{4+4t^2}\,dt \Rightarrow M = \int_C \delta(x,y,z)\,ds = \int_0^1 3t\sqrt{4+4t^2}\,dt = \left[\dfrac{1}{4}(4+4t)^{3/2}\right]_0^1$

$= 4\sqrt{2}-2$

(b) $M = \int_C \delta(x,y,z)\,ds = \int_0^1 \sqrt{4+4t^2}\,dt = \left[t\sqrt{1+t^2} + \ln\left(t+\sqrt{1+t^2}\right)\right]_0^1 = \sqrt{2} + \ln\left(1+\sqrt{2}\right)$

43. $\mathbf{r} = t\mathbf{i} + \left(\dfrac{2\sqrt{2}}{3}t^{3/2}\right)\mathbf{j} + \left(\dfrac{t^2}{2}\right)\mathbf{k}, 0 \le t \le 2 \Rightarrow x = t, y = \dfrac{2\sqrt{2}}{3}t^{3/2}, z = \dfrac{t^2}{2} \Rightarrow \dfrac{dx}{dt} = 1, \dfrac{dy}{dt} = \sqrt{2}t^{1/2}, \dfrac{dz}{dt} = t$

$\Rightarrow \sqrt{\left(\dfrac{dx}{dt}\right)^2 + \left(\dfrac{dy}{dt}\right)^2 + \left(\dfrac{dz}{dt}\right)^2}\,dt = \sqrt{1+2t+t^2}\,dt = \sqrt{(t+1)^2}\,dt = |t+1|\,dt = (t+1)\,dt$ on the domain given.

Then $M = \int_C \delta\,ds = \int_0^2 \left(\dfrac{1}{t+1}\right)(t+1)\,dt = \int_0^2 dt = 2; M_{yz} = \int_C x\delta\,ds = \int_0^2 t\left(\dfrac{1}{t+1}\right)(t+1)\,dt = \int_0^2 t\,dt = 2;$

$M_{xz} = \int_C y\delta\,ds = \int_0^2 \left(\dfrac{2\sqrt{2}}{3}t^{3/2}\right)\left(\dfrac{1}{t+1}\right)(t+1)\,dt = \int_0^2 \dfrac{2\sqrt{2}}{3}t^{3/2}\,dt = \dfrac{32}{15}; M_{xy} = \int_C z\delta\,ds$

$= \int_0^2 \left(\dfrac{t^2}{2}\right)\left(\dfrac{1}{t+1}\right)(t+1)\,dt = \int_0^2 \dfrac{t^2}{2}\,dt = \dfrac{4}{3} \Rightarrow \bar{x} = \dfrac{M_{yz}}{M} = \dfrac{2}{2} = 1; \bar{y} = \dfrac{M_{xz}}{M} = \dfrac{\left(\frac{32}{15}\right)}{2} = \dfrac{16}{15}; \bar{z} = \dfrac{M_{xy}}{M} = \dfrac{\left(\frac{4}{3}\right)}{2} = \dfrac{2}{3};$

$I_x = \int_C \left(y^2+z^2\right)\delta\,ds = \int_0^2 \left(\dfrac{8}{9}t^3 + \dfrac{t^4}{4}\right)dt = \dfrac{232}{45}; I_y = \int_C \left(x^2+z^2\right)\delta\,ds = \int_0^2 \left(t^2 + \dfrac{t^4}{4}\right)dt = \dfrac{64}{15};$

$I_z = \int_C \left(y^2+x^2\right)\delta\,ds = \int_0^2 \left(t^2 + \dfrac{8}{9}t^3\right)dt = \dfrac{56}{9}$

45. $\mathbf{r}(t) = \left(e^t \cos t\right)\mathbf{i} + \left(e^t \sin t\right)\mathbf{j} + e^t\mathbf{k}, \ 0 \le t \le \ln 2 \Rightarrow x = e^t \cos t, \ y = e^t \sin t, z = e^t \Rightarrow \frac{dx}{dt} = \left(e^t \cos t - e^t \sin t\right),$

$\frac{dy}{dt} = \left(e^t \sin t + e^t \cos t\right), \frac{dz}{dt} = e^t \Rightarrow \sqrt{\left(\frac{dx}{dt}\right)^2 + \left(\frac{dy}{dt}\right)^2 + \left(\frac{dz}{dt}\right)^2} \, dt$

$= \sqrt{\left(e^t \cos t - e^t \sin t\right)^2 + \left(e^t \sin t + e^t \cos t\right)^2 + \left(e^t\right)^2} \, dt = \sqrt{3e^{2t}} \, dt = \sqrt{3}\, e^t \, dt; M = \int_c \delta \, ds = \int_0^{\ln 2} \sqrt{3}\, e^t \, dt$

$= \sqrt{3}; M_{xy} = \int_c z\delta \, ds = \int_0^{\ln 2} \left(\sqrt{3}\, e^t\right)\left(e^t\right) dt = \int_0^{\ln 2} \sqrt{3}\, e^{2t} \, dt = \frac{3\sqrt{3}}{2} \Rightarrow \overline{z} = \frac{M_{xy}}{M} = \frac{\left(\frac{3\sqrt{3}}{2}\right)}{\sqrt{3}} = \frac{3}{2};$

$I_z = \int_c \left(x^2 + y^2\right)\delta \, ds = \int_0^{\ln 2}\left(e^{2t}\cos^2 t + e^{2t}\sin^2 t\right)\left(\sqrt{3}\, e^t\right)dt = \int_0^{\ln 2}\sqrt{3}\, e^{3t} \, dt = \frac{7\sqrt{3}}{3}$

47. Because of symmetry $\overline{x} = \overline{y} = 0$. Let $f(x, y, z) = x^2 + y^2 + z^2 = 25 \Rightarrow \nabla f = 2x\mathbf{i} + 2y\mathbf{j} + 2z\mathbf{k}$

$\Rightarrow |\nabla f| = \sqrt{4x^2 + 4y^2 + 4z^2} = 10$ and $\mathbf{p} = \mathbf{k} \Rightarrow |\nabla f \cdot \mathbf{p}| = 2z,$ since $z \ge 0 \Rightarrow M = \iint_R \delta(x, y, z) \, d\delta$

$= \iint_R z\left(\frac{10}{2z}\right) dA = \iint_R 5 \, dA = 5 \text{ (Area of the circular region)} = 80\pi; M_{xy} = \iint_R z\delta \, d\delta = \iint_R 5z \, dA$

$= \iint_R 5\sqrt{25 - x^2 - y^2} \, dx \, dy = \int_0^{2\pi}\int_0^4 \left(5\sqrt{25 - r^2}\right) r \, dr \, d\theta = \int_0^{2\pi} \frac{490}{3} d\theta = \frac{980}{3}\pi \Rightarrow \overline{z} = \frac{\left(\frac{980}{3}\pi\right)}{80\pi} = \frac{49}{12}$

$\Rightarrow (\overline{x}, \overline{y}, \overline{z}) = \left(0, 0, \frac{49}{12}\right); I_z = \iint_R \left(x^2 + y^2\right)\delta \, d\sigma = \iint_R 5\left(x^2 + y^2\right) dx \, dy = \int_0^{2\pi}\int_0^4 5r^3 dr \, d\theta = \int_0^{2\pi} 320 \, d\theta$

$= 640\pi$

49. $M = 2xy + x$ and $N = xy - y \Rightarrow \frac{\partial M}{\partial x} = 2y + 1, \frac{\partial M}{\partial y} = 2x, \frac{\partial N}{\partial x} = y, \frac{\partial N}{\partial y} = x - 1$

$\Rightarrow \text{Flux} = \iint_R \left(\frac{\partial M}{\partial x} + \frac{\partial N}{\partial y}\right) dx \, dy = \iint_R (2y + 1 + x - 1) \, dx \, dy = \int_0^1\int_0^1 (2y + x) \, dy \, dx = \frac{3}{2};$

$\text{Circ} = \iint_R \left(\frac{\partial N}{\partial x} - \frac{\partial M}{\partial y}\right) dx \, dy = \iint_R (y - 2x) \, dx \, dy = \int_0^1\int_0^1 (y - 2x) \, dy \, dx = -\frac{1}{2}$

51. $M = -\frac{\cos y}{x}$ and $N = \ln x \sin y \Rightarrow \frac{\partial M}{\partial y} = \frac{\sin y}{x}$ and $\frac{\partial N}{\partial x} = \frac{\sin y}{x} \Rightarrow \oint_c \ln x \sin y \, dy - \frac{\cos y}{x} dx$

$= \iint_R \left(\frac{\partial N}{\partial x} - \frac{\partial M}{\partial y}\right) dx \, dy = \iint_R \left(\frac{\sin y}{x} - \frac{\sin y}{x}\right) dx \, dy = 0$

53. $\frac{\partial}{\partial x}(2xy) = 2y, \frac{\partial}{\partial y}(2yz) = 2z, \frac{\partial}{\partial z}(2xz) = 2x \Rightarrow \nabla \cdot \mathbf{F} = 2y + 2z + 2x \Rightarrow \text{Flux} = \iiint_D (2x + 2y + 2z) \, dV$

$= \int_0^1\int_0^1\int_0^1 (2x + 2y + 2z) \, dx \, dy \, xz = \int_0^1\int_0^1 (1 + 2y + 2z) \, dy \, dz = \int_0^1 (2 + 2z) \, dz = 3$

55. $\frac{\partial}{\partial x}(-2x) = -2, \frac{\partial}{\partial y}(-3y) = -3, \frac{\partial}{\partial z}(z) = 1 \Rightarrow \nabla \cdot \mathbf{F} = -4; x^2 + y^2 + z^2 = 2$ and $x^2 + y^2 = z \Rightarrow z = 1$

$\Rightarrow x^2 + y^2 = 1 \Rightarrow \text{Flux} = \iiint_D -4 \, dV = -4\int_0^{2\pi}\int_0^1\int_{r^2}^{\sqrt{2-r^2}} dz \, r \, dr \, d\theta = -4\int_0^{2\pi}\int_0^1 \left(r\sqrt{2 - r^2} - r^3\right) dr \, d\theta$

$= -4\int_0^{2\pi}\left(-\frac{7}{12} + \frac{2}{3}\sqrt{2}\right) d\theta = \frac{2}{3}\pi\left(7 - 8\sqrt{2}\right)$

57. $\mathbf{F} = y\mathbf{i} + z\mathbf{j} + x\mathbf{k} \Rightarrow \nabla \cdot \mathbf{F} = 0 \Rightarrow \text{Flux} = \iint_S \mathbf{F} \cdot \mathbf{n} \, d\sigma = \iiint_D \nabla \cdot \mathbf{F} \, dV = 0$

59. $\mathbf{F} = xy^2\mathbf{i} + x^2 y\mathbf{j} + y\mathbf{k} \Rightarrow \nabla \cdot \mathbf{F} = y^2 + x^2 + 0 \Rightarrow \text{Flux} = \iint_S \mathbf{F} \cdot \mathbf{n} \, d\sigma = \iiint_D \nabla \cdot \mathbf{F} \, dV$

$= \iiint_D \left(x^2 + y^2 \right) dV = \int_0^{2\pi} \int_0^1 \int_{-1}^1 r^2 dz \, r \, dr \, d\theta = \int_0^{2\pi} \int_0^1 2r^3 dr \, d\theta = \int_0^{2\pi} \tfrac{1}{2} d\theta = \pi$

CHAPTER 15 ADDITIONAL AND ADVANCED EXERCISES

1. $dx = (-2\sin t + 2\sin 2t)\, dt$ and $dy = (2\cos t - 2\cos 2t)\, dt$; Area $= \tfrac{1}{2}\oint_C x\, dy - y\, dx$

$= \tfrac{1}{2}\int_0^{2\pi} \left[(2\cos t - \cos 2t)(2\cos t - 2\cos 2t) - (2\sin t - \sin 2t)(-2\sin t + 2\sin 2t) \right] dt$

$= \tfrac{1}{2}\int_0^{2\pi} \left[6 - (6\cos t \cos 2t + 6\sin t \sin 2t) \right] dt = \tfrac{1}{2}\int_0^{2\pi} (6 - 6\cos t)\, dt = 6\pi$

3. $dx = \cos 2t\, dt$ and $dy = \cos t\, dt$; Area $= \tfrac{1}{2}\oint_C x\, dy - y\, dx = \tfrac{1}{2}\int_0^{\pi} \left(\tfrac{1}{2}\sin 2t \cos t - \sin t \cos 2t \right) dt$

$= \tfrac{1}{2}\int_0^{\pi} \left[\sin t \cos^2 t - (\sin t)\left(2\cos^2 t - 1\right) \right] dt = \tfrac{1}{2}\int_0^{\pi} \left(-\sin t \cos^2 t + \sin t \right) dt = \tfrac{1}{2}\left[\tfrac{1}{3}\cos^3 t - \cos t \right]_0^{\pi} = -\tfrac{1}{3} + 1 = \tfrac{2}{3}$

5. (a) $\mathbf{F}(x, y, z,) = z\mathbf{i} + x\mathbf{j} + y\mathbf{k}$ is $\mathbf{0}$ only at the point $(0, 0, 0)$, and curl $\mathbf{F}(x, y, z) = \mathbf{i} + \mathbf{j} + \mathbf{k}$ is never $\mathbf{0}$.

   (b) $\mathbf{F}(x, y, z) = z\mathbf{i} + y\mathbf{k}$ is $\mathbf{0}$ only on the line $x = t, \ y = 0, \ z = 0$ and curl $\mathbf{F}(x, y, z) = \mathbf{i} + \mathbf{j}$ is never $\mathbf{0}$.

   (c) $\mathbf{F}(x, y, z) = z\mathbf{i}$ is $\mathbf{0}$ only when $z = 0$ (the xy-plane) and curl $\mathbf{F}(x, y, z) = \mathbf{j}$ is never $\mathbf{0}$.

7. Set up the coordinate system so that $(a, b, c) = (0, R, 0) \Rightarrow \delta(x, y, z) = \sqrt{x^2 + (y - R)^2 + z^2}$

$= \sqrt{x^2 + y^2 + z^2 - 2Ry + R^2} = \sqrt{2R^2 - 2Ry}$; let $f(x, y, z) = x^2 + y^2 + z^2 - R^2$ and $\mathbf{p} = \mathbf{i}$

$\Rightarrow \nabla f = 2x\mathbf{i} + 2y\mathbf{j} + 2z\mathbf{k} \Rightarrow |\nabla f| = 2\sqrt{x^2 + y^2 + z^2} = 2R \Rightarrow d\sigma = \dfrac{|\nabla f|}{|\nabla f \cdot \mathbf{i}|} dz\, dy = \dfrac{2R}{2x} dz\, dy$

$\Rightarrow \text{Mass} = \iint_S \delta(x, y, z)\, d\sigma = \iint_{R_{yz}} \sqrt{2R^2 - 2Ry} \left(\dfrac{R}{x} \right) dz\, dy = R \iint_{R_{yz}} \dfrac{\sqrt{2R^2 - 2Ry}}{\sqrt{R^2 - y^2 - z^2}} dz\, dy$

$= 4R\int_{-R}^{R}\int_0^{\sqrt{R^2 - y^2}} \dfrac{\sqrt{2R^2 - 2Ry}}{\sqrt{R^2 - y^2 - z^2}} dz\, dy = 4R\int_{-R}^{R} \sqrt{2R^2 - 2Ry}\, \sin^{-1}\left(\dfrac{z}{\sqrt{R^2 - y^2}} \right)\Big|_0^{\sqrt{R^2 - y^2}} dy$

$= 2\pi R\int_{-R}^{R} \sqrt{2R^2 - 2Ry}\, dy = 2\pi R\left(\dfrac{-1}{3R} \right)(2R^2 - 2Ry)^{3/2}\Big|_{-R}^{R} = \dfrac{16\pi R^3}{3}$

9. $M = x^2 + 4xy$ and $N = -6y \Rightarrow \frac{\partial M}{\partial x} = 2x + 4y$ and $\frac{\partial N}{\partial y} = -6 \Rightarrow$ Flux $= \int_0^b \int_0^a (2x + 4y - 6) \, dx \, dy$

$= \int_0^b \left(a^2 + 4ay - 6a \right) dy = a^2 b + 2ab^2 - 6ab$. We want to minimize

$f(a, b) = a^2 b + 2ab^2 - 6ab = ab(a + 2b - 6)$. Thus, $f_a(a,b) = 2ab + 2b^2 - 6b = 0$ and

$f_b(a,b) = a^2 + 4ab - 6a = 0 \Rightarrow b(2a + 2b - 6) = 0 \Rightarrow b = 0$ or $b = -a + 3$. Now $b = 0 \Rightarrow a^2 - 6a = 0 \Rightarrow a = 0$

or $a = 6 \Rightarrow (0,0)$ and $(6, 0)$ are critical points. On the other hand, $b = -a + 3 \Rightarrow a^2 + 4a(-a + 3) - 6a = 0$

$\Rightarrow -3a^2 + 6a = 0 \Rightarrow a = 0$ or $a = 2 \Rightarrow (0, 3)$ and $(2, 1)$ are also critical points. The flux at $(0, 0) = 0$, the flux at

$(6, 0) = 0$, the flux at $(0, 3) = 0$ and the flux at $(2, 1) = -4$. Therefore, the flux is minimized at $(2, 1)$ with

value -4.

11. (a) Partition the string into small pieces. Let $\Delta_i s$ be the length of the ith piece. Let (x_i, y_i) be a point in the

i^{th} piece. The work done by gravity in moving the ith piece to the x-axis is approximately

$W_i = (gx_i y_i \Delta_i s) y_i$ where $x_i y_i \Delta_i s$ is approximately the mass of the ith piece. The total work done by

gravity in moving the string to the x-axis is $\sum_i W_i = \sum_i g x_i y_i^2 \Delta_i s \Rightarrow$ Work $= \int_C gxy^2 \, ds$

(b) Work $= \int_C gxy^2 ds = \int_0^{\pi/2} g(2\cos t)(4\sin^2 t) \sqrt{4\sin^2 t + 4\cos^2 t} \, dt = 16g \int_0^{\pi/2} \cos t \sin^2 t \, dt$

$= \left[16g \left(\frac{\sin^3 t}{3} \right) \right]_0^{\pi/2} = \frac{16}{3} g$

(c) $\bar{x} = \frac{\int_C x(xy) ds}{\int_C xy \, ds}$ and $\bar{y} = \frac{\int_C y(xy) \, ds}{\int_C xy \, ds}$; the mass of the string is $\int_C xy \, ds$ and the weight of the string is

$g \int_C xy \, ds$. Therefore, the work done in moving the point mass at $(\bar{x}, \bar{y})$ to the x-axis is

$W = \left(g \int_C xy \, ds \right) \bar{y} = g \int_C xy^2 \, ds = \frac{16}{3} g$.

13. (a) Partition the sphere $x^2 + y^2 + (z - 2)^2 = 1$ into small pieces. Let $\Delta_i \sigma$ be the surface area of the ith piece

and let (x_i, y_i, z_i) be a point on the ith piece. The force due to pressure on the ith piece is approximately

$w(4 - z_i) \Delta_i \sigma$. The total force on S is approximately $\sum_i w(4 - z_i) \Delta_i \sigma$. This gives the actual force to be

$\iint_S w(4 - z) \, d\sigma$.

(b) The upward buoyant force is a result of the $\mathbf{k}$-component of the force on the ball due to liquid pressure.
The force on the ball at (x, y, z) is $w(4 - z)(-\mathbf{n}) = w(z - 4)\mathbf{n}$, where $\mathbf{n}$ is the outer unit normal at (x, y, z).
Hence the $\mathbf{k}$-component of this force is $w(z - 4)\mathbf{n} \cdot \mathbf{k} = w(z - 4)\mathbf{k} \cdot \mathbf{n}$. The (magnitude of the) buoyant
force on the ball is obtained by adding up all these $\mathbf{k}$-components to obtain $\iint_S w(z - 4)\mathbf{k} \cdot \mathbf{n} \, d\sigma$.

(c) The Divergence Theorem says $\iint_S w(z - 4)\mathbf{k} \cdot \mathbf{n} \, d\sigma = \iiint_D \text{div}\left(w(z - 4)\mathbf{k} \right) dV = \iiint_D w \, dV$, where D is

$x^2 + y^2 + (z - 2)^2 \le 1 \Rightarrow \iint_S w(z - 4)\mathbf{k} \cdot \mathbf{n} \, d\sigma = w \iiint_D 1 \, dV = \frac{4}{3} \pi w$, the weight of the fluid if it were to

occupy the region D.

15. $\oint_C f\,\nabla g\cdot d\mathbf{r} = \iint\limits_S \nabla\times(f\,\nabla\,g)\cdot\mathbf{n}\,d\sigma$ (Stokes' Theorem)

    $= \iint\limits_S (f\nabla\times\nabla\,g + \nabla f\times\nabla\,g)\cdot\mathbf{n}\,d\sigma$ (Section 16.8, Exercise 19b)

    $= \iint\limits_S [(f)(\mathbf{0}) + \nabla\,f\times\nabla\,g]\cdot\mathbf{n}\,d\sigma$ (Section 16.7, Equation 8)

    $= \iint\limits_S (\nabla f\times\nabla\,g)\cdot\mathbf{n}\,d\sigma$

17. False; let $\mathbf{F} = y\mathbf{i} + x\mathbf{j} \neq \mathbf{0} \Rightarrow \nabla\cdot\mathbf{F} = \frac{\partial}{\partial x}(y) + \frac{\partial}{\partial y}(x) = 0$ and $\nabla\times\mathbf{F} = \begin{vmatrix} \mathbf{i} & \mathbf{j} & \mathbf{k} \\ \frac{\partial}{\partial x} & \frac{\partial}{\partial y} & \frac{\partial}{\partial z} \\ x & y & 0 \end{vmatrix} = 0\mathbf{i} + 0\mathbf{j} + 0\mathbf{k} = \mathbf{0}$